Zinc Oxide Nanostructures: Synthesis and Characterization

Zinc Oxide Nanostructures: Synthesis and Characterization

Special Issue Editor

Sotirios Baskoutas

MDPI • Basel • Beijing • Wuhan • Barcelona • Belgrade

MDPI

Special Issue Editor
Sotirios Baskoutas
University of Patras
Greece

Editorial Office
MDPI
St. Alban-Anlage 66
Basel, Switzerland

This is a reprint of articles from the Special Issue published online in the open access journal *Materials* (ISSN 1996-1944) from 2017 to 2018 (available at: https://www.mdpi.com/journal/materials/special_issues/zinc_oxide_nanostructures_synthesis_and_characterization)

For citation purposes, cite each article independently as indicated on the article page online and as indicated below:

LastName, A.A.; LastName, B.B.; LastName, C.C. Article Title. *Journal Name* **Year**, *Article Number*, Page Range.

ISBN 978-3-03897-302-7 (Pbk)
ISBN 978-3-03897-303-4 (PDF)

Cover image courtesy of Ahmad Umar.

Contents

About the Special Issue Editor

Sotirios Baskoutas Professor Sotirios Baskoutas joined the Materials Science Department of the University of Patras in 2001, where he is currently a full Professor. Dr. Baskoutas has worked in several Universities and Research Institutes such as the Department of Physics, Universita di Roma La Sapienza (Italy); INT Institute for Nanotechnology, Karlsruhe (Germany); Max Planck Institute for Solid State Research, Stuttgart (Germany); and the Department of Chemistry, University of Hamburg (Germany). His research interests are focused mainly in theoretical but also experimental studies in semiconductor nanostructures, with an emphasis on their electronic and optical properties. Dr. Baskoutas has authored over 140 research articles in peer-reviewed journals in the fields of condensed matter physics and materials science

materials

MDPI

Editorial

Special Issue: Zinc Oxide Nanostructures: Synthesis and Characterization

Sotirios Baskoutas

Department of Materials Science, University of Patras, 26500 Patras, Greece; bask@upatras.gr;
Tel.: +30-261-096-9349

Received: 18 May 2018; Accepted: 23 May 2018; Published: 23 May 2018

Abstract: Zinc oxide (ZnO) is a wide band gap semiconductor with an energy gap of 3.37 eV at room temperature. It has been used considerably for its catalytic, electrical, optoelectronic, and photochemical properties. ZnO nanomaterials, such as quantum dots, nanorods, and nanowires, have been intensively investigated for their important properties. Many methods have been described in the literature for the production of ZnO nanostructures, such as laser ablation, hydrothermal methods, electrochemical deposition, sol–gel methods, Chemical Vapour Deposition, molecular beam epitaxy, the common thermal evaporation method, and the soft chemical solution method. The present Special Issue is devoted to the Synthesis and Characterization of ZnO nanostructures with novel technological applications.

Keywords: ZnO; synthesis; characterization; nanoparticles; nanorods; quantum wires; thin films

Among various metal oxide materials, ZnO presents itself as a multifunctional material due to its own properties and functionalities. The properties of ZnO include its wide band gap (3.37 eV), high exciton binding energy (60 meV) [1,2], biocompatibility, and ease of fabrication. Due to its excellent properties, ZnO is widely used for various potential applications, such as catalysis, solar cells, ultraviolet (UV) lasers, light-emitting diodes, photo-detectors, sensors (chemical, bio-, and gas), and optical and electrical devices [3]. Among various applications, the use of ZnO nanomaterials as a photocatalyst has attracted particular interest due to their large surface area; wide band gap; ease of fabrication and cost effective synthesis; and biocompatible and environmentally benign nature [4].

The synthesis of large-scale arrayed one-dimensional (1D) ZnO nanostructures, including nanowires, nanorods, nanobelts, and whiskers, is an important step for the fabrication of functional nano/microdevices. Recently, because of its high-temperature strength and rigidity, as well as excellent chemical stability, small-diameter ZnO whiskers have received great attention for industrial applications as reinforcement phase in composite materials. ZnO whiskers with a high aspect ratio have also been successfully used as a probing tip to develop new precise high-resolution imaging techniques for atomic force microscopy and scanning tunneling microscopy.

This Special Issue covers 3 review articles, 1 brief report, and 13 research articles. First of all, Chaudhary et al. [5] present an overview of the current advancements of ZnO-nanomaterial-based chemical sensors. Various operational factors, such as the effect of size, morphologies, compositions, and their respective working mechanisms, along with the selectivity, sensitivity, detection limit, and stability are discussed in this article. Scherzad et al. [6] in their review article summarize the existing data regarding the DNA damage that ZnO nanoparticles (NPs) induce, and focus on the possible molecular mechanisms underlying genotoxic events. Wang et al. [7] present a review on three-dimensional (3D) ZnO hierarchical nanostructures, and summarize major advances in solution phase synthesis and applications in the environment and electrical/electrochemical devices. They present the principles and growth mechanisms of ZnO nanostructures via different solution methods with an emphasis on rational control of the morphology and assembly. Then, they discuss the applications of 3D ZnO hierarchical nanostructures in photocatalysis, field emissions, electrochemical sensors, and lithium ion batteries. In the research articles, Giannouli et al. [8] present a comparative assessment of nanowire- versus nanoparticle-based ZnO dye-sensitized solar cells (DSSCs) in order to investigate the main parameters that affect device performance. Bittner at al. [9] perform a low-temperature fabrication of flexible ZnO photo-anodes for dye-sensitized solar cells (DSSCs) by templated electrochemical deposition of films in an enlarged and technically simplified deposition setup to demonstrate the feasibility of the scale up of the deposition process. Kwoka et al. [10] present the results of detailed X-ray photoelectron spectroscopy (XPS) studies combined with atomic force microscopy (AFM) investigation concerning the local surface chemistry and morphology of nanostructured ZnO thin films. Giuli et al. [11] report on the structural and electrochemical characterization of Fe-doped ZnO samples with varying dopant concentrations, which may potentially serve as anodes for rechargeable lithium-ion batteries (LIBs). Xia et al. [12] present two new functional materials based on zinc oxide (ZnO)—a legacy material in semiconductors but exceptionally novel to solid state ionics—that are developed as membranes in solid oxide fuel cells (SOFCs) for the first time. Sarwar et al. [13] in their work study an NH_4OH treatment to provide an optimum morphological trade-off to Ga-doped ZnO nanorods (n-GZR)/p-Si heterostructure characteristics. Quiñones et al. [14] in their paper use perfluorinated phosphonic acid modifications to modify zinc oxide (ZnO) nanoparticles because they create a more stable surface due to the electronegativity of the perfluoro head group. Umar et al. [15] report the growth of In-doped ZnO (IZO) nanomaterials, i.e., stepped hexagonal nanorods and nanodisks, by a thermal evaporation process using metallic zinc and indium powders in the presence of oxygen. The as-grown IZO nanomaterials were investigated by several techniques in order to examine their morphological, structural, compositional, and optical properties. From an application point of view, the grown IZO nanomaterials were used as a potential scaffold to fabricate sensitive phenyl hydrazine chemical sensors based on the I–V technique. In Pimentel et al.'s work [16], tracing and Whatman papers were used as substrates to grow zinc oxide (ZnO) nanostructures. The time-resolved photocurrent of the devices in response to UV being turned on/off was investigated and it has been observed that the ZnO nanorod arrays grown on the Whatman paper substrate present a responsivity that is 3 times greater than the ones grown on tracing paper. By using ZnO nanorods, the surface area-to-volume ratio will increase and will improve the sensor's sensibility, making these types of materials good candidates for low-cost and disposable UV sensors. Sagasti et al. [17] synthesized a nanostructured ZnO layer onto a Metglas magnetoelastic ribbon to immobilize hemoglobin (Hb) on it and study the Hb's electrochemical behavior towards hydrogen peroxide. Zhou et al. [18] fabricated a highly sensitive acetone chemical sensor using a ZnO nanoballs-modified silver electrode. Ibrahim et al. [19] reported a facile synthesis, characterization, and electrochemical-sensing application of ZnO nanopeanuts synthesized by a simple aqueous solution process and characterized by various techniques in order to confirm the compositional, morphological, structural, crystalline phase, and optical properties of the synthesized material. Beshkar et al. [20] have demonstrated a facile formation of CuO nanostructures on copper substrates by the oxidation of copper foil in ethylene glycol at 80 °C. The hydrophobic property of the products was characterized by means of water contact angle measurement. After simple surface modification with stearic acid and polydimethylsiloxane (PDMS), the resulting films showed hydrophobic and even superhydrophobic characteristics due to their special surface energy and nano-microstructure morphology. Finally, in their brief report, Hoffman et al. [21] use two-dimensional fluorescence difference spectroscopy (2-D FDS) to determine the unique spectral signatures of zinc oxide (ZnO), magnesium oxide (MgO), and a 5% magnesium zinc oxide nanocomposite (5% Mg/ZnO), which was then used to demonstrate the change in spectral signature that occurs when physiologically important proteins, such as angiotensin-converting enzyme (ACE) and ribonuclease A (RNase A), interact with ZnO nanoparticles (NPs).

Conflicts of Interest: The author declares no conflict of interest.

References

1. Baskoutas, S.; Bester, G. Conventional Optics from Unconventional Electronics in ZnO colloidal quantum dots. *J. Phys. Chem. C* **2010**, *114*, 9301–9307. [CrossRef]
2. Baskoutas, S.; Bester, G. Transition in the Optical Emission Polarization of ZnO Nanorods. *J. Phys. Chem. C* **2001**, *115*, 15862–15867. [CrossRef]
3. Chrissanthopoulos, A.; Baskoutas, S.; Bouropoulos, N.; Dracopoulos, V.; Poulopoulos, P.; Yannopoulos, S.N. Synthesis and characterization of ZnO/NiO p-n heterojunctions: ZnO nanorods grown on NiO thin film by thermal evaporation. *Photonic Nanostruct.* **2011**, *9*, 132–139. [CrossRef]
4. Dar, G.N.; Umar, A.; Zaidi, S.A.; Baskoutas, S.; Hwang, S.W.; Abaker, M.; Al-Hajry, A.; Al-Sayari, S.A. Ultra-high sensitive ammonia chemical sensor based on ZnO nanopencils. *Talanta* **2012**, *89*, 155–161. [CrossRef] [PubMed]
5. Chaudhary, S.; Umar, A.; Bhasin, K.K.; Baskoutas, S. Chemical Sensing Applications of ZnO Nanomaterials. *Materials* **2018**, *11*, 287. [CrossRef] [PubMed]
6. Scherzad, A.; Meyer, T.; Kleinsasser, N.; Hackenberg, S. Molecular Mechanisms of Zinc Oxide Nanoparticle-Induced Genotoxicity Short Running Title: Genotoxicity of ZnO NPs. *Materials* **2017**, *10*, 1427. [CrossRef] [PubMed]
7. Wang, X.; Ahmad, M.; Sun, H. Three-Dimensional ZnO Hierarchical Nanostructures: Solution Phase Synthesis and Applications. *Materials* **2017**, *10*, 1304. [CrossRef] [PubMed]
8. Giannouli, M.; Govatsi, K.; Syrrokostas, G.; Yannopoulos, S.N.; Leftheriotis, G. Factors Affecting the Power Conversion Efficiency in ZnO DSSCs: Nanowire vs. Nanoparticles. *Materials* **2018**, *11*, 411. [CrossRef] [PubMed]
9. Bittner, F.; Oekermann, T.; Wark, M. Scale-Up of the Electrodeposition of ZnO/Eosin Y Hybrid Thin Films for the Fabrication of Flexible Dye-Sensitized Solar Cell Modules. *Materials* **2018**, *11*, 232. [CrossRef] [PubMed]
10. Kwoka, M.; Lyson-Sypien, B.; Kulis, A.; Maslyk, M.; Borysiewicz, M.A.; Kaminska, E.; Szuber, J. Surface Properties of Nanostructured, Porous ZnO Thin Films Prepared by Direct Current Reactive Magnetron Sputtering. *Materials* **2018**, *11*, 131. [CrossRef] [PubMed]
11. Giuli, G.; Eisenmann, T.; Bresser, D.; Trapananti, A.; Asenbauer, J.; Mueller, F.; Passerini, S. Structural and Electrochemical Characterization of $Zn_{1-x}Fe_xO$—Effect of Aliovalent Doping on the Li^+ Storage Mechanism. *Materials* **2018**, *11*, 49. [CrossRef] [PubMed]
12. Xia, C.; Qiao, Z.; Feng, C.; Kim, J.-S.; Wang, B.; Zhu, B. Study on Zinc Oxide-Based Electrolytes in Low-Temperature Solid Oxide Fuel Cells. *Materials* **2018**, *11*, 40. [CrossRef] [PubMed]
13. Hassan, A.; Rana, S.; Kim, H.-S. NH_4OH Treatment for an Optimum Morphological Trade-off to Hydrothermal Ga-Doped n-ZnO/p-Si Heterostructure Characteristics. *Materials* **2018**, *11*, 37.
14. Quiñones, R.; Shoup, D.; Behnke, G.; Peck, C.; Agarwal, S.; Gupta, R.K.; Fagan, J.W.; Mueller, K.T.; Iuliucci, R.J.; Wang, Q. Study of Perfluorophosphonic Acid Surface Modifications on Zinc Oxide Nanoparticles. *Materials* **2017**, *10*, 1363. [CrossRef] [PubMed]
15. Umar, A.; Kim, S.H.; Kumar, R.; Al-Assiri, M.S.; Al-Salami, A.E.; Ibrahim, A.A.; Baskoutas, S. In-Doped ZnO Hexagonal Stepped Nanorods and Nanodisks as Potential Scaffold for Highly-Sensitive Phenyl Hydrazine Chemical Sensors. *Materials* **2017**, *10*, 1337. [CrossRef] [PubMed]
16. Pimentel, A.; Samouco, A.; Nunes, D.; Araújo, A.; Martins, R.; Fortunato, E. Ultra-Fast Microwave Synthesis of ZnO Nanorods on Cellulose Substrates for UV Sensor Applications. *Materials* **2017**, *10*, 1308. [CrossRef] [PubMed]
17. Sagasti, A.; Bouropoulos, N.; Kouzoudis, D.; Panagiotopoulos, A.; Topoglidis, E.; Gutiérrez, J. Nanostructured ZnO in a Metglas/ZnO/Hemoglobin Modified Electrode to Detect the Oxidation of the Hemoglobin Simultaneously by Cyclic Voltammetry and Magnetoelastic Resonance. *Materials* **2017**, *10*, 849. [CrossRef] [PubMed]
18. Zhou, Q.; Hong, C.X.; Yao, Y.; Ibrahim, A.M.; Xu, L.; Kumar, R.; Talballa, S.M.; Kim, S.H.; Umar, A. Fabrication and Characterization of Highly Sensitive Acetone Chemical Sensor Based on ZnO Nanoballs. *Materials* **2017**, *10*, 799. [CrossRef] [PubMed]
19. Ibrahim, A.A.; Tiwari, P.; Al-Assiri, M.S.; Al-Salami, A.E.; Umar, A.; Kumar, R.S.H.; Kim, Z.; Ansari, A.; Baskoutas, S. A Highly-Sensitive Picric Acid Chemical Sensor Based on ZnO Nanopeanuts. *Materials* **2017**, *10*, 795. [CrossRef] [PubMed]

20. Beshkar, F.; Khojasteh, H.; Salavati-Niasari, M. Flower-Like CuO/ZnO Hybrid Hierarchical Nanostructures Grown on Copper Substrate: Glycothermal Synthesis, Characterization, Hydrophobic and Anticorrosion Properties. *Materials* **2017**, *10*, 697. [CrossRef] [PubMed]
21. Hoffman, A.; Wu, X.; Wang, J.; Brodeur, A.; Thomas, R.; Thakkar, R.; Hadi, H.; Glaspell, G.P.; Duszynski, M.; Wanekaya, A.; DeLong, R.K. Two-Dimensional Fluorescence Difference Spectroscopy of ZnO and Mg Composites in the Detection of Physiological Protein and RNA Interactions. *Materials* **2017**, *10*, 1430. [CrossRef] [PubMed]

materials

MDPI

Article

Flower-Like CuO/ZnO Hybrid Hierarchical Nanostructures Grown on Copper Substrate: Glycothermal Synthesis, Characterization, Hydrophobic and Anticorrosion Properties

Farshad Beshkar, Hossein Khojasteh and Masoud Salavati-Niasari *

Institute of Nano Science and Nano Technology, University of Kashan, Kashan 87317-51167, Iran;
f.beshkar69@yahoo.com (F.B.); hn.khojasteh@gmail.com (H.K.)
* Correspondence: salavati@kashanu.ac.ir; Tel.: +98-315-591-2383; Fax: +98-315-591-3201

Received: 12 May 2017; Accepted: 21 June 2017; Published: 25 June 2017

Abstract: In this work we have demonstrated a facile formation of CuO nanostructures on copper substrates by the oxidation of copper foil in ethylene glycol (EG) at 80 °C. On immersing a prepared CuO film into a solution containing 0.1 g $Zn(acac)_2$ in 20 mL EG for 8 h, ZnO flower-like microstructures composed of hierarchical three-dimensional (3D) aggregated nanoparticles and spherical architectures were spontaneously formed at 100 °C. The as-synthesized thin films and 3D microstructures were characterized using XRD, SEM, and EDS techniques. The effects of sodium dodecyl sulphate (SDS), cetyltrimethylammonium bromide (CTAB), and polyethylene glycol (PEG) 6000 as surfactants and stabilizers on the morphology of the CuO and ZnO structures were discussed. Possible growth mechanisms for the controlled organization of primary building units into CuO nanostructures and 3D flower-like ZnO architectures were proposed. The hydrophobic property of the products was characterized by means of water contact angle measurement. After simple surface modification with stearic acid and PDMS, the resulting films showed hydrophobic and even superhydrophobic characteristics due to their special surface energy and nano-microstructure morphology. Importantly, stable superhydrophobicity with a contact angle of 153.5° was successfully observed for CuO-ZnO microflowers after modification with PDMS. The electrochemical impedance measurements proved that the anticorrosion efficiency for the CuO/ZnO/PDMS sample was about 99%.

Keywords: CuO film; flower-like ZnO; hydrophobicity; contact angle; anticorrosion

---

1. Introduction

Copper is widely applied in many applications such as the electronic industries and communications as a conductor in electrical power lines, pipelines for domestic and industrial water utilities including seawater, heat conductors, and heat exchangers [1]. Due to their special physical properties and high potential for various electronic and photonic device applications, semiconductors have attracted wide attention [2]. CuO is a p-type semiconductor that has a small band gap (Eg = 1.2 eV) and it is one of the most important and versatile semiconductors that is widely used as superconductors, catalysts, a potential field-emission material, and that has high-temperature durability [3,4]. Due to these properties, copper and its alloys have been one of the important materials in industry. Thus, the protection of used copper has particular importance in the industry. One of the effective methods for surface protection of metals against external threats is by making their surface hydrophobic to repulse water and its solved corrosion containing materials.

On the other hand, zinc oxide (ZnO) is a famous n-type semiconductor (Eg = 3.37 eV at 300 K) that has a high electron mobility and low recombination loss [5]. Because of its semiconducting, piezoelectric, and pyroelectric attributes, ZnO has become one of the most important semiconductor materials used for diverse applications such as optoelectronics, gas sensing, catalysis, solar cell, and actuators [6–9].

Producing hybrid nanostructured materials in order to achieve various properties such as high surface area, good conductivity, and permeability, have been considered to be one of the most important functional materials for the various research fields [10,11]. Some investigations have been reported for directly constructing ZnO nano or micro-structures on copper substrates. The combination of CuO and ZnO nanostructures has been reported, but at high temperature to oxidize a copper foil to CuO followed by the deposition of ZnO on the copper oxide surface [12]. As a result, many CuO/ZnO hybrid nanostructures have been fabricated so far, including CuO/ZnO core/shell heterostructures produced by thermal decomposition [13], CuO/ZnO nanocomposites for non-enzymatic glucose sensing applications [14], ZnO/CuO hetero-hierarchical nanotrees prepared by hydrothermal preparation [15], and flower-like CuO-ZnO structured nanowires by chemical deposition [16].

In addition, it has been reported that the hierarchical nanorod-like structure of the CuO/ZnO photoelectrode provides higher surface area, more reactive sites, effective light absorption, and reduced charge transfer resistance at the electrode/electrolyte interface that showed superior photoelectrochemical properties. Also, the photo conversion efficiency and stability of the CuO/ZnO photoelectrode were enhanced due to the formation of p-n junctions along the p-CuO core and n-ZnO protective shell, respectively [17].

As we know the performance of hybrid nanomaterials depends upon their size, morphology, composition, dispersion, structure, crystal phases, and crystal facets [18]. The surface of a material is the most important factor determining the compatibility and type of interaction with its environment [19]. In addition, the wetting behavior is one of the most important characteristic of a solid surface and is strongly influenced by the size and morphology of the surface particle structure. This property has a particular effect on the life span, energy consumption, and practical purposes of engineered materials [20]. Therefore, the controlled synthesis of hybrid metal oxide nanostructures is of prominent interest to obtain the required grain size, surface morphology, and structure of the hybrid nanostructure.

In the present work, flower-like ZnO micro/nanostructures fabricated on a copper substrate were produced by a facile, affordable, and environmentally-friendly glycothermal method in the presence of ethylene glycol as reduction agent and green solvent. Meanwhile, the effects of different surfactants on the final surface morphology and coarseness were investigated. The wetting and anticorrosion properties of the samples were also studied by the measurement of the contact angle and electrochemical impedance, respectively.

2. Experimental Section

2.1. Materials and Methods

Copper foils, hydrochloric acid (37%), glacial acetic acid, ethylene glycol (EG), cetyltrimethylammonium bromide (CTAB), sodium dodecyl sulphate (SDS), polyethylene glycol 6000 (PEG 6000), and zinc acetylacetonate (Zn(acac)$_2$) were purchased from Merck Company. Polydimethylsiloxane (PDMS) was provided by Sigma-Aldrich. Ammonia solution (25%), stearic acid, and ethanol were kindly provided by Ghatran Shimi Co (Tehran, Iran). All materials and solvents were used without further purification. Deionized water (DI water) was used throughout. GC-2550TG (Teif Gostar Faraz Company, Tehran, Iran) were used for all chemical analyses. The synthesized architecture materials were characterized using various analytical methods. XRD patterns were recorded by a Philips, X-ray diffractometer (Philips, Egham, England) using Ni-filtered Cu Kα radiation. The morphology of the products was measured using a Hitachi S-4160 field emission scanning electron microscope (FESEM, Hitachi, Ltd., Tokyo, Japan). Prior to taking images, the samples were coated with a very thin layer of Pt to make the sample surface conducting and prevent charge accumulation. The energy dispersive spectrometry (EDS) analysis was conducted using a Tescan mira3 microscope (Hitachi, Ltd., Tokyo, Japan). Contact angles (CAs) were measured using a contact-angle meter (Dataphysics, OCA 15 plus, DataPhysics Instruments GmbH, Filderstadt, Germany) equipped with a CCD camera at room temperature. The volume of water drip was approximately 4 µL. The average CA value was reported by measuring the right and left positions for the droplet.

2.2. Synthesis of Nanoscale CuO Arrays on the Cu Substrate

The CuO nanoscale array was prepared via in situ engraving of the Cu foil (1 cm × 1 cm) in alkaline conditions. Briefly, the Cu foils were immersed in 10 mL of 3 M HCl aqueous solution and ultrasonicated for about 10 min to refresh the surface and remove the surface impurities and oxide layers. After that, the foils were rinsed with ethanol and distilled water two times for 10 min in a sonication bath, respectively, and then were dried at 60 °C for 1 h in air atmosphere (Figure 1a). At the second step, the cleaned Cu foils were immersed into a sealed beaker containing 2.8 mL of concentrated HCl, 1.7 mL of acetic acid, and 10 mL of deionized water solution, and the reaction was performed at room temperature for 6 h. In this step, a rough surface for the modification and growth of the other material will be achieved. The change in the foil appearance during the reaction is shown in Figure 1b. As seen in the Figure 1a, the surface of the copper foil became bright and smooth after the cleaning treatment.

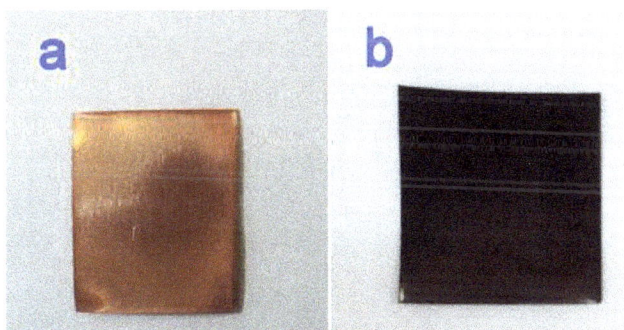

Figure 1. The appearance of the Cu foils between the different stages treatments. (**a**) the cleaned Cu foil; (**b**) the Cu foil reacted with EG at 80 °C for 5 days.

The prepared Cu foils were placed in 10 mL of EG at 80 °C for 5 days. The obtained samples were rinsed in DI water and then dried in air at 60 °C for 30 min. After drying, the uniformly black surface layer of the CuO nanosphere array growing on the Cu foil was obtained (Figure 1b). Then the surface morphology of CuO in the absence (C0 sample) or presence of 0.2 g of different stabilizers or surfactants (C1 = CTAB, C2 = SDS, and C3 = PEG6000) were investigated. The detailed preparation conditions are summarized in Table 1.

Table 1. Preparation conditions for all of the samples.

Sample	Stabilizer or Surfactant	Temperature (°C)	Time (h)	Morphology
C0	-	80	120	Rough surface with nanoscale particles
C1	PEG 6000	80	120	Coalesced particles and bulk structures
C2	SDS	80	120	Sponge-like
C3	CTAB	80	120	Agglomerated and impacted structures
Z0	-	100	8	Symmetry flower-like microstructures
Z1	PEG 6000	100	8	Symmetry cabbage-like microstructures
Z2	SDS	100	8	Asymmetry microstructures
Z3	CTAB	100	8	Non-sized cabbage-like microstructures

2.3. Preparation of 3D CuO-ZnO Hybrid Hierarchical Structures

To prepare the CuO-ZnO flower-like structures, a simple glycothermal method was used. A schematic illustration of the synthesis of flower-like ZnO is shown in Scheme 1. A solution containing 0.1 g of Zn(acac)$_2$ in 20 mL EG was prepared and stirred for several minutes to obtain a homogeneous solution. The needed amount of ammonia was added dropwise to the solution and the pH was adjusted to 10. After that, the previously prepared foils were immersed in the obtained solution. The prepared mixture was transferred to a 250 mL Teflon-lined autoclave and the substrate was hung in the Teflon container. The reaction was performed in glycothermal conditions at 100 °C for 8 h. Finally, the samples were rinsed in ethanol and DI water three times and then dried in air at 60 °C for 6 h. Here, the effects of the absence (Z0 sample) or presence of different stabilizer or surfactants (Z1 = CTAB, Z2 = SDS, and Z3 = PEG6000) with specific molar ratios (Zn:Surfactant = 1:2) on the final produced structures were investigated. The detailed preparation conditions are mentioned previously in Table 1.

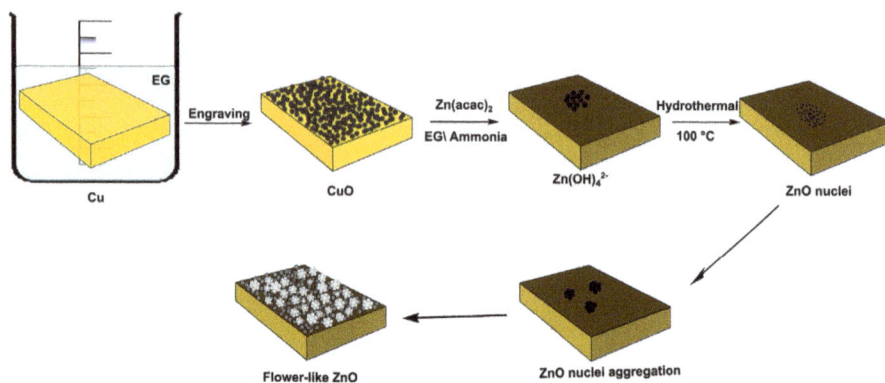

Scheme 1. Schematic illustration of the flower-like CuO-ZnO synthesis procedure.

2.4. Anticorrosion Behaviour of the Superhydrophobic CuO-ZnO Film

Electrochemical studies were carried out using an AUTOLAB model PGSTAT 30. To investigate the anticorrosion ability, Tafel plots and electrochemical impedance spectroscopy (EIS) tests were used by a conventional three-electrode cell (the platinum foil as the counter electrode, Ag/AgCl/KCl (3 mol·L^{-1}) as the reference electrode, and the foil with an exposed area of 1 cm^2 as the working electrode) with a capacity of 100 mL. All impedance curves were recorded at room temperature (25 ± 2 °C). The working electrode was immersed in the test solution (in aqueous NaCl solution (3.5%)) for 60 min until a steady state open-circuit potential was attained. The EIS measurements were performed over the frequency range from 100 kHz to 100 mHz at the open circuit potential by superimposing the alternating current (AC) signal of 0.01 V after immersion for 60 min in the corrosive media. NOVA software was used for fitting the impedance data in an equivalent circuit as well as for extrapolating the Tafel slopes. Experiments were carried out in triplicate to ensure the reproducibility of the results. The inhibition efficiencies (η) for each inhibitor concentration were calculated using Equation (1):

$$\eta\% = \frac{R_{Ct} - R_0}{R_{Ct}} \times 100 \tag{1}$$

where R_{ct} and R_0 are the charge transfer resistances in the presence and absence of the inhibitor, respectively.

3. Results and Discussion

X-ray powder diffraction (XRD) is a rapid, accurate, and nondestructive technique used for chemical and physical analysis of the materials. It is mostly used for phase identification of a crystalline material and can provide information on the unit cell dimensions. To compare and confirm the phase composition of the product, here the XRD experiment was carried out.

For evaluation of the crystal structure and purity of the as-prepared samples, wide-angle XRD patterns of clean Cu foil (C0 sample) and the as prepared final product (Z0 sample) were taken. The results are shown in Figure 2 as a comparison. As seen in Figure 2a, the Cu diffraction peaks are very strong and all the peaks corresponding to the face-centered cubic Cu were well matched with the database in JCPDS (File No. 02-1225).

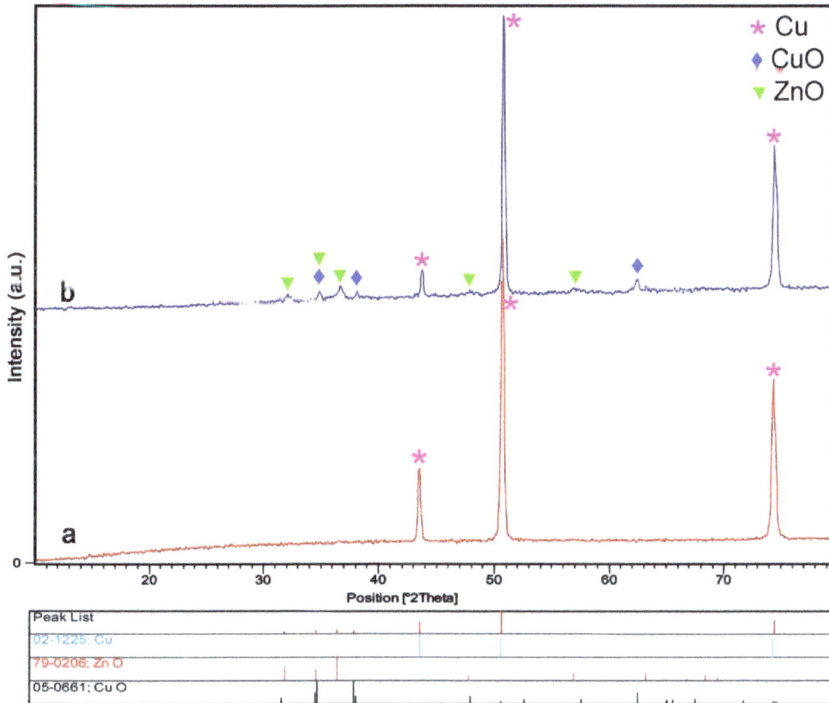

Figure 2. Wide-angle XRD patterns of the clean Cu foil obtained after cleaning treatment stage (**a**) and the as-prepared flower-like CuO-ZnO micro/nanostructures obtained after the engraving and glycothermal stages (**b**).

After interaction with acid and $Zn(acac)_2$, the XRD pattern of the product grown on the upward surface of the copper foil demonstrates the presence of phases, a monoclinic structure for CuO, and a hexagonal structure for flower-like ZnO (Figure 2b). It could be seen that the ZnO and CuO diffraction peaks are very weak, which is likely due to the small amount formed on the clean Cu substrate. The observed lattice constant values are in good agreement with the standard values with JCPDS file no. 79-0206 (a = 3.249 Å, c = 5.206 Å) for ZnO and JCPDS file no 05-066 (a = 4.688 Å, b = 3.423 Å, c = 5.132 Å) for the CuO crystals. The XRD pattern confirmed that the synthesis was successful.

For investigation of the chemical purity of the synthesized product, elemental analysis of 3D CuO-ZnO hybrid structures was performed by energy dispersive X-ray spectroscopy (EDS). The obtained EDS spectrum reveals the composition of the Cu, Zn, and O elements without any other impurity elements, as shown in Figure 3. The strong peaks of the Cu, Zn, and O elements are exhibited in the EDS spectrum. The Cu element originated from the CuO substrate and the Zn was contributed by the flower-like ZnO hierarchical structures. Furthermore, the atomic ratios of Zn to O and Cu to O are both 1:1, respectively, and this observation proves that the stoichiometric ratio is maintained.

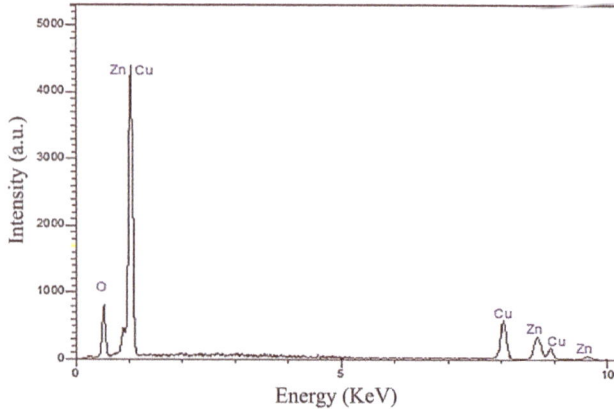

Figure 3. EDS spectrum of the flower-like CuO-ZnO hybrid hierarchical nanostructures grown on copper substrate.

Scanning Electron microscopy (SEM) was used to study the morphology of the products. In the SEM technique, a focused electron beam scans the conductive sample surface and reveals information about the sample including the external morphology (texture) and topography. Figure 4 shows the SEM images of the CuO nanostructures engraved on the Cu foil surface in different synthesis conditions. It is clearly observed from the SEM images that the copper foils were covered with films consisting of a large number of nanoparticles. Figure 4a indicates that the C0 sample synthesized from chemical oxidation has a sphere like structure. According to Figure 4a, the nanoscale roughness of the copper surface can clearly be seen and indicates that after interaction with the EG agent, the copper surface is covered with numerous nanoparticles which are typically 50–90 nm in size.

As shown in Figure 4b, when PEG 6000 was used as capping agent (C1 sample), a combination of coalesced particles and bulk structures are made. We propose that because PEG 6000 is a macromolecule with a high molecular weight, it acts as an obstacle by hindering the contact between the clean Cu foil surface and the EG solvent. Irregular distributions of the concentration of PEG 6000 form a disordered pattern of CuO structures on the Cu foil substrate. For the sample with SDS as the anionic surfactant (C2 sample), the SEM image shows that sponge-like CuO has been obtained (Figure 4c). It seems that after formation of the primary nucleus, this surfactant is adsorbed preferentially on the nuclei surface, and inhibits the accessibility of the surface for reactants by the steric hindrance mechanism [21]. Additionally, agglomerated and impacted structures were obtained by using CTAB as a cationic surfactant as shown in Figure 4d (sample C3). The formation of dense structures in Figure 4d using CTAB is associated with its cationic head group, and this surfactant easily interacts with free oxygen groups on the surface of the CuO substrate and increases the agglomeration of structures [22]. The SEM analysis indicates that the used surfactants operate as preventing agents for ethylene glycol to adhere to the Cu substrate. Therefore, using surfactants with high molecular weight and those which have strong interactions with the substrate is not recommended.

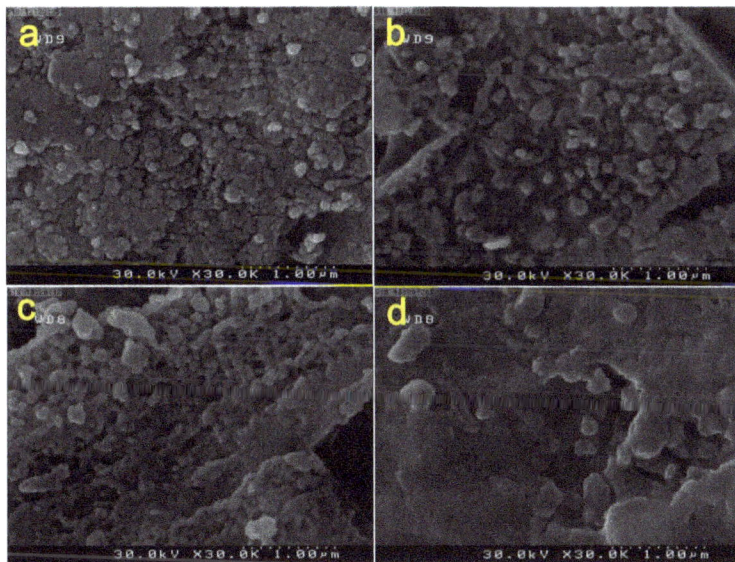

Figure 4. The SEM images of CuO nanostructures engraved on the Cu foil surface in different synthesis conditions. (**a**) Without any stabilizer or surfactants; and in the presence of (**b**) PEG 6000, (**c**) SDS; (**d**) CTAB.

3.1. The Mechanism of CuO Formation

The compound for starting the synthesis of CuO is hydroxide. The main feature of the reaction mechanism is that the reaction proceeds via the solution phase rather than the solid phase, and the metal particles are formed by nucleation and growth from the solution. As described in Figure 5 (Equations (1)–(3)), the formation of the main product, diacetyl, can be explained in terms of a double oxidation of acetaldehyde, previously produced by the dehydration of ethylene glycol [23].

In addition, the produced OH^- ions can react with metal ions for the formation of metal hydroxide (Equation (4)) [23,24]. Generally, the proposed reactions for the growth of CuO nanoparticles essentially consist of two stages of oxidation and dehydrogenation [25].

In the following, during the reaction, Cu foils as copper sources dissolved in the solutions and produced Cu^{2+} ions with two electrons left behind. Previous investigations have proven that the oxidation rate occurs quickly, and the Cu^{2+} ions are continuously released into the solution [26]. During the growth in a solution of EG, the Cu foil was slowly reacted and oxidized to $Cu(OH)_2$ in alkaline solution. It has been proven that Cu^{2+} ions prefer to arrange in a square-planar coordination with OH^-. Therefore, the transition $Cu(OH)_2$ product reacts with more hydroxyl groups to yield $Cu(OH)_4^{2-}$ species, which precipitate as $Cu(OH)_2$ on top of the copper foils, and then decompose into CuO (solid structures) after dehydration at 80 °C for 5 days [27–29]. As is well-known, during the formation of CuO nanoarrays in solution, $Cu(OH)_2$ usually serves as the precursor and template during the thermal dehydration process. Moreover, the $Cu(OH)_2$ precipitate is a thermodynamically metastable phase, which can be easily transformed into the more stable CuO solid film. This transformation is found to take place in aqueous media at a relatively low temperature [20]. The mechanism reactions for the formation of CuO nanoarrays are as follows (Figure 5 Equations (5)–(9)):

$$CH_2OH\text{-}CH_2OH \xrightarrow{\text{- }H_2O} CH_3\text{-}CHO \tag{1}$$

(2)

(3)

$$2\ HOCH_2CH_2OH + O_2 \longrightarrow 2\ HOCH_2CHO + 2H_2O \tag{4}$$

$$Cu^0 \longrightarrow Cu^{2+} + 2e^- \tag{5}$$

$$Cu^{2+} + 2OH^- \longrightarrow Cu(OH)_2\ \text{(Precipitate)} \tag{6}$$

$$Cu(OH)_2 + 2OH^- \longrightarrow Cu(OH)_4^{\ 2-}\ \text{(Complexions)} \tag{7}$$

$$2Cu(OH)_4^{\ 2-} \longrightarrow 2Cu(OH)_2\ \text{(Precipitate)} + H_2O + O_2 + 4e^- \tag{8}$$

$$Cu(OH)_2 \longrightarrow CuO\ \text{(Solid film)} + H_2O \tag{9}$$

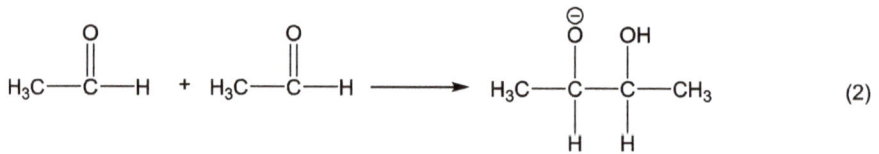

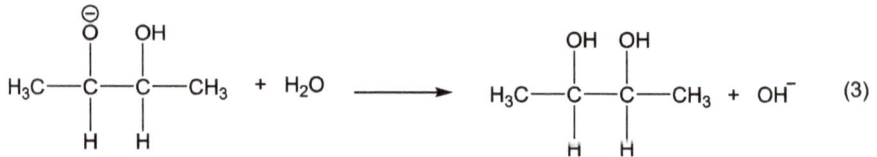

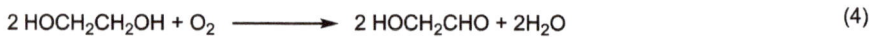

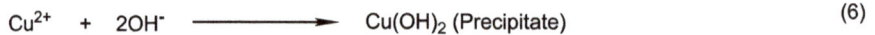

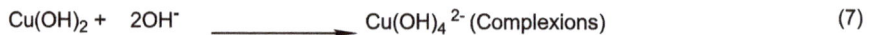

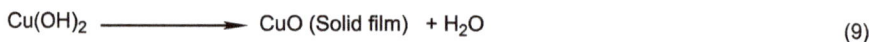

Figure 5. The proposed mechanism of CuO formation.

3.2. Studying the Morphology and Growth Mechanism of the Flower Like ZnO Products

Uniformly grown flower like ZnO hierarchical structures on the CuO substrate were confirmed using FE-SEM analysis. SEM images for the samples prepared without any surfactant or stabilizer (Z0 sample) at different magnifications are shown in Figure 6. The results indicate that flower like ZnO structures are partially grown on the CuO film (Figure 6a). The ZnO hybrid hierarchical structures are easily distinguishable from the CuO undercoat. The high magnification SEM image of the Z0 sample reveals that the flower-like ZnO has a special 3D structure that is stacked by seven mini-spheres. The average diameter of each petal of the ZnO microspheres is about 554 nm and the diameter of the total flower like structure is about 1.485 μm. Each mini-sphere contains numerous nanoparticles and the structures are symmetrical with specific shapes (Figure 6b).

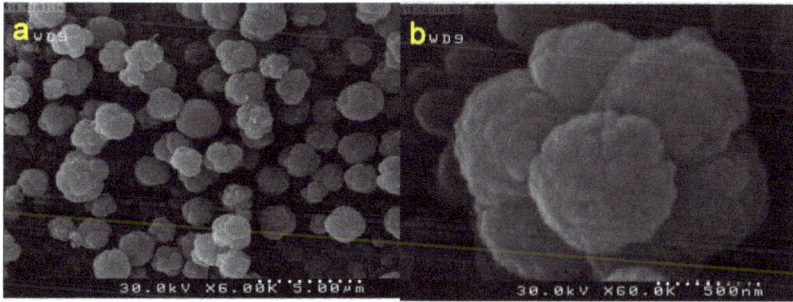

Figure 6. SEM images for the flower-like ZnO hierarchical nanostructures prepared without any surfactant or stabilizer at (**a**) 5 μm and (**b**) 500 nm magnifications.

Also, the effects of the surfactants and stabilizers on the final morphology of the 3D ZnO structures were studied. A representative SEM image of the 3D CuO-ZnO hierarchical structures is shown in Figure 7. According to Figure 7a, it is clear that when PEG 6000 is used as the capping agent (Z1 sample), the particles are aggregated and bulk structures are made. The flower like structure has also been destroyed and only micro spheres are obtained. In comparison with the Z0 sample, the microspheres are denser but the mean size of the spheres is approximately the same as the flower like structures (1.491 μm).

Figure 7. SEM images for the effects of surfactants and stabilizers on the final morphology of 3D CuO-ZnO hybrid hierarchical nanostructures. In the presence of (**a**) PEG 6000; (**b**) SDS and (**c**) CTAB.

The use of SDS as the surfactant had different impacts (Z2 sample). It seems that SDS leads to the formation of dome-shaped with flat base microstructures (Figure 7b). Some relics of the flower like symmetry still remain. Also in comparison with the Z0 sample, the mean size of the microstructures is increased ($\approx$1.7 μm). The surface of the CuO nanospheres is almost completely covered by ZnO materials. The results prove that the existence of SDS has real effects on the diffusion and crystallization of the flower-like ZnO structures. Thus, it is apparent that sodium dodecyl sulfate can induce morphological changes in the ZnO aggregates in polar solvents [30]. Although the effect of SDS on the ZnO architecture is obvious, the origin of the interaction is more complex. Sodium dodecyl sulfate (SDS), is an amphiphilic molecule that has an anionic head group composed of a sulfate ion and a sodium counterion, while the surfactant tail is a linear hydrophobic alkyl chain [31]. We propose that the effect of SDS on the morphology of the ZnO aggregates is a result of the electrostatic shielding properties of the sulfate and sodium ions. Because of electrostatic repulsion, the anionic head group of SDS has a repulsive interaction with the –OH groups of the EG solvent. On the other hand, the long hydrophobic chain is not soluble in polar solvents such as EG. So SDS does not undergo self-assembly in the solvent mixtures used throughout this study, and interacts in a different way with the Zn^{2+} ions to precipitate on the CuO substrate.

Figure 7c depicts the FE-SEM image of the sample Z3 synthesized using CTAB as the surfactant. As shown, the morphology of this sample is micro-particles with average sizes of about 1 μm. The morphology of the ZnO microstructures is changed and cabbage like structures are obtained. The formation of the small and dense structures is due to its cationic head group. CTAB easily interacts with the free oxygen groups on the surface of the formed ZnO nucleus and agglomeration and the final size of the products will decrease. On this basis, to obtain flower like ZnO architectures with better geometric symmetry, the use of these capping agents and surfactants are not recommended. Therefore, further investigations were continued with the Z0 sample. Figure 8 depicts the cross section view of the CuO-ZnO hierarchical structure. As is shown, a layer of flower like ZnO structures is uniformly covered on the CuO nanosphere substrate. The ZnO flowers have an average cross-section diameter and length of 1.8 μm and 1.9 μm, respectively. From the reaction process point of view, the reaction routes for the formation of flower like ZnO using $Zn(acac)_2$ as a precursor are shown below (Equations (10)–(12)):

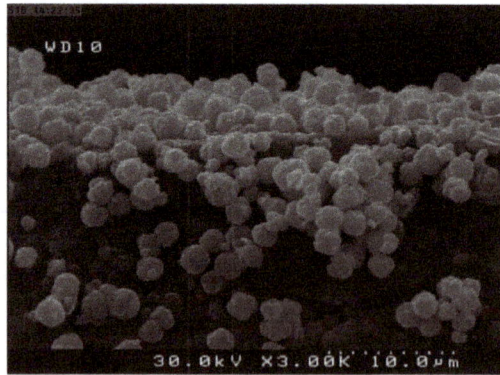

$$Zn^{2+} + 2OH^- \leftrightarrow Zn(OH)_2 \downarrow \tag{10}$$

$$Zn(OH)_2 + 4NH_3 \leftrightarrow Zn(NH_3)_4{}^{2+} + 2OH^- \tag{11}$$

$$Zn(OH)_2 \leftrightarrow ZnO + H_2O \tag{12}$$

Figure 8. Cross section view of the flower-like CuO-ZnO hybrid hierarchical nanostructures prepared without any surfactant or stabilizer.

Basically, zinc acetate dissolves to produce Zn^{2+} ions. Obviously, at the beginning of the reaction, $Zn(OH)_2$ precipitates were obtained (Equation (10)), evidenced by a white solid appearing in our experiment. After adding the NH_3 solution, the $Zn(OH)_2$ precipitate began to dissolve and a homogenous aqueous solution containing $Zn(NH_3)_4{}^{2+}$ ions is obtained [32]. This solution is stable and transparent with a large amount of complexing agents. According to the equilibrium in Equation (2), after increasing the temperature, the total concentration of NH_3 in the solution became lower and the reaction moves to the left to compensate [33]. Thereafter, $Zn(OH)_2$ would transform into ZnO nuclei by the dehydration reaction under hydrothermal conditions (Equation (12)). Here, EG as the stabilizer agent prevented the amalgamation of the ZnO nuclei in the supersaturated solvents during the reaction process. Furthermore, EG can serve as the director for the growth of ZnO nanoparticles along certain direction, which leads to dispersed flower like structures with the appearance of white precipitation [34].

It is commonly confirmed that the unusual wetting properties of superhydrophobic surfaces are dependent on both their chemical composition and geometric surface microstructures [35]. In addition, the characteristic of wettability of a surface can be well-controlled through the combination of surface roughness and different chemical modifications. It is expected that as-prepared flower-like ZnO-CuO architectures with special micro-nanostructures may result in a particular wettability. Considering the micro and nanoscale binary hierarchical architecture of the surfaces of CuO-ZnO covered Cu foil, the wetting properties of the CuO-ZnO microstructures obtained on the Cu foil during reactions were evaluated using contact angle (CA) measurements. Figure 9 shows the captured micrographs of a water droplet on the surface of the CuO-ZnO films. As can be seen in Figure 9a, for the CuO film, the corresponding CA is determined to be 117°, indicative of the surface hydrophobicity.

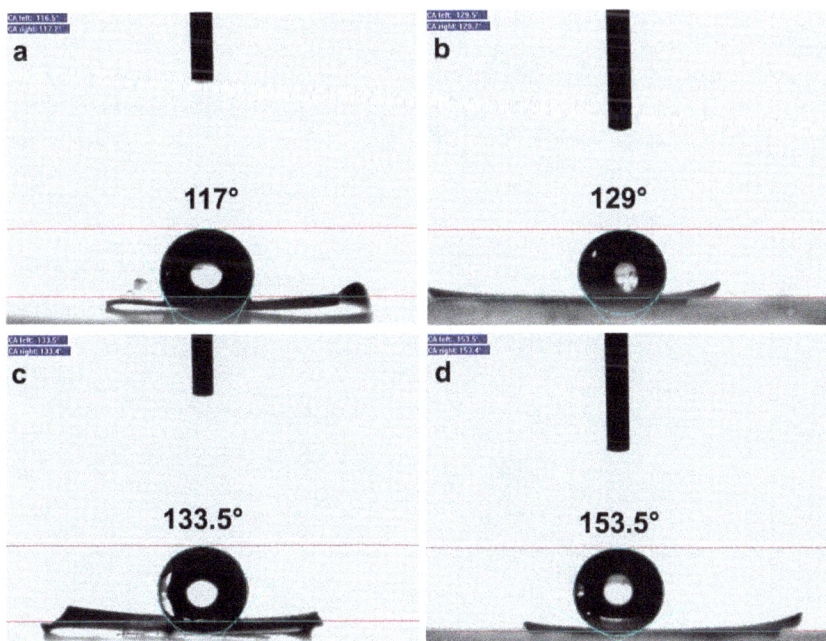

Figure 9. The shapes of a water droplet on different surfaces and the corresponding contact angle (CA) values: (**a**) modified CuO nanospheres surface; (**b**) modified underdeveloped CuO-ZnO micro-flower surfaces; (**c**) modified flat CuO-ZnO surface by SA; (**d**) modified flat CuO-ZnO surface by PDMS.

The water CA was found to increase from 117° for the nanostructured CuO film to 129° for the flower like CuO-ZnO microstructures (Figure 9b), indicative of the good hydrophobicity of the CuO-ZnO surface. Previously, our SEM results also confirmed that the ZnO microflowers with higher surface roughness demonstrate better hydrophobicity. This observation confirms that the persistence of the geometric structures of the ZnO micro flowers after the oxidation of the copper foil is essential for improving hydrophobicity. It is well known that for a specific surface, CA is strongly associated with both the surface roughness and surface energy, and the main factors that controls the surface energy is usually the type and properties of the surface functional groups [36]. For increasing the contact angle of the water droplet, the surface of the as prepared flower like ZnO hierarchical structures on the CuO substrate was further modified as follows; an as prepared CuO-ZnO foil was also modified with a stearic acid (SA) layer. Therefore, the as prepared CuO-ZnO substrate was immersed in 10 mL of 0.05 M SA ethanol solution for 1 h at room temperature. Finally the modified substrate was dried at 70 °C for 2 h and stored at room temperature. The result of the CA measurement is depicted in Figure 9c. As seen, the combination of the low surface energy SA and surface roughness (arising from the CuO nanoparticles and flower like ZnO micro structures) leads to obtaining a superhydrophobic surface with the static water contact angle of approximately 133.5°. In the other reaction, the surface of the as prepared CuO-ZnO substrate was modified with polydimethylsiloxane (PDMS). The intended substrate was put in a 3% (v/v) solution of PDMS in toluene for 1 h at room temperature. The foil was dried at 70 °C for 2 h and stored at room temperature and then its wettability was investigated. The results for the CA measurement confirmed that the as modified substrate showed excellent superhydrophobicity with a water contact angle as high as 153.5° (Figure 9d). For this sample, the water droplets can roll off the surface very quickly when the substrate is slightly crooked. Moreover, according to the previous studies, it should be noted that the rough surfaces without modification usually do not have a superhydrophobic property and this suggests that a suitable micro- nanoscale binary surface structure only is not enough to achieve a superhydrophobic surface [35].

One of the most advantageous characteristics of the superhydrophobic film was its tunable wettability when it is exposed to UV irradiation. Here, in order to study the UV-enhanced wettability conversion of the CuO/ZnO/PDMS coating, the sample (surface area of 1 cm^2) was placed under a UV light source (400 W mercury lamp) with a working distance of 10 cm in ambient conditions. The photodecomposition process and water contact angles as a function of the UV illumination time are shown in Figure 10. As shown, after exposing the sample to UV light for 30 min, the contact angle was decreased to about 4° and the sample nature was switched from superhydrophobic to superhydrophilic. However, when the UV-irradiated sample had been dried in a dark place at 80 °C for 24 h, the water contact angle (WCA) was increased again and the sample recovered its pristine superhydrophobic state with a contact angle of about 150°.

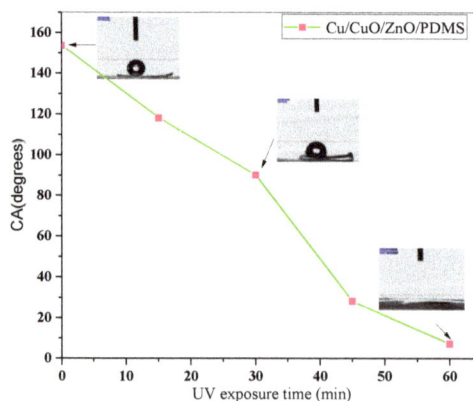

Figure 10. Water contact angle as a function of the UV irradiation time for the Cu/CuO/ZnO/PDMS sample.

For investigation of the corrosion resistance of the resulting surfaces, EIS was used as a nondestructive and useful technique to characterize the electrochemical reactions that occurred at the metal/salt solution boundary. Figure 11 shows the EIS results and typical Nyquist impedance plots of the bare copper and surface-treated copper oxide after 60 min of exposure in a 3.5 wt % aqueous NaCl solution. As we know, the low frequency impedance is known as the Warburg impedance (W). This impedance shows that the dissolution mechanism of copper is controlled by the mass transport rate. The diffusion step of copper dissolution in several papers is ascribed to the transport of the transportation of soluble cuprous chloride complexes from the surface of copper to the bulk solution. Moreover the depressed capacitive loop has been ascribed to roughness and in-homogeneities on the surface during corrosion. Figure 11 also shows that the Warburg impedance disappears at low frequencies when the surface of the bare copper is oxidized to CuO or when flower like ZnO structures are grafted on the surface. The Nyquist diagrams show a semicircle in the high frequency range related to the resistance of charge transfer (Rct) followed by a straight line in the low frequency region related to the double layer capacitance (C). As seen, the diameters of the Nyquist loop of the flower like ZnO structures on the Cu is significantly larger in compare with the bare copper and Cu/CuO sample. This observation demonstrates that the copper corrosion is controlled by the charge transfer process as the CuO/ZnO layer is added. The EIS results for further surface modification with SA and PDMS are shown in Figure 12. It is observed that the diameter of the semicircle Nyquist impedance plots increases obviously in the presence of the SA and PDMS as inhibitors, which increases the corrosion resistance of the sample. Obviously, in Table 2, the R_{ct} values increase with the modification of the surface. Consequently, the inhibition efficiencies (η) of the CuO, CuO/ZnO, CuO/ZnO/SA, and CuO/ZnO/PDMS calculated from Equation (1), are around 44.61, 87.49, 97.47, and 99.79%, respectively, after the initial 60 min of exposure. These results indicate that the superhydrophobic CuO/ZnO/PDMS surface has dramatically enhanced the corrosion resistance of the copper foil.

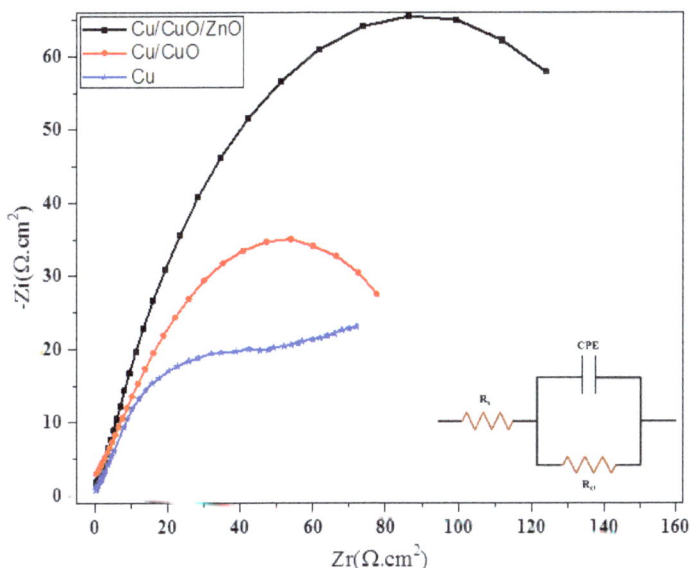

Figure 11. Nyquist plot for the bare Cu foil, Cu/CuO, and Cu/CuO/ZnO electrodes in 3.5% NaCl solution. The Cu/CuO/ZnO sample showed a better anticorrosion property.

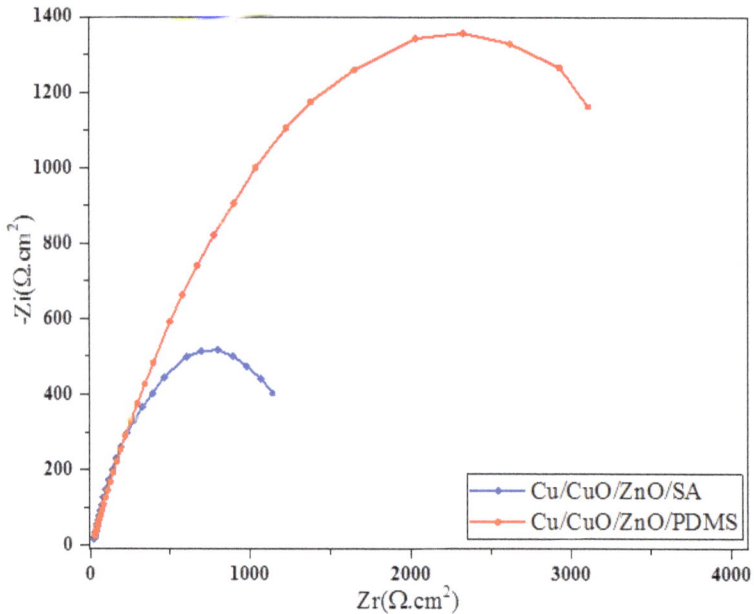

Figure 12. Nyquist plot for the modified Cu/CuO/ZnO/SA and Cu/CuO/ZnO/PDMS electrodes in 3.5% NaCl solution. The CuO/ZnO/PDMS sample showed a higher anticorrosion efficiency compared to the other samples.

Table 2. Impedance parameters and anticorrosion efficiencies of different samples in 3.5% NaCl solution.

Sample	R_s ($\Omega \cdot cm^2$)	R_{ct} ($\Omega \cdot cm^2$)	C ($\mu F \cdot cm^2$)	$\eta\%$
Cu	14 ± 4%	81.3 ± 4%	79.2 ± 6%	-
Cu/CuO	16.62 ± 4%	146.8 ± 4%	69.7 ± 7%	44.61
Cu/CuO/ZnO	23.58 ± 10%	650 ± 5%	27.4 ± 7%	87.49
Cu/CuO/ZnO/SA	35.9 ± 9%	3220 ± 10%	7.06 ± 8%	97.47
Cu/CuO/ZnO/PDMS	651 ± 8%	40,500 ± 9%	3.61 ± 7%	99.79

4. Conclusions

In this work, CuO-ZnO hierarchical nanostructures on copper substrates have been prepared by the oxidation of copper foil in EG solutions at a moderate temperature of 100 °C. The flowerlike ZnO microstructures are formed on copper oxide foils easily via a glycothermal method. The results indicate that EG acts as both an alkaline agent and reductant for the growth of CuO nanostructures and unusual flower like ZnO microstructures. The XRD pattern of the product grown on the upward surface of the copper foil confirms the true synthesis of the monoclinic structure for CuO and the hexagonal structure for flower-like ZnO. The SEM image results confirmed that the used surfactants (SDS, PEG 600, and CTAB) operate as an obstacle for ethylene glycol and prevent to reaction with the Cu substrate. Therefore, using surfactants with high molecular weight and those which have a strong interaction with the substrate is not recommended for the synthesis of both CuO and ZnO in the mentioned conditions. The surface hydrophobicity/superhydrophobicity was achieved on the modified Cu foils because of the combination of the uniform surface nano or microstructure and low surface energy. The results for the CA measurements confirmed that modifying the obtained rough surfaces of CuO and Cu-ZnO promotes the hydrophobic property, and using just a 3% solution (v/v) of PDMS in toluene leads to obtaining excellent superhydrophobicity with a water contact angle as high as 153.5°. Due to the high hydrophobicity of the surface, the obtained anticorrosion efficiency for the CuO/ZnO/PDMS sample was about 99%. These results provide valuable information for the design of the patterned superhydrophobic surfaces through a simple approach and multifunctional materials, which have many potential applications in corrosion protection, sensors, energy storage devices, and self-cleaning and anti-icing coatings.

Acknowledgments: The authors are grateful to the council of the Iran National Science Foundation (INSF) and the University of Kashan for supporting this work by Grant No. (159271/87990).

Author Contributions: Hossein Khojasteh and Farshad Beshkar conceived and designed the experiments; Farshad Beshkar performed the experiments; Masoud Salavati-Niasari and Hossein Khojasteh analyzed the data; Masoud Salavati-Niasari contributed reagents/materials/analysis tools; Masoud Salavati-Niasari and Hossein Khojasteh wrote the paper.

Conflicts of Interest: The authors declare no conflict of interest.

References

1. Núñez, L.; Reguera, E.; Corvo, F.; González, E.; Vazquez, C. Corrosion of copper in seawater and its aerosols in a tropical island. *Corros. Sci.* **2005**, *47*, 461–484. [CrossRef]
2. Wu, H.; Xue, M.; Ou, J.; Wang, F.; Li, W. Effect of annealing temperature on surface morphology and work function of ZnO nanorod arrays. *J. Alloys Compd.* **2013**, *565*, 85–89. [CrossRef]
3. Lu, C.; Qi, L.; Yang, J.; Zhang, D.; Wu, N.; Ma, J. Simple template-free solution route for the controlled synthesis of Cu(OH)$_2$ and CuO nanostructures. *J. Phys. Chem. B* **2004**, *108*, 17825–17831.
4. Xu, J.; Xue, D. Fabrication of malachite with a hierarchical sphere-like architecture. *J. Phys. Chem. B* **2005**, *109*, 17157–17161. [CrossRef] [PubMed]
5. Shi, X.; Yang, X.; Gu, X.; Su, H. CuO–ZnO heterometallic hollow spheres: Morphology and defect structure. *J. Solid State Chem.* **2012**, *186*, 76–80. [CrossRef]
6. Zhang, R.; Fan, L.; Fang, Y.; Yang, S. Electrochemical route to the preparation of highly dispersed composites of ZnO/carbon nanotubes with significantly enhanced electrochemiluminescence from ZnO. *J. Mater. Chem.* **2008**, *18*, 4964–4970. [CrossRef]
7. Rahman, M.M.; Jamal, A.; Khan, S.B.; Faisal, M. CuO codoped ZnO based nanostructured materials for sensitive chemical sensor applications. *ACS Appl. Mater. Interfaces* **2011**, *3*, 1346–1351. [CrossRef] [PubMed]
8. Xu, C.; Shin, P.; Cao, L.; Gao, D. Preferential growth of long zno nanowire array and its application in dye-sensitized solar cells. *J. Phys. Chem. C* **2010**, *114*, 125–129. [CrossRef]
9. Norton, D.P.; Heo, Y.W.; Ivill, M.P.; Ip, K.; Pearton, S.J.; Chisholm, M.F.; Steiner, T. ZnO: Growth, doping & processing. *Mater. Today* **2004**, *7*, 34–40.
10. Burda, C.; Chen, X.; Narayanan, R.; El-Sayed, M.A. Chemistry and properties of nanocrystals of different shapes. *Chem. Rev.* **2005**, *105*, 1025–1102. [PubMed]
11. Yin, Y.; Rioux, R.M.; Erdonmez, C.K.; Hughes, S.; Somorjai, G.A.; Alivisatos, A.P. Formation of hollow nanocrystals through the nanoscale kirkendall effect. *Science* **2004**, *304*, 711–714. [CrossRef] [PubMed]
12. Zhu, Y.; Sow, C.H.; Yu, T.; Zhao, Q.; Li, P.; Shen, Z.; Yu, D.; Thong, J.T.L. Co-synthesis of ZnO–CuO nanostructures by directly heating brass in air. *Adv. Funct. Mater.* **2006**, *16*, 2415–2422. [CrossRef]
13. Zhao, X.; Wang, P.; Li, B. CuO/ZnO core/shell heterostructure nanowire arrays: Synthesis, optical property, and energy application. *Chem. Commun.* **2010**, *46*, 6768–6770. [CrossRef] [PubMed]
14. Kumar, S.A.; Cheng, H.-W.; Chen, S.-M.; Wang, S.-F. Preparation and characterization of copper nanoparticles/zinc oxide composite modified electrode and its application to glucose sensing. *Mater. Sci. Eng. C* **2010**, *30*, 86–91. [CrossRef]
15. Guo, Z.; Chen, X.; Li, J.; Liu, J.-H.; Huang, X.-J. ZnO/CuO hetero-hierarchical nanotrees array: Hydrothermal preparation and self-cleaning properties. *Langmuir* **2011**, *27*, 6193–6200. [CrossRef] [PubMed]
16. Jung, S.; Yong, K. Fabrication of CuO-ZnO nanowires on a stainless steel mesh for highly efficient photocatalytic applications. *Chem. Commun.* **2011**, *47*, 2643–2645. [CrossRef] [PubMed]
17. Shaislamov, U.; Krishnamoorthy, K.; Kim, S.J.; Chun, W.; Lee, H.-J. Facile fabrication and photoelectrochemical properties of a CuO nanorod photocathode with a ZnO nanobranch protective layer. *RSC Adv.* **2016**, *6*, 103049–103056. [CrossRef]

18. Liu, P.; Wang, Y.; Zhang, H.; An, T.; Yang, H.; Tang, Z.; Cai, W.; Zhao, H. Vapor-phase hydrothermal transformation of $HTiOF_3$ intermediates into {001} faceted anatase single-crystalline nanosheets. *Small* **2012**, *8*, 3664–3673. [CrossRef] [PubMed]

19. Wang, S.; Zhang, Y.; Abidi, N.; Cabrales, L. Wettability and surface free energy of graphene films. *Langmuir* **2009**, *25*, 11078–11081. [CrossRef] [PubMed]

20. Xiao, F.; Yuan, S.; Liang, B.; Li, G.; Pehkonen, S.O.; Zhang, T. Superhydrophobic CuO nanoneedle-covered copper surfaces for anticorrosion. *J. Mater. Chem. A* **2015**, *3*, 4374–4388. [CrossRef]

21. Beshkar, F.; Zinatloo-Ajabshir, S.; Salavati-Niasari, M. Simple morphology-controlled fabrication of nickel chromite nanostructures via a novel route. *Chem. Eng. J.* **2015**, *279*, 605–614. [CrossRef]

22. Mousavi, Z.; Soofivand, F.; Esmaeili-Zare, M.; Salavati-Niasari, M.; Bagheri, S. $ZnCr_2O_4$ Nanoparticles: facile synthesis, characterization, and photocatalytic properties. *Sci. Rep.* **2016**, *6*, 20071. [CrossRef] [PubMed]

23. Patel, K.; Kapoor, S.; Dave, D.P.; Mukherjee, T. Synthesis of Pt, Pd, Pt/Ag and Pd/Ag nanoparticles by microwave-polyol method. *J. Chem. Sci.* **2005**, *117*, 311–316. [CrossRef]

24. Skrabalak, S.E.; Wiley, B.J.; Kim, M.; Formo, E.V.; Xia, Y. On the polyol synthesis of silver nanostructures: Glycolaldehyde as a reducing agent. *Nano Lett.* **2008**, *8*, 2077–2081. [CrossRef] [PubMed]

25. Wang, J.; Zhang, W.-D. Fabrication of CuO nanoplatelets for highly sensitive enzyme-free determination of glucose. *Electrochim. Acta* **2011**, *56*, 7510–7516. [CrossRef]

26. Jana, S.; Das, S.; Das, N.S.; Chattopadhyay, K.K. CuO nanostructures on copper foil by a simple wet chemical route at room temperature. *Mater. Res. Bull.* **2010**, *45*, 693–698. [CrossRef]

27. Hsu, Y.-K.; Chen, Y.-C.; Lin, Y.-G. Spontaneous formation of CuO nanosheets on Cu foil for H_2O_2 detection. *Appl. Surf. Sci.* **2015**, *354*, 85–89. [CrossRef]

28. Ekthammathat, N.; Thongtem, T.; Thongtem, S. Antimicrobial activities of CuO films deposited on Cu foils by solution chemistry. *Appl. Surf. Sci.* **2013**, *277*, 211–217. [CrossRef]

29. Fan, G.; Li, F. Effect of sodium borohydride on growth process of controlled flower-like nanostructured Cu_2O/CuO films and their hydrophobic property. *Chem. Eng. J.* **2011**, *167*, 388–396. [CrossRef]

30. Shen, H.; Eisenberg, A. Morphological phase diagram for a ternary system of block copolymer PS310-b-PAA52/Dioxane/H_2O. *J. Phys. Chem. B* **1999**, *103*, 9473–9487. [CrossRef]

31. Burke, S.E.; Eisenberg, A. Effect of Sodium Dodecyl sulfate on the morphology of polystyrene-b-poly(acrylic acid) aggregates in dioxane−water Mixtures. *Langmuir* **2001**, *17*, 8341–8347. [CrossRef]

32. Hui, Z.; Deren, Y.; Xiangyang, M.; Yujie, J.; Jin, X.; Duanlin, Q. Synthesis of flower-like ZnO nanostructures by an organic-free hydrothermal process. *Nanotechnology* **2004**, *15*, 622.

33. Sun, Y.; Hu, J.; Wang, N.; Zou, R.; Wu, J.; Song, Y.; Chen, H.; Chen, H.; Chen, Z. Controllable hydrothermal synthesis, growth mechanism, and properties of ZnO three-dimensional structures. *New J. Chem.* **2010**, *34*, 732–737. [CrossRef]

34. Yamabi, S.; Imai, H. Growth conditions for wurtzite zinc oxide films in aqueous solutions. *J. Mater. Chem.* **2002**, *12*, 3773–3778. [CrossRef]

35. Liu, J.; Huang, X.; Li, Y.; Sulieman, K.M.; He, X.; Sun, F. Hierarchical nanostructures of cupric oxide on a copper substrate: Controllable morphology and wettability. *J. Mater. Chem.* **2006**, *16*, 4427–4434. [CrossRef]

36. Wu, X.; Shi, G. Production and characterization of stable superhydrophobic surfaces based on copper hydroxide nanoneedles mimicking the legs of water striders. *J. Phys. Chem. B* **2006**, *110*, 11247–11252. [CrossRef] [PubMed]

materials

MDPI

Article

A Highly-Sensitive Picric Acid Chemical Sensor Based on ZnO Nanopeanuts

Ahmed A. Ibrahim [1,2,3], Preeti Tiwari [4], M. S. Al-Assiri [2,5], A. E. Al-Salami [6], Ahmad Umar [1,2,*], Rajesh Kumar [7], S. H. Kim [1,2], Z. A. Ansari [4] and S. Baskoutas [3]

[1] Department of Chemistry, Faculty of Science and Arts, Najran University, P.O. Box 1988, Najran 11001, Saudi Arabia; ahmedragal@yahoo.com (A.A.I.); semikim77@gmail.com (S.H.K.)
[2] Promising Centre for Sensors and Electronic Devices (PCSED), Najran University, P.O.Box-1988, Najran 11001, Saudi Arabia; msassiri@gmail.com
[3] Department of Materials Science, University of Patras, Patras GR-26504, Greece; bask@upatras.gr
[4] Centre for Interdisciplinary Research in Basic Sciences, Jamia Millia Islamia, New Delhi 110025, India; preet.19.pt@gmail.com (P.T.); zaansari@jmi.ac.in (Z.A.A.)
[5] Department of Physics, Faculty of Science and Arts, Najran University, P.O. Box 1988, Najran 11001, Saudi Arabia
[6] Department of Physics, Faculty of Science, King Khalid University, P.O. Box-9004, Abha 61413, Saudi Arabia; Salami11@gmail.com
[7] PG Department of Chemistry, JCDAV College, Dasuya, Punjab 144205, India; rk.ash2k7@gmail.com
* Corresponding: ahmadumar.asp@gmail.com or ahmadumar786@gmail.com; Tel.: +966-5-3457-4597

Received: 2 June 2017; Accepted: 8 July 2017; Published: 13 July 2017

Abstract: Herein, we report a facile synthesis, characterization, and electrochemical sensing application of ZnO nanopeanuts synthesized by a simple aqueous solution process and characterized by various techniques in order to confirm the compositional, morphological, structural, crystalline phase, and optical properties of the synthesized material. The detailed characterizations revealed that the synthesized material possesses a peanut-shaped morphology, dense growth, and a wurtzite hexagonal phase along with good crystal and optical properties. Further, to ascertain the useful properties of the synthesized ZnO nanopeanut as an excellent electron mediator, electrochemical sensors were fabricated based on the form of a screen printed electrode (SPE). Electrochemical and current-voltage characteristics were studied for the determination of picric acid sensing characteristics. The electrochemical sensor fabricated based on the SPE technique exhibited a reproducible and reliable sensitivity of ~1.2 μA/mM (9.23 μA·mM^{-1}·cm^{-2}), a lower limit of detection at 7.8 μM, a regression coefficient (R^2) of 0.94, and good linearity over the 0.0078 mM to 10.0 mM concentration range. In addition, the sensor response was also tested using simple *I-V* techniques, wherein a sensitivity of 493.64 μA·mM^{-1}·cm^{-2}, an experimental Limit of detection (LOD) of 0.125 mM, and a linear dynamic range (LDR) of 1.0 mM–5.0 mM were observed for the fabricated picric acid sensor.

Keywords: ZnO nanopeanuts; hydrothermal; electrochemical sensor; picric acid

---

1. Introduction

Electrochemical nanotechnology is an emerging combinational technique that involves electrochemical methods, and nanotechnology has been explored for many important applications in fields such as gas sensors, biosensors, electrochemical sensors, electronics, photovoltaic devices, supercapacitors, pH sensors, and humidity sensors [1,2]. Among the various applications, electrochemical detection and the sensing of hazardous and toxic chemicals are of utmost importance. Semiconductor metal oxide nanomaterials act as efficient electron mediators for the modification and fabrication of highly sensitive electrodes [3,4]. Among the various metal oxide nanomaterials, the

ZnO–II-VI semiconductor, with a low band gap energy of 3.37 eV and a large exciton binding energy of 60 MeV, is extensively studied [5]. Due to its tetrahedral structure and polar symmetry along the hexagonal axis of the wurtzite phase, ZnO with a variety of morphologies having high surface defect density can be synthesized [6–14]. These morphologies provide a large surface area for the adsorption of chemical species, which is a key factor for efficient electrochemical sensor applications [11,15]. As ZnO nanomaterials are *n*-type semiconductors, the adsorbed chemical species are reduced. Such redox changes on the surface of ZnO nanomaterials make these materials efficient electron mediators in the electrochemical sensor fabrication process [10,13].

ZnO nanostructures of different morphologies have been recently reported in the literature for their electrochemical applications for the detection of harmful, toxic, and even biologically important chemical substances [16–18]. Molaakbari et al. [19] fabricated a carbon paste electrode modified with ZnO nanorods and 5-(4′-amino-3′-hydroxy-biphenyl-4-yl)-acrylic acid (3,4′-AAZCPE) electrochemical sensors for levodopa, a precursor of the neurotransmitter dopamine, which is widely used in the clinical treatment of Parkinson's disease. The selective and sensitive determination of calcitonin was also reported by Patra et al. [20] in human blood serum samples using an electrochemical sensor comprising a medullary thyroid carcinoma marker imprinted polymer onto the surface of ZnO nanostructures. High sensitivity, a low detection limit, and a response time of ~26.58 $\mu A \cdot cm^{-2} \cdot mM^{-1}$, ~5 nM and 10 s, respectively were observed during the electrochemical sensing of ammonia at room temperature using ZnO nano pencil based electrochemical sensors [21,22]. Mehta et al. [23] reported an ultra-high sensitivity of ~97.133 $\mu A \cdot cm^{-2} \cdot \mu M^{-1}$ and a very low detection limit of 147.54 nM for hydrazine using a well-crystallized ZnO nanoparticle based amperometric sensor. A pristine ZnO nanorods array deposited on an inert alloy substrate were used as an efficient electron mediator for the fabrication of a hydrazine electrochemical sensor with a sensitivity of ~4.48 $\mu A \cdot \mu M^{-1} \cdot cm^{-2}$ [24]. One-dimensional (1D) ZnO nanorods and two-dimensional (2D) ZnO nanoflakes synthesized on an Au-coated substrate through a sonochemical approach showed 11.86 and 7.74 $\mu A \cdot M^{-1}$ sensitivities, respectively, for cortisol, a steroid hormone [25]. Additionally, other chemicals such as glucose [26–29], urea [30,31], uric acid [32], ethanol [33–36], nitrophenols [37], trinitrotoluene [38], nitrophenyl amine [39,40], and ethanolamine [41] are also detected through ZnO nanostructure-mediated electrochemical sensors.

Phenols and their derivatives, particularly 2,4,6-Trinitrophenol (Picric acid), find extensive application in many industries, such as pharmaceuticals, polymers, leathers, agriculture, fuel cells, and explosives [10,14]. Picric acid is a highly toxic and carcinogenic chemical and drastically affects the liver, kidney, eyes, and the respiratory tract [42,43]. A fast, reliable, and selective detection and sensing of even a low level of picric acid is thus required.

In this report, we present a facile, low-temperature solution method for ZnO nanoparticles with peanut shapes. Further, a highly sensitive picric acid electrochemical sensor based on ZnO nanopeanuts was fabricated for the sensing of picric acid.

2. Experimental Details

2.1. Materials

For the synthesis of the ZnO nanopeanuts and the fabrication of the picric acid electrochemical sensor, Zinc nitrate hexahydrate ($Zn(NO_3)_2 \cdot 6H_2O$), polyethylene glycol, butyl carbitol acetate (BCA), and ammonium hydroxide (NH_4OH) were purchased from Loba Chemie (Mumbai, India), and were used as received without any further purification. Triple deionized (DI) water was used as a solvent for the preparation of solutions.

2.2. Synthesis of ZnO Nanopeanuts

In a typical reaction process for the synthesis of ZnO nanopeanuts, 0.1 M Zinc nitrate hexahydrate ($Zn(NO_3)_2 \cdot 6H_2O$), was dissolved in 50 mL of DI water and mixed well, stirring with 0.2 g polyethylene glycol prepared in 50 mL of DI water. The stirring was continued for 30 min. After stirring, a few drops of ammonium hydroxide (NH_4OH) were mixed in the resultant solution to maintain the solution at a pH = 10.15. The resultant solution was again stirred for 20 min, and then transferred to a Teflon-lined autoclave for hydrothermal reaction at 170 °C for 7 h. On completion, the autoclave was cooled to room temperature, the white precipitates were decanted and washed with DI water to neutralize the pH, and finally dried at 80 °C in a convection oven.

2.3. Characterizations of ZnO Nanopeanuts

Detailed analytical and characterization techniques were used for the evaluation of the morphological, structural, crystalline, and optical properties of the hydrothermally synthesized ZnO nanopeanuts. Field emission scanning electron microscopy (FESEM; JEOL-JSM-7600F, JEOL Ltd., Tokyo, Japan) attached with Energy Dispersive Spectroscopy (EDS) was used to study the morphological, structural, and compositional features of the ZnO nanomaterials. Phase crystallinity and microstructural parameters were evaluated through X ray diffraction (XRD, PAN analytical Xpert Pro.) in the scan range of 10°–80° (2θ) angles using a Cu-Kα radiation source with a wavelength of 1.54 Å. The compositional, optical, and Raman scattering spectral properties of the ZnO nanopeanuts were analyzed through Fourier transform infrared spectroscopy (FTIR; Perkin Elmer-FTIR Spectrum-100, Perkin Elmer, Germany) in the scan range of 450–4000 cm^{-1}, a UV-visible spectrophotometer (Perkin Elmer-UV/VIS-Lambda 950, Perkin Elmer, Germany) in the absorption range of 200–800 nm, and Raman scattering spectroscopy (Perkin Elmer-Raman Station 400 series, Perkin Elmer, Germany) in the scan range of 200–700 cm^{-1}, respectively.

2.4. Fabrication of Electrochemical Sensor Based on Screen Printed Electrode (SPE)

Two sensor techniques i.e., electrochemical and simple current-voltage (*I-V*) techniques, were chosen to characterize the fabricated picric acid sensor using synthesized ZnO nanopeanuts. The in-house SPE was fabricated using Printed circuit board (PCB) technology on a glass epoxy substrate, which consisted of a three electrode system viz. working, counter, and reference electrode. All of the three electrodes were gold plated (Scheme 1). The working electrode (surface area 0.13 cm^2) was used for the coating of synthesized ZnO nanopeanuts by formulating a thick paste as reported elsewhere [44]. The paste was prepared by mixing a known and optimized amount of BCA (30%) in nanomaterial (70%), then printing it on the SPE, and drying it at 80 °C. BCA is known as an organic binder, and when mixed in an optimized ratio of 30:70 (BCA:nanomaterial), it results in a good thixotropic paste. The SPE is expected to play the role of conducting electrons from analyte/ZnO to the potentiostat. The cyclic voltammogram (CV) of the SPE was obtained using IVIUM's potentiostat/galvanostat at room-temperature. Different picric acid solutions of 0.0078 mM, 0.078 mM, 0.78 mM, 2 mM, 5 mM, and 10 mM were prepared in 0.1 M phosphate buffer solution (PBS) (pH = 7.4). The curves were obtained at a fix scan rate of 50 mV·s^{-1}, while the potential was varied from −1.0 to 1.0 V. CV measurements were also performed at various scan rates (50 to 500 m·Vs^{-1}) at one concentration of picric acid (2 mM).

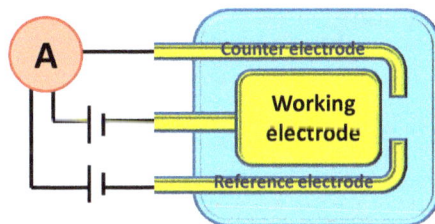

Scheme 1. Schematic representation of a screen printed electrode (SPE).

2.5. Fabrication of Picric Acid Sensor Based on Current-Voltage Technique Measurements

For sensors based on the *I-V* characteristic, a different set of electrodes were prepared. A cleaned silver electrode (AgE, surface area = 0.0214 cm^2, Purity supplier) was used as one of the electrodes (working), onto which a film of nanomaterial was coated using the paste prepared for the SPE's coating. A platinum wire was used as a counter electrode. Before coating, the Ag electrodes were rubbed against an alumina gel, followed by ultrasonic cleaning and repeated washings with deionized water. The electrode was dried in an air oven for 6 h at 70–75 °C. A Keithley electrometer, 6517A (USA) was used for measuring the current–voltage parameters at room temperature conditions. A platinum wire was used a counter electrode. The picric acid solutions were prepared in 0.1 M phosphate buffer solution (PBS) having a pH = 7.4 in the scan range of 0.0–4.0 V.

3. Results and Discussion

3.1. Characterizations and Properties of ZnO Nanopeanuts

The general morphologies of the synthesized material were examined by FESEM and the observed results are shown in Figure 1a,b. The observed FESEM images revealed that the prepared materials possess peanut shaped morphologies and grow in high density with almost uniform shape and size. The surface of the peanut shaped ZnO is highly rough, with swollen edges and a narrow central part. It is interesting to see that due to the high density growth, some nanopeanuts are linked to each other through one of their surfaces. The average diameter and length of ZnO nanopeanuts is ~110 ± 20 nm and ~220 ± 20 nm, respectively. Additionally, some dumb-bell and rod-shaped morphologies are also formed due to the aggregation of two ZnO nanopeanuts through their ends.

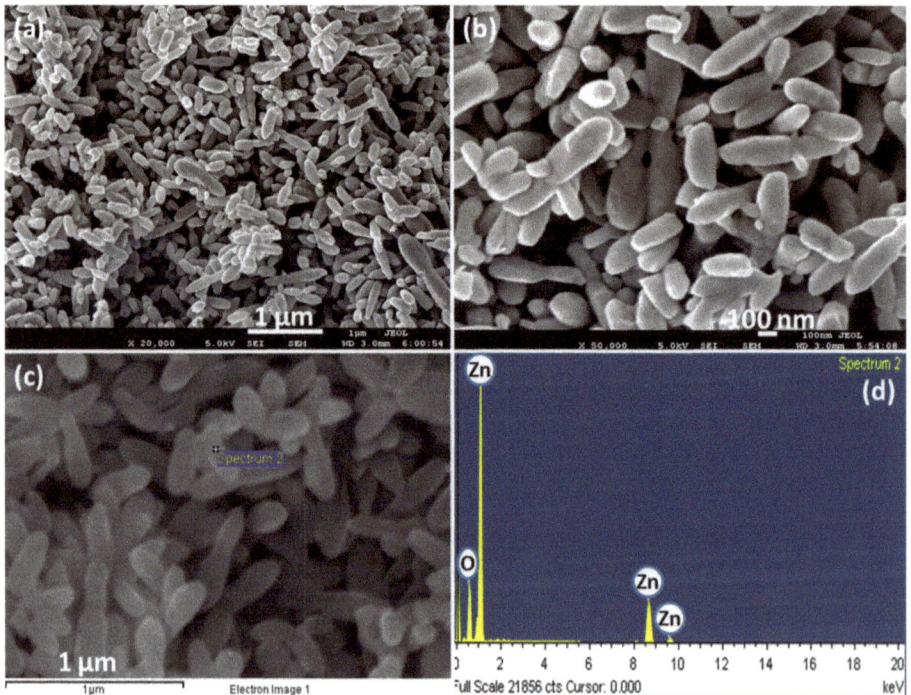

Figure 1. Typical (**a**,**b**) FESEM images; (**c**) EDS-SEM microscopic image and (**d**) EDS spectrum of the as-synthesized ZnO nanopeanuts.

The rough surface of the ZnO peanuts provides a sufficiently large surface area for intermolecular π stacking of the electron deficient picric acid benzene ring, which further facilitates a charge transfer from *n*-type semiconducting ZnO to picric acid molecules. The electron deficient nature of the picric acid benzene ring can be accounted for due to the presence of electron withdrawing nitro (–NO$_2$) groups. Additionally, the surface active sites of ZnO nanopeanuts attract the lone pairs of electrons present on the –OH group of the picric acid. The extent of chemisorptions is therefore increased, due to the high surface volume ratio of the as-synthesized ZnO peanuts. It has been reported that the chemisorption of the picric acid molecules amends the electronic states of the ZnO nanomaterials and improves the conductance [45]. These phenomena, such as charge transfer, altered electronic states, and improved conductance can be attributed to the excellent sensing performances of the ZnO nanomaterials.

To ascertain the elemental composition, the synthesized ZnO nanopeanuts were examined by energy dispersive spectroscopy (EDS) attached with FESEM (Figure 1c,d). As confirmed from the EDS spectrum, the synthesized material is made of zinc and oxygen, as no other peak related with any other element is seen in the observed EDS spectrum. The presence of only zinc and oxygen peaks in the EDS spectrum confirmed that the synthesized nanopeanuts are pure ZnO without any significant impurity.

It has been reported that Zn^{2+} ions combine with NH$_4^+$ and HO$^-$ ions to form $[\text{Zn}(\text{NH}_3)_4]^{2+}$ and $[\text{Zn}(\text{OH})_4]^{2-}$ growth units in the reaction medium [46,47]. The detailed growth mechanism for ZnO nanopeanuts has been elaborated elsewhere [47].

To further confirm the purity, crystal phases, and structure of the ZnO nanopeanuts, an X-ray diffraction pattern was recorded between 2θ = 10°–80°. Figure 2 represents the typical XRD pattern of the as-synthesized ZnO nanopeanuts, which indicates a hexagonal phase of pure ZnO in line with the reported literature [10,11,48,49]. The diffraction peaks observed at 31.6°, 34.3°, 36.3°, 47.4°, 56.6°, 62.8°, 66.5°, 68.5°, 69.7°, 74.3°, and 78.2° correspond to the lattice planes of ZnO (100), (002), (101), (102), (110), (103), (200), (112), (201), (004), and (202), respectively. No other diffraction peaks, except for a wurtzite hexagonal phase, are observed in the XRD pattern, which clearly confirmed that the synthesized nanopeanuts are only ZnO. The results of the XRD pattern matches that of the EDS observations.

The UV-Vis. absorption spectrum for the ZnO nanopeanuts is shown in Figure 3a. A well-defined single exciton absorption peak at 370 nm, corresponding to the pure wurtzite hexagonal phase, can be clearly seen, which is also in good agreement with the reported literature [48–51].

The band gap energy calculated using the well-known Planck's equation (Equation (1)) was found to be 3.35 eV [52]. No other absorption peak, except for 370 nm, confirms the fact that ZnO nanopeanuts possess excellent optical properties.

$$E_g = \frac{hc}{\lambda_{max}} = \frac{6.625 \times 10^{-34}\text{Js} \cdot 3 \times 10^8\text{ms}^{-1}}{370 \times 10^{-9}\text{m} \cdot 1.6 \times 10^{-19}\text{ J.eV}^{-1}} = 3.35\text{eV}, \qquad (1)$$

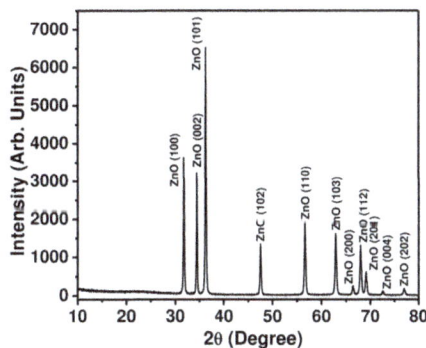

Figure 2. Typical XRD pattern for the as-synthesized ZnO nanopeanuts.

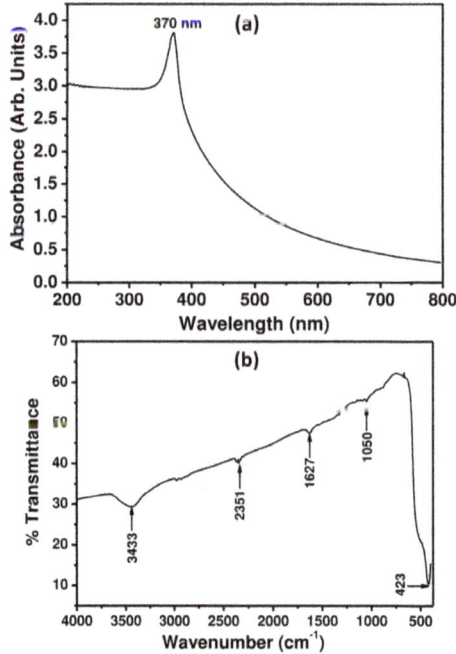

Figure 3. (a) UV-Vis and (b) FTIR spectra for the as-synthesized ZnO nanopeanuts.

Figure 3b depicts a typical FTIR spectrum of the as-synthesized ZnO nanoflakes, which exhibits various well-defined bands appearing at 423, 1050, 1627, 2351, and 3433 cm^{-1}. One sharp and well-defined peak at 423 cm^{-1} and another weak band at 1050 cm^{-1} may be assigned to the stretching and bending vibrational modes of the Zn-O bonds, respectively [49,53]. A weak band at 1627 cm^{-1} and a broad band 3433 cm^{-1} appear due to the bending and stretching vibration modes of the O–H groups, respectively, for the physisorbed water molecules on the surface of the ZnO nanopeanuts [52,54]. The low-intensity sharp band at 2351 cm^{-1} may be attributed to the asymmetric stretching of the C=O bonds of CO_2 molecules, adsorbed from the environment during KBr palletization [54].

To examine the scattering properties, the as-synthesized ZnO nanopeanuts were characterized by a Raman-scattering spectrum at room temperature, and the observed result is presented in Figure 4. The observed Raman-scattering spectrum exhibits various phonon peaks appearing at 330, 379, 437, and 581 cm^{-1}, which is consistent with the reported Raman-scattering spectrum of ZnO nanomaterials [41]. ZnO with a wurtzite hexagonal phase belongs to the C_{6v}^4 (P63mc) space group having four ZnO units per primitive cell. The peak corresponding to the non-polar optical phonon appeared at 437 cm^{-1}, and is assigned to the E_2^{High} mode. It is the characteristic peak for the wurtzite hexagonal phase of ZnO. The small but sharp peak at 330 cm^{-1} is assigned to the E_{2H}–E_{2L} multi-phonon process. The other weak bands at 379 and 581 cm^{-1} are associated with the A_{1T} and E_{1L} modes, respectively. All of these peaks also match well with the reported literature values [55–57].

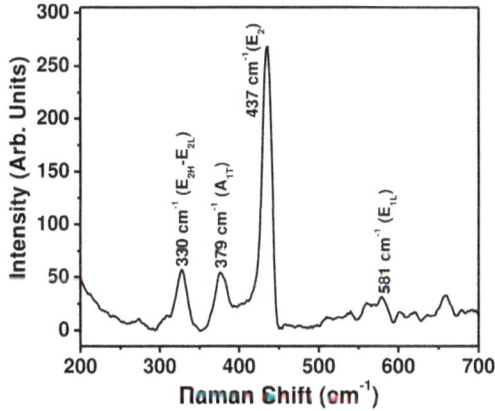

Figure 4. Typical Raman-scattering spectrum of the as-synthesized ZnO nanopeanuts.

3.2. Performance of Fabricated Picric Acid Sensors Based on ZnO Nanopeanuts

3.2.1. Sensing Properties of Picric Acid Sensor Based on ZnO Nanopeanuts Coated SPE

Figure 5a shows the average CV curves of two sets of measurements of a fabricated SPE at various concentrations of picric acid (0.0078, 0.078, 0.78, 2, 5, and 10 mM) prepared in 0.1M PBS (pH = 7.4) solution. The curves were acquired at the scan rate of 50 mV/s by varying the potential from −1.0 to 1.0 V. The peaks observed in the CV loop are related to the oxidation and reduction of the analyte that occurred at the potential of 0.04 V and −0.74 V, respectively. A systematic increase in peak current is noticed when increasing the concentrations of picric acid, which can be clearly seen in the inset of Figure 5a. At lower concentrations of picric acid, the increase in peak currents at a lower concentration is comparatively lower than those at higher concentrations. A slight increase in peak potential is observed with increasing concentration due to changes in the dielectric constant, with analyte being used as an electrolyte. The increase in the peak current can be attributed to the increased ionic strength at the electrode-electrolyte interface, and hence the extent of the electro-catalytic reaction occurring at the surface of the modified electrode [58–61] (schematically shown later in Figure 9).

Figure 5b shows the monotonic variation in the anodic peak current with concentration, which clearly shows the sensor's sensitivity to picric acid. A sudden increase in the peak current is noticed on addition of 7.8 µM picric acid to PBS, indicating that the developed device is able to detect change in the analyte concentration, and although the amount of change is less still a detectable change in peak current is noticed. It has been reported in the literature that electrochemical sensors are fairly sensitive to a level of pico-mole with a measureable peak current. A further increase of picric acid amount results in a systematic increase in the peak current, indicating a linear behavior of the developed sensor. This curve can be used as a calibration curve, and the sensitivity is therefore estimated as 0.12 µA/mM or 9.23 µA/mM·cm^{-2} in terms of per unit area of the working electrode. The lowest experimental limit is 7.8 µM, which is well below the safety limit (lethal dose, an indication of lethal toxicity of the material) of picric acid [62,63].

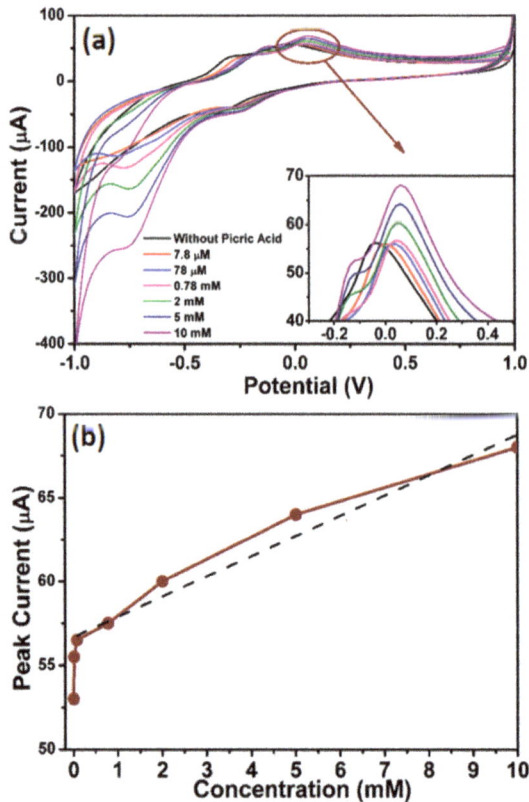

Figure 5. (a) Average cyclic voltammogram (CV) curves of the fabricated SPE at various concentrations of picric acid; and (b) Variation in the anodic peak current with concentration of picric acid.

Further, for the effect of scan rate (kinetics) on the electrochemical properties of the synthesized material, scan rate dependent CV curves were obtained. Figure 6a shows the average CV curves obtained at scan rates such as 50, 100, 150, 200, 250, 300, 350, 400, and 450 mV·s^{-1} at a particular concentration of 2 mM picric acid solution. It is known that increasing the scan rate would result in increased diffusion/depletion resulting in increased peak current. Thus, from the increase in the peak currents for anodic as well as cathodic processes, it could be confirmed that the electron transfer process is a diffusion confined electrode process.

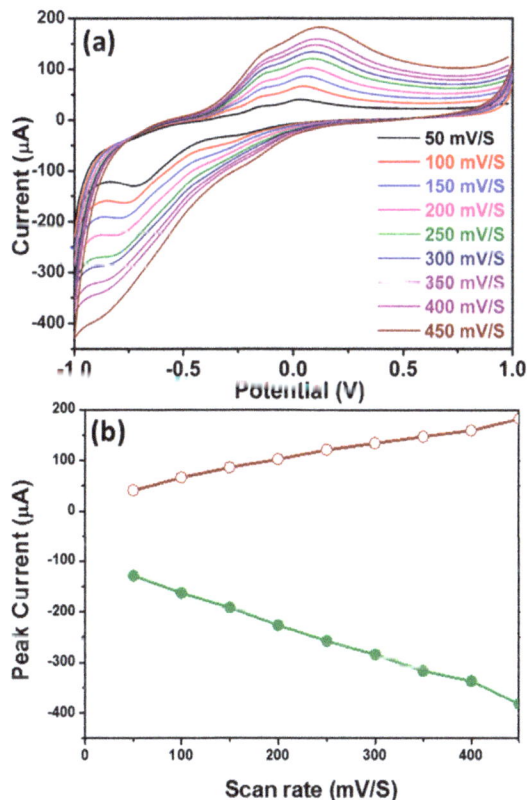

Figure 6. Typical (**a**) CV response curves at different scan rates (50, 100, 150, 200, 250, 300, 350, 400, and 450 mV/S) for a ZnO nanopeanut-modified SPE in 0.1 M PBS (pH = 7.4) containing 2 mM picric acid and (**b**) Variations of peak current with scan rate.

3.2.2. Sensing Properties of Picric Acid Sensor Based on ZnO Nanopeanuts Coated on AgE

A ZnO nanopeanut based sensor was fabricated to evaluate the sensing characteristics of picric acid in 0.1 M PBS with pH = 7.4 by a simple current-voltage (*I-V*) technique. Figure 7 represents the Current-Voltage (*I-V*) responses for a blank 0.1 M PBS and 0.125 mM picric acid solution prepared in PBS having pH = 7.4, in the scan range of 0.0–4.0 V, and using ZnO nanopeanut-modified AgE and Pt wire as a counter electrode. Prominent current variations for the picric acid solution, as compared to blank PBS, indicates that the ZnO nanopeanut-modified AgE is involved in redox changes, and hence can be used as an efficient electron mediator for electrochemical sensor applications. At 4.0 V potential, a current response of 16.18 µA and 4.804 µA were observed for the picric acid and blank solutions, respectively.

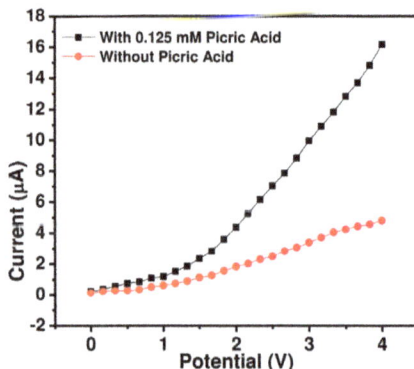

Figure 7. *I-V* performances for 0.125 mM picric acid and blank PBS using a ZnO nanopeanut-based AgE.

Figure 8a depicts the effect of picric acid concentration on *I-V* responses. For 0.125, 0.25, 0.5, 1.0, 2.1, 3.0, 4.0, and 5.0 mM solutions of picric acid solution prepared in PBS, the corresponding current responses were 16.18, 29.48, 38.63, 47.58, 68.09, 78.53, and 90.49 μA, respectively, measured at 4.0 V. This positive correlation can be explained on the basis of an increase in the ionic strength of the PBS buffer solution with the concentration of the picric acid. Further, it can also be postulated that the greater the extent of chemisorptions of picric acid molecules at higher concentrations (through to a saturation point), the greater are the changes in the electronic states and conduction at the electrode-electrolyte interface of the *n*-type semiconducting ZnO nanopeanuts, and thus higher is the current response [60].

The sensitivity of the ZnO nanopeanut-modified AgE was calculated from the ratio of the slope of the calibration graph plotted between molar concentrations (0.125–5.0 mM) of the picric acid solutions and the corresponding current responses measured at 4.0 V, and the active surface area of the AgE (Equation (2)) [61] (Figure 8b).

$$\text{Sensitivity} = \frac{\text{Slope of the calibration graph}}{\text{Active surface area of the modified AgE}} \tag{2}$$

A very high sensitivity of 493.64 μA·mM^{-1}·cm^{-2}, an experimental LOD of 0.125 mM, a linear dynamic range (LDR) of 1.0 mM–5.0 mM and a regression coefficient of $(R^2) = 0.9980$ were observed for the fabricated ZnO nanopeanut-modified AgE sensor. Due to the lower surface area of the electrode, a comparatively higher sensitivity is observed here than with the electrochemical sensing method. The sensitivity observed herein is high as compared to the reported sensitivities of some recently reported results. Huang et al. [64] reported a sensitivity of 0.00613 μA·μM^{-1} with a detection limit of 0.54 μM for picric acid through a reduced graphene oxide sensor modified with 1-pyrenebutyl-amino-β-cyclodextrin. It was proposed that β-cyclodextrin with the hydrophobic internal cavity and the hydrophilic external surface has a remarkable tendency to integrate with three hydrophobic -NO$_2$ groups of picric acid. However, there is a a low limit of detection of 0.6 μM for a copper-based electrochemical sensor [65,66] and 100 nM for a boron and nitrogen co-doped carbon nanoparticles based photoluminescent sensor [67] as compared to the ZnO peanut-based electrochemical sensors in this study.

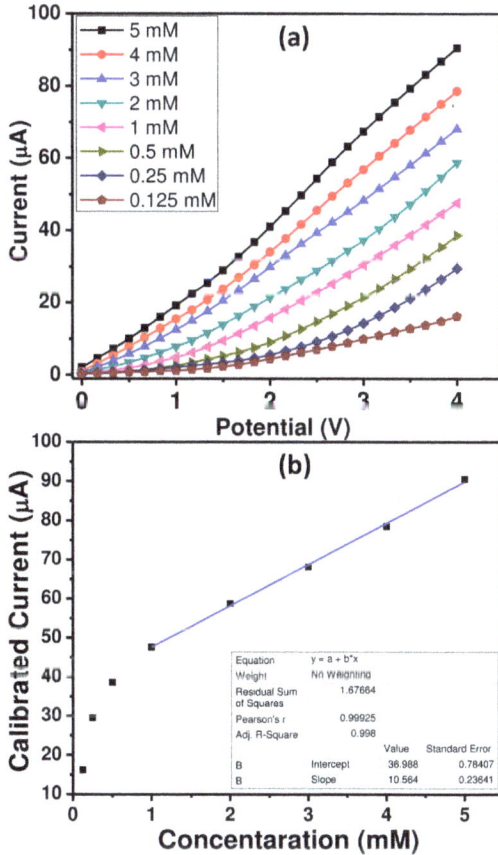

Figure 8. (a) *I-V* responses for various concentrations of picric acid solutions (b) Calibration graph for the fabricated ZnO nanopeanut-modified AgE electrochemical sensor.

4. Proposed Sensing Mechanism

On the basis of Frontier molecular orbital studies, it has been shown that the adsorption of the picric acid molecules on the surface of ZnO lowers the energy gap between the highest occupied molecular orbital (HOMO) and the lowest unoccupied molecular orbital (LUMO), resulting in the alteration of the electronic states and hence the charge transfer and conductance of the ZnO nanomaterials, as stated earlier in Section 3.1 [14,45]. Picric acid molecules with electron donor hydroxy (–OH) and electron withdrawing nitro (–NO$_2$) groups are adsorbed on the surface of the ZnO nanopeanuts through weak van der Waal interactions, where they undergo a series of redox changes [14]. Conduction band electrons of the ZnO nanopeanuts reduce the three –NO$_2$ groups to intermediate hydroxylamino (–HN–OH) groups, which are subsequently oxidized to nitroso (–NO) groups (Figure 9). The –NO groups undergo reversible reduction to release electrons back to the conduction band of the ZnO nanopeanuts, which are responsible for the increases in conductivity and current response [68].

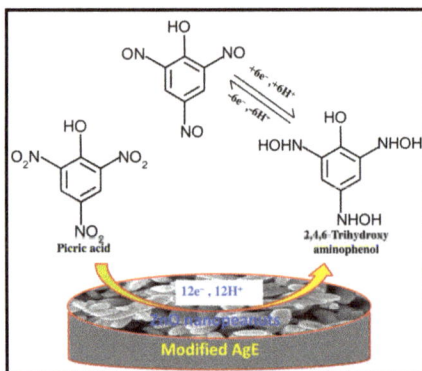

Figure 9. A sensing mechanism for picric acid sensing using modified AgE with ZnO nanopeanuts.

5. Conclusions

In summary, ZnO nanopeanuts were synthesized in a large quantity and characterized in detail, which revealed that the nanopeanuts possess high crystallinity and exhibit good optical and spectral properties. The synthesized nanopeanuts demonstrated high purity and were confirmed for the wurtzite hexagonal phase of pure ZnO. Further, the synthesized nanopeanuts were explored for their picric acid sensing applications in the form of an electrochemical and electrical sensor. Reliable, reproducible, and reversible CV and IV curves were obtained, indicating the versatility of the synthesized material for an efficient, sensitive, and reliable matrix for a picric acid sensor. Hence, ZnO nanopeanuts can be efficiently used as excellent electron mediators for the fabrication of sensors for other nitrophenols with very high sensitive and very low LOD.

Acknowledgments: The authors extend their appreciation to the Deanship of Scientific Research at King Khalid University for funding this work through research group program under grant number R.P.G.2/5/38.

Author Contributions: All the authors have contributed equally to the manuscript.

Conflicts of Interest: The authors declare no conflicts of interest.

References

1. Chen, A.; Chatterjee, S. Nanomaterials based electrochemical sensors for biomedical applications. *Chem. Soc. Rev.* **2013**, *42*, 5425–5438. [CrossRef] [PubMed]
2. Windmiller, J.R.; Wang, J. Wearable electrochemical sensors and biosensors: A Review. *Electroanalysis* **2013**, *25*, 29–46. [CrossRef]
3. Wang, F.; Hu, S. Electrochemical sensors based on metal and semiconductor nanoparticles. *Microchim. Acta* **2009**, *165*, 1–22. [CrossRef]
4. Park, G.C.; Lee, S.M.; Jeong, S.H.; Choi, J.H.; Lee, C.M.; Seo, T.Y.; Jung, S.-B.; Lim, J.H.; Joo, J. Enhanced photocatalytic activity of ZnO nanorods with tubular facet synthesized by hydrothermal method. *J. Nanosci. Nanotechnol.* **2016**, *16*, 11164–11168. [CrossRef]
5. Wei, A.; Pan, L.; Huang, W. Recent progress in the ZnO nanostructure-based sensors. *Mater. Sci. Eng. Solid-State Mater. Adv. Technol.* **2011**, *176*, 1409–1421. [CrossRef]
6. Katwal, G.; Paulose, M.; Rusakova, I.A.; Martinez, J.E.; Varghese, O.K. Rapid growth of zinc oxide nanotube-nanowire hybrid architectures and their use in breast cancer related volatile organics detection. *Nano Lett.* **2016**, *16*, 3014–3021. [CrossRef] [PubMed]
7. Jeong, S.H.; Park, G.C.; Choi, J.H.; Lee, C.M.; Lee, S.M.; Seo, T.Y.; Choi, D.H.; Jung, S.B.; Lim, J.H.; Joo, J. Effect of Al incorporation on morphology and electrical conductivity of ZnO nanorods prepared using hydrothermal method. *J. Nanosci. Nanotechnol.* **2016**, *16*, 11272–11276. [CrossRef]

8.	Caglar, M.; Gorgun, K. Characterization and heterojunction application of nanorod structure ZnO films prepared by microwave assisted chemical bath deposition method without any template. *J. Nanoelectron. Optoelectron.* **2016**, *11*, 769–776. [CrossRef]
9.	Kumar, R.; Al-Dossary, O.; Kumar, G.; Umar, A. Zinc oxide nanostructures for NO_2 gas-sensor applications: A review. *Nano-Micro Lett.* **2015**, *7*, 97–120. [CrossRef]
10.	Singh, K.; Chaudhary, G.R.; Singh, S.; Mehta, S.K. Synthesis of highly luminescent water stable ZnO quantum dots as photoluminescent sensor for picric acid. *J. Lumin.* **2014**, *154*, 148–154. [CrossRef]
11.	Ameen, S.; Akhtar, M.S.; Seo, H.K.; Shin, H.S. An electrochemical sensing platform based on hollow mesoporous ZnO nanoglobules modified glassy carbon electrode: Selective detection of piperidine chemical. *Chem. Eng. J.* **2015**, *270*, 564–571. [CrossRef]
12.	Katoch, A.; Choi, S.-W.; Sun, G.-J.; Kim, S.S. Low temperature sensing properties of Pt nanoparticle-functionalized networked ZnO nanowires. *J. Nanosci. Nanotechnol.* **2015**, *15*, 330–333. [CrossRef] [PubMed]
13.	Mani, G.K.; Rayappan, J.B.B. ZnO nanoarchitectures: Ultrahigh sensitive room temperature acetaldehyde sensor. *Sens. Actuators Chem.* **2016**, *223*, 2343–2351. [CrossRef]
14.	Farmanzadeh, D.; Tabari, L. DFT study of adsorption of picric acid molecule on the surface of single-walled ZnO nanotube; as potential new chemical sensor. *Appl. Surf. Sci.* **2015**, *324*, 864–870. [CrossRef]
15.	Kumar, R.; Rana, D.; Umar, A.; Sharma, P.; Chauhan, S.; Chauhan, M.S. Ag-doped ZnO nanoellipsoids: Potential scaffold for photocatalytic and sensing applications. *Talanta* **2015**, *137*, 204–213. [CrossRef] [PubMed]
16.	Ozden, P.B.; Caglar, Y.; Ilican, S.; Caglar, M. Effect of deposition time of electrodeposited ZnO nanorod films on crystallinity, microstructure and absorption edge. *J. Nanoelectron. Optoelectron.* **2016**, *11*, 244–249. [CrossRef]
17.	Zhu, L.-P.; Jiao, Y.-H.; Bing, N.-C.; Wang, L.-L.; Ye, Y.-K.; Wang, L.-J. Influences of Ni Doping on the morphology, optical and magnetic properties of ZnO nanostructures synthesized by solvothermal process. *J. Nanosci. Nanotechnol.* **2015**, *15*, 3234–3238. [CrossRef] [PubMed]
18.	Choi, H.J.; Lee, Y.-M.; Yu, J.-H.; Hwang, K.-H.; Boo, J.-H. Patterned well-aligned zno nanorods assisted with polystyrene monolayer by oxygen plasma treatment. *Materials (Basel)* **2016**, *9*, 656. [CrossRef]
19.	Molaakbari, E.; Mostafavi, A.; Beitollahi, H.; Alizadeh, R. Synthesis of ZnO nanorods and their application in the construction of a nanostructure-based electrochemical sensor for determination of levodopa in the presence of carbidopa. *Analyst* **2014**, *139*, 4356–4364. [CrossRef] [PubMed]
20.	Patra, S.; Roy, E.; Madhuri, R.; Sharma, P.K. Imprinted ZnO nanostructure-based electrochemical sensing of calcitonin: A clinical marker for medullary thyroid carcinoma. *Anal. Chim. Acta* **2015**, *853*, 271–284. [CrossRef] [PubMed]
21.	Dar, G.N.; Umar, A.; Zaidi, S.A.; Baskoutas, S.; Hwang, S.W.; Abaker, M.; Al-Hajry, A.; Al-Sayari, S.A. Ultra-high sensitive ammonia chemical sensor based on ZnO nanopencils. *Talanta* **2012**, *89*, 155–161. [CrossRef] [PubMed]
22.	Khan, A.; Khan, S.; Fawad, U.; Mujahid, M.; Khasim, S.; Hamdalla, T.; Kim, H.J. Three dimensional spherically evolved nanostructures of ZnO comprised of nanowires and nanorods for optoelectronic devices. *J. Nanoelectron. Optoelectron.* **2015**, *10*, 700–704. [CrossRef]
23.	Mehta, S.K.; Singh, K.; Umar, A.; Chaudhary, G.R.; Singh, S. Ultra-high sensitive hydrazine chemical sensor based on low-temperature grown ZnO nanoparticles. *Electrochim. Acta* **2012**, *69*, 128–133. [CrossRef]
24.	Liu, J.; Li, Y.; Jiang, J.; Huang, X. C@ZnO nanorod array-based hydrazine electrochemical sensor with improved sensitivity and stability. *Dalton Trans.* **2010**, *39*, 8693–8697. [CrossRef] [PubMed]
25.	Vabbina, P.K.; Kaushik, A.; Pokhrel, N.; Bhansali, S.; Pala, N. Electrochemical cortisol immunosensors based on sonochemically synthesized zinc oxide 1D nanorods and 2D nanoflakes. *Biosens. Bioelectron.* **2015**, *63*, 124–130. [CrossRef] [PubMed]
26.	Kim, J.Y.; Jo, S.Y.; Sun, G.J.; Katoch, A.; Choi, S.W.; Kim, S.S. Tailoring the surface area of ZnO nanorods for improved performance in glucose sensors. *Sens. Actuator Chem.* **2014**, *192*, 216–220. [CrossRef]
27.	Jin, Z.; Park, C.-I.; Hwang, I.-H.; Han, S.-W. Local structural properties and growth mechanism of ZnO nanorods on hetero-interfaces. *J. Nanosci. Nanotechnol.* **2015**, *15*, 5306–5309. [CrossRef] [PubMed]
28.	Lei, Y.; Yan, X.; Zhao, J.; Liu, X.; Song, Y.; Luo, N.; Zhang, Y. Improved glucose electrochemical biosensor by appropriate immobilization of nano-ZnO. *Colloids Surf. Bcointerfaces* **2011**, *82*, 168–172. [CrossRef] [PubMed]

29. Lee, C.T.; Chiu, Y.S.; Ho, S.C.; Lee, Y.J. Investigation of a photoelectrochemical passivated ZnO-based glucose biosensor. *Sensors* **2011**, *11*, 4648–4655. [CrossRef] [PubMed]

30. Ahmad, R.; Tripathy, N.; Hahn, Y.B. Highly stable urea sensor based on ZnO nanorods directly grown on Ag/glass electrodes. *Sens. Actuator Chem.* **2014**, *194*, 290–295. [CrossRef]

31. Ali, S.M.U.; Ibupoto, Z.H.; Salman, S.; Nur, O.; Willander, M.; Danielsson, B. Selective determination of urea using urease immobilized on ZnO nanowires. *Sens. Actuator Chem.* **2011**, *160*, 637–643. [CrossRef]

32. Ali, S.M.U.; Ibupoto, Z.H.; Kashif, M.; Hashim, U.; Willander, M. A potentiometric indirect uric acid sensor based on ZnO nanoflakes and immobilized uricase. *Sensors* **2012**, *12*, 2787–2797. [CrossRef] [PubMed]

33. Liang, Y.-C.; Chung, C.-C.; Lo, Y.-J.; Wang, C.-C. Microstructure-dependent visible-light driven photoactivity of sputtering-assisted synthesis of sulfide-based visible-light sensitizer onto ZnO nanorods. *Materials (Basel)* **2016**, *9*, 1014. [CrossRef]

34. Gençyilmaz, O.; Atay, F.; Akyüz, D. The effect of Co doping on ZnO films: Structural, morphological characterization and hall effect measurements. *J. Nanoelectron. Optoelectron.* **2015**, *10*, 799–805. [CrossRef]

35. Harraz, F.A.; Ismail, A.A.; Ibrahim, A.A.; Al-Sayari, S.A.; Al-Assiri, M.S. Highly sensitive ethanol chemical sensor based on nanostructured SnO2 doped ZnO modified glassy carbon electrode. *Chem. Phys. Lett.* **2015**, *639*, 238–242. [CrossRef]

36. Jeon, Y.S.; Seo, H.W.; Kim, S.H.; Kim, Y.K. Synthesis of Fe doped ZnO nanowire arrays that detect formaldehyde gas. *J. Nanosci. Nanotechnol.* **2016**, *16*, 4814–4819. [CrossRef]

37. Hu, Y.; Zhang, Z.; Zhang, H.; Luo, L.; Yao, S. Sensitive and selective imprinted electrochemical sensor for p-nitrophenol based on ZnO nanoparticles/carbon nanotubes doped chitosan film. *Thin Solid Films* **2012**, *520*, 5314–5321. [CrossRef]

38. Zhu, D.; He, Q.; Cao, H.; Cheng, J.; Feng, S.; Xu, Y.; Lin, T. Poly(phenylene ethynylene)-coated aligned ZnO nanorod arrays for 2,4,6-trinitrotoluene detection. *Appl. Phys. Lett.* **2008**, *93*, 11864. [CrossRef]

39. Ameen, S.; Akhtar, M.S.; Seo, H.K.; Shin, H.S. Deployment of aligned ZnO nanorod with distinctive porous morphology: Potential scaffold for the detection of p-nitrophenylamine. *Appl. Catal. Gen.* **2014**, *470*, 271–277. [CrossRef]

40. Selvam, N.C.S.; Jesudoss, S.K.; Rajan, P.I.; Kennedy, L.J.; Vijaya, J.J. Comparative investigation on the photocatalytic degradation of 2,4,6-trichlorophenol using pure and m-doped (M = Ba, Ce, Mg) ZnO spherical nanoparticles. *J. Nanosci. Nanotechnol.* **2015**, *15*, 5910–5917. [CrossRef]

41. Ameen, S.; Akhtar, M.S.; Shin, H.S. Low temperature grown ZnO nanotubes as smart sensing electrode for the effective detection of ethanolamine chemical. *Mater. Lett.* **2013**, *106*, 254–258. [CrossRef]

42. U.S. Environmental Protection Agency. *Nitrophenols, Ambient Water Qualify Criteria*; U.S. Environmental Protection Agency: Washington, DC, USA, 1980.

43. Kaur, N.; Sharma, S.K.; Kim, D.Y.; Sharma, H.; Singh, N. Synthesis of imine-bearing ZnO nanoparticle thin films and characterization of their structural, morphological and optical properties. *J. Nanosci. Nanotechnol.* **2015**, *15*, 8114–8119. [CrossRef] [PubMed]

44. Umar, A.; Al-Hajry, A.; Ahmad, R.; Ansari, S.; Al-Assiri, M.S.; Algarni, H. Fabrication and characterization of a highly sensitive hydroquinone chemical sensor based on iron-doped ZnO nanorods. *Dalton Trans.* **2015**, *44*, 21081–21087. [CrossRef] [PubMed]

45. Ibrahim, A.; Kumar, R.; Umar, A.; Kim, S.H.; Bumajdad, A.; Ansari, Z.A.; Baskoutas, S. Cauliflower-shaped ZnO nanomaterials for electrochemical sensing and photocatalytic applications. *Electrochim. Acta* **2017**, *222*, 463–472. [CrossRef]

46. Ahmed, F.; Arshi, N.; Jeong, Y.S.; Anwar, M.S.; Dwivedi, S.; Alsharaeh, E.; Koo, B.H. Novel biomimatic synthesis of ZnO nanorods using egg white (Albumen) and their antibacterial studies. *J. Nanosci. Nanotechnol.* **2016**, *16*, 5959–5965. [CrossRef] [PubMed]

47. Prabhu, M.; Mayandi, J.; Mariammal, R.N.; Vishnukanthan, V.; Pearce, J.M.; Soundararajan, N.; Ramachandran, K. Peanut shaped ZnO microstructures: Controlled synthesis and nucleation growth toward low-cost dye sensitized solar cells. *Mater. Res. Express* **2015**, *2*, 066202. [CrossRef]

48. Wang, A.J.; Liao, Q.C.; Feng, J.J.; Zhang, P.P.; Li, A.Q.; Wang, J.J. Apple pectin-mediated green synthesis of hollow double-caged peanut-like ZnO hierarchical superstructures and photocatalytic applications. *CrystEngComm* **2012**, *14*, 256–263. [CrossRef]

49. Kumar, R.; Kumar, G.; Akhtar, M.S.; Umar, A. Sonophotocatalytic degradation of methyl orange using ZnO nano-aggregates. *J. Alloys Compd.* **2015**, *629*, 167–172. [CrossRef]

50. Lee, C.-H.; Oh, S.-H. Preparation of Ga-doped ZnO thin films by metal-organic chemical vapor deposition with ultrasonic nebulization. *J. Nanosci. Nanotechnol.* **2016**, *16*, 11552–11557. [CrossRef]

51. Umar, A.; Kumar, R.; Kumar, G.; Algarni, H.; Kim, S.H. Effect of annealing temperature on the properties and photocatalytic efficiencies of ZnO nanoparticles. *J. Alloys Compd.* **2015**, *648*, 46–52. [CrossRef]

52. Umar, A.; Kumar, R.; Akhtar, M.S.; Kumar, G.; Kim, S.H. Growth and properties of well-crystalline cerium oxide (CeO2) nanoflakes for environmental and sensor applications. *J. Colloid Interface Sci.* **2015**, *454*, 61–68. [CrossRef] [PubMed]

53. Punnoose, A.; Dodge, K.; Rasmussen, J.W.; Chess, J.; Wingett, D.; Anders, C. Cytotoxicity of ZnO nanoparticles can be tailored by modifying their surface structure: A green chemistry approach for safer nanomaterials. *ACS Sustain. Chem. Eng.* **2014**, *2*, 1666–1673. [CrossRef] [PubMed]

54. Umar, A.; Lee, J.; Kumar, R.; Al-Dossary, O.; Ibrahim, A.; Baskoutas, S. Development of highly sensitive and selective ethanol sensor based on lance-shaped CuO nanostructures. *Mater. Des.* **2016**, *105*, 16–24. [CrossRef]

55. Kim, S.; Umar, A.; Kumar, R.; Algarni, H.; Kumar, G. Facile and rapid synthesis of ZnO nanoparticles for photovoltaic device application. *J. Nanosci. Nanotechnol.* **2015**, *15*, 6807–6812. [CrossRef] [PubMed]

56. Rana, U.H.S.; Kang, M.; Jeong, E.-S.; Kim, H.-S. Transition between ZnO nanorods and ZnO nanotubes with their antithetical properties. *J. Nanosci. Nanotechnol.* **2016**, *16*, 10772–10776. [CrossRef]

57. Kuriakose, S.; Satpati, B.; Mohapatra, S. Enhanced photocatalytic activity of Co doped ZnO nanodisks and nanorods prepared by a facile wet chemical method. *Phys. Chem. Chem. Phys.* **2014**, *16*, 12741–12749. [CrossRef] [PubMed]

58. Liu, S.; Li, H.; Li, S.; Li, M.; Yan, L. Synthesis, characterization and optical properties of ZnO nanostructure. *J. Nanosci. Nanotechnol.* **2016**, *16*, 8766–8771. [CrossRef]

59. Chang, W.-C.; Yu, W.-C.; Wu, C.-H.; Wang, C.-Y.; Hong, Z.-S.; Wu, R.-J. Flower-Like ZnO nanostructure for NO sensing at room temperature. *J. Nanosci. Nanotechnol.* **2016**, *16*, 9209–9214. [CrossRef]

60. Takeuchi, E.S.; Murray, R.W. Metalloporphyrin containing carbon paste electrodes. *J. Electroanal. Chem. Interfacial Electrochem.* **1985**, *188*, 49–57. [CrossRef]

61. Singh, K.; Ibrahim, A.A.; Umar, A.; Kumar, A.; Chaudhary, G.R.; Singh, S.; Mehta, S.K. Synthesis of CeO 2–ZnO nanoellipsoids as potential scaffold for the efficient detection of 4-nitrophenol. *Sens. Actuator Chem. B* **2014**, *202*, 1044–1050. [CrossRef]

62. Kim, K.Y.; Cho, C.H.; Le Shim, E. Effect of working pressure during ZnO thin-film layer deposition on transparent resistive random access memory device characteristics. *J. Nanosci. Nanotechnol.* **2016**, *16*, 10313–10318. [CrossRef]

63. Picric Acid, Human Health Effects, US Departement of Human & Health Services. Available online: https://toxnet.nlm.nih.gov/cgi-bin/sis/search/a?dbs+hsdb:@term+@DOCNO+2040 (accessed on 4 January 2011).

64. Huang, J.; Wang, L.; Shi, C.; Dai, Y.; Gu, C.; Liu, J. Selective detection of picric acid using functionalized reduced graphene oxide sensor device. *Sens. Actuator Chem.* **2014**, *196*, 567–573. [CrossRef]

65. Roy, N.; Chowdhury, A.; Paul, T.; Roy, A. Morphological, optical, and raman characteristics of ZnO nanoflowers on ZnO-seeded Si substrates synthesized by chemical method. *J. Nanosci. Nanotechnol.* **2016**, *16*, 9738–9745. [CrossRef]

66. Junqueira, J.R.C.; de Araujo, W.R.; Salles, M.O.; Paixão, T.R.L.C. Flow injection analysis of picric acid explosive using a copper electrode as electrochemical detector. *Talanta* **2013**, *104*, 162–168. [CrossRef] [PubMed]

67. Sadhanala, H.K.; Nanda, K.K. Boron and nitrogen Co-doped carbon nanoparticles as photoluminescent probes for selective and sensitive detection of picric acid. *J. Phys. Chem.* **2015**, *119*, 13138–13143. [CrossRef]

68. Sharma, P.; Rana, D.S.; Umar, A.; Kumar, R.; Chauhan, M.S.; Chauhan, S. Synthesis of cadmium sulfide nanosheets for smart photocatalytic and sensing applications. *Ceram. Int.* **2016**, *42*, 6601–6609. [CrossRef]

Article

Fabrication and Characterization of Highly Sensitive Acetone Chemical Sensor Based on ZnO Nanoballs

Qu Zhou [1,*], ChangXiang Hong [1], Yao Yao [2], Ahmed Mohamed Ibrahim [3], Lingna Xu [1], Rajesh Kumar [4], Sumaia Mohamed Talballa [5], S. H. Kim [6,7] and Ahmad Umar [6,7,*]

1   College of Engineering and Technology, Southwest University, Chongqing 400715, China; hcx111000@163.com (C.H.); lingnaxu@cqu.edu.cn (L.X.)
2   College of Communication Engineering, Chengdu University of Information Technology, Chengdu 610225, China; yaoyao386@yahoo.com
3   Department of Pharmaceutical Chemistry, Faculty of Pharmacy, Najran University, Najran 11001, Saudi Arabia; shakiroon4health@gmail.com
4   Department of Chemistry, Jagdish Chandra DAV College, Dasuya 144205, Punjab, India; rkuaah2k7@gmail.com
5   Department of Pathology, Faculty of Medicine, Najran University, Najran 11001, Saudi Arabia; omsuhieb@yahoo.com
6   Department of Chemistry, College of Science and Arts, Najran University, P.O. Box 1988, Najran 11001, Saudi Arabia; semikim77@gmail.com
7   Promising Centre for Sensors and Electronic Devices (PCSED), Najran University, P.O. Box 1988, Najran 11001, Saudi Arabia
*   Correspondence: zhouqul@swu.edu.cn (Q.Z.); ahmadumar786@gmail.com or umahmad@nu.edu.sa (A.U.); Tel.: +86-023-6825-1265 (Q.Z.)

Received: 2 June 2017; Accepted: 7 July 2017; Published: 14 July 2017

Abstract: Highly sensitive acetone chemical sensor was fabricated using ZnO nanoballs modified silver electrode. A low temperature, facile, template-free hydrothermal technique was adopted to synthesize the ZnO nanoballs with an average diameter of 80 ± 10 nm. The XRD and UV-Vis. studies confirmed the excellent crystallinity and optical properties of the synthesized ZnO nanoballs. The electrochemical sensing performance of the ZnO nanoballs modified AgE towards the detection of acetone was executed by simple current–voltage (I–V) characteristics. The sensitivity value of $\sim$472.33 $\mu A \cdot mM^{-1} \cdot cm^{-2}$ and linear dynamic range (LDR) of 0.5 mM–3.0 mM with a correlation coefficient (R^2) of 0.97064 were obtained from the calibration graph. Experimental limit of detection (LOD) for ZnO nanoballs modified AgE was found to be 0.5 mM.

Keywords: ZnO; nanoballs; acetone; current–voltage; electrochemical; sensor

1. Introduction

ZnO nanomaterials has received exceptional attention and interest worldwide among the research fraternity due to their unique properties such as large surface to volume ratio, non-toxicity, ease of synthesis, n-type semiconducting nature, wide band gap of ~3.30 eV, large exciton binding energy, high thermal stability, excellent electrical, magnetic, catalytic properties, etc. [1–5]. A large variety of methods for synthesis of ZnO nanomaterials is reported in the literature which results in the formations of different morphologies like nano-mushrooms, fluffy nanoballs, nanorods, nanoribbons, nanowires, nanoflakes, nano/microspheres, nanocones, nanopillars, nano/micro flowers, nanoneedles, nanosheets, nanoaggregates, etc.

Among the various potential applications, real-time and reliable electrochemical sensing of harmful, toxic and explosive chemicals using ZnO nanostructured based electrochemical sensing, is widely studied. Such sensors offer advantages such as ambient stability, resistivity towards toxic and hazardous chemicals, chemical inertness, electrocatalytic activity and ease of fabrication. It has been reported that *n*-type semiconducting metal oxide nanomaterials enhance the rate of electron transfer between electrode and analyte molecules, which drastically improves the current response for target molecules [6]. Additionally, inorganic metal oxide nanoparticles serve as supra-molecular assembling units which provide large surface area for electrochemical sensing interface [7,8]. Electrical signals resulted from the interaction of the target analyte molecules and the ZnO nanostructured transducer layers, coated on the surface of the modified electrode, provide the valuable analytical information [9].

Toxic and highly hazardous chemicals such as nitrophenols [10,11], ammonia [12], CO [13], hydrazines [14,15], nitroanilines [16–18], hydrogen sulfide [19], ethanolamine [20], picric acid [21], ethyl acetate [22], ethanol [23], synthetic antioxidants and dyes in food articles [24,25], some bio-molecules like glucose [26–28], uric acid [29,30], urea [31,32], aspartic acid [33], dopamine [34], pH sensors [35], etc. have been detected and analyzed through electrochemical sensing techniques using ZnO modified electrochemical sensors. Recently, Ahmad et al. [18] reported a binder-free, stable, and highly efficient hydrazine chemical sensor based on vertically aligned ZnO nanorods directly grown on the surface of Ag electrode through a low-temperature solution process. The average diameter and length of ZnO nanorods were ~50 nm and 2.2 μm with a high aspect ratio of about 44. Excellent sensitivity of 105.5 $\mu A \cdot \mu M^{-1} \cdot cm^{-2}$ with a linear dynamic range of 0.01–98.6 μM and low detection limit of 0.005 μM. was observed. Unique lotus-leaf-like ZnO nanostructures deposited on FTO substrate showed very low-level detection of ethyl acetate with high sensitivity of ~139.8 $\mu A \cdot mM^{-1} \cdot cm^{-2}$ and limit of detection of ~0.26 mM [22]. Ameen et al. [36] synthesized ZnO nanowhiskers through a hydrothermal method and utilized them as electron mediators for the fabrication of electrochemical sensors for detecting p-hydroquinone. As fabricated p-hydroquinone chemical sensor exhibited a substantially high sensitivity of ~99.2 $\mu A \cdot \mu M^{-1} \cdot cm^{-2}$ with a very low detection limit of ~4.5 μM and linear dynamic range of ~10–200 μM. Ibrahim et al. [21] observed a high sensitivity of 24.14 $\mu A \cdot mM^{-1} \cdot cm^{-1}$ with good LDR of 0.078–10.0 mM against picric acid using electrochemical sensor based on ZnO nanostructures with cauliflower shaped morphologies. Tailoring the ZnO morphologies for acquiring large surface to volume ratio for better adsorption of the analyte species and hence fast charge transfer during the electrochemical process is one of the most critical and desired aspects of electrochemical sensing applications.

In the present work, a simple, low cost, and template-free hydrothermal method was adopted for the synthesis of ZnO nanoballs with highly rough surfaces. Morphological, structural, optical, crystal phases, vibrational and scattering properties of the ZnO nanoballs were evaluated through different analytic techniques. ZnO nanoballs were further utilized for the fabrication of highly sensitive acetone electrochemical sensors through *I–V* techniques. The ZnO nanoballs modified AgE showed the high sensitivity towards acetone.

2. Results and Discussion

2.1. Morphological, Structural, Optical and Compositions Properties of ZnO Nanoballs

Figure 1 represents the field emission scanning electron microscopic (FESEM) images of the hydrothermally synthesized ZnO powders. Interestingly, almost ball shaped morphologies can be assigned to maximum of the ZnO particles from the low magnification (Figure 1a) as well as high magnification (Figure 1a,b) FESEM images. However, few ZnO structures with ellipsoidal and non-spherical shapes can also be seen. These ZnO nanoballs further form some agglomerated structures. The surface of the ZnO nanoballs is highly rough as confirmed from a close look at the high magnification FESEM image as shown in Figure 1c. The average diameter of the ZnO nanoballs is 80 ± 10 nm. The roughness of the ZnO nanoballs surface provides a high density of the active sites for the adsorption of the target analyte and O_2 from the air. In Figure 1d the energy dispersive spectroscopy (EDS) spectrum for the hydrothermally synthesized ZnO nanoballs is shown. The presence of peaks only for Zinc and oxygen atoms confirms the formation of the ZnO along with a high degree of purity for the synthesized ZnO nanoballs.

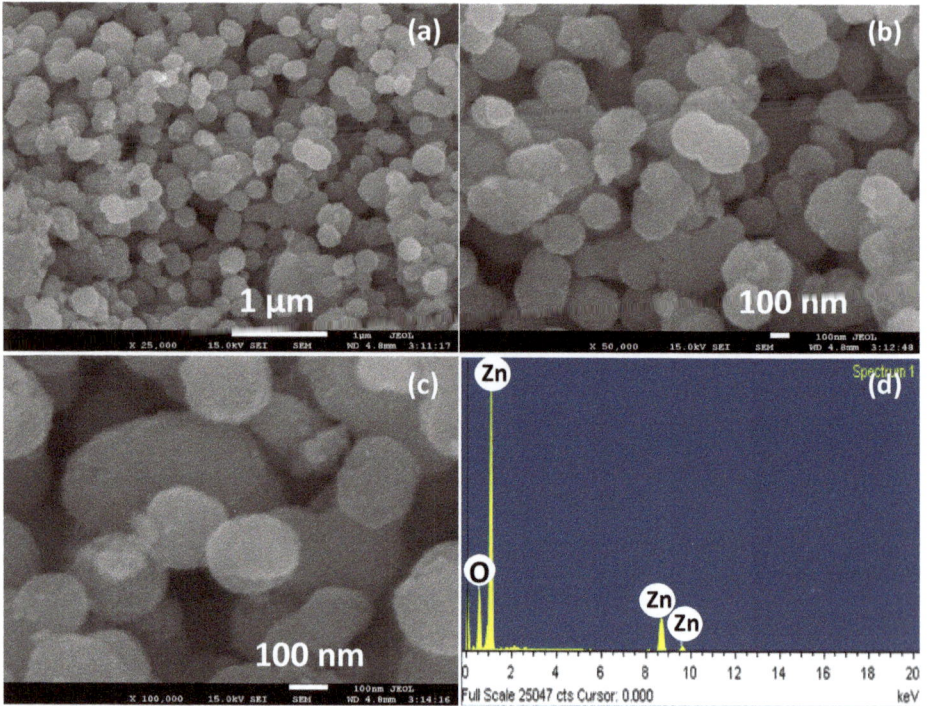

Figure 1. (**a**) Low magnification; and (**b**,**c**) high magnification FESEM images; and (**d**) EDS spectrum of ZnO nanoballs.

The crystallinity, crystalline size and microstructural phases for the ZnO nanoballs can be evaluated from the X-ray diffraction (XRD) spectrum as shown in Figure 2. Well-defined diffractions peaks corresponding to the diffraction planes (100), (002), (101), (102), (110), (103), (200), (112), (201), (004) and (202) at diffraction angles 31.78°, 34.43°, 36.23°, 47.63°, 56.61°, 62.91°, 66.40°, 67.95°, 69.14°, 72.59° and 76.71°, respectively, indicate the Wurtzite hexagonal phase for ZnO nanoballs. The results are supported by the JCPDS data card Nos. 36–1451 and reported literature [37–42]. No additional peak in the XRD spectrum related to any impurity, further confirms the results of EDS studies (Figure 1d).

Debye–Scherrer formula (Equation (1)) was used for calculating the crystallite size (d) of the ZnO nanoballs [43].

$$d = \frac{0.89\lambda}{\beta \cdot Cos\theta} \tag{1}$$

where λ = the wavelength of X-rays used (1.54 $\overset{o}{A}$), θ is the Bragg diffraction angle and β is the peak width at half maximum (FWHM). The FWHM values for the three most intense diffraction peaks corresponding to diffraction planes (100), (002) and (101) were taken into account. The corresponding results are given in Table 1. The average crystallite size of ZnO nanoballs was found to be 10.47 nm.

Table 1. The crystallite size of the hydrothermally synthesized ZnO nanoballs.

S.N	(hkl)	2θ (°)	FWHM (β)	Crystallite Size (nm)
1	(100)	31.78	0.71936	11.36
2	(002)	38.43	0.80871	10.18
3	(101)	36.23	0.83756	9.88

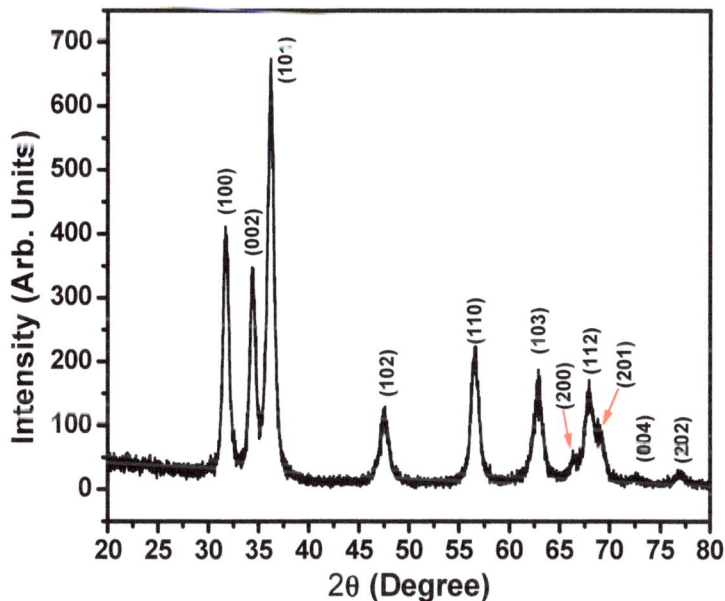

Figure 2. Typical XRD patterns for hydrothermally synthesized ZnO nanoballs.

Figure 3a represents the typical Fourier transform infrared (FTIR) spectrum of hydrothermally synthesized ZnO nanoballs. A sharp and well-defined peak at 476 cm^{-1} is the characteristic peak for metal-oxygen (M–O) bond and confirms the formation of the Zn–O bond. Another broad band at 3446 cm^{-1} is due to the O–H stretching vibrational modes of the water molecules physiosorbed on the surface of the ZnO nanoballs [16,44–46].

In Figure 3b, the UV-Vis. spectrum plotted in the range of 200–550 nm is shown. A single and sharp absorption peak at 390 nm is observed. The band gap energy (E_g) of 3.19 eV was calculated with the help of well-known Planck's quantum equation (Equation (2)) [47].

$$E_g = \frac{hc}{\lambda_{max}} = \frac{6.625 \times 10^{-34}\ Js \times 3 \times 10^8\ ms^{-1}}{390 \times 10^{-9}\ m \times 1.6 \times 10^{-19}} = 3.19\ eV \tag{2}$$

In order to evaluate the molecular vibrational, polarization and scattering information for the ZnO nanoballs, Raman-scattering analysis was performed at room temperature. Figure 4 represents the Raman scattering spectrum of the hydrothermally synthesized ZnO nanoballs.

Three distinct phonon peaks at 332, 382 and 438 cm^{-1} are the typical characteristic peaks of the ZnO wurtzite hexagonal phase and correspond to E$_{2H}$–E$_{2L}$ multiphonon process, A$_1$(TO) and E$_2^{High}$ modes, respectively [48]. Stronger E$_2^{High}$ indicates excellent crystal qualities and very low oxygen vacancies on the surface of the ZnO nanoballs [49].

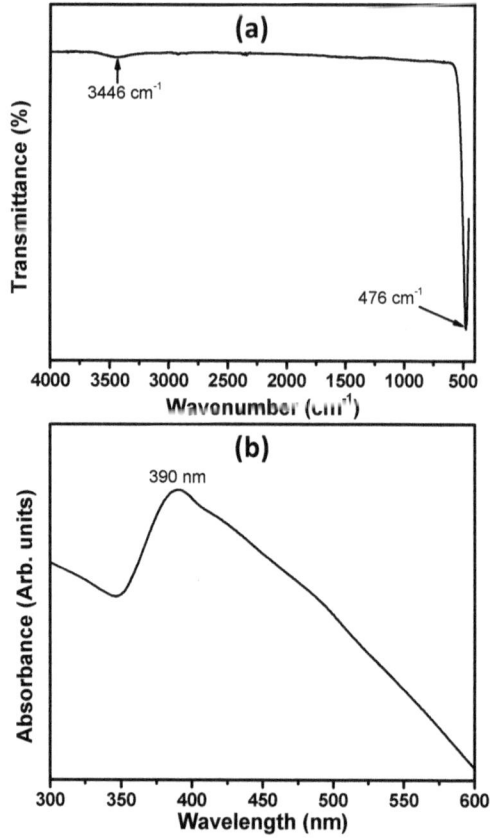

Figure 3. (**a**) FTIR; and (**b**) UV-Vis. spectra for hydrothermally synthesized ZnO nanoballs.

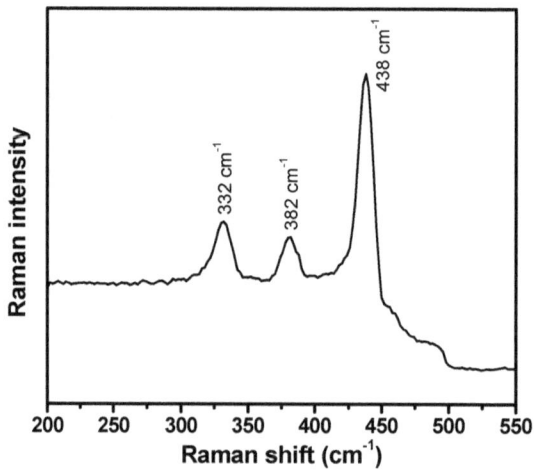

Figure 4. Raman spectrum for hydrothermally synthesized ZnO nanoballs.

2.2. Characterization of Acetone Sensor Fabricated Based on ZnO Nanoballs

The potential electro-catalytic sensing applications of ZnO nanoballs coated onto the surface of the AgE are demonstrated in this section. Initial experimentations involves the comparison of *I–V* responses of the ZnO nanoballs modified AgE for 0.5 mM acetone solution prepared in the 0.1 M PBS having pH 7.4 and blank PBS within the potential range of 0.0–2.5 V. As the applied potential increases, the current response increases remarkably for the PBS containing acetone as compared to blank PBS (Figure 5a). At an applied potential of 2.5 V, the maximum current responses of 6.84221 and 1.6182 µA were observed for 0.5 mM acetone solutions and blank PBS, respectively. This substantial response of the ZnO nanoballs modified AgE towards the sensing of acetone confirms the involvement of the ZnO nanostructures in the efficient electrocatalytic activities and fast electron exchange capabilities Figure 5b represents the effect of the acetone concentration on the current responses of the ZnO nanoballs modified AgE. Different solutions of acetone with concentration range of 0.5 mM–5.0 mM were prepared in 0.1 M PBS and were subjected to electrochemical analysis using ZnO nanoballs modified AgE as working electrode and a Pt wire as a counter electrode within the potential range of 0.0–2.5 V. It can be seen that the increase in the concentration of the acetone resulted in a marked increase in the current responses.

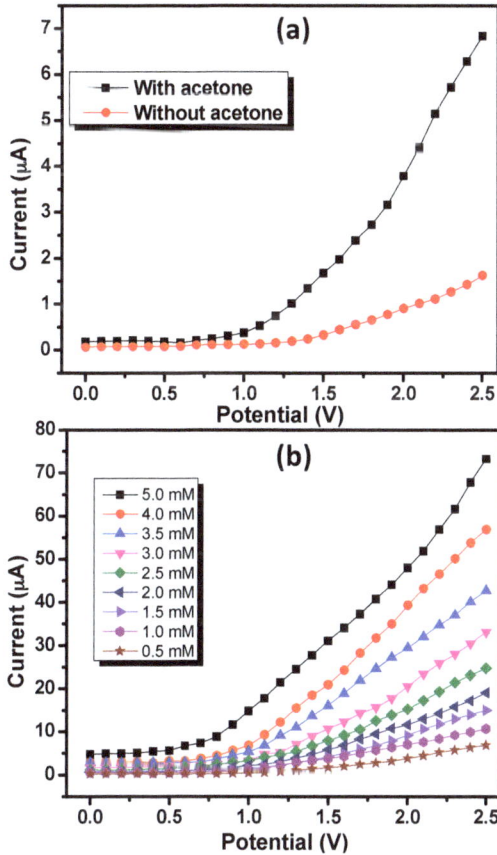

Figure 5. (**a**) *I–V* responses measured for 0.5 mM acetone in 0.1 M PBS solution and blank PBS solution using ZnO nanoballs modified AgE; and (**b**) *I–V* response variations for 0.5 mM–5.0 mM concentrations of acetone in 0.1 M PBS solution.

At an applied potential of 2.5 V, the current responses of 6.84221, 10.7272, 14.9831, 19.0696, 24.7101, 33.013, 42.7444, 56.8836 and 73.3094 μA were recorded for 0.5, 1.0, 1.5, 2.0, 2.5, 3.0, 3.5, 4.0 and 5.0 mM acetone solutions, respectively. Increased current responses with a concentration of the acetone can be attributed to the generation of a large number of ions and increased ionic strength of the analyte solutions [37].

Current vs. concentration calibration graph was plotted to determine the sensing parameters such as sensitivity, LOD and LDR (Figure 6). The sensitivity value of $\sim$472.33 μA·mM^{-1}·cm^{-2} and LDR of 0.5 mM–3.0 mM with a correlation coefficient (R^2) of 0.97064 were obtained from the calibration graph. Experimental LOD for ZnO nanoballs modified AgE was found to be 0.5 mM. As fabricated acetone sensors based on hydrothermally synthesized ZnO nanoballs exhibit better sensitivity compared to different sensors reported in the literature (Table 2).

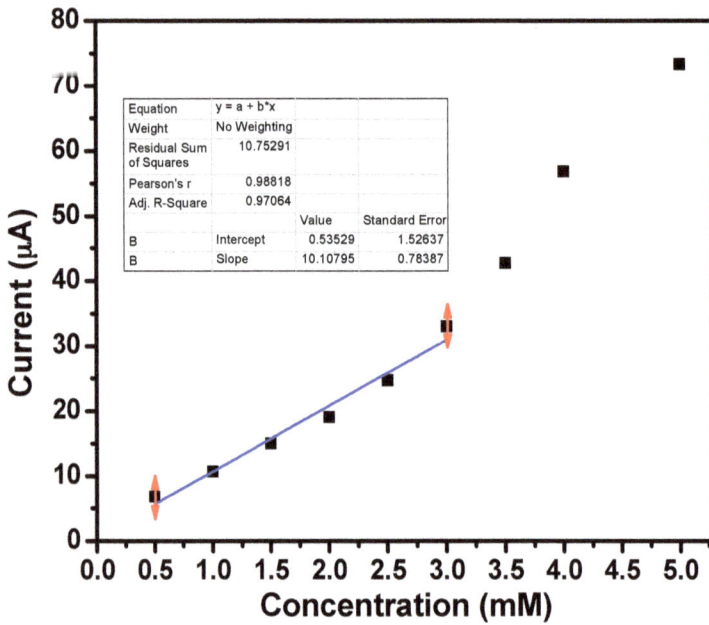

Figure 6. Calibration plot for ZnO nanoballs modified AgE towards acetone.

Table 2. Summary of the acetone sensing performances of different sensor materials.

Sensor	Method	Sensitivity	LDR	LOD	R^2	Ref.
ZnO-doped Co_3O_4 Nanorods/AgE	I–V	3.58 $\mu A \cdot mM^{-1} \cdot cm^{-2}$	66.8 μM–0.133 mM	14.7 ± 0.2 μM	0.5684	[50]
ZnO NPs/GCE	I–V	0.14065 $\mu A \cdot mM^{-1} \cdot cm^{-2}$	0.13 mM–0.13 M	0.068 ± 0.01 mM	-	[51]
Gd-ZnO-Nanopencils/AgE	I–V	208 ± 62 $\mu A \cdot mM^{-1} \cdot cm^{-2}$	750 μM–100 mM	0.5 mM	0.885	[52]
$ZnO/SnO_2/Yb_2O_3$/GCE	I–V	17.09 $\mu A \cdot mM^{-1} \cdot cm^{-2}$	3.34 nM–3.4 mM	0.05 ± 0.002 nM	0.9394	[53]
Lead foil electrode	Amperometric	2.07 $\mu A \cdot cm^{-2} \cdot ppm^{-1}$	50–250 ppm	50 ppm	0.998	[54]
Electro-deposited Pb electrode	Amperometric	4.16 $\mu A \cdot cm^{-2} \cdot ppm^{-1}$	100–400 ppm	-	0.99	[55]
Ag_2O microflower/GCE	I–V	1.699 $\mu A \cdot mM^{-1} \cdot cm^{-2}$	0.13 μM–0.67 M	0.1 μM	0.9462	[56]
ZnO nanoballs/AgE	*I–V*	*472.33 $\mu A \cdot mM^{-1} \cdot cm^{-2}$*	*0.5 mM–3.0 mM*	*0.5 mM*	*0.9706*	*This work*

2.3. Proposed Sensing Mechanism

It has been postulated in many studies that the adsorption of the molecular oxygen (O_2) from the PBS as well as from the surrounding environment onto the highly rough surface of the ZnO nanomaterials, is the key concern of the sensing applications. Surface reactions result in the formation of oxygenated anionic species such as superoxides (O_2^-), peroxides (O_2^{2-}), hydroxides (HO^-) and oxides (O^{2-}) [57]. The reduction is aided through the conduction band electrons of the ZnO nanomaterials coated onto the surface of AgE (Equations (3)–(6)).

$$O_{2\ (g)} \rightarrow O_{2\ (chemisorbed)} \tag{3}$$

$$O_{2\ (chemisorbed)} + e^- \rightarrow O_{2\ (Chemisorbed)}^- \tag{4}$$

$$O_{2\ (chemisorbed)} + 2e^- \rightarrow O_{2\ (Chemisorbed)}^{2-} \tag{5}$$

$$O_{2\ (chemisorbed)} + 4e^- \rightarrow 2O_{(Chemisorbed)}^{2-} \tag{6}$$

These chemisorbed oxygenated anionic species deplete the surface electron states of the ZnO and increase the resistance of the n-type semiconductor material due to the formation of an electron depletion layer at the ZnO nanoballs surfaces [58–61]. The increase in the current response is due to the release of the trapped electrons back into the conduction band during the catalytic oxidation the adsorbed acetone molecules into CO_2 and H_2O (Equations (7)–(10)) [18,44,62–64].

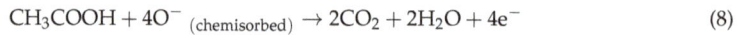

$$CH_3COCH_3 + 4O^-_{\ (chemisorbed)} \rightarrow CH_3COOH + CO_2 + H_2O + 4e^- \tag{7}$$

$$CH_3COOH + 4O^-_{\ (chemisorbed)} \rightarrow 2CO_2 + 2H_2O + 4e^- \tag{8}$$

$$CH_3COCH_3 + 4O_2^-_{\ (chemisorbed)} \rightarrow 3CO_2 + 3H_2O + 4e^- \tag{9}$$

$$CH_3COCH_3 + 8O^{2-}_{\ (chemisorbed)} \rightarrow 3CO_2 + 3H_2O + 16e^- \tag{10}$$

On the basis of above discussion, the proposed sensing mechanism for ZnO nanoballs modified AgE against acetone is represented in Figure 7.

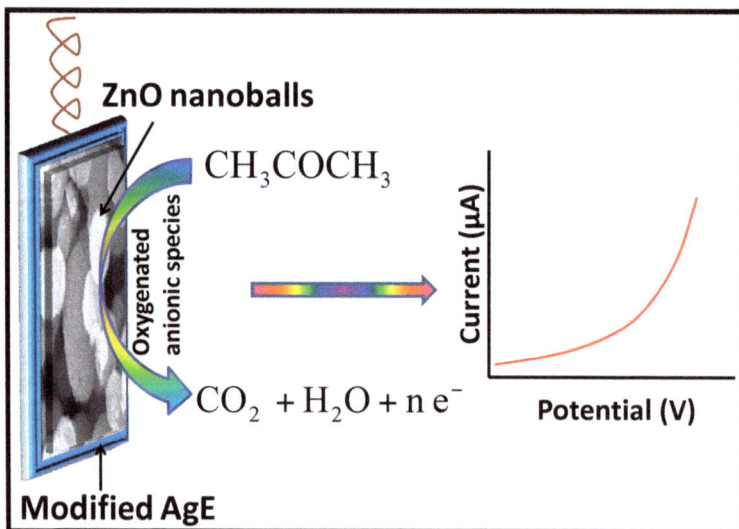

Figure 7. Proposed sensing mechanism for ZnO nanoballs modified AgE towards acetone in PBS.

Thus, ZnO nanoballs pose excellent electron mediator activities for the detection and sensing of very low level of acetone in PBS at room temperature.

3. Materials and Methods

3.1. Hydrothermal Synthesis of ZnO Nanoballs

For the synthesis of the ZnO nanoballs, all chemicals were purchased from Sigma–Aldrich (St. Louis, MO, USA) and were used as received without any further refinement. All solutions were prepared in DI water. A facile hydrothermal method was adopted for the synthesis of ZnO nanoballs in which 50 mL of 0.02 M Zinc nitrate hexahydrate [$Zn(NO_3)_2 \cdot 6H_2O$] was continuously stirred for 30 min along with aqueous NaOH solution, added dropwise in order to maintain a pH of 10. Thereafter the resulting solution was transferred to a Teflon-lined stainless steel autoclave which was heated to 150 °C. After heating the autoclave for the desired growth time oh for 5 h, it was slowly cooled to room temperature. The white product formed was filtered and washed with DI water and ethanol to remove any un-reacted reactants. Finally, the powder was dried at 70 ± 2 °C for 2 h in a hot air oven and characterized for its morphological, optical, structural, compositional and electrochemical sensing applications using different analytical techniques.

3.2. Fabrication of Acetone Sensor Based on ZnO Nanoballs

The silver electrode of active surface area of 0.0214 cm^2 was pre-cleaned using 0.05 μm alumina slurry followed by thorough washings with distilled water and ethanol and finally dried for 1 h in hot air oven at 70 °C. A homogeneous thin paste of ZnO nanoballs was prepared in butyl carbital acetate (BCA) conducting solvent and was coated in the form of a thin layer over the surface of the Ag electrode. The coated Ag electrode was dried at 70 °C in an air oven for 4 h. An electrochemical cell was then set up in which ZnO nanoballs modified AgE served as the working electrode and a Pt wire as the counter electrode. The current–voltage (I–V) measurements for solutions of acetone with different concentrations were measured at room temperature in the presence of 0.1 M phosphate buffer solution (PBS) with pH of 7.4 with the help of Keithley 6517A-USA electrometer (Tektronix, OR, USA) with computer interfacing. The acetone sensitivity was determined by generating a calibration curve of current vs. concentration. The sensitivity of the ZnO nanoballs modified AgE was determined from the ratio of the slope of the calibration graph plotted between current and concentration to the active surface area of modified AgE.

3.3. Characterization of ZnO Nanoballs

Field emission scanning electron microscopy (FESEM; JEOL-JSM-7600F, JEOL, Tokyo, Japan) integrated with EDS was examined in order to study the morphological, compositional, and structural properties of the hydrothermally synthesized ZnO nanoballs. X-ray diffraction (XRD; JDX-8030W, JEOL, Tokyo, Japan) studies were conducted between the diffraction angles (2θ) range of 20°–80° using Cu-Kα source radiation with a wavelength of 1.54 Å in order to explore the crystallinity, crystalline size, and microstructural phases. UV-visible spectrophotometer (Perkin Elmer-UV- -Vis. Lambda 950, PerkinElmer, MA, USA) analysis of the aqueous solution of ZnO nanoballs, sonicated for 15 min was carried out between the scan range of 300–600 nm to evaluate the band gap energy and optical properties. Fourier transform infrared spectroscopic (FTIR; Perkin Elmer-FTIR Spectrum-100, PerkinElmer, MA, USA) analysis was conducted in order to analyze the composition of the as-synthesized ZnO nanoballs. Raman-scattering spectroscopic (Perkin Elmer-Raman Station 400 series, PerkinElmer, MA, USA) technique was utilized for the examination of scattering properties of the ZnO nanoballs in the scan range of 200–550 cm^{-1}.

4. Conclusions

A simple, low cost, template-free hydrothermal method was adopted for the synthesis of ZnO nanoballs with highly rough surfaces. Morphological, structural, optical, crystal phases, vibrational and scattering properties of the ZnO nanoballs were evaluated through different analytic techniques such as FESEM, EDS, UV-Vis, FTIR and Raman scattering spectroscopy. ZnO nanoballs were further utilized for the fabrication of highly sensitive acetone electrochemical sensors through $I–V$ techniques. All the observations were recorded at room temperature and in the presence of the 0.1 M PBS with pH of 7.4. High sensitivity of ~472.33 $\mu A \cdot mM^{-1} \cdot cm^{-2}$ and LDR of 0.5 mM–3.0 mM were obtained. Hence, ZnO nanoballs based electrochemical sensors may introduce a relatively new avenue for the fabrication of efficient sensor for hazardous and carcinogenic chemicals and in environmental and healthcare monitoring.

Acknowledgments: This work has been supported in part by the National Natural Science Foundation of China (No. 51507144), China Postdoctoral Science Foundation funded project (Nos. 2015M580771 and 2016T90832); the Chongqing Science and Technology Commission (CSTC) (No. cstc2016jcyjA0400); Postdoctoral Science Funded Project of Chongqing (No. Xm2015016); Visiting Scholarship of State Key Laboratory of Power Transmission Equipment & System Security and New Technology (No. 2007DA10512716423); and Fundamental Research Funds for the Central Universities (No. XDJK2015B005).

Author Contributions: Qu Zhou and Ahmad Umar conceived and designed the experiments; Qu Zhou, ChangXiang Hong, Ahmad Umar, and Ahmed Mohamed Ibrahim performed the experiments; ChangXiang Hong, Ahmad Umarand Yao Yao analyzed the data; Qu Zhou and Ahmad Umar wrote the paper; and Qu Zhou, Ahmed Mohamed Ibrahim, Rajesh Kumar, Sumaia Mohamed Talballa, S. H. Kim and Ahmad Umar reviewed and revised the manuscript. All authors read and approved the manuscript.

Conflicts of Interest: The authors declare no conflict of interest.

References

1. Kumar, R.; Al-Dossary, O.; Kumar, G.; Umar, A. Zinc oxide nanostructures for NO_2 gas sensor applications: A review. *Nano-Micro Lett.* **2015**, 7, 97–120. [CrossRef]
2. Schmidt-Mende, L.; MacManus-Driscoll, J.L. ZnO—Nanostructures, defects, and devices. *Mater. Today* **2007**, 10, 40–48. [CrossRef]
3. Kumar, R.; Umar, A.; Kumar, G.; Nalwa, H.S. Antimicrobial properties of ZnO nanomaterials: A review. *Ceram. Int.* **2017**, 43, 3940–3961. [CrossRef]
4. Kumar, R.; Umar, A.; Kumar, G.; Nalwa, H.S.; Kumar, A.; Akhtar, M.S. Zinc oxide nanostructure-based dye-sensitized solar cells. *J. Mater. Sci.* **2017**, 52, 4743–4795. [CrossRef]
5. Chongsri, K.; Sinornate, W.; Boonyarattanakalin, K.; Pecharapa, W. Growth and characterization of Ga/F Co-doped ZnO nanorods/nanodisks via hydrothermal process. *J. Nanosci. Nanotechnol.* **2016**, 16, 12962–12966. [CrossRef]
6. Park, G.C.; Lee, S.M.; Jeong, S.H.; Choi, J.H.; Lee, C.M.; Seo, T.Y.; Jung, S.-B.; Lim, J.H.; Joo, J. Enhanced photocatalytic activity of ZnO nanorods with tubular facet synthesized by hydrothermal method. *J. Nanosci. Nanotechnol.* **2016**, 16, 11164–11168. [CrossRef]
7. Nouira, W.; Barhoumi, H.; Maaref, A.; Renault, N.J.; Siadat, M. Tailoring of analytical performances of urea biosensors using nanomaterials. *J. Phys. Conf. Ser.* **2013**, 416, 12010. [CrossRef]
8. Zhang, F.; Yang, P.; Matras-Postolek, K. Au catalyst decorated silica spheres: Synthesis and high-performance in 4-nitrophenol reduction. *J. Nanosci. Nanotechnol.* **2016**, 16, 5966–5974. [CrossRef] [PubMed]
9. Hieu, N.M.; Kim, H.; Kim, C.; Hong, S.-K.; Kim, D. A hydrogen sulfide gas sensor based on pd-decorated ZnO nanorods. *J. Nanosci. Nanotechnol.* **2016**, 16, 10351–10355. [CrossRef]
10. Thirumalraj, B.; Rajkumar, C.; Chen, S.-M.; Lin, K.-Y. Determination of 4-nitrophenol in water by use of a screen-printed carbon electrode modified with chitosan-crafted ZnO nanoneedles. *J. Colloid Interface Sci.* **2017**, 499, 83–92. [CrossRef] [PubMed]
11. Hu, Y.; Zhang, Z.; Zhang, H.; Luo, L.; Yao, S. Sensitive and selective imprinted electrochemical sensor for p-nitrophenol based on ZnO nanoparticles/carbon nanotubes doped chitosan film. *Thin Solid Films* **2012**, 520, 5314–5321. [CrossRef]

12. Dar, G.N.; Umar, A.; Zaidi, S.A.; Baskoutas, S.; Hwang, S.W.; Abaker, M.; Al-Hajry, A.; Al-Sayari, S.A. Ultra-high sensitive ammonia chemical sensor based on ZnO nanopencils. *Talanta* **2012**, *89*, 155–161. [CrossRef] [PubMed]

13. Chen, W.; Li, Q.; Xu, L.; Zeng, W. Gas sensing properties of ZnO/SnO$_2$ nanostructures. *J. Nanosci. Nanotechnol.* **2015**, *15*, 1245–1252. [CrossRef] [PubMed]

14. Mehta, S.K.; Singh, K.; Umar, A.; Chaudhary, G.R.; Singh, S. Ultra-high sensitive hydrazine chemical sensor based on low-temperature grown ZnO nanoparticles. *Electrochim. Acta* **2012**, *69*, 128–133. [CrossRef]

15. Hu, J.; Zhao, Z.; Sun, Y.; Wang, Y.; Li, P.; Zhang, W.; Lian, K. Controllable synthesis of branched hierarchical ZnO nanorod arrays for highly sensitive hydrazine detection. *Appl. Surf. Sci.* **2016**, *364*, 434–441. [CrossRef]

16. Ibrahim, A.A.; Umar, A.; Kumar, R.; Kim, S.H.; Bumajdad, A.; Baskoutas, S. Sm$_2$O$_3$-doped ZnO beech fern hierarchical structures for nitroaniline chemical sensor. *Ceram. Int.* **2016**, *42*, 16505–16511. [CrossRef]

17. Ahmad, N.; Umar, A.; Kumar, R.; Alam, M. Microwave-assisted synthesis of ZnO doped CeO$_2$ nanoparticles as potential scaffold for highly sensitive nitroaniline chemical sensor. *Ceram. Int.* **2016**, *42*, 11562–11567. [CrossRef]

18. Ahmad, R.; Tripathy, N.; Ahn, M.-S.; Hahn, Y.-B. Highly stable hydrazine chemical sensor based on vertically-aligned ZnO nanorods grown on electrode. *J. Colloid Interface Sci.* **2017**, *494*, 153–158. [CrossRef] [PubMed]

19. Park, N.-K.; Lee, T.H.; Choi, H.Y.; Lee, T.J. Changing electric resistance of ZnO nano-rods by sulfur compounds for chemical gas sensor. *J. Nanosci. Nanotechnol.* **2015**, *15*, 1752–1755. [CrossRef] [PubMed]

20. Ameen, S.; Shaheer Akhtar, M.; Shin, H.S. Low temperature grown ZnO nanotubes as smart sensing electrode for the effective detection of ethanolamine chemical. *Mater. Lett.* **2013**, *106*, 254–258. [CrossRef]

21. Ibrahim, A.A.; Kumar, R.; Umar, A.; Kim, S.H.; Bumajdad, A.; Ansari, Z.A.; Baskoutas, S. Cauliflower-shaped ZnO nanomaterials for electrochemical sensing and photocatalytic applications. *Electrochim. Acta* **2016**, *222*, 463–472. [CrossRef]

22. Ameen, S.; Park, D.-R.; Akhtar, M.S.; Shin, H.S. Lotus-leaf like ZnO nanostructures based electrode for the fabrication of ethyl acetate chemical sensor. *Mater. Lett.* **2016**, *164*, 562–566. [CrossRef]

23. Jianjiao, Z.; Hongyan, Y.; Erjun, G.; Shaolin, Z.; Liping, W.; Chunyu, Z.; Xin, G.; Jing, C.; Hong, Z. Novel gas sensor based on ZnO nanorod circular arrays for C$_2$H$_5$OH gas detection. *J. Nanosci. Nanotechnol.* **2015**, *15*, 2468–2472. [CrossRef] [PubMed]

24. Gan, T.; Zhao, A.; Wang, S.; Lv, Z.; Sun, J. Hierarchical triple-shelled porous hollow zinc oxide spheres wrapped in graphene oxide as efficient sensor material for simultaneous electrochemical determination of synthetic antioxidants in vegetable oil. *Sens. Actuators B Chem.* **2016**, *235*, 707–716. [CrossRef]

25. Ya, Y.; Jiang, C.; Li, T.; Liao, J.; Fan, Y.; Wei, Y.; Yan, F.; Xie, L. A zinc oxide nanoflower-based electrochemical sensor for trace detection of sunset yellow. *Sensors* **2017**, *17*. [CrossRef] [PubMed]

26. Zhou, F.; Jing, W.; Wu, Q.; Gao, W.; Jiang, Z.; Shi, J.; Cui, Q. Effects of the surface morphologies of ZnO nanotube arrays on the performance of amperometric glucose sensors. *Mater. Sci. Semicond. Process.* **2016**, *56*, 137–144. [CrossRef]

27. Rodrigues, A.; Castegnaro, M.V.; Arguello, J.; Alves, M.C.M.; Morais, J. Development and surface characterization of a glucose biosensor based on a nanocolumnar ZnO film. *Appl. Surf. Sci.* **2017**, *402*, 136–141. [CrossRef]

28. Kitture, R.; Chordiya, K.; Gaware, S.; Ghosh, S.; More, P.A.; Kulkarni, P.; Chopade, B.A.; Kale, S.N. ZnO nanoparticles-red sandalwood conjugate: A promising anti-diabetic agent. *J. Nanosci. Nanotechnol.* **2015**, *15*, 4046–4051. [CrossRef] [PubMed]

29. Lei, Y.; Liu, X.; Yan, X.; Song, Y.; Kang, Z.; Luo, N.; Zhang, Y. Multicenter uric acid biosensor based on tetrapod-shaped ZnO nanostructures. *J. Nanosci. Nanotechnol.* **2012**, *12*, 513–518. [CrossRef] [PubMed]

30. Ahmad, R.; Tripathy, N.; Jang, N.K.; Khang, G.; Hahn, Y.B. Fabrication of highly sensitive uric acid biosensor based on directly grown ZnO nanosheets on electrode surface. *Sens. Actuators B Chem.* **2015**, *206*, 146–151. [CrossRef]

31. Ansari, S.G.; Wahab, R.; Ansari, Z.A.; Kim, Y.S.; Khang, G.; Al-Hajry, A.; Shin, H.S. Effect of nanostructure on the urea sensing properties of sol-gel synthesized ZnO. *Sens. Actuators B Chem.* **2009**, *137*, 566–573. [CrossRef]

32. Ali, S.M.U.; Ibupoto, Z.H.; Salman, S.; Nur, O.; Willander, M.; Danielsson, B. Selective determination of urea using urease immobilized on ZnO nanowires. *Sens. Actuators B Chem.* **2011**, *160*, 637–643. [CrossRef]

33. Liu, H.; Gu, C.; Hou, C.; Yin, Z.; Fan, K.; Zhang, M. Plasma-assisted synthesis of carbon fibers/ZnO core–shell hybrids on carbon fiber templates for detection of ascorbic acid and uric acid. *Sens. Actuators B Chem.* **2016**, *224*, 857–862. [CrossRef]

34. Ghanbari, K.; Moloudi, M. Flower-like ZnO decorated polyaniline/reduced graphene oxide nanocomposites for simultaneous determination of dopamine and uric acid. *Anal. Biochem.* **2016**, *512*, 91–102. [CrossRef] [PubMed]

35. Copa, V.C.; Tuico, A.R.; Mendoza, J.P.; Ferrolino, J.P.R.; Vergara, C.J.T.; Salvador, A.A.; Estacio, E.S.; Somintac, A.S. Development of resistance-based pH sensor using zinc oxide nanorods. *J. Nanosci. Nanotechnol.* **2016**, *16*, 6102–6106. [CrossRef]

36. Ameen, S.; Akhtar, M.S.; Shin, H.S. Highly dense ZnO nanowhiskers for the low level detection of p-hydroquinone. *Mater. Lett.* **2015**, *155*, 82–86. [CrossRef]

37. Ameen, S.; Akhtar, M.S.; Shin, H.S. Highly sensitive hydrazine chemical sensor fabricated by modified electrode of vertically aligned zinc oxide nanorods. *Talanta* **2012**, *100*, 377–383. [CrossRef] [PubMed]

38. Umar, A.; Akhtar, M.S.; Al-Hajry, A.; Al-Assiri, M.S.; Dar, G.N.; Saif Islam, M. Enhanced photocatalytic degradation of harmful dye and phenyl hydrazine chemical sensing using ZnO nanourchins. *Chem. Eng. J.* **2013**, *262*, 588–596. [CrossRef]

39. Zheng, L.; Wan, Y.; Qi, P.; Sun, Y.; Zhang, D.; Yu, L. Lectin functionalized ZnO nanoarrays as a 3D nano-biointerface for bacterial detection. *Talanta* **2017**, *167*, 600–606. [CrossRef] [PubMed]

40. Kumar, R.; Kumar, G.; Umar, A. ZnO nano-mushrooms for photocatalytic degradation of methyl orange. *Mater. Lett.* **2013**, *97*, 100–103. [CrossRef]

41. Cullity, B.D. *Elements of X-ray Diffraction*, 2nd ed.; Addison-Wesley Publishing Co.: Reading, MA, USA, 1978.

42. Dogar, S.; Kim, S.M.; Kim, S.D. Ultraviolet photonic response of AlGaN/GaN high electron mobility transistor-based sensor with hydrothermal ZnO nanostructures. *J. Nanosci. Nanotechnol.* **2016**, *16*, 10175–10181. [CrossRef]

43. Patterson, A.L. The scherrer formula for X-ray particle size determination. *Phys. Rev.* **1939**, *56*, 978–982. [CrossRef]

44. Umar, A.; Alshahrani, A.A.; Algarni, H.; Kumar, R. CuO nanosheets as potential scaffolds for gas sensing applications. *Sens. Actuators B Chem.* **2017**, *250*, 24–31. [CrossRef]

45. Jeong, E.-S.; Kang, M.; Kim, H.-S. Surface acoustic wave propagation properties with ZnO thin film for thermo-electric sensor applications. *J. Nanosci. Nanotechnol.* **2016**, *16*, 10219–10224. [CrossRef]

46. Kim, S.; Park, S.; Kheel, H.; Lee, W.I.; Lee, C. Enhanced ethanol gas sensing performance of the networked Fe2O3-functionalized ZnO nanowire sensor. *J. Nanosci. Nanotechnol.* **2016**, *16*, 8585–8588. [CrossRef]

47. Umar, A.; Kumar, R.; Akhtar, M.S.; Kumar, G.; Kim, S.H. Growth and properties of well-crystalline cerium oxide (CeO2) nanoflakes for environmental and sensor applications. *J. Colloid Interface Sci.* **2015**, *454*, 61–68. [CrossRef] [PubMed]

48. Silambarasan, M.; Saravanan, S.; Soga, T. Effect of Fe-doping on the structural, morphological and optical properties of ZnO nanoparticles synthesized by solution combustion process. *Physica E* **2015**, *71*, 109–116. [CrossRef]

49. Dong, Y.; Feng, C.; Jiang, P.; Wang, G.; Li, K.; Miao, H. Simple one-pot synthesis of ZnO/Ag heterostructures and the application in visible-light-responsive photocatalysis. *RSC Adv.* **2014**, *4*, 7340–7346. [CrossRef]

50. Rahman, M.M.; Khan, S.B.; Asiri, A.M.; Alamry, K.A.; Khan, A.A.P.; Khan, A.; Rub, M.A.; Azum, N. Acetone sensor based on solvothermally prepared ZnO doped with Co3O4 nanorods. *Microchim. Acta* **2013**, *180*, 675–685. [CrossRef] [PubMed]

51. Khan, S.B.; Faisal, M.; Rahman, M.M.; Jamal, A. Low-temperature growth of ZnO nanoparticles: Photocatalyst and acetone sensor. *Talanta* **2011**, *85*, 943–949. [CrossRef] [PubMed]

52. Ibrahim, A.A.; Hwang, S.W.; Dar, G.N.; Kim, S.H.; Abaker, M.; Ansari, S.G. Synthesis and characterization of Gd-doped ZnO nanopencils for acetone sensing application. *Sci. Adv. Mater.* **2015**, *7*, 1241–1246. [CrossRef]

53. Rahman, M.M.; Alam, M.M.; Asiri, A.M.; Islam, M.A. Fabrication of selective chemical sensor with ternary ZnO/SnO2/Yb2O3 nanoparticles. *Talanta* **2017**, *170*, 215–223. [CrossRef] [PubMed]

54. Wang, C.-C.; Weng, Y.-C.; Chou, T.-C. Acetone sensor using lead foil as working electrode. *Sens. Actuators B Chem.* **2007**, *122*, 591–595. [CrossRef]

55. Chou, T.-C. An amperometric acetone sensor by using an electro-deposited Pb-modified electrode. *Z. Naturforsch. B* **2006**, *61*, 560–564.

56. Rahman, M.M.; Khan, S.B.; Jamal, A.; Faisal, M.; Asiri, A.M. Fabrication of highly sensitive acetone sensor based on sonochemically prepared as-grown Ag_2O nanostructures. *Chem. Eng. J.* **2012**, *192*, 122–128. [CrossRef]

57. Saravanan, T.; Raj, S.G.; Chandar, N.R.K.; Jayavel, R. Synthesis, optical and electrochemical properties of Y_2O_3 nanoparticles prepared by co-precipitation method. *J. Nanosci. Nanotechnol.* **2015**, *15*, 4353–4357. [CrossRef] [PubMed]

58. Alharbi, N.D.; Shahnawaze Ansari, M.; Salah, N.; Khayyat, S.A.; Khan, Z.H. Zinc oxide-multi walled carbon nanotubes nanocomposites for carbon monoxide gas sensor application. *J. Nanosci. Nanotechnol.* **2016**, *16*, 439–447. [CrossRef] [PubMed]

59. Ahmad, R.; Tripathy, N.; Jung, D.-U. J.; Hahn, Y.-B. Highly sensitive hydrazine chemical sensor based on ZnO nanorods field-effect transistor. *Chem. Commun.* **2014**, *50*, 1890–1893. [CrossRef] [PubMed]

60. Fan, Z.; Wang, D.; Chang, P.-C.; Tseng, W.-Y.; Lu, J.G. ZnO nanowire field-effect transistor and oxygen sensing property. *Appl. Phys. Lett.* **2004**, *85*, 5923–5925. [CrossRef]

61. Gujarati, T.P.; Ashish, A.G.; Kal, M.; Chaijumon, M.M. Highly ordered vertical arrays of TiO_2/ZnO hybrid nanowires: Synthesis and electrochemical characterization. *J. Nanosci. Nanotechnol.* **2015**, *15*, 5000–5829. [CrossRef] [PubMed]

62. Majumder, S. Synthesis and characterisation of SnO_2 films obtained by a wet chemical process. *Mater. Sci.* **2009**, *27*, 123–129.

63. Ahmad, R.; Tripathy, N.; Ahn, M.-S.; Hahn, Y.-B. Development of highly-stable binder-free chemical sensor electrodes for p-nitroaniline detection. *J. Colloid Interface Sci.* **2017**, *494*, 300–306. [CrossRef] [PubMed]

64. Behera, B.; Chandra, S. Catalyst-free synthesis of ZnO nanowires on oxidized silicon substrate for gas sensing applications. *J. Nanosci. Nanotechnol.* **2015**, *15*, 4534–4542. [CrossRef] [PubMed]

materials

MDPI

Article

Nanostructured ZnO in a Metglas/ZnO/Hemoglobin Modified Electrode to Detect the Oxidation of the Hemoglobin Simultaneously by Cyclic Voltammetry and Magnetoelastic Resonance

Ariane Sagasti [1], Nikolaos Bouropoulos [2,3,*], Dimitris Kouzoudis [4], Apostolos Panagiotopoulos [2], Emmanuel Topoglidis [2] and Jon Gutiérrez [1,5]

[1] BCMaterials, Ibaizabal Bidea, Edificio 500, Parque Tecnológico de Bizkaia, 48160 Derio, Spain; ariane.sagasti@bcmaterials.net (A.S.); jon.gutierrez@ehu.eus (J.G.)
[2] Department of Materials Science, University of Patras, 26504 Patras, Greece; apostolospanas@gmail.com (A.P.), etop@upatras.gr (E.T.)
[3] Foundation for Research and Technology Hellas, Institute of Chemical Engineering and High Temperature Chemical Processes, 26504 Patras, Greece
[4] Department of Chemical Engineering, University of Patras, 26504 Patras, Greece; kouzoudi@upatras.gr
[5] Department of Electricity and Electronics, Universidad del País Vasco/Euskal Herriko Unibertsitatea, 48080 Bilbao, Spain
* Correspondence: nbouro@upatras.gr; Tel.: +30-2610-997-164

Received: 26 June 2017; Accepted: 21 July 2017; Published: 25 July 2017

Abstract: In the present work, a nanostructured ZnO layer was synthesized onto a Metglas magnetoelastic ribbon to immobilize hemoglobin (Hb) on it and study the Hb's electrochemical behavior towards hydrogen peroxide. Hb oxidation by H_2O_2 was monitored simultaneously by two different techniques: Cyclic Voltammetry (CV) and Magnetoelastic Resonance (MR). The Metglas/ZnO/Hb system was simultaneously used as a working electrode for the CV scans and as a magnetoelastic sensor excited by external coils, which drive it to resonance and interrogate it. The ZnO nanoparticles for the ZnO layer were grown hydrothermally and fully characterized by X-Ray Diffraction (XRD), Scanning Electron Microscopy (SEM) and photoluminescence (PL). Additionally, the ZnO layer's elastic modulus was measured using a new method, which makes use of the Metglas substrate. For the detection experiments, the electrochemical cell was performed with a glass vial, where the three electrodes (working, counter and reference) were immersed into PBS (Phosphate Buffer Solution) solution and small H_2O_2 drops were added, one at a time. CV scans were taken every 30 s and 5 min after the addition of each drop and meanwhile a magnetoelastic measurement was taken by the external coils. The CV plots reveal direct electrochemical behavior of Hb and display good electrocatalytic response to the reduction of H_2O_2. The measured catalysis currents increase linearly with the H_2O_2 concentration in a wide range of 25–350 µM with a correlation coefficient 0.99. The detection limit is 25–50 µM. Moreover, the Metglas/ZnO/Hb electrode displays rapid response (30 s) to H_2O_2, and exhibits good stability and reproducibility of the measurements. On the other hand, the magnetoelastic measurements show a small linear mass increase versus the H_2O_2 concentration with a slope of 152 ng/µM, which is probably due to H_2O_2 adsorption in ZnO during the electrochemical reaction. No such effects were detected during the control experiment when only PBS solution was present for a long time.

Keywords: ZnO nanostructures; Metglas; magnetoelastic resonance; Hemoglobin; synthesis; characterizations; sensors

Materials **2017**, *10*, 849

1. Introduction

In order to fabricate a good biosensor, the supporting material for the target biomolecule of choice has to be biocompatible in order to immobilize it in a stable and functional way. Hydrogels, surfactants, biopolymers, conducting polymers, metal oxides and ionic liquids have been successfully used in the past as substrates for the immobilization of proteins or enzymes: they are nontoxic and provide favorable environmental conditions to examine and study the direct electrochemical activity of redox biomolecules as they allow the active redox center of the immobilized biomolecules to come into direct contact with the material [1,2].

In the last years, nanostructured materials are being intensively studied for applications in many different nanoscale functional devices. Semiconductor nanomaterials such as ZnO, SnO_2, TiO_2 and ZnS are good examples where the nano-structure affects their intrinsic physical and chemical properties [3–5] and the subsequently derived applications. In particular, the semiconductor ZnO is a good candidate for the construction of nanostructured functional devices because of its low toxicity, good biocompatibility and biodegradability, good thermal stability and oxidation resistance, large specific surface area and high electron mobility [6]. ZnO is a transparent semiconductor with a direct band gap (E_g = 3.37 eV) and a large exciton binding energy (60 meV), exhibiting near UV emission [7]. It is widely used in the chemical industry [8,9], for biomedical applications [10,11], or food technology [12] among others. ZnO nanostructures are being widely used as chemical gas sensors (based on conductance changes) and for biological agents detection purposes (by profiting of biocompatibility and low toxicity) [12–14]. ZnO nanoparticles and nanostructures with different size and growth morphologies can be prepared using a variety of techniques. The most popular fabrication processes include thermal evaporation, thermal decomposition and hydrothermal growth [15–17].

For materials scientists and engineers, the knowledge of the elastic properties of such nanostructured materials is a key factor, since elastic moduli are closely linked to the internal structure of solids at the atomic level [18,19]. The capability to measure those elastic moduli and their dependence under external influences, as temperature or mass load, unveils the utility of such nanostructured materials for oriented applications, as thin-film deposition growth control or deposition control of specific targets.

On the other hand, over the last 30 years there has been a great effort to develop new H_2O_2 electrochemical biosensors in order to understand the redox processes of enzymes and proteins and if these are maintained after their immobilization on electrodes surface. The protein's structure and redox transformation of protein molecules are actually a preferential task devoted to give a deep insight into physiological electron transfer processes. H_2O_2 presents cytotoxic effects and associated tissue injury, but also plays a role in physiological and biomedical studies as well as when monitoring biological processes. H_2O_2 is also a side product of many oxidative biological reactions catalyzed by enzymes such as glucose oxidase (GO_x), lactate oxidase (LO_x, cholesterol oxidase ($ChoO_x$) and many others [20]. Therefore, it is of high importance to be able to achieve sensitive determination of H_2O_2 presence in many biological processes and related applications. Additionally, it is also well known that, due to its intrinsic peroxidase activity, Hb is an excellent protein to fabricate H_2O_2 electrochemical biosensors (see, for example, works by Chen et al. [20] or Shamsipur et al. [21]).

Hb is a physiologically oxygen transfer protein with a well known and documented structure, of low cost and exhibiting relatively higher stability and intrinsic peroxidase activity [22]. Hb has four polypeptide chains, each with one electroactive iron heme group [23], as can be seen in Figure 1. It is a prototype molecule for studying biological electron transfer processes and therefore it has been extensively used as an ideal model enzyme to study biological electron transfer reactions, to evaluate materials for their choice to be used as substrates for the immobilization of biomolecules in an active configuration and it has already been used in the past for the fabrication of electrochemical biosensors and bioreactors (see, for example, works by Zhang et al. [24] or Duan et al. [1]).

Nevertheless, many of the fabricated sensors exhibit slow electron exchange due to the unfavorable orientation of Hb molecules onto electrode surfaces, and so efforts point towards the development of new immobilization methods and supporting materials to promote the direct electron transfer of Hb while maintaining its enzymatic activity. Among other possibilities, ZnO nanoparticles are good candidates for such purposes [2,21].

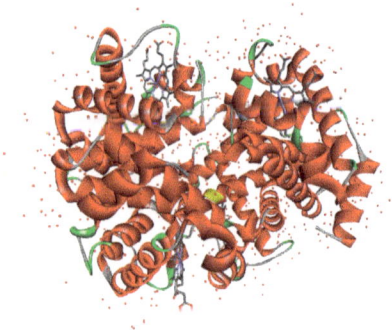

Figure 1. Ribbon diagram of Bovine Hemoglobin showing the position of the four hemes (blue) taken from the RCSB Protein Data Bank and plotted on BIOvia Discovery Studio Visualizer.

The experimental technique of CV can be used to monitor the electrochemical behavior of modified electrodes. On the other hand, magnetoelastic materials working in resonant conditions are known to be extremely sensitive to external parameters, such as mass load [25]. Magnetoelasticity is a property of ferromagnetic materials, which describes the efficient conversion of magnetic energy into elastic, and vice versa [26]. Amorphous metallic glass ribbons are among the best magnetoelastic materials known for such energy conversion processes, due to their almost null internal magnetocrystalline anisotropy and internal stresses appearing during their fabrication process [27]. Freestanding magnetoelastic ribbons can easily be induced to vibrate by exposure to an external, time varying magnetic field. Such ribbons exhibit a resonance frequency that depends upon factors such as sample geometry, mass load and elastic constants, which are (magnetic) field-dependent. The appearance of any mass load onto a magnetoelastic ribbon will immediately cause a decrease in its resonant frequency and this decrease can be used to determine the loaded mass value by comparing to calibration curves of known mass loads. Another important advantage of such a device is that the whole detection process is remote, thus eliminating the need for direct electrical connections which sometimes is a nuisance. These magnetoelastic resonant platforms can be converted into very selective microbalances by depositing on them nanostructured materials that can work as selective adsorbing layers. Thus, the detection of specific targets of biological or chemical origin is possible [28]. Additionally, the dependence of the resonance frequency of the magnetoelastic platform on the elastic modulus, allows the study of the elastic properties of the deposited nanostructured coatings.

With this purpose in mind, we have fabricated a biosensor to detect the oxidation of Hb by H_2O_2. The biosensor is composed of a thin-film of nanostructured ZnO deposited onto a magnetoelastic strip of commercial magnetoelastic material Metglas 2826MB ($Fe_{40}Ni_{38}Mo_4B_{18}$). The ZnO nanoparticles for the ZnO layer were prepared using the hydrothermal method and a layer of Hb was successfully immobilized on the ZnO layer. Adsorption of Hb on ZnO film results in the yellow-brown coloration of the film indicating the even distribution of the Hb molecules on its surface. As demonstrated in the past, the binding of Hb on metal oxide films such as ZnO and TiO_2 is mainly electrostatic and controlled by the buffer pH, the protein surface charge, and the solution ionic strength [29]. The resultant three-layer sensor was used in two simultaneous detection techniques, as a working electrode (Metglas/ZnO/Hb) in CV and as the resonant platform in MR. The detection experiment consisted of a standard electrochemical cell composed of three electrodes, the sensor as the working electrode (WE), the Pt counter electrode (CE) and the Ag/AgCl reference electrode (RE). The cell was immersed in a PBS buffer solution where drops of H_2O_2 were added. An external coil was wrapped around the glass vial, which contained the electrolyte solution and the cell. A detailed scheme of the detection system is shown in Figure 2. The voltage V was scanned during the CV scans and the resulting current I was recorded. The external coils were controlled by a magnetoelastic resonator in order to drive the sensor to resonance.

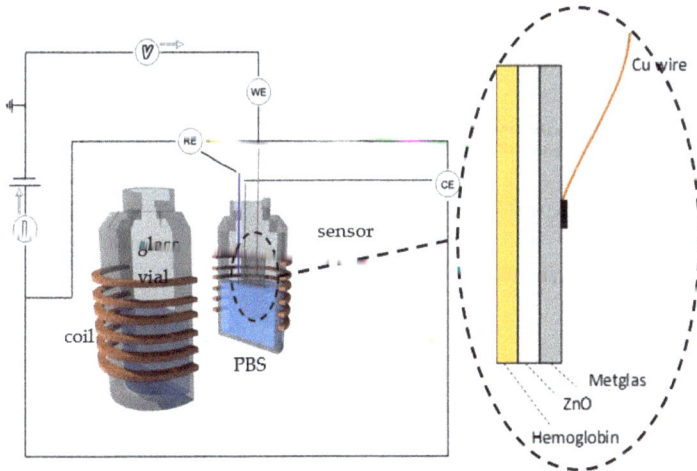

Figure 2. The detection experiment consisted of three electrodes, the sensor working electrode (WE), the Pt counter electrode (CE) and the Ag/AgCl reference electrode (RE), all immersed in a PBS buffer solution inside a glass vial on which a coil was wrapped externally.

The resulted CV scans exhibit the direct electrochemical behavior of the immobilized Hb and display good electrocatalytic responses to the reduction of H_2O_2. The catalysis currents increase linearly to the H_2O_2 concentration in a wide range of 25–350 µM with a correlation coefficient 0.99. The detection limit is 25–50 µM. Moreover, the Metglas/ZnO/Hb electrode displays rapid response (30 s) to H_2O_2, and possesses good stability and reproducibility. The magnetoelastic measurements show that the mass load of the sensor increases linearly with the concentration of the H_2O_2 reaching a mass of about 57 µg when the molar concentration of H_2O_2 was 375 µM. The corresponding slope is equal to 152 ng/µM. To our knowledge, this is the first time that the two methods of CV and MR have been used simultaneously for biodetection.

2. Materials and Methods

2.1. Reagents and Materials

For the magnetoelastic measurements, a commercial ribbon of Metglas 2826MB ($Fe_{40}Ni_{38}Mo_4B_{18}$) purchased from Hitachi Metals Europe GmbH (Dusseldorf, Germany) was used as resonant platform. Magnetoelastic sensors are made of this ribbon by cutting strips of 2 cm length after applying a cleaning treatment with analytical grade acetone purchased from Sigma-Aldrich Chemie GmbH (Taufkirchen, Germany).

For the ZnO nanoparticle synthesis and later film deposition the reagents used were analytical-grade without further purification. Lithium hydroxide monohydrate [$Li(OH)·H_2O$], zinc acetate dihydrate [$Zn(CH_3COO)_2·H_2O$] and ethanol were purchased from Sigma.

Hb (MW 65,000), from Bovine blood was purchased from Sigma and was used without further purification. Sodium dihydrogen orthophosphate (0.01 M) from Sigma was used to prepare the supporting electrolyte, and its pH was adjusted to 7 using NaOH and was thoroughly deaerated by bubbling with Argon prior to the experiments. H_2O_2 (30% *w/v* solution) was purchased from Lach-Ner (Neratovice, Czech Republic), and was diluted. All solutions were prepared with deionized water.

2.2. Apparatus

The crystalline structure of the synthesized ZnO nanoparticles as well as the deposited ZnO layers were analyzed by X-ray diffraction (XRD) with a Bruker D8 advanced diffractometer (Bruker AXS GmbH, Karlsruhe, Germany) operated at 40 kV and 40 mA using CuKα radiation, with a scanning speed of 0.35 sec/step for 2θ in a range from 15° to 70°.

Morphology of the deposited ZnO layers was obtained by using scanning electron microscopy (SEM) and obtaining images with a Zeiss SUPRA 35VP instrument operated at 10 kV (Carl Zeiss SMT, Oberköchen, Germany). As a further proof of the quality of the ZnO layers obtained, the photoluminescence (PL) spectra were also recorded at room temperature with a Hitachi F2500 Fluorescence Spectrophotometer (Hitachi Ltd, Tokyo, Japan) from 350 to 600 nm at an excitation wavelength of 325 nm.

For the magnetoelastic resonance measurements, a microcontroller-controlled frequency generator drove a current amplifier connected to a single coil. The resulting alternating magnetic field induced elastic waves on the sensor due to its magnetoelastic properties, causing a mechanical vibration. When the frequency of this vibration matches with the natural frequency of the sensor, resonance occurs, and the maximum measured at that point is easily followed by our automated set-up. Extensive information about this can be found in a previous work of ours [28].

The detection experimental set-up was described in the Introduction and is shown in Figure 2. The glass vial is cylindrical with a height of 4.6 cm and a diameter of 2.3 cm. The electrochemical measurements were performed on an Autolab PGStat 101 Potentiostat (Metrohm Autolab, Utrecht, The Netherlands) with a conventional three-electrode system. The Metglas/ZnO/Hb was used as the working electrode, a platinum wire as a counter electrode, and a Ag/AgCl as a reference electrode. Simultaneously, magnetoelastic resonance detection was performed by using a magnetoelastic resonator made by Sentec which was driving a homemade coil ($N = 24$ turns, $R = 0.6\ \Omega$, $L = 6.9\ \mu H$) which was wrapped around the glass vial.

3. Experimental

3.1. ZnO Nanoparticle Synthesis and Film Deposition

3.1.1. Synthesis of the ZnO Nanoparticles

The preparation of the ZnO nanoparticles for the seeding procedure has been performed by using all chemicals as analytical-grade reagents without further purification. This synthesis was carried out by following a standard hydrothermal procedure: 0.4 g of lithium hydroxide monohydrate [Li(OH)·H_2O] was suspended in 100 mL absolute ethanol under magnetic stirring. This suspension was added into 50 mL ethanoic solution of zinc acetate dihydrate [Zn(CH_3COO)$_2$·H_2O] 0.1 M, again under magnetic stirring. The obtained solution was then sealed in an autoclave reactor and kept at 100 °C for 3 h, followed by normal cooling down to room temperature. The obtained particles were centrifuged at 4000 rpm for 10 min, washed after resuspension in water, and centrifugation (those two last steps repeated three times) and finally dried at 80 °C. Afterwards and for the subsequent ZnO layer deposition, a suspension was prepared with 100 mg of those ZnO nanoparticles in ethanol under magnetic stirring and sonication.

3.1.2. ZnO Film Deposition onto the Magnetoelastic Resonant Platform

A commercial ribbon of Metglas 2826MB ($Fe_{40}Ni_{38}Mo_4B_{18}$) was used as the magnetoelastic resonant platform. Equal strips of 2 cm length were cut and cleaned in acetone for 15 min under sonication. For the ZnO layer deposition procedure, the Metglas strips were placed in a petri dish with the rough side facing upwards and 2 mL of the above-mentioned ZnO nanoparticle solution were added. Finally, the petri dish was left in the oven at 85 °C until all the solvent was totally evaporated, giving as a result the Metglas + ZnO layer product. This procedure was repeated several times for each sensor, and after each step the structure and morphology of the deposited product was analyzed. In addition, at each step, a measurement of the total mass and the resonance frequency of the composite strip was taken to determine the elastic modulus of the deposited ZnO thin-film, as it will be shown below.

3.2. Hemoglobin Immobilization

The Metglas/ZnO/Hb electrode was fabricated following the procedure described above for the Metglas/ZnO film, plus the immobilization of Hb on its surface. For the Hb immobilization, a 20 μM Hb solution was prepared using 0.01 M Phosphate Buffer Solution (PBS), pH 7 and stored at 4 °C. Hb was deposited on the surface of the Metglas/ZnO by dropping 5 μL of Hb solution on the surface of the material and allowing it to dry at 30 °C for 30 min. Prior to all electrochemical measurements, the Metglas/ZnO/Hb electrode was rinsed with PBS to remove any non-immobilized Hb from its surface.

4. Results and Discussion

4.1. Characterization of ZnO Nanoparticles and Film

Figure 3a shows the XRD results for the synthetic ZnO nanoparticles. The measured diffraction pattern is compared to the standard JCPDS card for ZnO (No 36-1451) which corresponds to the wurtzite crystal No 36-1451 structure of ZnO. The observed experimental peaks are fitted to the standard card values corresponding to the ZnO reflections from (100), (002), (101), (102), (110) and (103) planes.

The average particle size was estimated using the Scherrer's formula:

$$d = \frac{k \cdot \lambda}{\beta \cdot cos\theta} \tag{1}$$

where d is the average crystallite size, k is the Scherrer constant taken equal to 0.9, λ is the wavelength of the X-ray radiation, β is the full width at half-maximum and θ is the diffraction angle. It was found that the average particle size is 9.0, 20.7 and 9.3 nm corresponding to the (100), (002) and (101) diffraction lines, respectively.

Figure 3b shows the different XRD patterns observed for the ZnO nanoparticles, the Metglas substrate and the same substrate with the ZnO nanoparticles deposited on it. The comparison of the three XRD patterns proves that no impurities were involved during the synthesis process, which confirms the purity of our obtained product. As the Metglas strip is an amorphous material, the XRD pattern gives a noisy and broad signal with a wide peak from 40° to 50°. While peak positions for ZnO nanoparticles and film are coincident (reflections (100), (002), and (101)), the intensity is much lower for this last one, which make us affirm that we have actually a quite thin film of ZnO nanoparticles deposited onto the Metglas strip. Further discussion about the thickness of the deposited film will be given in a following section.

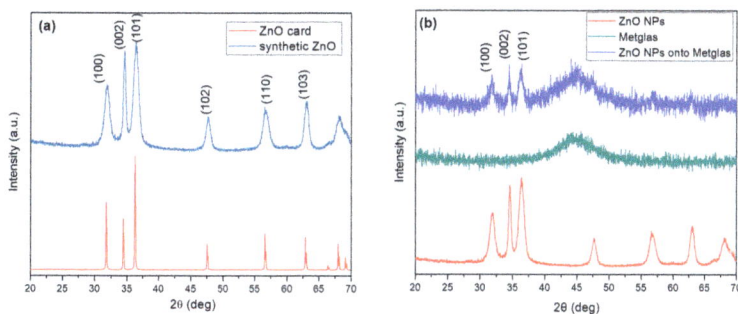

Figure 3. XRD patterns: (**a**) Synthetic ZnO nanoparticles and the standard ZnO wurtzite structure; and (**b**) synthetic ZnO nanoparticles (NPs), a clean strip of Metglas and the ZnO coated Metglas strip.

The SEM micrographs in Figure 4 show the morphology of the ZnO layer onto the Metglas substrate. It can be seen that the particles are found as aggregates composed of individual nanoparticles of spherical shape (Figure 4a). The size of nanoparticles was measured using the ImageJ software (National Institutes of Health, Bethesda, USA). The size ranged from 11 to 32 nm with a mean value of 18.3 ± 4.0 nm. The deposited ZnO layer covers entirely the surface of the metallic ribbon (Figure 4b).

| (a) | (b) |

Figure 4. SEM images: (**a**) high magnification showing individual ZnO nanocrystals; and (**b**) a surface view of the obtained ZnO layer, after six depositions. Insert in image (**a**) shows particle size distribution.

Figure 5 shows photoluminescence spectra of the ZnO layer. A small peak appears at 380 nm, which can be attributed to the near band edge emission, arising from the recombination of free excitons. However, the spectrum is dominated by a broad band with a maximum around $\lambda_{max} = 545$ nm (2.32 eV) which is known as green emission and has a full width at half maximum of $\Delta E_{1/2} = 330$ meV.

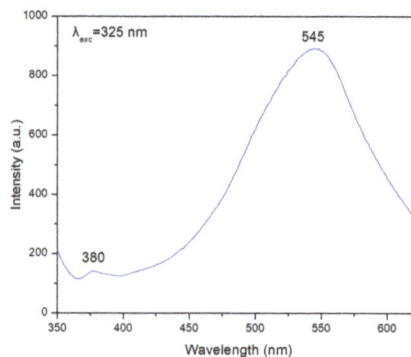

Figure 5. PL spectrum recorded by using an excitation wavelength at 325 nm over the ZnO film deposited onto the Metglas 2826MB strip.

Previous works [30,31] have shown that the green luminescence is caused by electronic transitions between shallow donors and deep acceptors (VZn), or transitions from the conduction band to VZn-levels. That is, the maximum of the green luminescence band located at 2.35 ± 0.05 eV corresponds to the case of zinc vacancies being responsible for the observed luminescence. Since the synthesis of our ZnO nanoparticles and subsequent films occurs in air atmosphere, there is an excess of oxygen, which probably causes zinc vacancies in the ZnO structure [32].

4.2. ZnO Deposited Film Elastic Modulus Determination

In this section, we will show that it is possible to determine the elastic Young's modulus of the deposited ZnO film onto the Metglas substrate, following a method that has been described in detail previously [33,34]. The knowledge of the elastic parameters of such a thin film turns out to be of great importance in the design and fabrication of sensing devices that use this kind of material. The usual technique of the uniaxial tensile testing to measure Young's modulus of bulk materials is almost impossible to apply when dealing with thin films and nanoscales, where manipulation of the material and application of the force and accurate measurement of the displacement is extremely difficult. Thus, the possibility to perform in situ experiments at the nanoscale becomes a necessary tool in order to obtain not only quantitative but also qualitative information about nanosized materials [35]. The fundamental resonance frequency of a single flat layer, stress free ribbon of length L, density ρ, and elastic Young modulus E is given by the well-known relationship [36,37]:

$$f_R = \frac{1}{2L}\sqrt{\frac{E}{\rho}} \tag{2}$$

According to this method, when dealing with two layers, as in our case with the Metglas and ZnO layers shown schematically in Figure 6, the above formula needs to be modified to:

$$f_R'' = \frac{1}{2L}\sqrt{\frac{E''}{\rho''}} = \frac{1}{2L}\sqrt{\frac{E + E'\frac{h'}{h}}{\rho + \rho'\frac{h'}{h}}} \tag{3}$$

where $E'' = E + E'(h'/h)$, $\rho'' = \rho + \rho'(h'/h)$, the un-primed parameters refer to the Metglas 2826MB strip alone and the primed parameters refer to the ZnO layer. As shown in the Figure 6, h' and h are the thicknesses of the two layers.

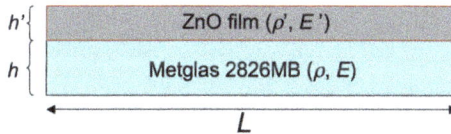

Figure 6. Schematic representation of the Metglas 2826MB and ZnO layers layout in our resonant devices.

According to the method, if a series of similar films with different thicknesses h' can be synthesized, then a plot of the E'' parameter of the bilayer system (extracted numerically from the resonance frequency) versus h' will be a straight line with a slope equal to the value of the Young's modulus E' of the film. Additionally, the Young's modulus E of the substrate layer can be extracted from the y-intercept. In our case, a thickness of $h \approx 30$ μm was measured for the Metglas 2826MB layer ($\rho = 7900$ kg/m³), while the thickness h' of the ZnO layer was estimated assuming a uniform film, given its mass, dimensions and the (bulk) density value of ZnO $\rho' = 5606$ kg/m³. For this purpose, six different and successive depositions of the ZnO solution were performed onto the Metglas strips, with the final one resulting to the thickest ZnO solid film of about 1 μm thickness. Figure 7 shows the obtained results and the corresponding linear fit for one of our resonant platforms. The good linearity reveals the validity of the aforementioned method. From the y-intercept, a Young's modulus of 160 GPa is estimated for the bare Metglas strip, in good agreement with previous results [34]. From the slope, a Young's modulus value of 60 GPa is calculated for the ZnO film (estimated error from mean Young's modulus values about ±2%).

The Young's modulus of bulk ZnO is ≈140 GPa, a value that is generally accepted and was calculated by Kobiakov [38] starting from elastic constants for ZnO crystal. The range of experimental values measured when dealing with ZnO at the nanoscale, is quite diverse depending not only on the geometry of the material but also on the experimental process used for its measurement (see, for example, Table 1 in [35]). Thus, for ZnO nanowires, Song et al. [39] gave a value of 29 GPa determined by AFM bending measurements and Desai et al. [40] obtained a value of 21 GPa measured by using a MEMS test-bed to perform uniaxial tensile experiments. On the other hand, Ji et al. [41] have reported values of Young's modulus as high as 117 GPa and 232 GPa for ZnO nanowires with diameters of 100 nm and 30 nm, respectively, by studying the buckling of the nanowires with nanoindentation. For ZnO nanobelts, Bai et al. [42] gave a Young's modulus value of 50 GPa, and Wang obtained a value of 52 GPa [43], in both cases by measuring the dynamic response of the specimen in an alternating electrostatic field inside a TEM. Considering all the previously reported values, our observations agree most with those obtained for nanobelt shaped samples, and we can infer that our $\sim$1 μm thickness ZnO film on the Metglas 2826MB strip, behaves like a wide nanobelt with Young's modulus of about 60 GPa.

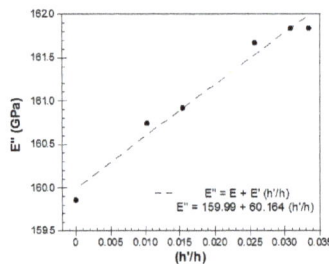

Figure 7. Total Young's modulus (E'') measured as a function of the ratio h'/h (width of the deposited ZnO layer/width of Metglas 2826MB strip).

4.3. Simultaneous Electrochemical and Magnetoelastic Detection of H_2O_2 Using a Metglas/ZnO/Hemoglobin Electrode-Sensor

Hb is an auto-oxidating protein where heme iron atoms easily oxidize from ferrous Fe (II) to ferric Fe (III) and reduce from Fe (III) to Fe (II). The reaction scheme for the electrochemical reduction and oxidation of Hb can be written as follows:

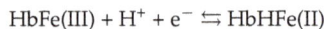

$$HbFe(III) + H^+ + e^- \leftrightarrows HbHFe(II)$$

An excellent and complete graphical representation of all involved reactions in this reduction and oxidation of Hb can be found in Figure 1 of [44]. It is well known that the Hb molecule can catalyze the reduction of H_2O_2 [1,2,5–7,44] and accordingly Hb has been extensively used to construct H_2O_2 biosensors. As shown in Figure 8, the enzymatic reaction mechanism can be described as follows [45–47]:

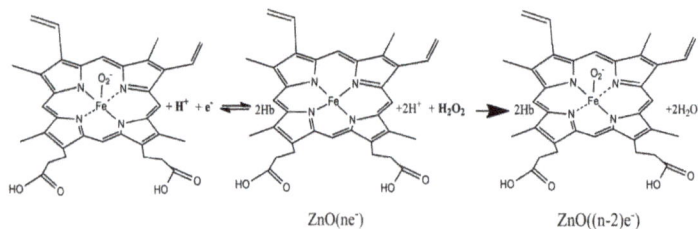

$$2HbHFe(II) + H_2O_2 + 2H^+ \rightarrow 2HbFe(III) + 2H_2O$$

Figure 8. Reaction scheme for the direct reduction and oxidation of the immobilized hemes of Hb and the electrocatalytic reduction of H_2O_2 on the sensor (created on Chemdraw).

4.3.1. Electrochemical Behavior of Metglas, Metglas/ZnO, Metglas/ZnO/Hb Film Electrodes

When using CV, the potential is scanned from a certain initial voltage to a certain final potential to charge the capacitor and again scanned back in the reverse direction in order to discharge it. This allows the tracking of the electrochemical properties of the modified electrodes. The resulting current is plotted against the applied potential with respect to a reference electrode. The CV curve of an electric double layer capacitor (such as Metglas/ZnO) would be of a rectangular shape, in absence of a faradic reaction. In the presence of faradic redox reactions, the CV curve should exhibit peak currents, which are due to the effect of pseudo-capacitance exhibited by the electrode material.

All CV experiments were carried out in a Hb-free, 10 mM aqueous PBS electrolyte solution of pH 7 at room temperature. Figure 9 shows the CV curves of (a) Metglas 2826MB; (b) Metglas/ZnO and (c) Metglas/ZnO/Hb electrodes at a scan rate of 0.1 V/s.

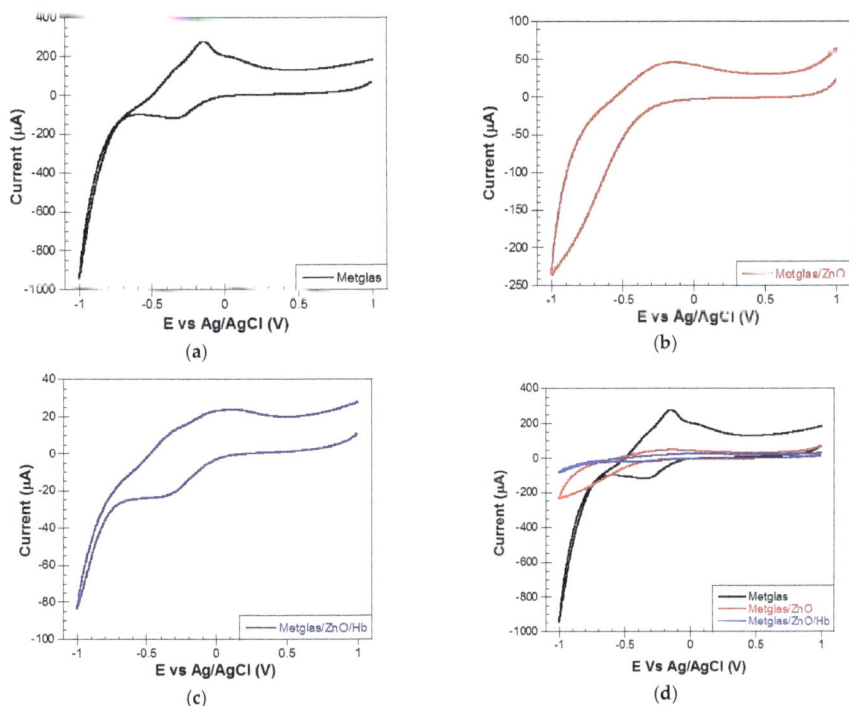

Figure 9. Cyclic voltammetry curves of different electrodes at a scan rate of 0.1 V/s: (**a**) Metglas 2826MB; (**b**) Metglas/ZnO; (**c**) Metglas/ZnO/Hb; and (**d**) all three curves plotted together, for comparison purposes.

The Metglas monolayer is exhibiting high peak redox currents, which are shown in the CV (Figure 9a). This Metglas is exhibiting a very interesting electrochemical behavior, seems likely to be a diffusion-controlled system with charge transfer phenomena in play. The Metglas film shows a characteristic reduction peak at -0.35 V and a (re)oxidation peak at -0.15 V in aqueous electrolyte solution. The voltage range is taken from -1 V to $+1$ V because in this range the Metglas electrode is effectively working without any breakdown. These peaks (reversible process) could be possibly due to the high content of iron in Metglas which is an amorphous metallic material and thus the iron atoms can occur in both oxidizing states Fe(II) and Fe(III), depending on their local neighborhood in the amorphous atomic framework. Thus, depending on the applied potential, iron can be oxidized and reduced easily. These peaks could be a sum of contribution of various oxidation processes of iron to form divalent or trivalent species [47]. In addition, the metallic Metglas gives back a greater and broader current when compared to the semiconducting Metglas/ZnO, due to its conducting nature.

The Metglas/ZnO film (Figure 9b) shows the characteristic charging/discharging currents assigned to electron injection into sub-band gap/conduction band states of the ZnO film. The charging of the ZnO film as seen in Figure 9b starts at −0.16 V, which is around the same value reported at previous studies in literature [48]. Over the potential range examined, for potentials more positive to −0.16 V, the ZnO is insulating and serves only as a support for the immobilization of biomolecules.

The CV of the Metglas/ZnO/Hb electrode in PBS solution is shown in Figure 9c. It was used to estimate the midpoint redox potential of the immobilized Hb. As the applied potential was ramped from 1.0 V to −1.0 V and reversed vs Ag/AgCl, the Metglas/ZnO/Hb electrode exhibits in addition to the film charging currents, nearly reversible, but not equivalent well-defined reduction (−0.35 V) and oxidation(+0.15 V) peaks. These peaks are assigned to Hb reduction Fe (II) and re-oxidation Fe (III). The Fe (III)/Fe (II) redox chemistry of heme is termed quasi-reversible as the peak to peak separation was > 60 mV and the peak oxidation current was typically much less than the reduction peak current [48]. These peaks are clearly absent from the CVs of the same electrode before the immobilization of Hb. It clearly demonstrates that the immobilized protein is electroactive and could be used for the sensing of H_2O_2.

Figure 9d shows all the above-mentioned different curves of the three electrodes, plotted together for direct comparison. It is evident in the plot that the metallic Metglas gives back a greater and broader current with respect to the other two electrodes.

4.3.2. Simultaneous Electrochemical and Magnetoelastic Resonant Detection of H_2O_2

- Control curves for the two methods

To be certain that the detection signal which was obtained in our measurements (see next section) was due to the oxidation of immobilized Hb by H_2O_2, we performed control experiments without the presence of H_2O_2, in plain PBS solution. Shown in Figure 10 are the CV scans taken every 5 min in a total time period of 50 min. It is evident that there is no much activity during these 50 min as expected, except for a small initial 10–15 min transition to the final stable state. It is not clear to us what caused this transition but similar control experiments with bare Metglas strips in PBS solution did not show such a behavior (the signal was stable within some noise). In subsequent measurements, the system was given enough time in the stable state of Figure 11b before H_2O_2 was added in PBS and no transition states were observed. From long-term stability data, we know that the detection limit of our magnetoelastic resonator is ±0.01 kHz, as shown by the error bars in Figure 11b.

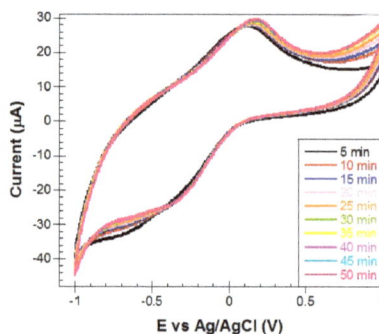

Figure 10. CVs of a Metglas/ZnO/Hb electrode at a scan rate of 0.1 V/s at specific time intervals (5–50 min) (no H_2O_2 present).

Shown in Figure 11 are the magnetoelastic resonance measurements. Figure 11a is a typical resonance signal (amplified coil voltage in mV) which is received when the Metglas/ZnO/Hb electrode is immersed in the PBS solution. The continuous line is a Gaussian fit to the data and the resonance frequency is extracted by the x-value at the peak. In Figure 11b, different resonance frequencies are received for different times with the Metglas/ZnO/Hb electrode immersed in PBS solution, in the absence of H_2O_2. It is clear that the curve is quite flat with an error of about 0.02 kHz.

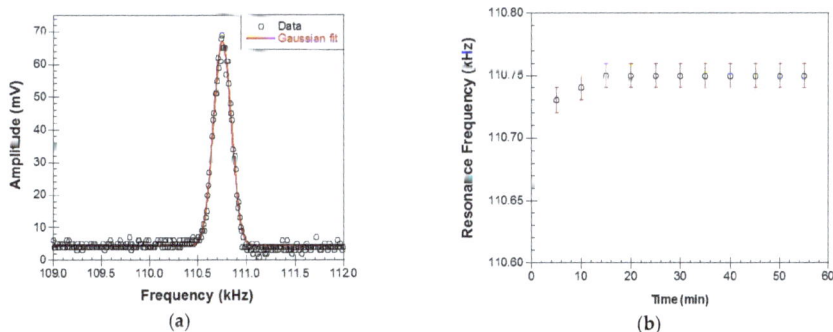

Figure 11. Magnetoelastic resonance measurements: (**a**) a typical resonance signal received when the Metglas/ZnO/Hb electrode is immersed in the PBS solution. The continuous line is a Gaussian fit to the data; and (**b**) resonance frequency versus time when the electrode is immersed in PBS solution.

- Performance in the detection of H_2O_2

To test the electrochemical reaction between Hb and H_2O_2, we added successively in the cell where the Metglas/ZnO/Hb electrode was immersed in PBS buffer, 5 µL aliquots of 30 µM H_2O_2 solution each time to increase the H_2O_2 concentration by 25 µM at each step. Each amount was added at time intervals of 5 min and CV scans were obtained right afterwards (30 seconds) and shortly before the end of the interval (5 min). CV scans at 50 µM step additions of H_2O_2 are shown in Figure 12a. This plot reveals an intense electrochemical activity (electrocatalytic responses), as expected, since it is well known that Hb can catalyze the reduction of H_2O_2 [1,2,44]. Hb was immobilized in the mesopores of the ZnO film in a stable and functional way and was able to interact with the semi-conducing substrate as well as the aqueous electrolyte solution. When the H_2O_2 molecules were added to the electrolyte solution in the cell, they could easily enter the mesopores of the ZnO film, interacted there with the immobilized molecules of Hb and were reduced by the four bound iron atoms on each of the heme molecules of Hb. In Figure 12a these interactions between the immobilized Hb and the added H_2O_2 are displayed by the gradual increase of the current peaks and their gradual shift to the right (to less negative biases) Figure 12b shows the peak current versus the H_2O_2 concentration with a good linear correlation (R = 0.99). This plot proves that the CV method is not only sensitive enough to detect the electrochemical changes that take place between Hb and H_2O_2 but also that the corresponding signals produce a linear calibration plot that could be used as a H_2O_2 biosensor.

For comparison, the peak current of these curves, is plotted in Figure 13 as solid circles, together with the corresponding signal (solid triangles), which is received when H_2O_2 is added in the solution. The control experiment produces a flat response with a small error of about 0.5 µA and it is obvious that the changes brought up by the electrochemical reaction of H_2O_2 with the Hb, produce a big enough sensing signal of about 8 µA in variation, much larger than the above error. Thus, CV method is a sensitive enough method to detect the electrochemical reactions caused by the addition of H_2O_2.

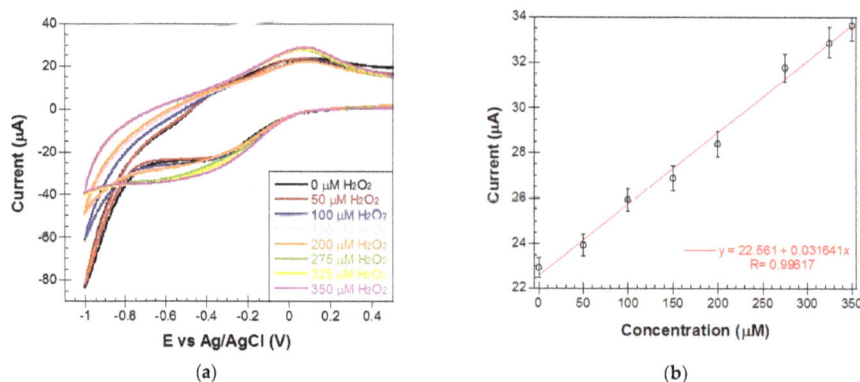

(a)

(b)

Figure 12. (a) CVs obtained for a Metglas/ZnO/Hb electrode in PBS buffer before and after the addition of increasing amounts (50–350 μM) of H_2O_2 at a scan rate of 0.1 V/s (sensing signals measured 30 s after each addition of H_2O_2); and (b) a plot of peak current values vs. H_2O_2 concentration. Error bars were determined from repeating the measurements on the same electrode at least three times.

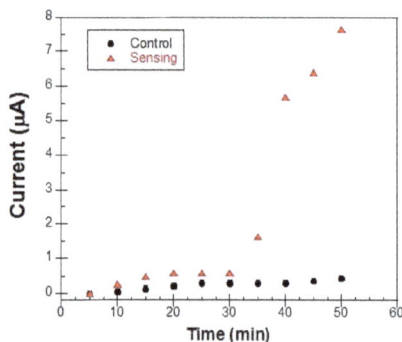

Figure 13. Comparison of the control peak current (circles) obtained from the CVs of a Metglas/ZnO/Hb electrode in PBS solution and the corresponding sensing current (triangles) when H_2O_2 is added in the solution.

Shown in Figure 14 are the magnetoelastic data, which show that the resonance frequency of the sensor has a linear drop versus time as the H_2O_2 concentration increases. Additionally, the total change of 0.075 KHz is larger than the error of 0.02 kHz observed at the control experiment and thus the change should be related to the H_2O_2 concentration. As it was mentioned in the introduction, the magnetoelastic sensors are used as microbalances since the resonance frequency depends on the mass load. For the particular Metglas ribbon used, calibration with known small mass loads gives a calibration factor of −1.4 kHz/mg. From this factor and the maximum H_2O_2 concentration of 350 μM, we conclude that there was a corresponding mass increase on the sensor of 152 ng/μM, which is probably due to H_2O_2 adsorption in the mesopores of the ZnO film during the electrochemical reaction.

We have also tested if the time interval between the addition of H_2O_2 aliquots and the electrochemical detection has any influence in the sensing results. We have tested three different cases (three different concentrations of H_2O_2 and two different time intervals, 30 s and 5 min, after each addition of H_2O_2). The results are shown in Figure 15. Comparing the obtained CV scans at each time interval, it can be clearly seen that the detection happens instantly, so we can affirm that there is no time dependence in the electrochemical detection process.

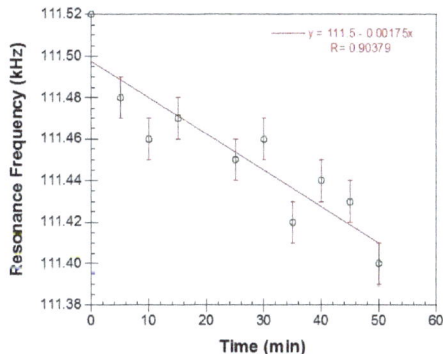

Figure 14. Magnetoelastic resonance data of a Metglas/ZnO/Hb electrode measured 5 min after the addition of increasing aliquots of H_2O_2.

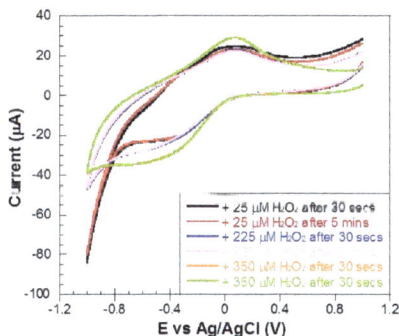

Figure 15. CVs of a Metglas/ZnO/Hb electrode at a scan rate of 0.1 V/s after the addition of three different concentrations of H_2O_2 measured after 30 s and 5 min after each addition.

5. Conclusions

For the first time, we have shown the fabrication of a simultaneous electrochemical magnetoelastic biosensor for studying Hb electrochemical behavior towards H_2O_2. To achieve this, we first succeeded in fabricating good quality ZnO nanoparticles, and depositing them as a film onto a magnetoelastic ribbon of Metglas 2826MB, which helped us determine the ZnO film Young's modulus of about 60 GPa. Next, Hb molecules were deposited onto the surface of the Metglas/ZnO bilayer, thus making both a sensitive modified voltammetry electrode and a magnetoelastic biosensor. This way we were able to monitor the reaction of the immobilized protein with specific aliquots of H_2O_2 by using simultaneously cyclic voltammetry and magnetoelastic detection procedures, which reveal a mass increase of about 152 ng/μM, which is probably due to H_2O_2 adsorption in the mesopores of the ZnO film during the electrochemical reaction.

Acknowledgments: Ariane Sagasti wants to thank BCMaterials and the European Erasmus+ Program for financial support. A. Sagasti and J. Gutiérrez also would like to thank the financial support provided by the Basque Government under the MICRO4FAB (ELKARTEK program) and Research Groups IT711-13 projects.

Author Contributions: Nikolaos Bouropoulos, Dimitris Kouzoudis, Jon Gutiérrez and Emmanuel Topoglidis conceived and designed the experiments; Ariane Sagasti and Apostolos Panagiotopoulos performed the experiments; Nikolaos Bouropoulos, Dimitris Kouzoudis, Emmanuel Topoglidis, Apostolos Panagiotopoulos, Ariane Sagasti and Jon Gutiérrez analyzed the data; Nikolaos Bouropoulos, Dimitris Kouzoudis and Emmanuel Topoglidis contributed reagents/materials/analysis tools; Jon Gutiérrez, Ariane Sagasti, Dimitris Kouzoudis, Emmanuel Topoglidis and Nikolaos Bouropoulos wrote the paper.

Conflicts of Interest: The authors declare no conflict of interest.

References

1. Duan, G.; Li, Y.; Wen, Y.; Ma, X.; Wang, Y.; Ji, J.; Wu, P.; Zhang, Z.; Yang, H. Direct electrochemistry and electrocatalysis of Hemoglobin/ZnO-Chitosan/nano-Au modified glassy carbon electrode. *Electroanalysis* **2008**, *20*, 2454–2459. [CrossRef]
2. Rifkind, J.; Nagababu, E.; Ramasamy, S.; Ravi, L.B. Hemoglobin redox reactions and oxidative stress. *Redox Rep.* **2003**, *8*. [CrossRef] [PubMed]
3. Liu, J.F.; Roussel, C.; Lagger, G.; Tacchini, P.; Girault, H.H. Antioxidant sensors based on DNA-modified electrodes. *Anal. Chem.* **2005**, *77*, 7687–7694. [CrossRef] [PubMed]
4. Jing, Z.H.; Zhan, J.H. Fabrication and gas sensing properties of porous ZnO nanoplates. *Adv. Mater.* **2008**, *20*, 4547–4551. [CrossRef]
5. Fang, X.; Bando, Y.; Liao, M.; Gautam, U.K.; Zhi, C.; Dierre, B.; Liu, B.; Zhai, T.; Sekiguchi, T.; Koide, Y.; et al. Single-crystalline ZnS nanobelts as ultraviolet-light sensors. *Adv. Mater.* **2009**, *21*, 2034–2039. [CrossRef]
6. Wei, A.; Xu, C.Y.; Sun, X.W.; Huang, W.; Lo, G.Q. Field emission from hydrothermally grown ZnO nanoinjectors. *J. Display Technol.* **2008**, *4*, 9–12.
7. Wang, Z.L. Towards self-powered nanosystems: From nanogenerators to nanopiezotronics. *Adv. Funct. Mater.* **2008**, *18*, 3553–3567. [CrossRef]
8. Qi, Q.; Zhang, T.; Liu, L.; Zheng, X.J.; Yu, Q.J.; Zeng, Y.; Yang, H.B. Selective acetone sensor based on dumbbell-like ZnO with rapid response and recovery. *Sens. Actuators B* **2008**, *134*, 166–170. [CrossRef]
9. Solanki, P.R.; Kaushik, A.; Ansari, A.A.; Sumana, G.; Malhotra, B.D. Zinc oxide-chitosan nanobiocomposite for urea sensor. *Appl. Phys. Lett.* **2008**, *93*, 163903. [CrossRef]
10. Zhang, Y.; Nayak, T.R.; Hong, H.; Cai, W. Biomedical applications of zinc oxide nanomaterials. *Cur. Mol. Med.* **2013**, *13*, 1633–1645. [CrossRef]
11. Mirzaei, H.; Darroudi, M. Zinc oxide nanoparticles: Biological synthesis and biomedical applications. *Ceram. Int.* **2017**, *43*, 907–914. [CrossRef]
12. Wei, A.; Pan, L.; Huang, W. Recent progress in the ZnO nanostructure-based sensors. *Mater. Sci. Eng. B* **2011**, *176*, 1409–1421. [CrossRef]
13. Dar, G.N.; Umar, A.; Zaidi, S.A.; Baskoutas, S.; Hwang, S.W.; Abaker, M.; Al-Hajry, A.; Al-Sayari, S.A. Ultra-high sensitive ammonia chemical sensor based on ZnO nanopencils. *Talanta* **2012**, *89*, 155–161. [CrossRef] [PubMed]
14. Mehta, S.K.; Singh, K.; Umar, A.; Chaudhary, G.R.; Singh, S. Ultra-high sensitive hydrazine chemical sensor based on low-temperature grown ZnO nanoparticles. *Electrochim. Acta* **2012**, *69*, 128–133. [CrossRef]
15. Chrissanthopoulos, A.; Baskoutas, S.; Bouropoulos, N.; Dracopoulos, V.; Tasis, D.; Yannopoulos, S.N. Novel ZnO nanostructures grown on carbon nanotubes by thermal evaporation. *Thin Solid Films* **2007**, *515*, 8524–8528. [CrossRef]
16. Baskoutas, S.; Giabouranis, P.; Yannopoulos, S.N.; Dracopoulos, V.; Toth, L.; Chrissanthopoulos, A.; Bouropoulos, N. Preparation of ZnO nanoparticles by thermal decomposition of zinc alginate. *Thin Solid Films* **2007**, *515*, 8461–8464. [CrossRef]
17. Baruah, S.; Dutta, J. Hydrothermal growth of ZnO nanostructures. *Sci. Technol. Adv. Mater.* **2009**, *10*, 013001. [CrossRef] [PubMed]
18. Morkoç, H.; Özgur, Ü. Chapter I: General properties of ZnO. In *Zinc Oxide: Fundamentals, Materials and Device Technology*; WILEY-VCH Verlag GmbH & Co.: Weinheim, Germany, 2009; pp. 1–76.
19. Polarz, S.; Roy, A.; Lehmann, M.; Driess, M.; Kruis, F.E.; Hoffmann, A.; Zimmer, P. Structure–property–function relationships in nanoscale oxide sensors: A case study based on zinc oxide. *Adv. Funct. Mater.* **2007**, *17*, 1385–1391. [CrossRef]
20. Chen, W.; Cai, S.; Ren, Q.Q.; Wen, W.; Zhao, Y.D. Recent advances in electrochemical sensing for hydrogen peroxide: A review. *Analyst* **2012**, *137*, 49–58. [CrossRef] [PubMed]
21. Shamsipur, M.; Pashabadi, A.; Molaabasi, F. A novel electrochemical hydrogen peroxide biosensor based on hemoglobin capped gold nanoclusters-chitosan composite. *RSC Adv.* **2015**, *5*, 61725–61734. [CrossRef]
22. Xu, Y.; Hu, C.; Hu, S. A hydrogen peroxide biosensor based on direct electrochemistry of hemoglobin in Hb-Ag sol films. *Sens. Actuators B* **2008**, *130*, 816–822. [CrossRef]

23. Weissbluth, M. *Hemoglobins (Cooperativity and Electronic Properties)*; Springer: Berlin/Heidelberg, Germany, 1974; ISBN 3-540-06582-2.

24. Zhang, C.L.; Liu, M.C.; Li, P.; Xian, Y.Z.; Chang, Y.X.; Zhang, F.F.; Wang, X.L.; Jin, L.T. Fabrication of ZnO nanorods modified electrode and its application to the direct electrochemical determination of hemoglobin and cytochrome c. *Chin. J. Chem.* **2005**, *23*, 144–148. [CrossRef]

25. Stoyanov, P.G.; Grimes, C.A. A remote query magnetostrictive viscosity sensor. *Sens. Actuators A* **2000**, *80*, 8–14. [CrossRef]

26. Van der Burgt, C.M. Dynamical physical parameters of the magnetostrictive excitation of extensional and torsional vibrations in ferrites. *Philips Res. Rep.* **1953**, *8*, 91–132.

27. Luborsky, F.E. Chapter 6: Amorphous ferromagnets. In *Handbook of Magnetic Materials*; North-Holland Publishing: Amsterdam, The Netherlands, 1980; ISBN 978-0-444-85311-0.

28. Bouropoulos, N.; Kouzoudis, D.; Grimes, C. The real-time, in situ monitoring of calcium oxalate and brushite precipitation using magnetoelastic sensors. *Sens. Actuators B* **2005**, *109*, 227–232. [CrossRef]

29. Topoglidis, E.; Astuti, Y.; Duriaux, F.; Grätzel, M., Durrant, J.R. Direct electrochemistry and nitric oxide interaction of heme proteins adsorbed on nanocrystalline tin oxide electrodes. *Langmuir* **2003**, *19*, 6894–6900. [CrossRef]

30. Reynolds, D.C.; Look, D.C.; Jogai, B. Fine structure on the green band in ZnO. *J. Appl. Phys.* **2001**, *89*, 6189–6191. [CrossRef]

31. Chen, H.; Gu, S.; Tang, K.; Zhu, S.; Zhu, Z.; Ye, J.; Zhang, R.; Zheng, Y. Origins of green band emission in high-temperature annealed N-doped ZnO. *J. Lumin.* **2011**, *131*, 1189–1192. [CrossRef]

32. Borseth, T.M.; Svenson, B.G.; Kuznetsov, A.Y.; Klason, P.; Zhao, Q.X.; Willander, M. Identification of oxygen and zinc vacancy optical signals in ZnO. *Appl. Phys. Lett.* **2006**, *89*, 262112. [CrossRef]

33. Schmidt, S.; Grimes, C.A. Elastic modulus measurement of thin films coated onto magnetoelastic ribbons. *IEEE Trans. Mag.* **2001**, *37*, 2731–2733. [CrossRef]

34. Baimpos, T.; Giannakopoulos, I.G.; Nikolakis, V.; Kouzoudis, D. Effect of gas adsorption on the elastic properties of faujasite films measured using magnetoelastic sensors. *Chem. Mater.* **2008**, *20*, 1470–1475. [CrossRef]

35. Manoharan, M.P.; Desai, A.V.; Neely, G.; Haque, M.A. Synthesis and elastic characterization of zinc oxide nanowires. *J. Nanomater.* **2008**, *1*. [CrossRef]

36. Landau, L.D.; Lifshitz, E.M. Theory of Elasticity. In *A Course of Theoretical Physics*, 2nd ed.; Pergamon Press: Oxford, UK, 1970.

37. Livingston, J.D. Magnetomechanical properties of amorphous metals. *Phys. Stat. Sol. A* **1982**, *70*, 591–596. [CrossRef]

38. Kobiakov, I.B. Elastic, piezoelectric and dielectric properties of ZnO and CdS single crystals in a wide range of temperatures. *Solid State Commun.* **1980**, *35*, 305–310. [CrossRef]

39. Song, J.; Wang, X.; Riedo, E.; Wang, Z.L. Elastic property of vertically aligned nanowires. *Nano Lett.* **2005**, *5*, 1954–1958. [CrossRef] [PubMed]

40. Desai, A.V.; Haque, M.A. Mechanical properties of ZnO nanowires. *Sens. Actuators A* **2007**, *134*, 169–176. [CrossRef]

41. Ji, L.W.; Young, S.J.; Fang, T.H.; Liu, C.H. Buckling characterization of vertical ZnO nanowires using nanoindentation. *Appl. Phys. Lett.* **2007**, *90*, 033109. [CrossRef]

42. Bai, X.D.; Gao, P.X.; Wang, Z.L.; Wang, E.G. Dual-mode mechanical resonance of individual ZnO nanobelts. *Appl. Phys. Lett.* **2003**, *82*, 4806–4808. [CrossRef]

43. Wang, Z.L. Zinc oxide nanostructures: Growth, properties and applications. *J. Phys. Condens. Matter* **2004**, *16*, R829–R858. [CrossRef]

44. Mollan, T.L.; Alayash, A.I. Redox reactions of Hemoglobin: Mechanism of toxicity and control. *Antioxid. Redox Signal.* **2013**, *18*, 2251–2253. [CrossRef] [PubMed]

45. Gonga, C.; Shena, Y.; Chena, J.; Songa, Y.; Chena, S.; Songa, Y.; Wanga, L. Microperoxidase-11@PCN-333 (Al)/three-dimensional microporous carbon electrode for sensing hydrogen peroxide. *Sens. Actuators B-Chem.* **2017**, *239*, 890–897. [CrossRef]

46. Topoglidis, E.; Cass, A.E.G.; O'Regan, B.; Durrant, J.R. Immobilization and bioelectrochemistry of proteins on nanoporous TiO2 and ZnO films. *J. Electroanal. Chem.* **2001**, *517*, 20–27. [CrossRef]

47. Altube, A.; Pierna, A.R. Thermal and electrochemical properties of cobalt containing Finemet type alloys. *Electrochem. Acta* **2004**, *49*, 303–311. [CrossRef]

48. Bard, A.J.; Faulkner, L.R. *Electrochemical Methods Fundamentals and Applications*, 2nd ed.; John Wiley & Sons: New York, NY, USA, 2001.

materials

MDPI

Review

Three-Dimensional ZnO Hierarchical Nanostructures: Solution Phase Synthesis and Applications

Xiaoliang Wang [1,*]**, Mashkoor Ahmad** [2] **and Hongyu Sun** [3,*]

1   College of Science, Hebei University of Science and Technology, Shijiazhuang 050018, China
2   Nanomaterials Research Group, Physics Division, Pakistan Institute of Nuclear Science and Technology,
    P.O. Nilore, Islamabad 44000, Pakistan; mashkoorahmad2003@yahoo.com
3   Department of Micro- and Nanotechnology, Technical University of Denmark,
    2800 Kongens Lyngby, Denmark
*   Correspondence: wxlsr@126.com (X.W.); hsun@nanotech.dtu.dk (H.S.); Tel.: +45-45-25-68-40 (H.S.)

Received: 17 October 2017; Accepted: 10 November 2017; Published: 13 November 2017

Abstract: Zinc oxide (ZnO) nanostructures have been studied extensively in the past 20 years due to their novel electronic, photonic, mechanical and electrochemical properties. Recently, more attention has been paid to assemble nanoscale building blocks into three-dimensional (3D) complex hierarchical structures, which not only inherit the excellent properties of the single building blocks but also provide potential applications in the bottom-up fabrication of functional devices. This review article focuses on 3D ZnO hierarchical nanostructures, and summarizes major advances in the solution phase synthesis, applications in environment, and electrical/electrochemical devices. We present the principles and growth mechanisms of ZnO nanostructures via different solution methods, with an emphasis on rational control of the morphology and assembly. We then discuss the applications of 3D ZnO hierarchical nanostructures in photocatalysis, field emission, electrochemical sensor, and lithium ion batteries. Throughout the discussion, the relationship between the device performance and the microstructures of 3D ZnO hierarchical nanostructures will be highlighted. This review concludes with a personal perspective on the current challenges and future research.

Keywords: zinc oxide; hierarchical nanostructures; solution phase synthesis; photocatalysis; field emission; sensor; lithium ion batteries

---

1. Introduction

Advanced nanomaterials, which are earth abundant and environmentally compatible, show the potential to solve the serious energy and environment problems. As an important and widely used wide bandgap (3.0–3.2 eV) oxide semiconductor, ZnO shows unique physical and chemical properties [1]. The applications of ZnO materials range from room temperature nanolasers, nanogenerators, solar cells, lithium ion batteries and photocatalysts.

ZnO crystal shows the stable structure as hexagonal wurtzite under the condition of normal temperature and atmospheric pressure. The crystal structure of ZnO can be viewed as a number of alternating planes composed of tetrahedrally coordinated oxygen and zinc ions stacked alternately along the [0001] direction (Figure 1). There are two important structural characteristics in wurtzite ZnO, i.e., the absence of inversion symmetry of the positive and negative charge centers and polar surfaces, which are the origin of piezoelectric properties and unique growth behaviors in the synthesized ZnO nanostructures. The most typical polar surface in ZnO structure is the basal plane (Figure 1), in which positively charged Zn-(0001) and negatively charged O-(000-1) polar surfaces are produced. These net charges on the polar surfaces are ionic charges and non-mobile. The distribution of surface charges is responsible for the minimizing the electrostatic energy of the system, which is also one of the important driving forces for growing nanostructures with the domination of polar surfaces [2]. In principle,

Materials **2017**, *10*, 1304

the equilibrium morphology of a crystal is determined by the standard Wulff construction, which depends on the relaxation energies [3]. For ZnO crystals, the kinetic parameters vary with different crystal planes and growth direction due to different relaxation energies, which are emphasized under a given growth condition [4]. Therefore, from the viewpoint of decreasing the system total energy, a ZnO crystallite will commonly develop into a three-dimensional morphology with well-defined and low-index crystallographic faces (Figure 1). In addition, the surface energy can be modified by selective adsorption of additives or surfactants on specific planes, and the morphology can be controlled accordingly by adding suitable agents during the synthesis.

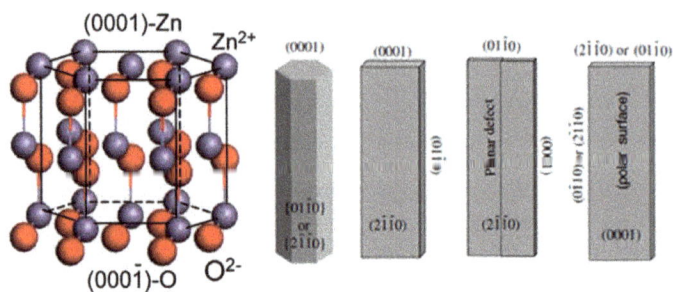

Figure 1. Crystal structure model of wurtzite ZnO (reprinted from [5] with permission, Copyright—Elsevier B.V., 2004), and typical morphologies of 1D ZnO nanostructures with exposed facets (reprinted from [2] with permission, Copyright—Elsevier B.V., 2009).

Inspired by the size- or morphology-dependent properties or device performances, numerous efforts have been devoted to the synthesis of ZnO nanostructures with 1D morphologies, such as nanowires, nanobelts, nanorings, nanohelices and so on [2,5,6]. Recently, more attention has been paid to assembling low-dimensional nano-sized building blocks into three-dimensional (3D) complex hierarchical structures [7]. Compared to mono-morphological structures, 3D ZnO hierarchical structures usually exhibit high surface to volume ratios, a large accessible surface area and better permeability. In addition, the hierarchical structures can increase the number of light traveling paths and thereby facilitate light absorption. Finally, such 3D hierarchical structures not only inherit the excellent properties of the single nano-sized building blocks but also provide potential applications in the bottom-up fabrication of functional devices including photocatalysts, sensors and drug release systems [8]. It is, therefore, important to develop facile approaches to synthesize 3D hierarchical structures with controlled fashion.

Herein, we summarize the most recent progress in the synthesis of 3D ZnO hierarchical nanostructures by using solution phase routes, and discuss the related applications. Firstly, typical solution phase synthesis methods towards 3D ZnO hierarchical nanostructures are reviewed, including direct precipitation, microemulsions, hydrothermal/solvothermal, sol-gel, electrochemical deposition, and chemical bath deposition. The basic principles of the synthesis and main factors that influence the structure and morphology of the products are analyzed. Then, different applications based on the 3D hierarchical architectures are discussed in the context of photocatalysis, field emission, electrochemical sensors, and electrodes for lithium ion batteries (Figure 2). Finally, current challenges and future outlooks of the synthesis and applications of 3D ZnO hierarchical nanostructures are briefly outlined.

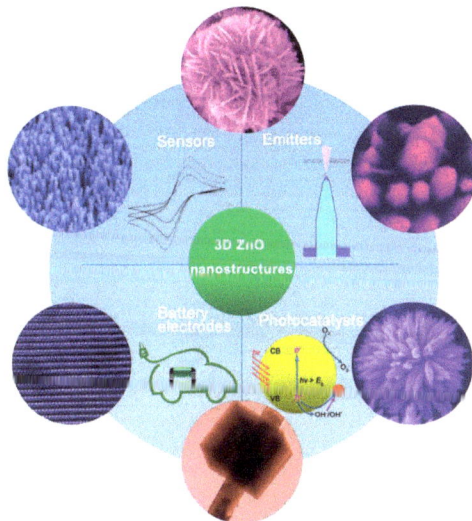

Figure 2. Typical 3D ZnO hierarchical nanostructures and their applications as photocatalysts, field electron emitters, electrochemical sensors, and electrodes for batteries.

2. Solution Phase Synthesis of 3D ZnO Hierarchical Nanostructures

The above discussion shows that the hierarchical assembly in ZnO nanostructures is related to the electronic/optical properties and thus a wide range of potential applications. So far, many synthesis strategies based on physical (physical and chemical vapor deposition, laser ablation, ball milling, lithographic, etc.), chemical (gas phase reaction, various solution phase synthesis), or biological methods have been well established to obtain 3D ZnO hierarchical nanostructures [9]. Comparing to other methods, solution phase route shows unique advantages, such as low cost (low in energy consumption and equipment costs), scalability, and ease of handling. Most of the solution phase reactions occur under mild condition with a relatively low temperature (<200 °C). Therefore, solution phase synthesis has attracted increasing interest. Typical solution phase synthesis includes precipitation, microemulsions, hydrothermal/solvothermal, sol-gel, electrochemical deposition, chemical bath deposition, and so on. There are several excellent reviews describing the synthesis of ZnO nanostructures [10–13]. In this paper, we mainly focus on ZnO hierarchical nanostructures synthesized by different solution phase methods.

2.1. Precipitation

In a typical precipitation process, different kinds of alkalis (NaOH, KOH, ammonium, urea, hexamethylene tetramine (HMT), etc.) and zinc sources (zinc salts, zinc foil, etc.) are used (Figure 3a). The synthesis starts with a reaction between zinc and hydroxide ions followed by the process of aggregation. A resultant precipitate is collected by filtration or centrifugation. The morphology and assembly of the ZnO products can be controlled by adjusting the reaction conditions, including the concentration of reacted solutions, temperature, time, and additives to tune the reaction steps [14]. Kołodziejczak-Radzimska and co-workers [15] optimized the precipitation conditions that would ensure getting the uniform particles of ZnO with the minimum diameter. Sepulveda-Guzman et al. [16] synthesized submicron ZnO arrays by a simple one-step aqueous precipitation method with zinc nitrate ($Zn(NO_3)_2$) and sodium hydroxide (NaOH) as the reagents. The effect of reaction temperature on the morphology change was studied. Snowflake-like and flower-like morphologies were obtained at 60 °C and 70 °C. The ZnO arrays are formed through self-aggregation process, and that such an oriented aggregation is enhanced by increasing the reaction temperature. In another work, Oliveira and co-workers [17] systematically investigated the influence of zinc salts ($Zn(NO_3)_2·6H_2O$, zinc sulfate ($ZnSO_4·7H_2O$)), pH value, temperature, additives (sodium sulfate, sodium dodecyl sulfate) on the final ZnO morphology and size. The results show the importance of nucleation of nanometric primary particles followed by oriented aggregation to produce uniform submicrometric particles. Recently, López et al. [18] reported an interesting work on exploring the synthesis of ZnO by employing zinc sources leached from alkaline batteries, paving a new way to recycling useful nanostructures from various chemical wastes.

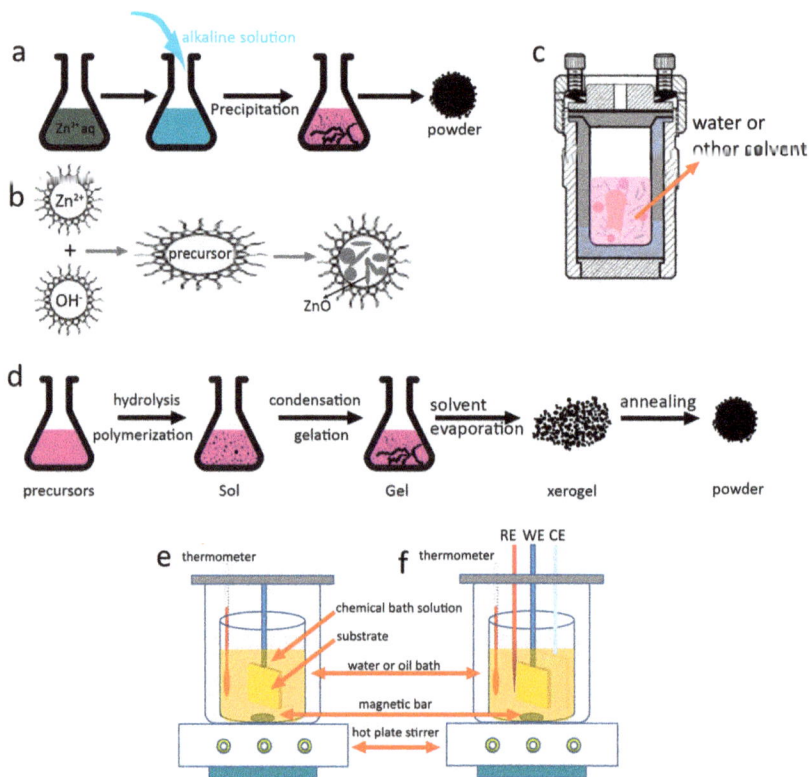

Figure 3. Typical solution phase methods for the synthesis of 3D ZnO hierarchical nanostructures. (**a**) Precipitation; (**b**) microemulsions; (**c**) hydrothermal/solvothermal; (**d**) sol-gel; (**e**) chemical bath deposition; and (**f**) electrochemical deposition.

It should be noted that ZnO is a typical amphoteric oxide, which can be etched in either an acid or alkali environment. The reaction between the zinc ions and alkaline ions results in the formation of ZnO crystals. Meanwhile, the newly formed ZnO further reacts with alkaline in growth solution until an equilibrium is achieved. Thus, tuning the precipitation (growth) and in situ etching provides a facile approach to control the morphology of nanocrystals [19–21]. For example, Xi et al. [19] fabricated large-scale arrays of highly oriented single-crystal ZnO nanotubes by an in situ chemical etching of the ZnO nanorods. Yang et al. [20] have demonstrated a novel anisotropic etching methodology for the synthesis of complex Ag nanoparticles shapes by controlling the concentrations of etching solution, and highlighted their important applications as highly sensitive surface enhanced Raman spectroscopy substrates. In our previous studies [8], we reported a simple precipitation to synthesize ZnO 3D hierarchical structures by using $Zn(NO_3)_2 \cdot 6H_2O$, Zn foil, and KOH as the zinc source and alkalis. Importantly, the morphology of the hierarchical structures can be simply selected by changing the KOH concentrations. The morphology of the products are sheet, flower-like microsphere assembled by nanosheets, flower-like microsphere assembled by nanoneedles, and flower-like microsphere assembled by thinner nanoneedles when the concentration of KOH increased from 0.5 to 2, 4, and 8 M, respectively (Figure 4a). A possible formation mechanism based on preferential etching of KOH along the [001] direction of ZnO sheets was proposed (Figure 4a). The growth or/and etching can be further controlled by adding suitable capping agents that prefer to absorb on specific crystal planes, making it possible to have fine control over the morphology. Tian et al. [??] demonstrated that citrate ions can selectively adsorb on the (001) surfaces of ZnO crystals and thus inhibit the growth along [001] orientation. Based on this knowledge, they synthesized large arrays of oriented ZnO nanorods with controlled aspect ratios and a series of complex morphologies. In our studies, we combined the selective capping (citrate ions) and etching (KOH) together to capture the intermediate of the morphology change, and proposed a possible formation mechanism based on capping–etching competitive interactions (Figure 4b) [23]. The strategy can also be applied to the synthesis and modification of other materials by carefully selecting suitable capping agents and etchants.

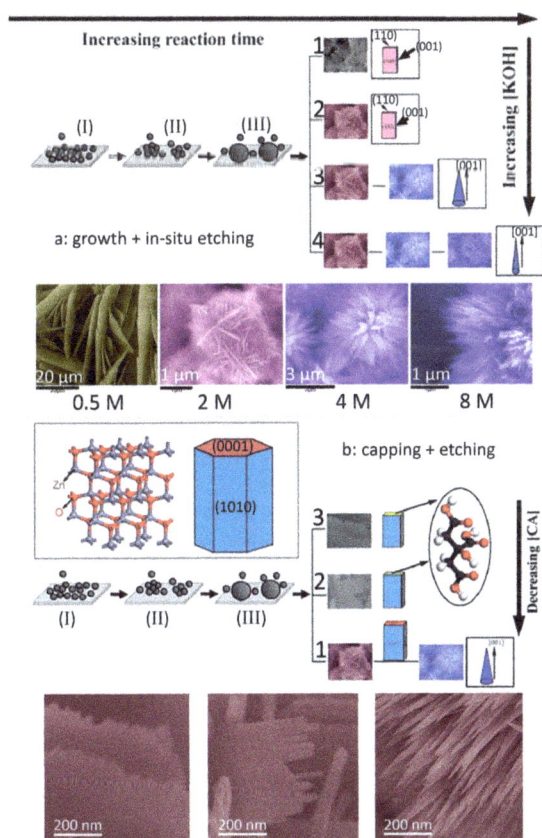

Figure 4. Schematic illustration of the formation process of ZnO 3D hierarchical structures via the combination of (**a**) growth and in situ etching (reprinted from [8] with permission, Copyright—The Royal Society of Chemistry, 2012); (**b**) capping and etching (reprinted from [23] with permission, Copyright—The Royal Society of Chemistry, 2015).

2.2. Microemulsions

Materials synthesis via microemulsion is occurred in a stable mixed solution that contains water, oil, and surfactant. Depending on the properties of immiscible liquid–liquid interface, the formation of microemulsion can be in the form of oil-swollen micelles dispersed in water (oil-in-water, or O/W) or the reverse case, i.e., water swollen micelles dispersed in oil (water-in-oil, W/O). The two configurations are usually named as microemulsion and reverse microemulsion, respectively [24]. Figure 3b shows a typical process for the synthesis of ZnO via microemulsion method. The typical size of the microemulsion or reverse microemulsion is smaller than 100 nm; therefore, the microemulsions can be used as nanoreactors for materials synthesis, which provides a bottom-up route to control the size and morphology by adjusting the microemulsion properties. By studying the phase diagrams of oil-water-surfactant system containing toluene, zinc acetate solution, cetyltrimethylammonium bromide and butanol, Lin et al. [25] obtained the desired size and shape of ZnO particles (Figure 5a–d). This study demonstrates the importance of investigating the intrinsic properties of this multi-liquid phase system. Jesionowski et al. [26] made emulsion that was composed of cyclohexane as the organic phase, zinc acetate as the water phase and appropriate emulsifiers. The obtained ZnO showed a narrow particle size distribution and large specific surface area. In their further work [27], they synthesized a series of ZnO materials in a similar emulsion system. ZnO structures with the morphologies of solids, ellipsoids, rods and flakes were successfully obtained by applying modifications of the ZnO precipitation process.

2.3. Hydrothermal and Solvothermal

Hydrothermal synthesis refers to the materials synthesis by chemical reactions of substances in a sealed heated aqueous solution above ambient temperature and pressure, while solvothermal is very similar to the hydrothermal route except the precursor solution is usually non-aqueous (Figure 3c) [28–31]. Well control over the hydrothermal/solvothermal synthetic conditions is a key to the synthesis of ZnO nanomaterials with defined structure, morphology, composition, and assembly. Typical control experimental parameters include reagents, solvent, additives, filling degree, temperature, time, and so on [32,33]. In addition, substrate is an important and interesting parameter that can affect the ZnO morphology, especially the aligned ZnO nanostructures. In a study, ZnO nanostructures on Zn foil was synthesized by hydrothermal synthesis [34]. The morphology and assembly of ZnO arrays are dependent on the solvent properties. Specifically, ZnO nanorod arrays and randomly scattered nanorods are obtained in the mixed solvent containing ammonia aqueous solution (1%) and pure water, and pure water system, respectively. Moreover, repetitive hydrothermal or solvothermal treatment can yield more complex and hierarchical configurations. By employing this simple strategy, Ko et al. [35] synthesized high density and long branched tree-like ZnO nanoforests (Figure 5e). In a report of Joo et al. [36], single-crystalline ZnO nanowires were grown on substrates with zinc oxide seed layers in aqueous solutions by hydrothermal method. The effect of specific additives on the ZnO morphology was studied in detail. The results show that the addition of positively charged complex ions (Cd, Cu, Mg, Ca) and negatively charged complexes (Al, In, Ga) promote low and high aspect ratio growth of ZnO nanostructures (Figure 5f). They demonstrated that face-selective electrostatic crystal growth inhibition mechanism governed this selective synthesis. Wysokowski et al. [37] used β-Chitinous scaffolds as a template during the hydrothermal synthesis of ZnO under mild conditions (70 °C). The obtained samples showed unique film-like morphology, which showed good antibacterial properties against Gram-positive bacteria. Very recently, Van Thuan and Kim et al. [38] synthesized oval-shaped graphene/ZnO quantum hybrids by employing a facile chemical-hydrothermal method. The samples exhibited excellent catalytic properties for the selective reduction of nitroarenes.

2.4. Sol-Gel Processing

The sol-gel process contains the formation of solid material from a solution by using a sol or a gel as an intermediate step. The synthesis of metal oxide materials often involves controlled hydrolysis and condensation of the alkoxide precursors or salts. Figure 3d illustrates the main steps of preparation of metal oxides powder by the sol-gel process: (i) preparation of the precursor solution; (ii) hydrolysis of the molecular precursor and polymerization via successive bimolecular additions of ions, forming oxo-, hydroxyl, or aquabridges; (iii) condensation by dehydration; (iv) solvent evaporation and organic compounds removal to form xerogel; and (v) heat treatment of the xerogel to form powers. Properties of the final products, including the particle size, surface area, crystallinity, and agglomeration, are highly dependent on the reaction parameters, especially the precursors, solvents, additives, evaporation, drying, and post-treatment conditions [39] (Figure 5g). Tseng et al. [40] employed the sol-gel process to synthesize ZnO polycrystalline nanostructures using $Zn(CH_3COO)_2 \cdot 2H_2O$ as the zinc solute and different alcohols as solvents (glycol, glycerol, and diethylene glycol). The morphology of the final ZnO was fiber, rhombic flakes, and spherical particles. The formation of thorn-like ZnO nanostructures in the sol-gel process was reported by Khan and co-workers [41]. They further modified this method by mechanical stirring during the sol generation, and found the agitation speed was a critical value in determining the size and aspect ratio of the particles. Higher stirring speed is favorable for anisotropic growth of ZnO nanoparticles.

2.5. Electrochemical and Chemical Bath Deposition

Electrochemical deposition (ECD) and chemical bath deposition (CBD) methods are facile and can produce materials or nanostructures that cannot be obtained by other deposition methods. The CBD process only requires suitable solution containers and substrate mounting devices (Figure 3e), while for the ECD method, additional power supplies, electrodes (counter electrode/CE, reference electrode/RE) are necessary, and the substrate (working electrode/WE) must be conductive (Figure 3f). More importantly, the morphology and orientation of the deposited samples can be tuned by controlling the reaction thermodynamics and kinetics, including solution properties, additives, substrate, temperature, and electrochemical parameters (applied potential, current density, etc.) [42–47].

The synthesis of ZnO nanostructures via the ECD process includes the reduction of precursor at the electrode, the supersaturation at the vicinity of the electrode, and subsequent precipitation. Theoretically, factors that affect any step should be considered to achieve the good control of final structures. In this regard, Illy et al. [48] systematically studied the effect of various experimental parameters (electrolyte concentration, pH value, reaction temperature, and overpotential) on the morphology, thickness, transparency, roughness and crystallographic orientation of the ZnO materials. They found that ZnO nanostructures with (002) preferential orientation and controlled thickness can be grown by using optimized parameters, which are important for organic photovoltaic applications. Different additives can either interact with the ions in the electrolyte, absorb on specific sites on the deposited ZnO structure, or change the electrolyte itself (conductivity, viscosity, etc.), resulting in a different deposition pathways and thus final products. In a study by Oekermann and co-workers [49], the addition of water soluble tetrasulfonated metallophthalocyanines (TSPcMt), in which Mt = Zn(II), Al(III)[OH] or Si(IV)[OH], in the electrolyte containing zinc nitrate yields completely different morphology and assembly, which is ascribed to the preferential adsorption of the additive molecules onto the crystal planes of ZnO (Figure 5h,i).

Synthesis of ZnO nanostructures via CBD is based on a direct chemical reaction involving dissolved zinc ions and oxygen precursors in the solution. Different from ECD where the deposition only occurs on the conductive substrate, the growth of ZnO in CBD process can take place either in the solution or on the substrate surface. The morphology and assembly of ZnO products can also be controlled by the solution, additives, and so on [50]. Moreover, patterned or flexible ZnO hierarchical nanostructures can be obtained by applying corresponding substrates [51] (Figure 5j).

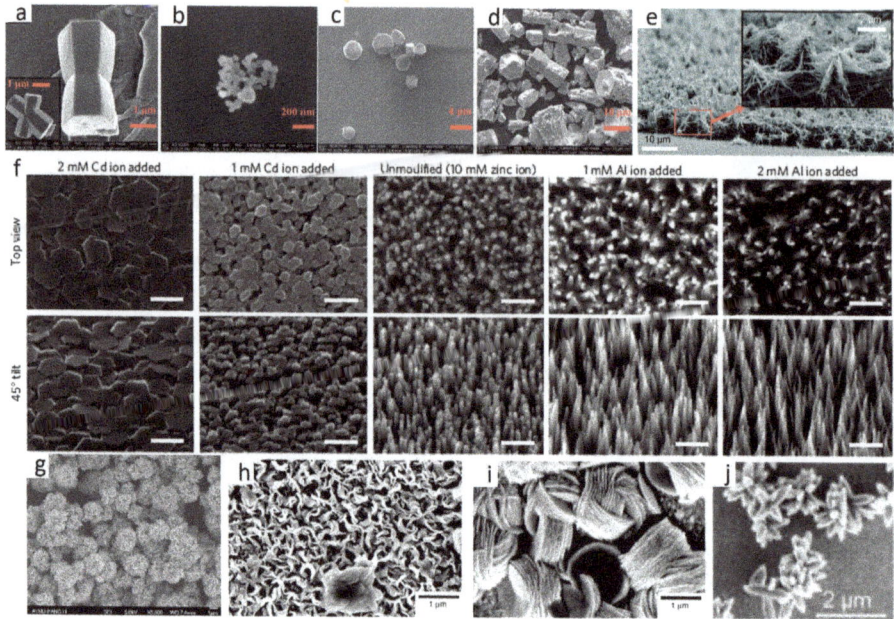

Figure 5. Typical 3D ZnO hierarchical nanostructures synthesized by solution phase methods: (**a–d**) microemulsion process (reprinted from [25] with permission, Copyright—The Royal Society of Chemistry, 2012); (**e**) repetitive hydrothermal (reprinted from [35] with permission, open access, American Chemical Society, 2011); (**f**) ion-mediated hydrothermal (scale bars = 500 nm, reprinted from [36] with permission, Copyright—Macmillan Publishers Limited, 2011); (**g**) sol-gel method (reprinted from [39] with permission, Copyright—Elsevier Ltd and Techna Group S.r.l., 2013); (**h,i**) electrochemical deposition (reprinted from [49] with permission, Copyright—The Owner Societies, 2011); and (**j**) chemical bath deposition method (reprinted from [51] with permission, Copyright—American Chemical Society, 2015).

The above-mentioned solution phase synthesis methods have their own advantages and disadvantages. Table 1 summarizes and compares the methods for the synthesis of 3D ZnO hierarchical nanostructures. In addition, the different solution phase methods can be combined together or with other treatments, such as microwave heating, sonochemistry, etc., to achieve even more complex and interesting hierarchical nanostructures with useful applications. For example, sol-gel processing or microemulsions are often employed with hydrothermal/solvothermal treatment to prepare various nanostructures [52]. The pre-synthesized ZnO nanostructures via hydrothermal or ECD methods can be further etched to form needle or tube arrays [53]. The assembly of ZnO architectures can be tuned by means of ECD via deformation and coalescence of soft colloidal templates in reverse microemulsion [54]. It is worth noting that a wide range of ZnO hierarchical nanostructures can also be synthesized from different biomass, such as microorganisms, enzymes, bacteria, and plant extracts, which are eco-friendly alternatives compared to the conventional synthesis methods [55–57]. In a word, solution phase synthesis provides plenty of room to control and optimize hierarchical ZnO architectures for diverse applications.

Table 1. The comparison of solution phase methods for synthesis of 3D ZnO nanostructures.

Synthesis Methods	Advantages	Disadvantages
Precipitation	Simplicity, low cost, and rapid	The nucleation and growth occur simultaneously due to the rapid reaction, making it difficult to study the detail growth process; sometimes, further thermal treatment is needed
Microemulsions	Novel morphology can be obtained by selecting suitable microemulsion system as a reactor (template)	Surfactants are difficult to remove; upscale synthesis may be hindered by the high price of some surfactants
Hydrothermal/solvothermal	Simple equipment (autoclave), low cost, and large area uniform production	Higher pressure and reaction temperature; organic solvents are needed for solvothermal method
Sol-gel	Simplicity, low cost, and relatively mild conditions of synthesis	Sol-gel matrix components may involved in the samples and additional purification is needed
Electrochemical deposition	Low synthesis temperature, low cost, and rapid; the structure and morphology can be easily controlled by electrochemical parameters	The growth substrate must be conductive
Chemical bath deposition	Simple cost, effective, and the samples can be deposited at arbitrary substrates	heterogeneous growth at the growth substrate and homogeneous formation in the bath take place at the same time; wastage of solution after every deposition

3. Applications of 3D ZnO Hierarchical Nanostructures

3D ZnO hierarchical nanostructures show unique advantages of high surface area, porous structures, and synergistic interactions of the constituted nano building blocks. Therefore, 3D ZnO nanostructures possess improved physical/chemical properties, such as enhanced light harvesting, increased reaction sites, and improved electron and ion transportation, which are highly needed for optical, electrical, and electrochemical applications. In this paper, we will review the most recent progress in the research activities on 3D ZnO hierarchical nanostructures used for photocatalysis, field emission, electrochemical sensors, and electrodes for lithium ion batteries.

3.1. Photocatalysis

ZnO nanostructures have attracted much attention to the fields of photocatalysis, including photocatalytic degradation of organic contaminants, photocatalytic water splitting, and so on, due to the notable merits such as nontoxicity, biological compatibility, and universality. Typical steps involved in heterogeneous photocatalysis process are as follows (Figure 6) [58]: (1) light absorption; (2) the generation and separation of photoexcited electrons and holes; (3) the migration, transport and recombination of carriers; and (4) surface electrocatalytic reduction and oxidation reactions. The overall catalysis efficiency is related to the cumulative effects of these consecutive steps. For ZnO photocatalysts, the activity is limited by the intrinsic wide bandgap (3.0–3.2 eV) and the high electron–hole recombination rate, which can be tuned by optimizing the structural parameters of the photocatalysts, such as size, morphology, assembly, specific surface area, and the defect density [59]. Compared to the 0D, 1D, or 2D counterparts, and 3D ZnO hierarchical nanostructures show advantages of high surface area and porous structures, enhanced light harvesting, and synergistic effects between the nano building blocks. All of these characters are beneficial for the photocatalysis enhancement.

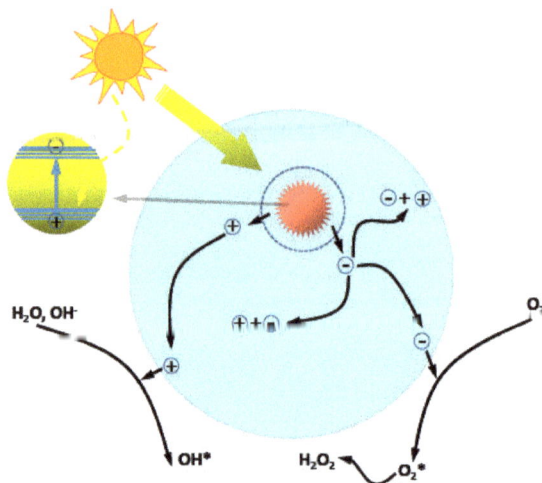

Figure 6. Schematic illustration on the photocatalytic processes in ZnO.

The comparison of the photodegradation of organic dye Rhodamine B (RhB) under UV-irradiation with different 3D ZnO hierarchical nanostructures yielded by facile solution phase method is shown in (Figure 7a,b) [8]. Under the same experimental conditions, the relative photocatalytic activity is thin needle flowers > needle flowers > sheet flowers > nanosheets. The significant improvement in the photocatalytic activity of the thin needle flowers structure can be attributed to the following reasons: (1) optical quality and special structural features; (2) the large active surface area and interspaces of the flower structure, which facilitate the diffusion and mass transportation of RhB molecules and hydroxyl radicals; (3) the improved efficiency of electron–hole separation. The morphology of ZnO nanostructures dependent photocatalysis is later demonstrated in the degradation of methylene blue in aqueous solution [60]. By coupling the strategies of elemental doping [61], defect engineering [62], modifying the surface with visible light active materials [63] or plasmonic-metal nanostructures (Ag, Pt, Au, etc.) [64], and the photodegradation properties of 3D ZnO hierarchical nanostructures can be further improved. For example, we evaluated photocatalytic performance of ZnO needle flowers and Au nanoparticles/ZnO needle flowers composite (Au/ZnO) by degradating organic dye RhB under UV irradiation (Figure 7c–e) [65]. The pure ZnO showed observable photocatalytic activity but with rather slow kinetics. Only ~60% of the RhB was decomposed within 90 min. In contrast, when the Au/ZnO composites were applied as the photocatalyst, a significant synergistic enhancement effect was observed, i.e., RhB was decomposed thoroughly within 90 min. The Au nanoparticles enhanced photodegradation is also observed in ZnO sheet flowers. In these studies, besides the hierarchical morphology of ZnO nanostructures, the improvement of photocatalytic properties can be ascribed to the presence of noble metal nanoparticles: (1) the light absorption is increased due to the strong surface plasmon resonance of the noble metal nanoparticles; and (2) the efficiency of charge separation of the photo-generated electron–hole pairs is increased due to the strong electronic interaction between strong electronic interaction and ZnO.

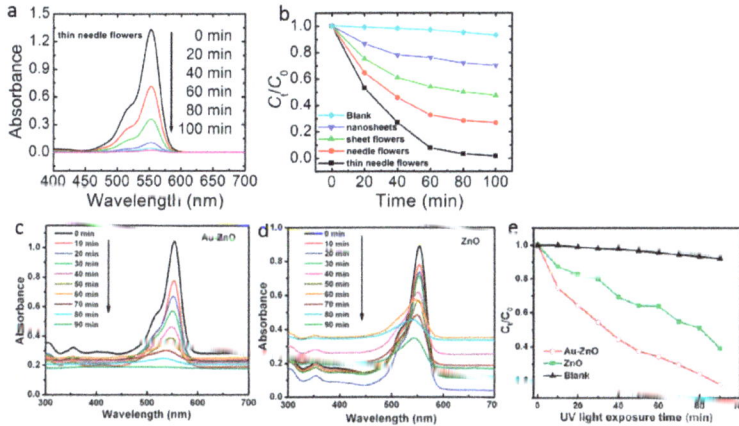

Figure 7. Photocatalytic degradation of RhB via (**a**,**b**) 3D ZnO hierarchical nanostructures with different morphologies (reprinted from [8] with permission, Copyright—The Royal Society of Chemistry, 2012); (**c**–**e**) ZnO needle flowers and Au nanoparticles/ZnO needle flowers composite (reprinted from [65] with permission, Copyright—The Royal Society of Chemistry, 2011).

The above routes to improve the degradation properties in ZnO hierarchical photocatalysts can also be applied to photosplit water to generate hydrogen. For example, elemental doping and defect engineering are effective to narrow the bandgap of ZnO materials, which results in an extension of light absorption range from UV into visible light range, while generating interface structures by depositing plasmonic noble metals separates photogenerated carriers, improves visible and near-infrared photo-absorption, and thus achieves high-performance photocatalytic hydrogen evolution.

3.2. Field Emission

Field emission devices show several advantages, such as the resistance to temperature fluctuation and radiation, less power consumption, low thermionic noise, low energy spread, miniature volume and nonlinear, and the exponential current–voltage (I–V) relationship in which a small variation in the voltage results in a large change in the emission current instantaneously [66]. Theoretical calculations show that the external filed induces a decrease of the surface barrier height by a value of $\Delta\phi \sim 3.8\,F^{1/2}$ (for ϕ in eV and F in V/Å) (Figure 8a), and the field is off the order of $\sim 10^9$ V/m. If the emitter surface is sharp configuration as shown in Figure 8b, electrons can be extracted at a considerably lower applied field. The relationship between the field emission current density (J) and the applied electric field (E) is described by Fowler–Nordheim equation: $J = (A\beta^2 E^2/\phi)\exp\lfloor-B\phi^{3/2}(\beta E)^{-1}\rfloor$, where A ($1.54 \times 10^{-6}$ A eV V^{-2}) and B (6.83×10^3 eV$^{-3/2}$ V μm^{-1}) are the first and second F-N constants, ϕ is the work function, and β is the field enhancement factor, which reflects the magnitude of electric field at the emitting surface. The field enhancement factor β can be defined as the ratio of local electric field divided to the applied electric field. Therefore, β is a dimensionless quantity. The emission current density at a constant cathode-anode distance is strongly dependent on the work function ϕ and the field enhancement factor β that is related to geometric configuration of the emitter, crystal structure, conductivity, and so on.

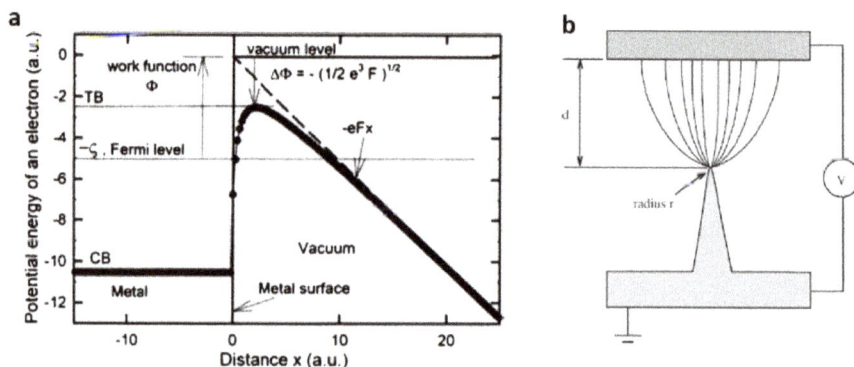

Figure 8. (a) Potential energy of an electron near the cathode surface (reprinted from [67] with permission, Copyright American Vacuum Society, 2007); (b) illustration of field electron emission from a tip (reprinted from [68] with permission, Copyright—Elsevier B.V., 2004).

With inherent properties of being thermally stable and oxidation resistant, ZnO nanostructures show the potential to be good candidates for field emission. Moreover, a variety of ZnO nanostructures can be synthesized by facile solution phase methods as discussed above, which not only reduce the cost, but also make it possible to fabricate flexible field emission devices based on polymers or other metal substrate materials [69,70].

Field emission studies on ZnO nanorod arrays synthesized on zinc foils by the solvothermal route are presented by Dev et al. [71]. The effect of solvothermal parameters including the solvent (distilled water and ethylenediamine), temperature, and time on the morphology and field emission properties of ZnO nanostructures are studied. It was observed that, with the increase in ethylenediamine concentrations, the alignment of the nanorods gets better, corresponding to the increase of field enhancement factor from 850 to 1044. ZnO nanotube arrays were prepared by hydrothermal reaction in ammonia and zinc chloride solutions by Wei et al. [72]. The turn-on field of the ZnO nanotube arrays was extrapolated to be about 7.0 V m^{-1} at a current density of 0.1 A cm^{-2}, the emission current densities reached 1 mA cm^{-2} at a bias field of 17.8 V m^{-1}, and the field enhancement factor was estimated to be 910. Cao et al. [73] reported the field emission of wafer-scale ZnO nanoneedle arrays synthesized by template-free electrochemical deposition method. The field enhancement factor of the ZnO nanoneedle arrays was 657 with a working distance of 250 µm between the cathode and anode. In our studies [74], we compared the field emission properties of ZnO nanowire arrays with flat ends and nanoneedle arrays with sharp ends (Figure 9a). The ZnO nanowires were synthesized by the hydrothermal method at 70 °C. Then, solution etching was employed to form ZnO nanoneedles at room temperature. The turn-on electronic fields of ZnO nanoneedles and nanowires are 2.7 and 5.3 V µm^{-1} at a current density of 10 µA cm^{-2}. The threshold electronic fields, which were defined as the field value at the emission current density J of 0.1 mA cm^{-2}, of ZnO nanoneedles and nanowires are 3.9 and 6.1 V µm^{-1}, respectively (Figure 9b). The field enhancement factors were estimated to be 4939.3 for ZnO nanoneedles and 1423.6 for ZnO nanowires (Figure 9c). In addition, there is no obvious degradation of the current density, demonstrating the excellent emission stability of the ZnO array materials (Figure 9d). This study highlights the important effect of emitter geometry on the field emission.

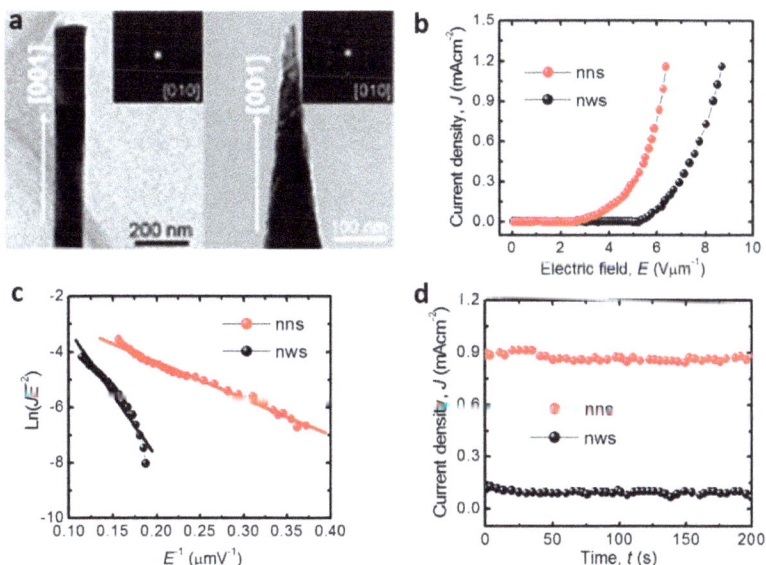

Figure 9. The comparison of field emission properties of ZnO nanowires with flat ends and nanoneedles with sharp ends. (**a**) Transmission electron microscopy (TEM) images; (**b**) Current density (*J*)-applied electric field (*E*) curves; (**c**) Fowler–Nordheim plots; (**d**) stability of the emission current density under a constant electric field of 6.0 V μm^{-1} (reprinted from [74] with permission, Copyright—Elsevier B.V., 2017).

Besides sample geometry, intrinsic electric conductivity in ZnO material also influences the field emission properties. Using a simple solution reduction method, oxygen-deficient ZnO nanorod arrays were synthesized by Su et al. [75]. The concentration of oxygen vacancies can be effectively controlled by adjusting the reduction temperature ranging from 30 to 110 °C, resulting in a controlled tailoring of the band structure of the ZnO. The final oxygen-deficient ZnO nanorod arrays with optimized topography show excellent field emission properties, the threshold electronic field was as low as 0.67 V μm^{-1}, the field enhancement factor was as large as 64,601, and the stability was also favorable. In addition, doping with metal [76] or non-metal elements [77] through facile solution phase methods is also an effective method to improve the field emission properties of ZnO nanostructures. Doping induced conductivity enhancement and electron increase in the conduction band are the possible reasons for the emission properties improvement.

3.3. Electrochemical Sensors

Continuous monitoring of biological molecules and metal ions has attracted much interest due to the significant use in biotechnology, medicines, food and processing industry, and as a valuable biological marker for many oxidative biological reactions. In this regard, electrochemical sensors show unique advantages of high sensitivity, wide range of detection, real-time monitoring, ease of fabrication and control, reproducibility, and low cost, which can not be simultaneously achieved by other techniques, such as radioisotope tracing and nuclear magnetic resonance. The principle of electrochemical sensors is based on electroanalytical chemistry techniques in which quantitative investigating sensing is made by varying the potential and measuring the resulting current as an analyte reacts electrochemically with the working electrodes surface (nanostructures modified glass carbon electrode, (GCE), Figure 10). The frequently used electrochemical techniques employed in sensors include cyclic voltammetry (CV), linear scan voltammograms (LSV), differential pulse voltammetry (DPV), electrochemical impedance spectroscopy (EIS), and so on.

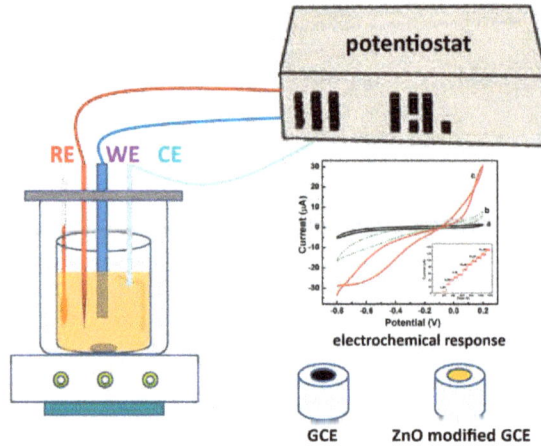

Figure 10. Schematic illustration of the electrochemical sensor testing.

Thanks to the biocompatible and nontoxic nature and high sensitivity to chemical species, ZnO nanostructures have been intensively studied as different kinds of electrochemical sensors to monitor important biological molecules and metal ions in organism, typically including glucose, dopamine, uric acid, L-lactic acid, L-Cysteine, hydrogen peroxide, potassium, sodium, calcium, magnesium, and iron ions. Due to the large surface area and porous nature, 3D ZnO hierarchical nanostructures based electrochemical sensors generally show good sensitivity, low detection limit, long-term stability, and repeatability.

Yang et al. [78] prepared ordered single-crystal ZnO nanotube arrays on indium-doped tin oxide coated glass by combining electrochemical deposition and subsequent chemical etching methods. The samples were used as a working electrode to fabricate an enzyme-based glucose biosensor, which exhibited high sensitivity of 30.85 μA cm^{-2} mM^{-1} at an applied potential of +0.8 V vs. saturated calomel electrode (SCE), wide linear calibration ranges from 10 μM to 4.2 mM, and a low limit of detection at 10 μM for sensing of glucose. We studied electrochemical sensing of hydrogen peroxide by using noble metal nanoparticle-functionalized ZnO nanoflowers. Firstly, the hierarchical flower-like ZnO structures were synthesized by a co-precipitation method in a solution containing Zn $(NO_3)_2 \cdot 6H_2O$ and KOH. Au or Ag nanoparticles were decorated on the surface of ZnO nanoflowers by subsequent hydrothermal treatment. Au/ZnO, Ag/ZnO and bare ZnO nanostructures modified GCE were fabricated and used as H_2O_2 sensors. The electrodes were tested in 0.05 M Phosphate buffered saline at pH = 7.2 with a platinum counter electrode and a saturated calomel electrode (SCE) reference electrode. CV results show that the electrochemical oxidation of H_2O_2 started at about −0.68 to −0.1 V versus SCE (Figure 11a), and the CV response for Ag/ZnO electrode was much higher than Au/ZnO and bars ZnO electrodes. In addition, the Ag/ZnO electrode also exhibited rapid and sensitive response to the change in concentration of H_2O_2 and the amperometric current is noticeably increased upon successive addition of H_2O_2 (Figure 11b). The linear range of calibration curve for these modified electrodes was from 1 to 20 μM (correction factor, R = −0.998) with a low limit of detection (LOD) of about −2.5 μM (Figure 11c). The sensitivity of the H_2O_2 sensor for Ag/ZnO modified electrode is 50.8 μA cm^{-2} μM^{-1}, which is much higher than that of Au/ZnO and bare ZnO electrodes. Stability test showed that Ag/ZnO modified electrode was more stable as compared to Au/ZnO and bare ZnO showing higher value of current with steady state current loss of 1.5% after 300 s (Figure 11d). This work demonstrated that noble metal-integrated ZnO nanostructures provided a new platform for applications in designing enzymeless biosensors. By decorating ZnO nanostructures with optimized alloy clusters, the electrochemical activity can be further improved [79].

Due to the mild condition of solution phase synthesis, ZnO nanostructures can be directly grown on a wide range of flexible and conductive substrates, making it possible to fabricate free-standing and flexible electrochemical sensors. For example, ZnO nanorods were uniformly anchored on the surface of carbon cloth by a simple hydrothermal method [80]. The product was directly used as an electrode for the simultaneous determination of dihydroxybenzene isomers. The electrodes showed good electrochemical stability, high sensitivity, and high selectivity. The linear ranges of concentration for hydroquinone, catechol, and resorcinol were 2–30, 2–45, and 2–385 µM, respectively, and the corresponding limits of detection (S/N = 3) were 0.57, 0.81, and 7.2 µM.

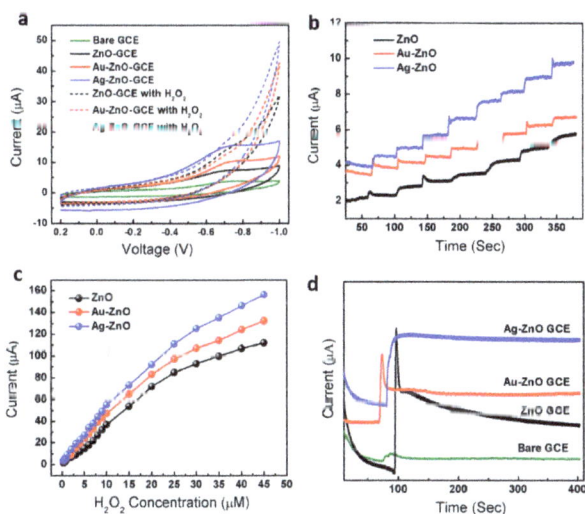

Figure 11. (**a**) cyclic voltammograms of bare and modified GCE with pure ZnO, Au/ZnO, and Ag/ZnO in the absence of H_2O_2 and in the presence of H_2O_2; (**b**) amperometric response of three modified GCE at constant voltage of −0.45 V with successive addition of 1 µM H_2O_2 in 0.05 M PBS under stirring; (**c**) corresponding calibration curves of the three modified electrodes and (**d**) stability plot of the three modified GCE at constant potential of −0.45 V in the presence of 1 µM H_2O_2 (reprinted from [81] with permission, Copyright—Springer Science+Business Media Dordrecht, 2016).

3.4. Lithium Ion Batteries

With a growing world population and increasing industrialization, energy and environment become the two main factors that restrict the society sustainability. It is thus of urgent need to develop renewable energy conversion and storage techniques. Lithium ion batteries are one of the most important energy storage devices, which dominate the market of portable electronic devices, and also show potential in hybrid/electric vehicles. A typical lithium ion batteries system mainly includes anode, cathode, electrolyte, and separator, of which active materials used in both electrodes play pivotal roles in determining the overall performance of batteries. For example, the specific capacity of graphite anode in current commercial lithium ion batteries is as low as 372 mA·h·g^{-1}, which is insufficient for many applications. Searching for high-performance electrode materials remains one of the most important focuses in the battery community. Among many potential electrode candidates, ZnO nanostructures have attracted much attention due to the abundance of raw materials, environmental benignity as well as facile synthesis [82–86]. In principle, the reaction between lithium and ZnO anodes occurs through the so-called mechanism of "conversion reaction". During lithiation, the ZnO anode undergoes a conversion reaction to form Li_2O embedded with nanosized metallic zinc clusters. This step is followed by an alloying reaction between lithium and the formed Zn NPs. The reaction processes are described as the following equations:

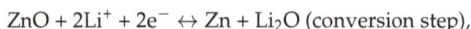

$$ZnO + 2Li^+ + 2e^- \leftrightarrow Zn + Li_2O \text{ (conversion step)},$$

$$Zn + Li^+ + e^- \leftrightarrow LiZn \text{ (alloying step)},$$

which yields a higher theoretical capacity (987 mA·h·g^{-1}) than that of graphite [87–89]. However, the practical use of ZnO based anodes mainly suffer from low Columbic efficiency (especially in the first cycle), severe capacity fading, and poor electrochemical kinetics [90,91]. Firstly, the conversion step in the lithiation reaction represents the largely irreversible reduction process of ZnO to metallic Zn. This irreversible chemical transformation is partly responsible for the large initial irreversible capacity loss. Secondly, the alloying step is accompanied by a large volume change (~228%) upon cycling, which results in material pulverization, electrode failure, and thus rapid capacity fading. The volume change of the anodes also results in the formation of an unstable solid-electrolyte interphase (SEI) layer, which continuously traps Li ions, leading to capacity loss. Thirdly, the low intrinsic electronic conductivity of ZnO materials causes a moderate lithium ion diffusion coefficient and limits the high-rate applications. To enhance the lithium storage properties of ZnO anodes, the construction of 3D ZnO hierarchical nanostructures with proper morphology, composition, and assembly has been proven to be an effective approach to overcome the above limitations. (1) Capacity—Compared to the corresponding nanobuilding blocks, hierarchical structures possess larger surface area, which increases the contact area between electrode and electrolyte and thus the number of active sites for electrode reactions with lithium ions. In addition, the hierarchical electrodes can lead to new lithium storage mechanisms, such as surface, interface, and nanopore storage, which lead to excess capacity. (2) Stability—The low-dimensional ZnO nanobuilding blocks have high mechanical strength, more resistance to mechanical damage, and can be engineered to allow volume change, and the assembled hierarchical structures can also prevent the possible agglomeration during the continuous cycling. Both are essential to ensure the structural integrity of the electrodes and long-term stability. (3) Rate performance—The rate of battery operation is related to the solid-state diffusion of lithium ions in the electrodes, which can be reduced in the nanoscale electrodes.

The assembly of ZnO hierarchical nanostructures shows great influence on the battery performance. Zhang et al. [92] synthesized ZnO nanostructures with different morphology by a facile hydrothermal and subsequent annealing treatment. The ZnO particles anode delivers the largest initial discharge capacity of 1815.8 mA·h·g^{-1}, and a reversible charge capacity of 870.0 mA·h·g^{-1} at the current density of 50 mA·g^{-1}, while cabbage-like ZnO nanosheets' electrodes displays better cycling stability. In other work, ZnO nanorod arrays with dandelion-like morphology were grown on copper substrates by a hydrothermal synthesis method [81]. The samples can be directly used as electrodes without any additives or binders. Cycling performance was performed at a current density of 0.1 mA cm^{-2}. The charge capacity of the dandelion-like ZnO electrode decreases to 596, 481 and 419 mA·h·g^{-1} in the second, third and fifth cycles, respectively. The ZnO arrays keep a capacity larger than 310 mA·h·g^{-1} even after 40 cycles, which is about four times higher than the stabilized capacity of the bulk ZnO electrode. The unique dandelion-like binary-structure played an important role in the electrochemical performance of the array electrodes.

Besides the architecture design, the electrochemical properties of ZnO anodes can further be improved by composting with electronically conductive agents (such as carbon nanofibers, carbon nanotubes, graphene, metals, metal compounds and so on) [93–96]. Those additives can not only enhance the conductivity of the electrodes but also modify the chemistry at the electrode/electrolyte interface. Therefore, 3D ZnO hierarchical nanostructures with suitable surface or interface composition modification show unique advantageous as improved lithium storage properties. In our previous studies, we synthesized hierarchical flower-like ZnO nanostructure by a facile solution phase approach. Au nanoparticles were functionalized on the surface of ZnO by subsequent electrochemical deposition treatment. The diameter of the pristine ZnO microflower is about 6–10 µm, and the length of an individual nanoneedle varies by 2–3 µm (Figure 12a). After electrodeposition, Au nanoparticles with an average diameter of 4–6 nm are decorated on the surface of each ZnO nanoneedle (Figure 12b,c). Comparing to the bare ZnO material, the Au-ZnO hybrid hierarchical structures possess large specific surface area, abundant void spaces, stable structure and strong electronic interaction between Au nanoparticles and ZnO. Those structural characters are beneficial for lithium storage enhancement. The initial discharge and charge capacity of Au-ZnO electrode are 1280 and 660 mA·h·g^{-1}, respectively, yielding a Coulombic efficiency of 79% (Figure 12d). In comparison, the initial discharge and charge capacity of pure ZnO electrode are 958 and 590 mA·h·g^{-1}, respectively (Figure 12e). The initial Coulombic efficiency of ZnO electrode is 52%, which is 27% lower than that of the Au–ZnO hybrids electrode. The stability test results show that the charge capacity of the Au–ZnO electrode decreases to 519 and 485 mA·h·g^{-1} after the second and third cycle, and stabilizes at 392 mA·h·g^{-1} after 50 cycles (Figure 12f). In contrast, the capacity of the ZnO electrode decays rapidly to 252 mA·h·g^{-1} (Figure 11f). The better lithium storage properties, including improved capacity and cycle life of the Au–ZnO electrode, can be attributed to the Au nanoparticles, which act as good electronic conductors and serve as a good catalyst during the lithiation/delithiation process. Due to the strong electronic interaction between Au nanoparticles and ZnO, electrons can easily reach all the positions where lithium ions' intercalation takes place. This feature is very important when the battery is cycled at high current density.

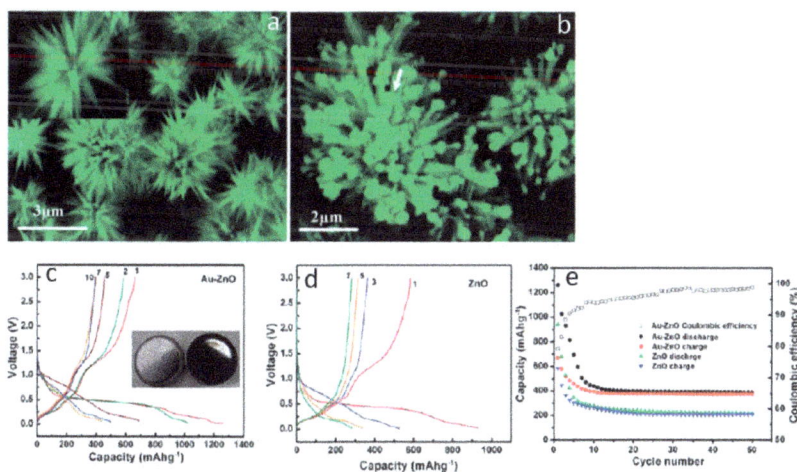

Figure 12. (**a,b**) Electron microscopy images and (**c–e**) lithium storage properties of the ZnO needle flowers and the Au nanoparticles/ZnO needle flowers composite (reprinted from [65] with permission, Copyright—The Royal Society of Chemistry, 2011).

Anchoring ZnO nanostructures on various carbon materials, such as graphite, mesoporous carbon, mesoporous carbon bubble, hierarchical porous carbon, vertically aligned graphene, and graphene aerogels, also facilitates the electron and lithium ion transport during charge/discharge cycles [97–99]. Those composite anodes show improved lithium storage properties, especially high rate capacity, which are highly dependent on the strong interaction between ZnO and carbon nanostructures. To strengthen this adhesion, one strategy is in situ formation carbon modification and ZnO nanostructures from metal organic compounds containing carbon and zinc elements. For example, Zhang and co-workers designed a facile and scalable strategy to synthesize integrated, binder-free, and lightweight ZnO nanoarray-based electrode as shown in Figure 13a [100]. Firstly, aligned and ordered ZnO nanorods were grown on carbon cloth via a low temperature solution deposition method. The ZnO nanorods were then served as the template as well as the Zn source to induce the formation of zeolitic imidazolate frameworks-8 (ZIF-8), a typical metal-organic framework, on the surface of ZnO nanorods (Figure 13b). Finally, unique ZnO@ZnO QDs/C core–shell nanorod arrays were obtained by thermal treatment in N_2. Structure studies show that the shell of each nanorod is constituted by amorphous carbon framework and ultrafine ZnO quantum dots (Figure 13c), resulting in a stronger adhesive force between carbon and the active ZnO materials, which is important in order to accelerate the charge transfer in the electrode. Benefitting from this structure design, the resultant ZnO@ZnO QDs/C anode not only exhibits remarkable cycling performance (Figure 13d), but also provides a remarkable rate capability, i.e., a reversible capability of 1055, 913, 762, 591, and 530 mA·h·g^{-1} at the current density of 100, 200, 400, 800, and 1000 mA·g^{-1} (Figure 13e).

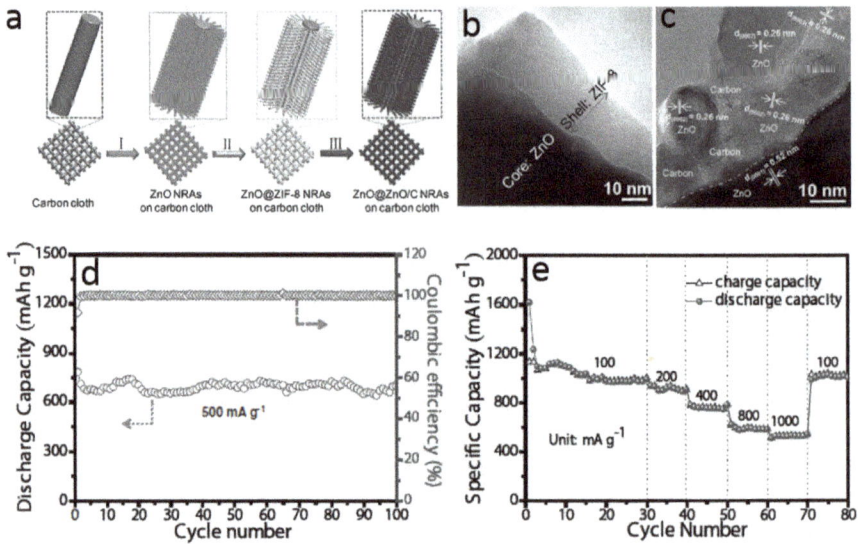

Figure 13. (**a**) Schematic illustrating the synthesis procedures of ZnO@ZnO QDs/C core-shell nanorod arrays on carbon cloth; (**b**,**c**) TEM and high-resolution transmission electron microscopy (HRTEM) images; (**d**,**e**) lithium storage properties of ZnO@ZnO QDs/C core-shell structures (reprinted from [100] with permission, Copyright—John Wiley and Sons, 2015).

It is worth mentioning that the ZnO nanostructures can also be used as anodes for sodium ion batteries [101,102], which are important complementarities of current lithium ion batteries. Due to the fact that sodium chemistry is similar to the case of lithium, the established electrode-design strategies for ZnO materials in lithium ion batteries system can be transferred to and expedite the sodium ion battery studies.

4. Conclusions

In recent years, there have been explosive research and development efforts on ZnO materials, ranging from facile synthesis to advanced characterizations and device applications. 3D ZnO hierarchical nanomaterials possess a high surface area with porous structures, and facilitate multiple physical and chemical processes. In addition, the hierarchical materials not only inherit the excellent properties of an individual nanostructure but also generate new properties due to the interactions between the nano building blocks. Therefore, 3D ZnO hierarchical nanostructures provide a wide range of applications. This review article summarized the main progress in the synthesis of 3D ZnO hierarchical nanostructures via different solution phase methods, such as precipitation, microemulsions, hydrothermal/solvothermal, sol-gel, electrochemical deposition, and chemical bath deposition. In each method, the synthesis principle, factors that affect the final structure and morphology, and typical examples are briefly discussed. Then, the applications of the 3D hierarchical architectures in the fields of photocatalysis, field emission, electrochemical sensors, and lithium ion batteries are analyzed, especially the effect of hierarchical morphology on the performance are evaluated. Despite these impressive advances, several challenges still remain.

(1)   Although great success has been made on the controllable synthesis of 3D ZnO hierarchical architectures, there is still room for improvement in terms of quality and scale of the products. Moreover, new synthesis methods also provide opportunities to explore novel morphology and understand the formation mechanism of the nanostructures.

(2)   Besides the hierarchical architectures, the device performance is also related to the size, composition and defects of ZnO nanostructures. Coupling the 3D nanostructures with the ability of fine control over geometry and chemistry can further optimize the chemical and physical properties.

(3)   Direct study of the dynamic morphological and chemical evolution of ZnO nanostructures during the practical applications is of significance to study the performance degradation and develop strategies to improve the stability.

Acknowledgments: This work was financially supported by the National Natural Science Foundation of China (Grants No. 51401114, 51701063).

Author Contributions: Hongyu Sun and Xiaoliang Wang wrote the first draft of the manuscript. Editing and revising were carried out by all the authors.

Conflicts of Interest: The authors declare no conflict of interest.

References

1.   Sun, Y.; Chen, L.; Bao, Y.; Zhang, Y.; Wang, J.; Fu, M.; Wu, J.; Ye, D. The applications of morphology controlled ZnO in catalysis. *Catalysts* **2016**, *6*, 188. [CrossRef]

2.   Wang, Z.L. ZnO nanowire and nanobelt platform for nanotechnology. *Mater. Sci. Eng. R* **2009**, *64*, 33–71. [CrossRef]

3.   Baruah, S.; Dutta, J. Hydrothermal growth of ZnO nanostructures. *Sci. Technol. Adv. Mater.* **2009**, *10*, 013001. [CrossRef] [PubMed]

4.   Tang, C.; Spencer, M.J.S.; Barnard, A.S. Activity of ZnO polar surfaces: An insight from surface energies. *Phys. Chem. Chem. Phys.* **2014**, *16*, 22139–22144. [CrossRef] [PubMed]

5.   Wang, Z.L. Nanostructures of zinc oxide. *Mater. Today* **2004**, *7*, 26–33. [CrossRef]

6.   Xu, S.; Wang, Z.L. One-dimensional ZnO nanostructures: Solution growth and functional properties. *Nano Res.* **2011**, *4*, 1013–1098. [CrossRef]

7.   Burke-Govey, C.P.; Plank, N.O.V. Review of hydrothermal ZnO nanowires: Toward FET applications. *J. Vac. Sci. Technol. B* **2013**, *31*, 06F101. [CrossRef]

8.   Sun, H.; Yu, Y.; Luo, J.; Ahmad, M.; Zhu, J. Morphology-controlled synthesis of ZnO 3D hierarchical structures and their photocatalytic performance. *CrystEngComm* **2012**, *14*, 8626–8632. [CrossRef]

9.   Kołodziejczak-Radzimska, A.; Jesionowski, T. Zinc oxide—From synthesis to application: A review. *Materials* **2014**, *7*, 2833–2881. [CrossRef] [PubMed]

10.   Zhang, Y.; Ram, M.K.; Stefanakos, E.K.; Goswami, D.Y. Synthesis, characterization, and applications of ZnO nanowires. *J. Nanomater.* **2012**, *2012*, 624520. [CrossRef]

11.   Naveed Ul Haq, A.; Nadhman, A.; Ullah, I.; Mustafa, G.; Yasinzai, M.; Khan, I. Synthesis Approaches of Zinc Oxide Nanoparticles: The Dilemma of Ecotoxicity. *J. Nanomater.* **2017**, *2017*, 8510342. [CrossRef]

12.   Cargnello, M.; Gordon, T.R.; Murray, C.B. Solution-phase synthesis of titanium dioxide nanoparticles and nanocrystals. *Chem. Rev.* **2014**, *114*, 9319–9345. [CrossRef] [PubMed]

13.   Znaidi, L. Sol-gel-deposited ZnO thin films: A review. *Mater. Sci. Eng. B* **2010**, *174*, 18–30. [CrossRef]

14.   Król, A.; Pomastowski, P.; Rafińska, K.; Railean-Plugaru, V.; Buszewski, B. Zinc oxide nanoparticles: Synthesis, antiseptic activity and toxicity mechanism. *Adv. Colloid Interface Sci.* **2017**. [CrossRef] [PubMed]

15. Kołodziejczak-Radzimska, A.; Jesionowski, T.; Krysztafkiewicz, A. Obtaining zinc oxide from aqueous solutions of KOH and Zn(CH₃COO)₂. *Physicochem. Probl. Miner. Process.* **2010**, *44*, 93–102.

16. Sepulveda-Guzman, S.; Reeja-Jayan, B.; de la Rosa, E.; Torres-Castro, A.; Gonzalez-Gonzalez, V.; Jose-Yacaman, M. Synthesis of assembled ZnO structures by precipitation method in aqueous media. *Mater. Chem. Phys.* **2009**, *115*, 172–178. [CrossRef]

17. Oliveira, A.P.A.; Hochepied, J.-F.; Grillon, F.; Berger, M.-H. Controlled precipitation of zinc oxide particles at room temperature. *Chem. Mater.* **2003**, *15*, 3202–3207. [CrossRef]

18. López, F.A.; Cebriano, T.; García-Díaz, I.; Fernández, P.; Rodríguez, O.; Fernández, A.L. Synthesis and microstructural properties of zinc oxide nanoparticles prepared by selective leaching of zinc from spent alkaline batteries using ammoniacal ammonium carbonate. *J. Clean. Prod.* **2017**, *148*, 795–803. [CrossRef]

19. Xi, Y.; Song, J.; Xu, S.; Yang, R.; Gao, Z.; Hu, C.; Wang, Z.L. Growth of ZnO nanotube arrays and nanotube based piezoelectric nanogenerators. *J. Mater. Chem.* **2009**, *19*, 9260–9264. [CrossRef]

20. Mulvihill, M.J.; Ling, X.Y.; Henzie, J.; Yang, P. Anisotropic etching of silver nanoparticles for plasmonic structures capable of single-particle SERS. *J. Am. Chem. Soc.* **2010**, *132*, 268–274. [CrossRef] [PubMed]

21. Li, F.; Ding, Y.; Gao, P.; Xin, X.; Wang, Z.L. Single-crystal hexagonal disks and rings of ZnO: Low-temperature, large-scale synthesis and growth mechanism. *Angew. Chem. Int. Ed.* **2004**, *43*, 5238–5242. [CrossRef] [PubMed]

22. Tian, Z.R.; Voigt, J.A.; Liu, J.; Mckenzie, B.; Mcdermott, M.J.; Rodriguez, M.A.; Konishi, H.; Xu, H. Complex and oriented ZnO nanostructures. *Nat. Mater.* **2003**, *2*, 821–826. [CrossRef] [PubMed]

23. Mujtaba, J.; Sun, H.; Fang, F.; Ahmad, M.; Zhu, J. Fine control over the morphology and photocatalytic activity of 3D ZnO hierarchical nanostructures: Capping vs. etching. *RSC Adv.* **2015**, *5*, 56232–56238. [CrossRef]

24. Malik, M.A.; Wani, M.Y.; Hashim, M.A. Microemulsion method: A novel route to synthesize organic and inorganic nanomaterials. *Arabian J. Chem.* **2012**, *5*, 397–417. [CrossRef]

25. Lin, J.-C.; Lee, C.-P.; Ho, K.-C. Zinc oxide synthesis via a microemulsion technique: Morphology control with application to dye-sensitized solar cells. *J. Mater. Chem.* **2012**, *22*, 1270–1273. [CrossRef]

26. Jesionowski, T.; KołodziejczakRadzimska, A.; Ciesielczyk, F.; Sójka-Ledakowicz, J.; Olczyk, J.; Sielski, J. Synthesis of Zinc Oxide in an Emulsion System and its Deposition on PES Nonwoven Fabrics. *Fibres Text. East. Eur.* **2011**, *19*, 70–75.

27. Kołodziejczak-Radzimska, A.; Markiewicz, E.; Jesionowski, T. Structural Characterisation of ZnO Particles Obtained by the Emulsion Precipitation Method. *J. Nanomater.* **2012**, *2012*, 1–9. [CrossRef]

28. Demazeau, G. Solvothermal reactions: An original route for the synthesis of novel materials. *J. Mater. Sci.* **2008**, *43*, 2104–2114. [CrossRef]

29. Demazeau, G. Solvothermal processes: Definition, key factors governing the involved chemical reactions and new trends. *Z. Naturforsch. B* **2014**, *65*, 999–1006. [CrossRef]

30. Feng, S.; Xu, R. New materials in hydrothermal synthesis. *Acc. Chem. Res.* **2001**, *34*, 239–247. [CrossRef] [PubMed]

31. Ong, C.B.; Ng, L.Y.; Mohammad, A.W. A review of ZnO nanoparticles as solar photocatalysts: Synthesis, mechanisms and applications. *Renew. Sustain. Energy Rev.* **2018**, *81*, 536–551. [CrossRef]

32. Shi, W.; Song, S.; Zhang, H. Hydrothermal synthetic strategies of inorganic semiconducting nanostructures. *Chem. Soc. Rev.* **2013**, *42*, 5714–5743. [CrossRef] [PubMed]

33. Hayashi, H.; Hakuta, Y. Hydrothermal synthesis of metal oxide nanoparticles in supercritical water. *Materials* **2010**, *3*, 3794–3817. [CrossRef] [PubMed]

34. Yang, H.; Song, Y.; Li, L.; Ma, J.; Chen, D.; Mai, S.; Zhao, H. Large-scale growth of highly oriented ZnO nanorod arrays in the Zn-NH₃·H₂O hydrothermal system. *Cryst. Growth Des.* **2008**, *8*, 1039–1043. [CrossRef]

35. Ko, S.H.; Lee, D.; Kang, H.W.; Nam, K.H.; Yeo, J.Y.; Hong, S.J.; Grigoropoulos, C.P.; Sung, H.J. Nanoforest of hydrothermally grown hierarchical ZnO nanowires for a high efficiency dye-sensitized solar cell. *Nano Lett.* **2011**, *11*, 666–671. [CrossRef] [PubMed]

36. Joo, J.; Chow, B.Y.; Prakash, M.; Boyden, E.S.; Jacobson, J.M. Face-selective electrostatic control of hydrothermal zinc oxide nanowire synthesis. *Nat. Mater.* **2011**, *10*, 596–601. [CrossRef] [PubMed]

37. Wysokowski, M.; Motylenko, M.; Stöcker, H.; Bazhenov, V.V.; Langer, E.; Dobrowolska, A.; Czaczyk, K.; Galli, R.; Stelling, A.L.; Behm, T.; et al. An extreme biomimetic approach: Hydrothermal synthesis of β-chitin/ZnO nanostructured composites. *J. Mater. Chem. B* **2013**, *1*, 6469–6476. [CrossRef]

38. Van Thuan, D.; Nguyen, T.K.; Kim, S.-W.; Chung, J.S.; Hur, S.H.; Kim, E.J.; Hahn, S.H.; Wang, M. Chemical-hydrothermal synthesis of oval-shaped graphene/ZnO quantum hybrids and their photocatalytic performances. *Catal. Commun.* **2017**, *101*, 102–106. [CrossRef]

39. Zhao, X.; Lou, F.; Li, M.; Lou, X.; Li, Z.; Zhou, J. Sol-gel-based hydrothermal method for the synthesis of 3D flower-like ZnO microstructures composed of nanosheets for photocatalytic applications. *Ceram. Int.* **2014**, *40*, 5507–5514. [CrossRef]

40. Khan, M.F.; Ansari, A.H.; Hameedullah, M.; Ahmad, E.; Husain, F.M.; Zia, Q.; Baig, U.; Zaheer, M.R.; Alam, M.M.; Khan, A.M.; et al. Sol-gel synthesis of thorn-like ZnO nanoparticles endorsing mechanical stirring effect and their antimicrobial activities: Potential role as nano-antibiotics. *Sci. Rep.* **2016**, *6*, 27689. [CrossRef] [PubMed]

41. Tseng, Y.-K.; Chuang, M.-H.; Chen, Y.-C.; Wu, C.-H. Synthesis of 1D, 2D, and 3D ZnO polycrystalline nanostructures using the sol-gel method. *J. Nanotechnol.* **2012**, *2012*, 1–8. [CrossRef]

42. Skompska, M.; Zarębska, K. Electrodeposition of ZnO nanorod arrays on transparent conducting substrates A review. *Electrochim. Acta* **2014**, *127*, 467–488. [CrossRef]

43. Kumar, M.; Sasikumar, C. Electrodeposition of nanostructured ZnO thin film: A review. *Am. J. Mater. Sci. Eng.* **2014**, *2*, 18–23. [CrossRef]

44. Switzer, J.A.; Hodes, G. Electrodeposition and chemical bath deposition of functional nanomaterials. *MRS Bull.* **2010**, *35*, 743–750.

45. Bhattacharya, R. Chemical bath deposition, electrodeposition, and electroless deposition of semiconductors, superconductors, and oxide materials. In *Solution Processing of Inorganic Materials*; Mitzi, D.B., Ed.; John Wiley & Sons, Inc.: Hoboken, NJ, USA, 2008; pp. 199–237.

46. Choi, K.-S.; Jang, H.S.; McShane, C.M.; Read, C.G.; Seabold, J.A. Electrochemical synthesis of inorganic polycrystalline electrodes with controlled architectures. *MRS Bull.* **2010**, *35*, 753–760. [CrossRef]

47. Lincot, D. Solution growth of functional zinc oxide films and nanostructures. *MRS Bull.* **2010**, *35*, 778–789. [CrossRef]

48. Illy, B.N.; Cruickshank, A.C.; Schumann, S.; Da Campo, R.; Jones, T.S.; Heutz, S.; McLachlan, M.A.; McComb, D.W.; Riley, D.J.; Ryan, M.P. Electrodeposition of ZnO layers for photovoltaic applications: Controlling film thickness and orientation. *J. Mater. Chem.* **2011**, *21*, 12949–12957. [CrossRef]

49. Oekermann, T.; Yoshida, T.; Schlettwein, D.; Sugiura, T.; Minoura, H. Photoelectrochemical properties of ZnO/tetrasulfophthalocyanine hybrid thin films prepared by electrochemical self-assembly. *Phys. Chem. Chem. Phys.* **2001**, *3*, 3387–3392. [CrossRef]

50. Shi, Z.; Walker, A.V. Chemical Bath Deposition of ZnO on Functionalized self-assembled monolayers: Selective deposition and control of deposit morphology. *Langmuir* **2015**, *31*, 1421–1428. [CrossRef] [PubMed]

51. Hodes, G. Semiconductor and ceramic nanoparticle films deposited by chemical bath deposition. *Phys. Chem. Chem. Phys.* **2007**, *9*, 2181–2196. [CrossRef] [PubMed]

52. She, G.-W.; Zhang, X.-H.; Shi, W.-S.; Fan, X.; Chang, J.C.; Lee, C.-S.; Lee, S.-T.; Liu, C.-H. Controlled synthesis of oriented single-crystal ZnO nanotube arrays on transparent conductive substrates. *Appl. Phys. Lett.* **2008**, *92*, 053111. [CrossRef]

53. Xu, L.; Liao, Q.; Zhang, J.; Ai, X.; Xu, D. Single-crystalline ZnO nanotube arrays on conductive glass substrates by selective disolution of electrodeposited ZnO nanorods. *J. Phys. Chem. C* **2007**, *111*, 4549–4552. [CrossRef]

54. Hu, L.; Hu, Z.; Liu, C.; Yu, Z.; Cao, X.; Han, Y.; Jiao, S. Electrochemical assembly of ZnO architectures via deformation and coalescence of soft colloidal templates in reverse microemulsion. *RSC Adv.* **2014**, *4*, 24103–24109. [CrossRef]

55. Mirzaei, H.; Darroudi, M. Zinc oxide nanoparticles: Biological synthesis and biomedical applications. *Ceram. Int.* **2017**, *43*, 907–914. [CrossRef]

56. Tseng, Y.-H.; Lin, H.-Y.; Liu, M.-H.; Chen, Y.-F.; Mou, C.-Y. Biomimetic synthesis of nacrelike faceted mesocrystals of ZnO—Gelatin composite. *J. Phys. Chem. C* **2009**, *113*, 18053–18061. [CrossRef]

57. Morin, S.A.; Amos, F.F.; Jin, S. Biomimetic assembly of zinc oxide nanorods onto flexible polymers. *J. Am. Chem. Soc.* **2007**, *129*, 13776–13777. [CrossRef] [PubMed]

58. Zhai, B.; Huang, Y.-M. A review on recent progress in ZnO based photocatalysts. *Optoelectron. Mater.* **2016**, *1*, 22–36.

59. Xia, Y.; Wang, J.; Chen, R.; Zhou, D.; Xiang, L. A review on the fabrication of hierarchical ZnO nanostructures for photocatalysis application. *Crystals* **2016**, *6*, 148. [CrossRef]

60. Hezam, A.; Namratha, K.; Drmosh, Q.A.; Chandrashekar, B.N.; Sadasivuni, K.K.; Yamani, Z.H.; Cheng, C.; Byrappa, K. Heterogeneous growth mechanism of ZnO nanostructures and the effects of their morphology on optical and photocatalytic properties. *CrystEngComm* **2017**, *19*, 3299–3312. [CrossRef]

61. Wu, X.L.; Ding, S.H.; Peng, Y.; Xu, Q.; Lu, Y. Study of the photocatalytic activity of Na and Al-doped ZnO powders. *Ferroelectrics* **2013**, *455*, 90–96.

62. Kurbanov, S.; Yang, W.C.; Kang, T.W. Kelvin probe force microscopy of defects in ZnO nanocrystals associated with emission at 3.31 eV. *Appl. Phys. Express* **2011**, *4*, 021101. [CrossRef]

63. Mohd, A.M.A.; Julkapli, N.M.; Abd, H.S.B. Review on ZnO hybrid photocatalyst: Impact on photocatalytic activities of water pollutant degradation. *Rev. Inorg. Chem.* **2016**, *36*, 77–104.

64. Kumar, S.G.; Rao, K.S.R.K. Zinc oxide based photocatalysis: Tailoring surface-bulk structure and related interfacial charge carrier dynamics for better environmental applications. *RSC Adv.* **2015**, *5*, 3306–3351. [CrossRef]

65. Ahmad, M.; Shi, Y.; Nisar, A.; Sun, H.; Shen, W.; Wei, M.; Zhu, J. Synthesis of hierarchical flower-like ZnO nanostructures and their functionalization by Au nanoparticles for improved photocatalytic and high performance Li-ion battery anodes. *J. Mater. Chem.* **2011**, *21*, 7723–7729. [CrossRef]

66. Sui, M.; Gong, P.; Gu, X. Review on one-dimensional ZnO nanostructures for electron field emitters. *Front. Optoelectron.* **2013**, *6*, 386–412. [CrossRef]

67. Semet, V.; Adessi, C.; Capron, T.; Thien Binh, V. Electron emission from low surface barrier cathodes. *J. Vac. Sci. Technol. B* **2007**, *25*, 513–516. [CrossRef]

68. Xu, N.S.; Huq, S.E. Novel cold cathode materials and applications. *Mater. Sci. Eng. R* **2005**, *48*, 47–189. [CrossRef]

69. Cui, J.B.; Daghlian, C.P.; Gibson, U.J.; Püsche, R.; Geithner, P.; Ley, L. Low-temperature growth and field emission of ZnO nanowire arrays. *J. Appl. Phys.* **2005**, *97*, 044315. [CrossRef]

70. Liu, J.; Huang, X.; Li, Y.; Ji, X.; Li, Z.; He, X.; Sun, F. Vertically aligned 1D ZnO nanostructures on bulk alloy substrates: Direct solution synthesis, photoluminescence, and field emission. *J. Phys. Chem. C* **2007**, *111*, 4990–4997. [CrossRef]

71. Dev, A.; Kar, S.; Chakrabarti, S.; Chaudhuri, S. Optical and field emission properties of ZnO nanorod arrays synthesized on zinc foils by the solvothermal route. *Nanotechnology* **2006**, *17*, 1533–1540. [CrossRef]

72. Wei, A.; Sun, X.W.; Xu, C.X.; Dong, Z.L.; Yu, M.B.; Huang, W. Stable field emission from hydrothermally grown ZnO nanotubes. *Appl. Phys. Lett.* **2006**, *88*, 213102. [CrossRef]

73. Cao, B.; Cai, W.; Duan, G.; Li, Y.; Zhao, Q.; Yu, D. A template-free electrochemical deposition route to ZnO nanoneedle arrays and their optical and field emission properties. *Nanotechnology* **2005**, *16*, 2567–2574. [CrossRef]

74. Ma, H.; Qin, Z.; Wang, Z.; Ahmad, M.; Sun, H. Enhanced field emission of ZnO nanoneedle arrays via solution etching at room temperature. *Mater. Lett.* **2017**, *206*, 162–165. [CrossRef]

75. Su, X.-F.; Chen, J.-B.; He, R.-M.; Li, Y.; Wang, J.; Wang, C.-W. The preparation of oxygen-deficient ZnO nanorod arrays and their enhanced field emission. *Mater. Sci. Semicond. Process.* **2017**, *67*, 55–61. [CrossRef]

76. Mahmood, K.; Park, S.B.; Sung, H.J. Retracted Article: Enhanced photoluminescence, Raman spectra and field-emission behavior of indium-doped ZnO nanostructures. *J. Mater. Chem. C* **2013**, *1*, 3138–3149. [CrossRef]

77. Shao, D.; Gao, J.; Xin, G.; Wang, Y.; Li, L.; Shi, J.; Lian, J.; Koratkar, N.; Sawyer, S. Cl-doped ZnO nanowire arrays on 3D graphene foam with highly efficient field emission and photocatalytic properties. *Small* **2015**, *11*, 4785–4792. [CrossRef] [PubMed]

78. Yang, K.; She, G.-W.; Wang, H.; Ou, X.-M.; Zhang, X.-H.; Lee, C.-S.; Lee, S.-T. ZnO nanotube arrays as biosensors for glucose. *J. Phys. Chem. C* **2009**, *113*, 20169–20172. [CrossRef]

79. Cosentino, S.; Fiaschi, G.; Strano, V.; Hu, K.; Liao, T.-W.; Hemed, N.M.; Yadav, A.; Mirabella, S.; Grandjean, D.; Lievens, P.; et al. Role of Au_xPt_{1-x} clusters in the enhancement of the electrochemical activity of ZnO nanorod electrodes. *J. Phys. Chem. C* **2017**, *121*, 15644–15652. [CrossRef]

80. Meng, S.; Hong, Y.; Dai, Z.; Huang, W.; Dong, X. Simultaneous detection of dihydroxybenzene isomers with ZnO nanorod/carbon cloth electrodes. *ACS Appl. Mater. Interfaces* **2017**, *9*, 12453–12460. [CrossRef] [PubMed]

81. Hussain, M.; Sun, H.; Karim, S.; Nisar, A.; Khan, M.; Ul Haq, A.; Iqbal, M.; Ahmad, M. Noble metal nanoparticle-functionalized ZnO nanoflowers for photocatalytic degradation of RhB dye and electrochemical sensing of hydrogen peroxide. *J. Nanopart. Res.* **2016**, *18*, 95. [CrossRef]

82. Wang, H.; Pan, Q.; Cheng, Y.; Zhao, J.; Yin, G. Evaluation of ZnO nanorod arrays with dandelion-like morphology as negative electrodes for lithium-ion batteries. *Electrochim. Acta* **2009**, *54*, 2851–2855. [CrossRef]

83. Xiao, C.; Zhang, S.; Wang, S.; Xing, Y.; Lin, R.; Wei, X.; Wang, W. ZnO nanoparticles encapsulated in a 3D hierarchical carbon framework as anode for lithium ion battery. *Electrochim. Acta* **2016**, *189*, 245–251. [CrossRef]

84. Zhang, J.; Gu, P.; Xu, J.; Xue, H.; Pang, H. High performance of electrochemical lithium storage batteries: ZnO-based nanomaterials for lithium-ion and lithium-sulfur batteries. *Nanoscale* **2016**, *8*, 18578–18595. [CrossRef] [PubMed]

85. Sun, Y.; Liu, N.; Cui, Y. Promises and challenges of nanomaterials for lithium-based rechargeable batteries. *Nat. Energy* **2016**, *1*, 16071. [CrossRef]

86. Yu, S.-H.; Lee, S.H.; Lee, D.J.; Sung, Y.-E.; Hyeon, T. Conversion reaction-based oxide nanomaterials for lithium ion battery anodes. *Small* **2016**, *12*, 2146–2172. [CrossRef] [PubMed]

87. Xu, X.; Cao, K.; Wang, Y.; Jiao, L. 3D hierarchical porous ZnO/ZnCo$_2$O$_4$ nanosheets as high-rate anode material for lithium-ion batteries. *J. Mater. Chem. A* **2016**, *4*, 6042–6047. [CrossRef]

88. Chen, H.; Ding, L.-X.; Xiao, K.; Dai, S.; Wang, S.; Wang, H. Highly ordered ZnMnO$_3$ nanotube arrays from a "self-sacrificial" ZnO template as high-performance electrodes for lithium ion batteries. *J. Mater. Chem. A* **2016**, *4*, 16318–16323. [CrossRef]

89. Kim, J.; Hong, S.-A.; Yoo, J. Continuous synthesis of hierarchical porous ZnO microspheres in supercritical methanol and their enhanced electrochemical performance in lithium ion batteries. *Chem. Eng. J.* **2015**, *266*, 179–188. [CrossRef]

90. Chen, Y.; Du, N.; Zhang, H.; Yang, D. Porous Si@C coaxial nanotubes: Layer-by layer assembly on ZnO nanorod templates and application to lithium-ion batteries. *CrystEngComm* **2017**, *19*, 1220–1229. [CrossRef]

91. Yuan, C.; Cao, H.; Zhu, S.; Hua, H.; Hou, L. Core-shell ZnO/ZnFe$_2$O$_4$@C mesoporous nanospheres with enhanced lithium storage properties towards high-performance Li-ion batteries. *J. Mater. Chem. A* **2015**, *3*, 20389–20398. [CrossRef]

92. Zhang, L.; He, W.; Liu, Y.; Zheng, P.; Yuan, X.; Guo, S. Temperature effect on morphology and electrochemical properties of nanostructured ZnO as anode for lithium ion batteries. *IET Micro Nano Lett.* **2016**, *11*, 535–538. [CrossRef]

93. Lu, S.; Wang, H.; Zhou, J.; Wu, X.; Qin, W. Atomic layer deposition of ZnO on carbon black as nanostructured anode materials for high-performance lithium-ion batteries. *Nanoscale* **2017**, *9*, 1184–1192. [CrossRef] [PubMed]

94. Gao, S.; Fan, R.; Li, B.; Qiang, L.; Yang, Y. Porous carbon-coated ZnO nanoparticles derived from low carbon content formic acid-based Zn(II) metal-organic frameworks towards long cycle lithium-ion anode material. *Electrochim. Acta* **2016**, *215*, 171–178. [CrossRef]

95. Xie, Q.; Lin, L.; Ma, Y.; Zeng, D.; Yang, J.; Huang, J.; Wang, L.; Peng, D.-L. Synthesis of ZnO-Cu-C yolk-shell hybrid microspheres with enhanced electrochemical properties for lithium ion battery anodes. *Electrochim. Acta* **2017**, *226*, 79–88. [CrossRef]

96. Ahmad, M.; Yingying, S.; Sun, H.; Shen, W.; Zhu, J. SnO$_2$/ZnO composite structure for the lithium-ion battery electrode. *J. Solid State Chem.* **2012**, *196*, 326–331. [CrossRef]

97. Yang, X.; Xue, H.; Yang, Q.; Yuan, R.; Kang, W.; Lee, C.-S. Preparation of porous ZnO/ZnFe$_2$O$_4$ composite from metal organic frameworks and its applications for lithium ion batteries. *Chem. Eng. J.* **2017**, *308*, 340–346. [CrossRef]

98. Quartarone, E.; Dall'Asta, V.; Resmini, A.; Tealdi, C.; Tredici, I.G.; Tamburini, U.A.; Mustarelli, P. Graphite-coated ZnO nanosheets as high-capacity, highly stable, and binder-free anodes for lithium ion batteries. *J. Power Sources* **2016**, *320*, 314–321. [CrossRef]

99. Ge, X.; Li, Z.; Wang, C.; Yin, L. Metal-organic frameworks derived porous core/shell structured ZnO/ZnCo$_2$O$_4$/C hybrids as anodes for high-performance lithium-ion battery. *ACS Appl. Mater. Interfaces* **2015**, *7*, 26633–26642. [CrossRef] [PubMed]

100. Zhang, G.; Hou, S.; Zhang, H.; Zeng, W.; Yan, F.; Li, C.C.; Duan, H. High-performance and ultra-stable lithium-ion batteries based on MOF-derived ZnO@ZnO quantum dots/C core-shell nanorod arrays on a carbon cloth anode. *Adv. Mater.* **2015**, *27*, 2400–2405. [CrossRef] [PubMed]
101. Xu, F.; Li, Z.; Wu, L.; Meng, Q.; Xin, H.L.; Sun, J.; Ge, B.; Sun, L.; Zhu, Y. In situ TEM probing of crystallization form-dependent sodiation behavior in ZnO nanowires for sodium-ion batteries. *Nano Energy* **2016**, *30*, 771–779. [CrossRef]
102. Asayesh-Ardakani, H.; Yao, W.; Yuan, Y.; Nie, A.; Amine, K.; Lu, J.; Shahbazian-Yassar, R. In Situ TEM Investigation of ZnO Nanowires during Sodiation and Lithiation Cycling. *Small Methods* **2017**, *1*, 1700202. [CrossRef]

![materials logo] *materials*

MDPI

Article

Ultra-Fast Microwave Synthesis of ZnO Nanorods on Cellulose Substrates for UV Sensor Applications

Ana Pimentel *, Ana Samouco, Daniela Nunes, Andreia Araújo, Rodrigo Martins and Elvira Fortunato *

i3N/CENIMAT, Department of Materials Science, Faculty of Sciences and Technology,
Universidade NOVA de Lisboa, Campus de Caparica, 2829-516 Caparica, Portugal;
a.samouco@campus.fct.unl.pt (A.S.); daniela.gomes@fct.unl.pt (D.N.);
andreiajoiaraujo@hotmail.com (A.A.); rm@uninova.pt (R.M.)
*   Correspondence: acgp@campus.fct.unl.pt (A.P.); emf@fct.unl.pt (E.F.);
    Tel.: +351-21-294-8562 (A.P. & E.F.); Fax: +351-21-294-0558 (A.P. & E.F.)

Received: 31 October 2017; Accepted: 12 November 2017; Published: 15 November 2017

Abstract: In the present work, tracing and Whatman papers were used as substrates to grow zinc oxide (ZnO) nanostructures. Cellulose-based substrates are cost-efficient, highly sensitive and environmentally friendly. ZnO nanostructures with hexagonal structure were synthesized by hydrothermal under microwave irradiation using an ultrafast approach, that is, a fixed synthesis time of 10 min. The effect of synthesis temperature on ZnO nanostructures was investigated from 70 to 130 °C. An Ultra Violet (UV)/Ozone treatment directly to the ZnO seed layer prior to microwave assisted synthesis revealed expressive differences regarding formation of the ZnO nanostructures. Structural characterization of the microwave synthesized materials was carried out by scanning electron microscopy (SEM) and X-ray diffraction (XRD). The optical characterization has also been performed. The time resolved photocurrent of the devices in response to the UV turn on/off was investigated and it has been observed that the ZnO nanorod arrays grown on Whatman paper substrate present a responsivity 3 times superior than the ones grown on tracing paper. By using ZnO nanorods, the surface area-to-volume ratio will increase and will improve the sensor sensibility, making these types of materials good candidates for low cost and disposable UV sensors. The sensors were exposed to bending tests, proving their high stability, flexibility and adaptability to different surfaces.

Keywords: ZnO; nanorod substrates; microwave irradiation; UV sensors

---

1. Introduction

In the recent years, a huge effort has been made to produce materials that can be used in different applications, such as nanoelectronics, optoelectronics, photonics, gas sensors, solar cells, photocatalysis, lab-on-paper for rapid diagnostic tests and antibacterial applications using flexible, biodegradable and green substrates, like cellulosic fiber-based substrates. The use of cellulosic substrates in these types of applications bring some advantages as cellulose is the Earth major biopolymer being suitable for low-cost applications, besides being flexible, lightweight, biocompatible and biodegradable [1,2].

Regular paper is composed by cylindrical cellulosic fibers, with diameters ranging from 20 to 50 μm and lengths that can reach 2 to 5 mm [3]. The larger roughness and porosity of the surface are intrinsic barriers to the development of electronic devices on the surface of this type of substrates. Nevertheless, the development of some devices on cellulosic fibers based substrates have already been reported [1,4–10].

Zinc Oxide (ZnO) is an *n*-type semiconductor with a wide and direct band gap of about 3.37 eV and a large free exciton binding energy of 60 meV at room temperature which allows it to act as an efficient semiconductor material [11]. ZnO possesses unique electrical, optical, photocatalytic and antibacterial properties, also being a low-cost material with a high surface reactivity. The physical and chemical properties of ZnO nanomaterials vary as a function of size, shape, morphology and crystalline structures.

Many efforts have been devoted to the development of different ZnO nanostructures with improved properties. The shape and the aspect ratio of this nanostructures are key factors that greatly influence the electrical and optical properties of ZnO material. Different techniques, precursors and solvents can be used to prepare a vast variety of nanostructures. Thus, new green synthesis strategies are vital for the development of novel nanomaterials [6,12].

Hydrothermal synthesis is a simple method that allows the production of uniform and well distributed ZnO nanorod arrays, and is generally associated with conventional heating, which is known to be inefficient, besides being time and energy consuming. Microwave irradiation, on other hand is a low cost technology due to its unique features such as short reaction time thus energy saving, enhanced reaction selectivity, homogeneous volumetric heating and high reaction rate [13]. In fact, microwave synthesis has been successfully employed for several sources of nanomaterials, shortening the synthesis reaction time [6,8,14–18].

As it is largely known, to synthesize a continuous ZnO layer using the hydrothermal synthesis method, it is imperative to use a seed layer [19,20]. In this sense, the use of different techniques for surface treatments have been reported by some authors to improve the surface wettability and adhesion. Examples of techniques employed include plasma treatment [21,22], wet chemical [23] and UV/Ozone treatment [24,25]. The UV/Ozone treatment is a simple, inexpensive and low-temperature method that allows simultaneously the removal of some surface contaminants and the polarization of ZnO seed layer surface.

The direct growth of ZnO nanorods on paper substrates has so far been reported by very few authors [20,26], however to the best of the author's knowledge, this is the first time that ZnO nanorod arrays grown on different sources of paper substrates under microwave irradiation, with short synthesis time (10 min) and having UV/Ozone treatments inflicted to the seed layer have been reported.

Zinc oxide is frequently employed in a wide range of applications, from optoelectronics to biological fields [20,27–31]. Nevertheless, one of the most investigated areas for ZnO is UV/Ozone sensing, mainly due to its capacity of absorbing UV light, being transparent to visible light, presenting a high sensitivity and selectivity [32–34]. UV sensors are mostly used in UV source and environmental monitoring, space technologies, as well as in chemical and biological detection [35]. The use of ZnO nanostructures instead of thin films allows an increase in sensor responsivity and sensitivity due to a higher aspect ratio of length to diameter and higher surface area of nanostructures [36].

Many authors have reported the direct growth of ZnO nanorods to act as the photoactive layer in UV sensors on rigid (glass and silicon) [35,37] and flexible substrates (polymeric, textile and cellulosic substrates [26,36,38]). These reports showed responsivities ranging from 15 to 24 mA/W for glass and silicon substrates, while 0.022 µA/W was reported for other types of flexible substrates [35–37]. Nowadays, the scientific community seeks inexpensive, adaptable and flexible devices, which makes paper an appealing option for the next generation UV sensors devices.

The present work reports the synthesis of ZnO nanorods on tracing and Whatman papers by the hydrothermal synthesis method under microwave irradiation. The influence of synthesis temperature with a fixed time and the influence of UV/Ozone treatment on the ZnO seed layer have been studied. After an extensive structural, morphological and optical characterization of the synthesized nanorod arrays, the materials were tested as UV sensors.

2. Results and Discussion

2.1. Characterization of Paper Substrates: SEM, Thermal Analysis, XRD

As previously mentioned, the fact of cellulosic paper substrates present high roughness and porosity can be disadvantageous to the development of electronic devices. Nevertheless, two different types of substrates were chosen to investigate these effects, tracing paper and Whatman paper.

Prior to the seed layer deposition and growth of ZnO nanorod arrays, both pristine papers were fully characterized to understand their physical characteristics and limits (smoothness, impurities and temperature degradation).

Figure 1 shows the scanning electron microscopy (SEM) images and the 3D profilometry measurements showing the roughness profiles of tracing and Whatman paper substrates. In the case of the tracing paper, it is almost impossible to distinguish the cellulose fibers. The Whatman paper revealed a high-density structure of intertwined cellulose fibers with a cylindrical and flat shape in the micrometer range. Thus, it is possible to observe that the tracing paper substrate presents a smoother surface when compared to the Whatman paper. This assumption is confirmed by the surface 3D profilometry measurements (see Figure 1c,f), in which it is possible to observe that tracing paper presents a root mean square (RMS) of 3.9 μm, while Whatman paper has a rougher surface with a RMS value of 12.6 μm. The energy dispersive spectrometry (EDS) images (see Figure 1b,e) confirm the absence of calcium carbonate ($CaCO_3$) or any other contaminants.

Figure 1. Scanning electron microscopy (SEM) images of (**a**) tracing paper and (**d**) Whatman paper substrates; (**b,e**) SEM images (artificial colored) together with the corresponding X-ray maps of C and O species of tracing and Whatman papers respectively; (**c,f**) Surface 3D profilometry of tracing and Whatman paper, respectively.

To infer the maximum working temperature of tracing and Whatman paper substrates, differential scanning calorimetric (DSC) and thermogravimetry (TG) measurements were carried out. The results are presented on Figure 2a,b, for tracing and Whatman papers, respectively. It is well known that cellulosic fibers undergo for a rapid thermal degradation at low/moderate temperatures (below 400 °C). The thermal decomposition of cellulosic fibers consists on the degradation of several components that are on its composition—decomposition of hemicelluloses, pyrolysis of lignin, depolymerization of cellulose, active flaming combustion and finally char oxidation [39]. So, as mentioned by other authors [39–41], the decomposition of a cellulosic substrate can be divided in four steps. The first step occurs at temperatures between 40 and 120 °C and is related with the extraction of water or moistures presented on paper. The second step, usually is accompanied with a major mass loss, and can be correlated with the main degradation reaction of cellulose fibers due to depolymerization and carbonization of glycosyl units (being characterized by an endothermic peak approximately at 350 °C). The third step corresponds to the oxidation of the char produced with the fibers decomposition (this stage can be absent on some types of papers), occurring between 400 and 500 °C. The fourth stage usually occurs at temperatures above 630 °C and corresponds to the decomposition of calcium carbonate. The thermal decomposition of cellulosic fibers is greatly influenced by their structure and chemical composition. So as the chemical structure of the cellulosic fibers are arranged differently, they will decompose at different temperatures ranges and possess different decomposition profiles [39,41].

The DSC curve of tracing paper presents a small endothermic peak at 85 °C, accompanied by a small weight loss (about 6.40%), which corresponded to desorption or water evaporation from cellulose fibers. Between 280 and 400 °C an enhanced weight loss of about 60% is observed, correlated with two endothermic peaks at 295 and 368 °C. These two peaks can be associated to a stage of the decomposition step, corresponding to the oxidative decomposition of cellulose fibers. Relative to Whatman paper, although without any peaks present, a small mass loss is observed until 120 °C, corresponding to water evaporation. One endothermic peak at 336 °C is also detected, accompanied by one decomposition step, with a weight loss of approximately 80%. This peak is correlated to the thermal decomposition of cellulosic fibers.

So, by observing the results obtained, it is possible to ensure that the substrates can be heated up to 200 °C, without damage and without losing their properties (temperature at which the mass of the sample starts to decrease due to decomposition, indicating the maximum working temperature for this type of substrate).

The X-ray diffraction (XRD) results (see Figure 2c,d, for tracing and Whatman paper, respectively) show that both types of paper present the characteristic peaks of native cellulosic fibers: $(1\bar{1}0)$, (110), (200) and (004) at $2\theta = 14.9°$, $2\theta = 16.6°$, $2\theta = 22.7°$ and $2\theta = 35°$, respectively; which are in accordance with that reported in the literature [39,41]. Due to the high intensity of peak (200) is possible to conclude that both types of paper are highly crystalline. No impurities or other crystallographic phases were detected.

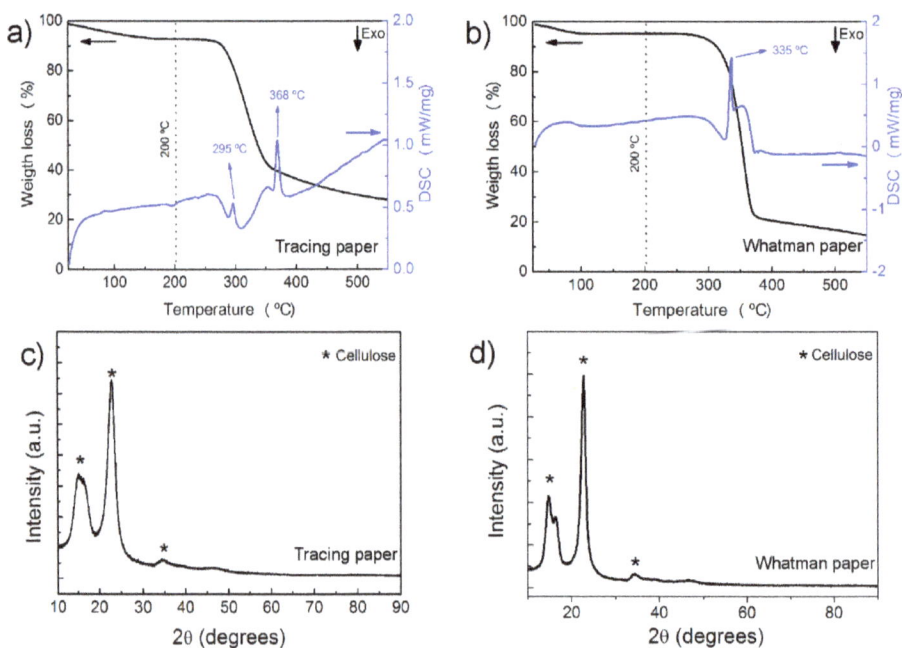

Figure 2. (**a**,**b**) Differential Scanning Calorimetry (DSC) and (**c**,**d**) X-ray diffraction (XRD) diffractograms of tracing and Whatman paper substrates, respectively.

The crystallinity index, based on the "Segal peak-height method" can be inferred by using the XRD data and calculating the ratio between intensity of the crystalline peak ($I_{002} - I_{AM}$) and the total intensity of peak (002) (I_{002}) [42,43]. It was estimated a crystallinity index of 78% and 89% for tracing and Whatman paper, respectively, indicating that Whatman paper have less amorphous fibres and more type 1 cellulose.

2.2. UV/Ozone ZnO Seed Layer Treatment

After the deposition of ZnO seed layer on both cellulosic substrates, an UV/Ozone treatment was used in order to improve the ZnO surface polarity. In Figure 3 it is possible to observe the XRD measurement and SEM images of the ZnO seed layer, deposited by sputtering technique. It is possible to observe that the ZnO seed layer is very smooth with a preferable orientation along the (002) crystallographic plane.

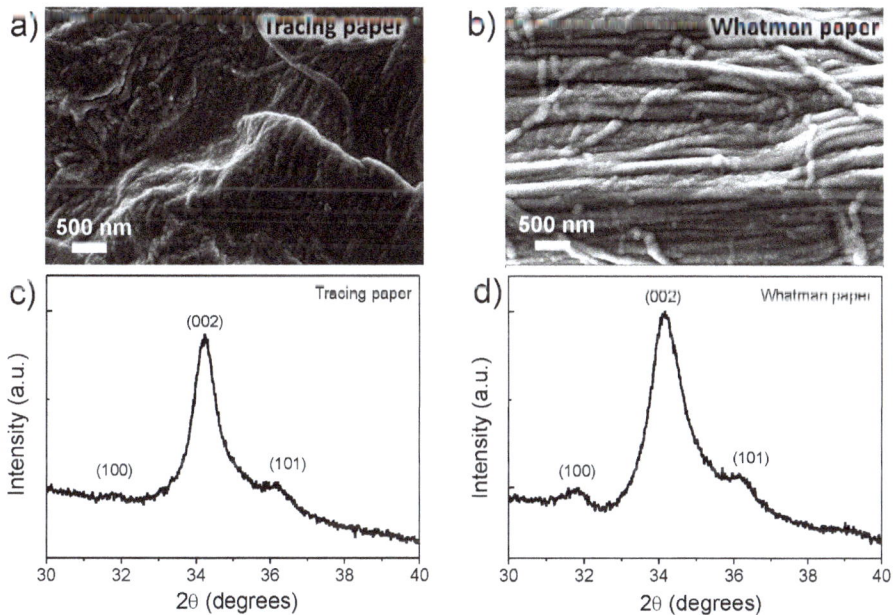

Figure 3. (**a**,**b**) SEM images and (**c**,**d**) XRD diffractograms of zinc oxide (ZnO) seed layer, deposited by sputtering technique on tracing and Whatman paper substrates, respectively.

The use of UV/Ozone systems for surface treatment have already been used by other authors in order to increase the surface oxygen, polarity and wettability [44,45]. This UV/Ozone system presents some advantages when compared with other systems: no vacuum is required (no need of any sophisticated apparatus), and the absence of a wet chemistry treatment gives the advantage of no residual solvents or other contaminants left at the substrate/sample surface; also, it can be used at room temperature [46].

ZnO is a crystal with a hexagonal structure that grows along the *c* axis and possesses both polar and nonpolar surfaces, arising from the anisotropy of the wurtzite structure. It presents high energy polar surfaces, with a Zn^{2+} terminate (0001) plane and an O^{2-} terminated (000$\bar{1}$) plane (this surfaces reconstruct to lower the surface energy) [45,47]. As reported by Talebian et al., the solvents play an important role on hydrothermal/solvothermal synthesis [48]. When synthesizing ZnO nanoparticles by the hydrothermal method, it will originate a very strong interaction between the polar terminate plane (0001) and (000$\bar{1}$) of ZnO surfaces. When a ZnO nucleus is formed, due to the high energy of polar surfaces, the incoming precursor molecules will tend to adsorb on the polar surface of the ZnO seed layer. After adsorbing one layer of precursor molecules, the polar surface will transform into another polar surface, but with an inverted polarity (Zn^{2+} terminated surface will change into O^{2-} terminated surface, or vice versa). This process will be repeated over time, promoting the increase of the rate of crystal growth perpendicular to this surface (in the c-direction) and exposing the non-polar ($1\bar{1}00$) and ($2\bar{1}\bar{1}0$) surfaces [47–49]. So, by exposing the ZnO seed layer to UV light, the O_2 adsorbed to the surface will be decomposed and the surface will become more polar, with a Zn terminate plane (0001), which will promote the growth of ZnO nanorods by hydrothermal synthesis method assisted by microwave irradiation. A more polar ZnO seed layer will improve the growth of ZnO nanorods with a direction perpendicular to the seed layer surface. Figure 4 shows a schematic of surface modification prior to ZnO nanorod arrays grown by hydrothermal method under microwave irradiation [50].

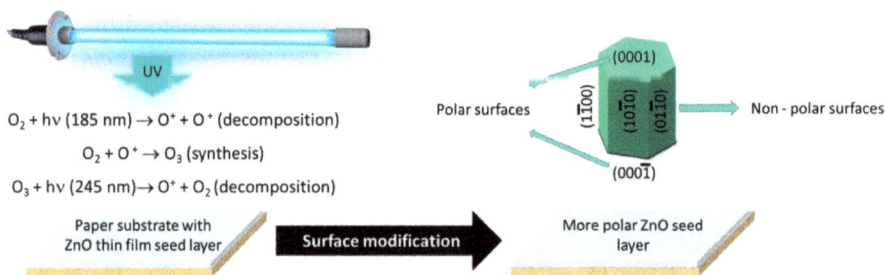

Figure 4. Schematic of surface treatment/modification of ZnO seed layer with an UV/Ozone system.

2.3. Crystallographic Structure and Morphology Analysis of ZnO Nanorods

To infer the crystallographic structure and the morphology of the synthesized ZnO nanorod arrays, XRD and SEM analysis were carried out to all materials produced on this study. The XRD diffractograms of the ZnO nanorods grown on cellulosic substrates are presented in Figure 5. For all of the materials produced, the observed peaks can be fully indexed to the hexagonal wurtzite ZnO structure, with lattice constants of a = b = 0.3296 nm and c = 0.52065 nm, which is in accordance with the literature [51]. The three peaks observed are fully assigned to the crystallographic planes (100), (002) and (101). The results confirm that it was possible to grow pure ZnO nanocrystals on cellulosic substrates (tracing and Whatman papers). Nevertheless, is possible to observe that the crystallinity of ZnO nanorods increases with the increase of synthesis temperature and with UV/Ozone treatments prior to synthesis. These results corroborate the previous assumption that the UV/Ozone treatment favor the growth of ZnO nanorods on the surface of cellulosic substrates.

On the tracing paper condition, a splitting of the XRD peak corresponding to the crystallographic plane (002) is observed, and it might be due to the peak doublet of K-alpha 1 and K-alpha 2 [52].

SEM analysis for ZnO nanorod arrays synthesized on tracing paper substrate is shown on Figure 6. The effect of temperature and UV treatment prior to microwave synthesis has been investigated. The synthesis time was constant for all samples (10 min). It is possible to observe that without UV treatment, the ZnO nanorods grow in an inhomogeneous way, with a non-uniform shape and size (regardless the synthesis temperature). Moreover, it is easily observed that the ZnO nanorods do not cover fully the substrate surface. With a UV treatment prior to ZnO nanorod array synthesis, the growth of these nanostructures becomes more homogeneous, covering all the substrate surface. For low temperature synthesis (70 °C), no ZnO nanoparticle growth is observed, but with the increase of synthesis temperature the ZnO nanorods start to cover all of the substrate surface, becoming higher in length and well aligned, especially in the case of the tracing paper substrate, mainly due to its very smooth surface. Thus, with the increase of temperature, it is possible to observe that the nanorods become higher and thicker, for synthesis with UV treatment. The ZnO nanorods present an average length of 130 nm, 300 nm and 500 nm for synthesis temperatures of 90, 110 and 130 °C, respectively. Regarding the thickness, they present an average value of 85, 100 and 110 nm for the same range of temperatures.

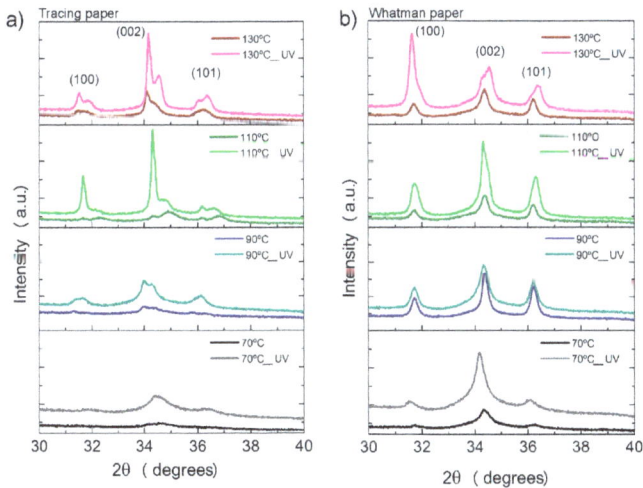

Figure 5. XRD diffractograms of ZnO nanorods arrays produced by an ultrafast hydrothermal method assisted by microwave irradiation, grown on (**a**) tracing paper substrate and (**b**) Whatman paper substrate. All samples were produced with a synthesis time of 10 min.

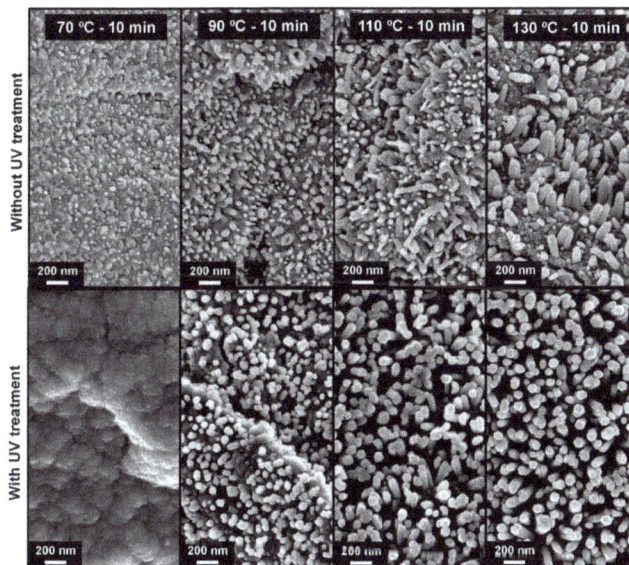

Figure 6. Surface SEM images of ZnO nanorods produced by hydrothermal method assisted by microwave irradiation on tracing paper, with a synthesis time of 10 min, with different synthesis temperatures (70, 90, 110 and 130 °C). On top are the images of samples produced without UV treatment and on the bottom samples produced with UV treatment.

Figure 7 shows the SEM images of the synthesis of ZnO nanorod arrays using Whatman paper as substrate. Also in this case the growth as a function of synthesis temperature is observed, with and without an UV treatment prior to synthesis. As observed in tracing paper, it is possible to see that the use of an UV treatment prior to ZnO growth will favor the growth of this type of nanostructures. Without UV treatment, the growth is very heterogeneous, not completely covering the substrate surface. With UV treatment, the ZnO nanorod arrays grew uniformly, covering all of the surface. By increasing the synthesis temperature, the ZnO nanorods increased their length, presenting average values of 120, 340 and 480 nm for synthesis temperatures of 90, 100 and 130 °C, respectively. Regarding the thickness, they present an average value of 55, 66 and 75 nm for the same range of temperatures.

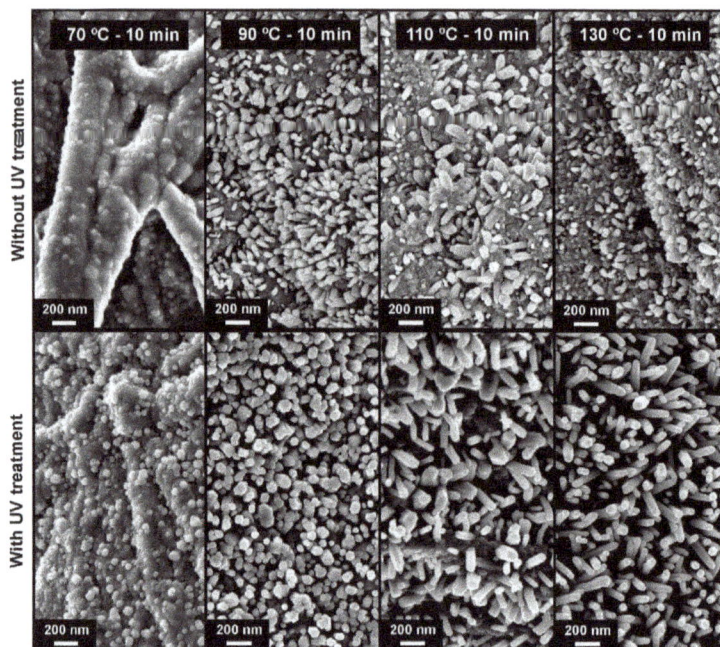

Figure 7. Surface SEM images of ZnO nanorods produced by hydrothermal method assisted by microwave irradiation on Whatman paper, with a synthesis time of 10 min, with different synthesis temperatures (70, 90, 110 and 130 °C). On top are the images of samples produced without UV treatment and on the bottom samples produced with UV treatment.

The misalignment of nanorods with the increase of temperature is due to the high roughness of the surface.

So, the top view SEM images (from both type of substrates) indicate that well-defined crystallographic planes of the hexagonal single crystalline ZnO nanorods can be identified and that they grow along the [0001] direction. Nevertheless, the growth in length is more expressive than in thickness, which implies that the growth rate must be along [0001] directions, and this latter direction is more temperature-sensitive when compared to those along [10$\bar{1}$1] and [10$\bar{1}$0] directions [53].

Comparing the two types of papers used, it is possible to conclude that tracing paper needs a higher temperature for ZnO nanorods start to nucleate at the surface. On Figure 8 is possible to observe a comparison between the length and the diameter of the ZnO nanorods as a function of temperature, for both type of cellulosic substrates, with the corresponding SEM images.

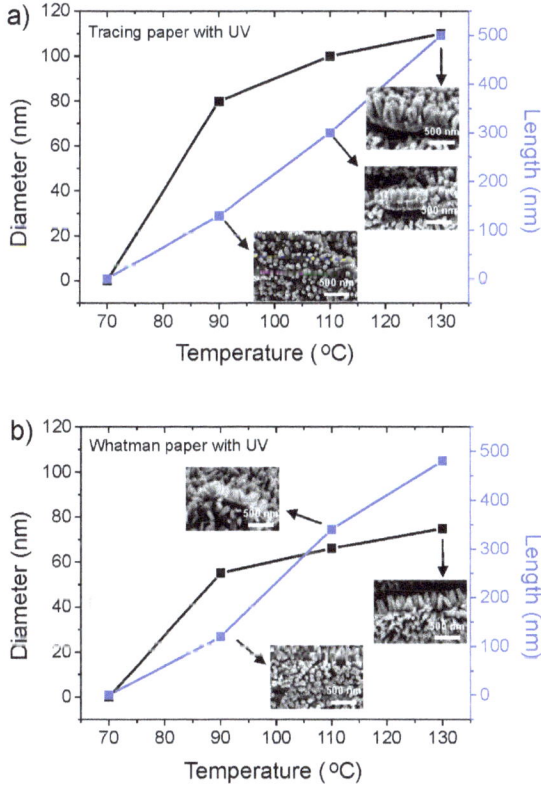

Figure 8. Behavior of length and diameter of the ZnO nanorods as a function of temperature for (**a**) tracing paper and (**b**) Whatman paper. The insets reveal the corresponding SEM images.

The small nanorods with and a relatively large length result in a high specific area, which is an important parameter in UV sensor application.

2.4. Optical Properties

The optical band gap of ZnO nanorods grown on cellulosic substrates was evaluated from reflectance results. It was applied the Tauc equation to reflectance values, for direct band semiconductors (see Equation (1)) [54]:

$$(\alpha h v)^m = A(h v - E_g) \tag{1}$$

where E_g is the material optical band gap, h is the Plank constant ($h = 4.135 \times 10^{-15}$ eV s), v is the frequency, α is the material absorption coefficient, m is a constant that depends on the type of the optical transition ($m = 2$ for allowed indirect transitions and $m = 1/2$ for allowed direct transition) and A is an energy-independent constant and.

Figure 9 shows the optical band gap calculation for ZnO nanorods synthesis with and without UV treatment and as a function of synthesis temperature. The optical band gap was calculated by extrapolating $(\alpha h v)^2$ as a function of $h v$. On the insets, it is possible to observe the reflectance behavior of each sample, that is, the ZnO nanorod arrays produced absorb almost all the light in the UV region of the spectra.

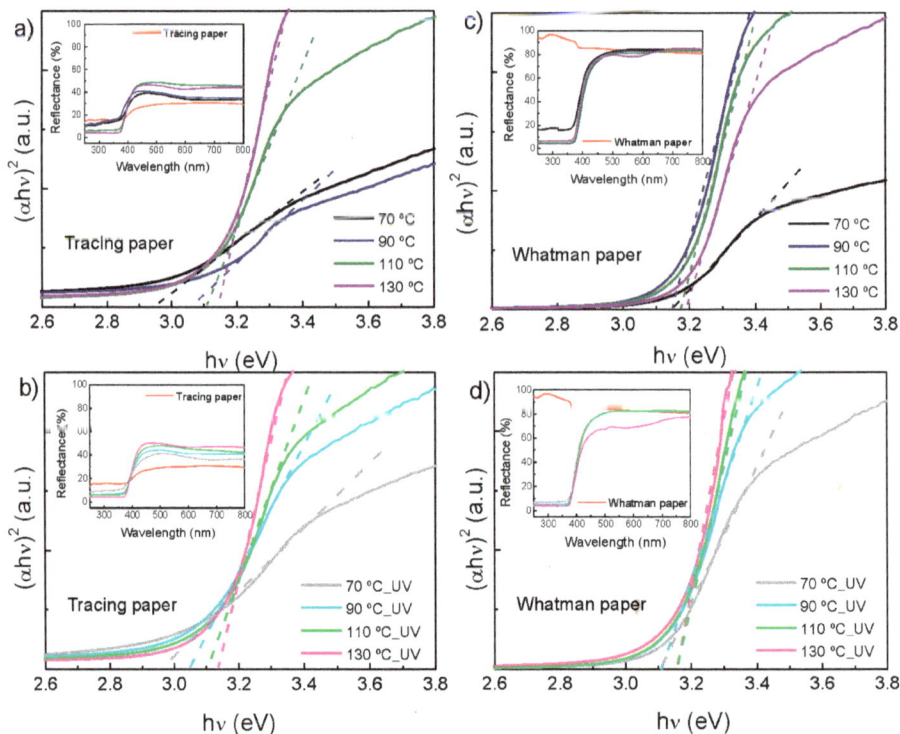

Figure 9. Absorbance spectra of ZnO nanorods arrays produced by an ultrafast hydrothermal method assisted by microwave irradiation, grown on (**a,b**) tracing paper substrate and (**c,d**) Whatman paper substrate, without and with UV/Ozone treatment. All samples were produced with a synthesis time of 10 min.

The estimated optical band gap values, calculated from the Tauc equation, are discriminated on Table 1. By observing the values obtained it is possible to see that for both types of cellulosic substrates the band gap value increased with the increase of synthesis temperature, with or without the use of UV treatment prior to ZnO nanorods synthesis. This result is probably due to the increase of temperature, a more homogeneous nanorod arrays cover all the substrate surface, thus the measured band gap value become closer to the theoretical value ZnO band gap of 3.2–3.4 eV [55–57].

Table 1. Optical band gap of ZnO nanorods, produced with different synthesis time and temperature, obtained by extrapolating $(\alpha h\nu)^2$ vs. $h\nu$.

Synthesis Temperature		70 °C	90 °C	110 °C	130 °C
Tracing paper	Without UV treatment	2.95 eV	3.07 eV	3.10 eV	3.14 eV
	With UV treatment	2.96 eV	3.05 eV	3.11 eV	3.14 eV
Whatman paper	Without UV treatment	3.15 eV	3.15 eV	3.18 eV	3.19 eV
	With UV treatment	3.10 eV	3.10 eV	3.16 eV	3.16 eV

It is also possible to conclude that the same range of values was obtained, regardless of the type of cellulosic substrate used or if a UV treatment was used prior to ZnO nanorod synthesis.

2.5. Application of ZnO Nanorods in Paper-Based UV Sensors

The ZnO nanorods arrays grown on cellulosic substrates were applied as UV sensors, and for that only one condition from each substrate has been selected. By observing the SEM images (see Figures 6 and 7) it was decided to test the ones with higher surface coverage and larger ZnO nanorods—Synthesis condition of 130 °C.

So, for ZnO nanorod UV sensor production, carbon interdigital contacts were screen printed on paper substrates after ZnO nanorods growth, like is exemplified in Figure 10. A polymeric mask was used with the interdigital contacts design that allow us to ensure that the contacts were equal for both the samples.

Carbon ink screen printing

Figure 10. Schematic of the carbon interdigital contacts deposited by inkjet printing on cellulosic substrates.

The samples were then subjected to UV radiation cycles, with a bias voltage of +10 V and on/off cycles of 2 min.

The sensing mechanism is based on the oxygen vacancies that exist on the surface and that will influence the properties of ZnO nanoparticles. In the dark, molecules containing high concentrations of O_2 are adsorbed at vacancy sites that accept electrons, which will be withdrawn and effectively depleted from the conduction band. This mechanism will lead to a conductivity reduction. When exposed to UV light, reducing molecules will react with the adsorbed oxygen, leaving behind an electron, inducing an increase in the electrical conductivity [19,58,59]. This phenomena can be described by the following equations [60]:

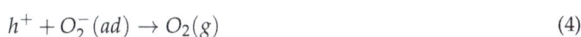

$$O_2(g) + e^- \rightarrow O_2^- \ (ad) \tag{2}$$

$$h\nu \rightarrow e^- + h^+ \tag{3}$$

$$h^+ + O_2^- (ad) \rightarrow O_2(g) \tag{4}$$

In Figure 11, it is possible to observe the time resolved photocurrent of ZnO nanorods paper UV sensor, produced for 10 min and 130 °C on tracing and Whatman substrates, in response to UV radiation turn on/off. The ZnO synthesized on Whatman paper substrates displayed enhanced sensing performance when compared to tracing paper substrate. Under the bias voltage of +10 V, the photocurrent exponential increase from 16.6 nA to 3.8 μA for tracing paper while for Whatman paper increased from 0.76 μA to 10.36 μA, within about 60 s, with saturation in the on-state. After the UV radiation was turned off, the current decreased to the initial value stage of current. The photocurrent of ZnO UV sensors was completely reproducible during several cycles of photocurrent switching. This behavior was obtained for both types of paper. The low values of photocurrent observed in both UV sensors may be related to the use of paper as a substrate. They are rough and present a high porosity. Moreover, after ZnO nanorods synthesis the tracing paper becomes slightly wrinkled, which may make its use as an UV sensor device difficult. These characteristics will make the current flow through the sample difficult. Regarding the nanorods' influence, no significant structural differences (in length and diameter) have been observed between the nanorod arrays produced in both paper substrates, confirming that the major contribution must come from the substrate.

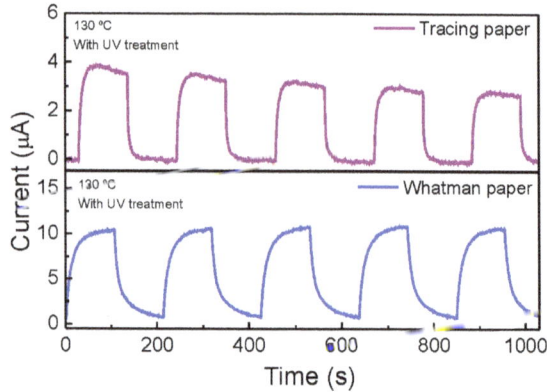

Figure 11. Cycling behavior of cellulosic ZnO nanorods UV sensors grown on tracing paper and on Whatman paper.

In order to determine the responsivity, *R*, of each UV ZnO paper sensor it was used the following equation [61]:

$$R = \frac{I_{ph} - I_{dark}}{P_{UV}} \tag{5}$$

where P_{UV} is the power of the UV source lamps, I_{ph} is the UV sensor photocurrent and I_{dark} is the UV sensor dark current. The obtained responsivity was 0.39 µA/W and 1.19 µA/W, for the ZnO nanorods on tracing paper and on Whatman paper, respectively. The responsivity was calculated taking into account the current value when the sensor reaches 95% of its stable value [62].

So, the ZnO nanorods UV sensor produced on Whatman substrate presents a photo response 3 times superior to the one produced in tracing substrate. It is expected to occur due the grain size effect. It is well known that the sensitivity of a nanostructured sensor is related to grain size, with the particle geometry, oxygen absorption and lattice defects. Smaller grain size will induce a higher sensor sensitivity due to an increase of the specific surface area and oxygen adsorption quantity [63]. In order to infer the grain size of the ZnO nanorods grown on tracing and Whatman paper substrates, it was used the Scherres's equation, D = 0.94 λ/β cosθ, where λ is the X-ray radiation wavelength, θ is Bragg's angle β is the full width at half maximum [64]. A grain size of 69 nm and 26 nm was estimated for ZnO nanorods grown on tracing and Whatman paper, respectively, which may justify the higher value of responsivity obtained with UV sensor on Whatman substrate.

The flexibility of the UV sensors produced was also tested, by placing them on round mods with curvature radius of 45, 25 and 15 cm. The results are shown on Figure 12. Both sensors produced on tracing and Whatman paper substrates show a decrease in the responsivity for smaller bending radius and consequently higher strains. This can be explained both by the device resistance increment, and by alteration of the light interaction with the sensor, which may be less efficient for larger angular scattering. When strain is induced to the ZnO nanorod arrays, it will induce the formation of piezoelectric polarizations charges that can promote the oxygen adsorption/re-adsorption processes, reducing the sensor responsivity [65].

In Table 2, it is possible to observe the measured parameters, as a function of the curvature radius, of ZnO nanorods UV sensors produced on tracing and Whatman papers. It is also possible to observe that the response time and the recovery time does not change much with the curvature radius, presenting a higher value for Whatman paper that may be due to the high porosity observed for this type of substrate.

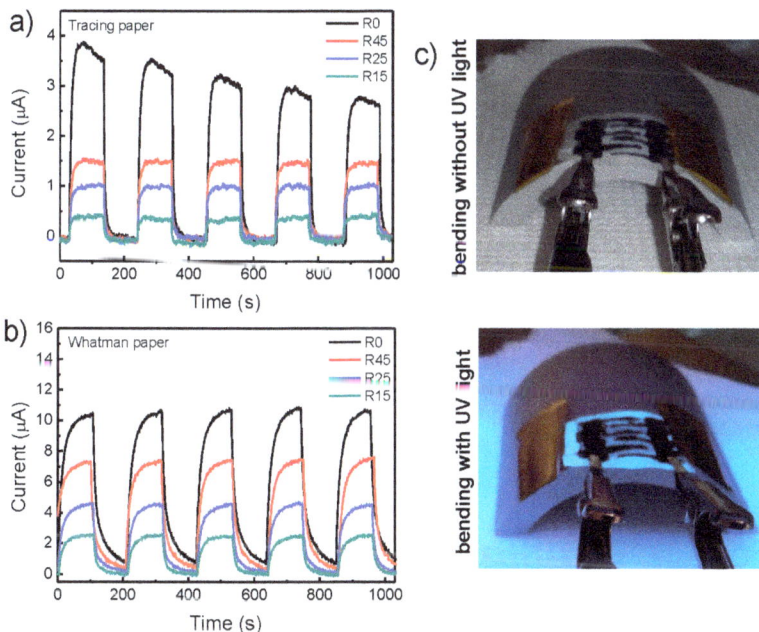

Figure 12. Flexibility cycling behavior of cellulosic ZnO nanorods UV sensors grown on (**a**) tracing paper and (**b**) Whatman paper as a function of a curvature radius of 45, 25 and 15 cm (R45, R25 and R15, respectively). R0 is the measurement with no bending. (**c**) Photograph of cellulosic ZnO nanorods UV sensor, in a bending state, without and with UV light.

Table 2. UV Sensor parameters measured as a function of a curvature radius of 45, 25 and 15 cm.

Paper	Radius	Response Time (s)	Recovery Time (s)	Responsivity (μA/W)
Tracing	R0	30	27	0.39
	R45	27	25	0.19
	R25	24	26	0.12
	R15	21	30	0.044
Whatman	R0	57	65	1.20
	R45	57	62	0.84
	R25	62	57	0.57
	R15	61	48	0.32

3. Experimental Details

3.1. Synthesis of ZnO Nanostructures

An ultrafast method based on hydrothermal synthesis assisted by microwave irradiation have been used for the synthesis of ZnO nanorod arrays on paper substrate. Two distinct types of paper substrates coated with a ZnO seed thin film layer have been used: tracing (Canson, Annonay, France) and Whatman (Sigma-Aldrich, St. Louis, MO, USA) no. 2 papers.

The ZnO seed layer was deposited on the two types of paper substrates by radio frequency (RF) sputtering, at room temperature. A ceramic oxide target of ZnO with a purity of 99.99% was used for the deposition. For the depositions, the chamber was evacuated to a base pressure of 10^{-6} mbar. A shutter between the substrate and the target enabled the protection of the targets from cross contamination. For the deposition of ZnO seed layer, it was used a power density of 12.30 Wcm^{-2}

and a deposition pressure of 4×10^{-3} mbar. The distance between the target and substrate was fixed at 15 cm. The deposition was carried out for 90 min allowing the formation of a 200 nm ZnO layer.

After uniformly coating the two types of paper substrates with ZnO thin films, ZnO nanorod arrays were grown by hydrothermal synthesis under microwave irradiation with a Discover SP microwave system, from CEM (Matthews, NC, USA). Two different approaches were tested, UV/ozone treated or without any previous treatment. For UV treatment, the substrates (tracing and Whatman no. 2) were placed for 5 min in a UV/Ozone system from Novascan (Bonne, MO, USA), equipped with two UV lamps with wavelengths of 185 nm and 254 nm. The distance between the paper and UV lamps were kept at 10 cm. For the microwave-assisted synthesis, the ZnO seeded substrates (20×20 mm) were placed at an angle against the Pyrex vessel, with the seed layer facing down [16] and filled with an aqueous solution of 25 mM zinc nitrate hexahydrate ($Zn(NO_3)_2 \cdot 6H_2O$; 98%, CAS: 10196-18-6) and 25 mM hexamethylenetetramine ($(C_6H_{12}N_4)_2$; 99%, CAS: 100-97-0) [12], both from Sigma Aldrich (St. Louis, MO, USA). Microwave synthesis was fixed at 10 min.

The use of a UV/Ozone treatment was tested before the nanorods synthesis and was optimized the growth temperature to obtain a uniformly coated paper with ZnO nanorods. The synthesis was done at different temperatures: 70, 90, 110 and 130 °C. After each synthesis process, the paper substrates were cleaned with deionized water and ethanol and dried with nitrogen compressed air. Figure 13 represents a schematic of the production process for ZnO nanorods arrays synthesis on paper substrates (with and without an UV/Ozone treatment).

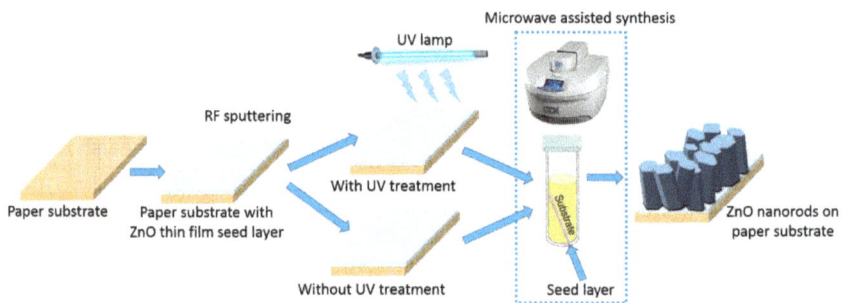

Figure 13. Schematic of ZnO nanorods hydrothermal synthesis assisted by microwave irradiation on paper substrate with a ZnO thin film seed layer deposited by sputtering method.

3.2. Characterization Techniques

Differential scanning calorimetric (DSC) measurements of tracing and Whatman paper substrates were carried out using a Simultaneous Thermal Analyser (TGA-DSC-STA 449 F3 Jupiter, Netzsch-Geratebau GmnH, Selb, Germany). Approximately 5–7 mg of each sample was loaded into an open aluminium crucible and heated from room temperature to 550 °C with a heating rate of 5 °C min^{-1}, in air atmosphere.

Surface 3D profilometry of the paper substrate was performed using an Ambios XP-200 (Ambios Technology, Santa Cruz, CA, USA) profiler for an area of 1×1 μm^2.

The crystallinity of the ZnO nanorod arrays has been determined X-ray diffraction (XRD), using a PANalytical's X'Pert PRO MRD X-ray diffractometer, (PANalytical B.V., Almero, The Netherlands) with a monochromatic CuKα radiation source (wavelength 1.540598 Å). XRD measurements have been carried out from 10° to 90° (2θ), for paper analysis, and from 30° to 40°, for ZnO nanorods measurements, with a scanning step size of 0.016°. The morphology of papers substrates and ZnO nanorods has been characterized by Scanning Electron Microscopy (SEM) using a Carl Zeiss AURIGA CrossBeam Workstation instrument (Carl Zeiss Microscopy GmbH, Oberkochen, Germany) equipped with an Oxford X-ray Energy Dispersive Spectrometer (Carl Zeiss Microscopy GmbH, Oberkochen,

Germany). The length and width of the ZnO nanorods were determined from SEM micrographs using 20 individual nanorods, using ImageJ software [66].

Room temperature diffuse reflectance measurements of were performed using a Perkin Elmer lambda 950 UV/VIS/NIR spectrophotometer (Perkin Elmer, Inc., Waltham, MA, USA) with a diffuse reflectance module (150 mm diameter integrating sphere, internally coated with Spectralon). The calibration of the system was achieved by using a standard reflector sample (reflectance, R = 1.00 from Spectralon disk). The reflectance (R) was obtained from 250 to 800 nm.

3.3. Characterization of ZnO Nanorods on Tracing and Whatman Substrate as a UV Sensor

The synthesized ZnO nanorod arrays on tracing and Whatman paper substrates were characterized as a UV sensor, using a potentiostat model 600, from Gramy Instruments, Inc. (Warminster, PA, USA), in a chronoamperiometry configuration, with a constant applied voltage of 10 V. For interdigital electrical contacts, a carbon resistive ink, PE-C-774, from Conductive Compounds (Hudson, NH, USA), was used. The ZnO nanorod arrays were subjected to UV irradiation with an EL-Series Twin Tube UV lamps, from UVP (Upland, CA, USA) with an intensity of 8 W at a wavelength of 365 nm. The sensor produced was irradiated for 2 min, followed by 2 min in the off state.

4. Conclusions

In the present work, the synthesis of ZnO nanorod arrays was studied on two different cellulosic substrates, tracing and Whatman papers. An ultra-fast chemical synthesis method, based on microwave irradiation, was employed requiring just 10 min to produce ZnO nanorod arrays in both substrates. The influence of a UV treatment prior to synthesis was studies together with the synthesis temperature effect on the growth of ZnO nanorod arrays on paper substrates. It was observed that without UV treatment, the growth on ZnO was heterogeneous, not covering all the substrate surface. With the use of UV treatment and increase of synthesis temperature, larger ZnO nanorods fully covering both cellulosic substrates could be observed. The XRD analysis confirmed the formation of pure and crystalline wurtzite ZnO, with no other impurities detected. The samples produced at 130 °C, with UV treatment prior to ZnO synthesis, were characterized as a UV sensor, revealing an increase of 3 times in the responsivity with the use Whatman paper, when compared with tracing paper. Bending tests revealed that decreasing the curvature radius will also decrease the responsivity of the UV sensor. Nevertheless, the results show that these types of sensor are stable when working in a bending mode.

Acknowledgments: This work was partially financed by FEDER funds through the COMPETE 2020 Programme and National Funds through FCT (Portuguese Foundation for Science and Technology) through BPD/76992/2011, BD/85587/2012 and BPD/84215/2012 and under the project number POCI-01-0145-FEDER, reference UID/CTM/50025/2013.

Author Contributions: Ana Pimentel and Andreia Araújo performed the experiments; optical characterization was performed by Ana Pimentel; the DSC-TG and XRD characterization was performed by Ana Pimentel; Daniela Nunes performed the SEM; the surface 3D profilometry was executed by Andreia Araujo; Ana Samouco performed the ZnO UV sensors tests; the work and paper was under the supervision of Rodrigo Martins and Elvira Fortunato.

Conflicts of Interest: The authors declare no conflict of interest.

References

1. Fortunato, E.; Correia, N., Barquinha, P.; Pereira, L.; Goncalves, G.; Martins, R. High-Performance Flexible Hybrid Field-Effect Transistors Based on Cellulose Fiber Paper. *IEEE Electron Device Lett.* **2008**, *29*, 988–990. [CrossRef]
2. Shah, J.; Brown, R.M. Towards electronic paper displays made from microbial cellulose. *Appl. Microbiol. Biotechnol.* **2005**, *66*, 352–355. [CrossRef] [PubMed]
3. Tobjörk, D.; Österbacka, R. Paper Electronics. *Adv. Mater.* **2011**, *23*, 1935–1961. [CrossRef] [PubMed]

4. Zhou, Y.; Fuentes-Hernandez, C.; Khan, T.M.; Liu, J.-C.; Hsu, J.; Shim, J.W.; Dindar, A.; Youngblood, J.P.; Moon, R.J.; Kippelen, B. Recyclable organic solar cells on cellulose nanocrystal substrates. *Sci. Rep.* **2013**, *3*, 1536. [CrossRef] [PubMed]

5. Wang, B.; Kerr, L.L. Dye sensitized solar cells on paper substrates. *Sol. Energy Mater. Sol. Cells* **2011**, *95*, 2531–2535. [CrossRef]

6. Pimentel, A.; Nunes, D.; Duarte, P.; Rodrigues, J.; Costa, F.M.; Monteiro, T.; Martins, R.; Fortunato, E. Synthesis of Long ZnO Nanorods under Microwave Irradiation or Conventional Heating. *J. Phys. Chem. C* **2014**, *118*, 14629–14639. [CrossRef]

7. Liana, D.D.; Raguse, B.; Gooding, J.J.; Chow, E. Recent Advances in Paper-Based Sensors. *Sensors* **2012**, *12*, 11505–11526. [CrossRef] [PubMed]

8. Marques, A.C.; Santos, L.; Costa, M.N.; Dantas, J.M.; Duarte, P.; Gonçalves, A.; Martins, R.; Salgueiro, C.A.; Fortunato, E. Office paper platform for bioelectrochromic detection of electrochemically active bacteria using tungsten trioxide nanoprobes. *Sci. Rep.* **2015**, *5*, 9910. [CrossRef] [PubMed]

9. Costa, M.N.; Veigas, B.; Jacob, J.M.; Santos, D.S.; Gomes, J.; Baptista, P.V.; Martins, R.; Inácio, J.; Fortunato, E. A low cost, safe, disposable, rapid and self sustainable paper-based platform for diagnostic testing: Lab-on-paper. *Nanotechnology* **2014**, *25*, 94006. [CrossRef] [PubMed]

10. Oliveira, M.J.; Quaresma, P.; de Almeida, M.P.; Araújo, A.; Pereira, E.; Fortunato, E.; Martins, R.; Franco, R.; Águas, H. Office paper decorated with silver nanostars—An alternative cost effective platform for trace analyte detection by SERS. *Sci. Rep.* **2017**, *7*, 1–14. [CrossRef] [PubMed]

11. Morkoc, H.; Ozgur, Ü. Zinc Oxide: Fundamentals, Materials and Device Technology. Available online: http://eu.wiley.com/WileyCDA/WileyTitle/productCd-3527408134.html (accessed on 14 March 2016).

12. Pimentel, A.; Rodrigues, J.; Duarte, P.; Nunes, D.; Costa, F.M.; Monteiro, T.; Martins, R.; Fortunato, E. Effect of solvents on ZnO nanostructures synthesized by solvothermal method assisted by microwave radiation: A photocatalytic study. *J. Mater. Sci.* **2015**, *50*, 5777–5787. [CrossRef]

13. Hayes, B.L. *Microwave Synthesis: Chemistry at the Speed of Light*; CEM Publishing: Matthews, NC, USA, 2002.

14. Nunes, D.; Pimentel, A.; Barquinha, P.; Carvalho, P.A.; Fortunato, E.; Martins, R. Cu₂O polyhedral nanowires produced by microwave irradiation. *J. Mater. Chem. C* **2014**, *2*, 6097. [CrossRef]

15. Gonçalves, A.; Resende, J.; Marques, A.C.; Pinto, J.V.; Nunes, D.; Marie, A.; Goncalves, R.; Pereira, L.; Martins, R.; Fortunato, E. Smart optically active VO₂ nanostructured layers applied in roof-type ceramic tiles for energy efficiency. *Sol. Energy Mater. Sol. Cells* **2016**, *150*, 1–9. [CrossRef]

16. Nunes, D.; Pimentel, A.; Pinto, J.V.; Calmeiro, T.R.; Nandy, S.; Barquinha, P.; Pereira, L.; Carvalho, P.A.; Fortunato, E.; Martins, R. Photocatalytic behavior of TiO₂ films synthesized by microwave irradiation. *Catal. Today* **2015**, *278*, 262–270. [CrossRef]

17. Pimentel, A.; Nunes, D.; Pereira, S.; Martins, R.; Fortunato, E. Photocatalytic Activity of TiO₂ Nanostructured Arrays Prepared by Microwave-Assisted Solvothermal Method. In *Semiconductor Photocatalysis—Materials, Mechanisms and Applications*; Cao, W.B., Ed.; InTech: Rijeka, Croatia, 2016.

18. Nunes, D.; Pimentel, A.; Santos, L.; Barquinha, P.; Fortunato, E.; Martins, R. Photocatalytic TiO₂ Nanorod Spheres and Arrays Compatible with Flexible Applications. *Catalysts* **2017**, *7*, 60. [CrossRef]

19. Pimentel, A.; Ferreira, S.; Nunes, D.; Calmeiro, T.; Martins, R.; Fortunato, E. Microwave Synthesized ZnO Nanorod Arrays for UV Sensors: A Seed Layer Annealing Temperature Study. *Materials* **2016**, *9*, 299. [CrossRef] [PubMed]

20. Araújo, A.; Pimentel, A.; Oliveira, M.J.; Mendes, M.J.; Franco, R.; Fortunato, E.; Águas, H.; Martins, R. Direct growth of plasmonic nanorod forests on paper substrates for low-cost flexible 3D SERS platforms. *Flex. Print. Electron.* **2017**, *2*, 14001. [CrossRef]

21. Major, S.; Kumar, S.; Bhatnagar, M.; Chopra, K.L. Effect of hydrogen plasma treatment on transparent conducting oxides. *Appl. Phys. Lett.* **1986**, *49*, 394. [CrossRef]

22. Park, J.-S.; Jeong, J.K.; Mo, Y.-G.; Kim, H.D.; Kim, S.-I. Improvements in the device characteristics of amorphous indium gallium zinc oxide thin-film transistors by Ar plasma treatment. *Appl. Phys. Lett.* **2007**, *90*, 262106. [CrossRef]

23. Angermann, H.; Korte, L.; Rappich, J.; Conrad, E.; Sieber, I.; Schmidt, M.; Hübener, K.; Hauschild, J. Optimisation of electronic interface properties of a-Si:H/c-Si hetero-junction solar cells by wet-chemical surface pre-treatment. *Thin Solid Films* **2008**, *516*, 6775–6781. [CrossRef]

24. Ip, K.; Gila, B.; Onstine, A.; Lambers, E.; Heo, Y.; Baik, K.; Norton, D.; Pearton, S.; Kim, S.; LaRoche, J.; et al. Effect of ozone cleaning on Pt/Au and W/Pt/Au Schottky contacts to n-type ZnO. *Appl. Surf. Sci.* **2004**, *236*, 387–393. [CrossRef]
25. Cho, J.M.; Kwak, S.-W.; Aqoma, H.; Kim, J.W.; Shin, W.S.; Moon, S.-J.; Jang, S.-Y.; Jo, J. Effects of ultraviolet–ozone treatment on organic-stabilized ZnO nanoparticle-based electron transporting layers in inverted polymer solar cells. *Org. Electron.* **2014**, *15*, 1942–1950. [CrossRef]
26. Manekkathodi, A.; Lu, M.-Y.; Wang, C.W.; Chen, L.-J. Direct growth of aligned zinc oxide nanorods on paper substrates for low-cost flexible electronics. *Adv. Mater.* **2010**, *22*, 4059–4063. [CrossRef] [PubMed]
27. Fortunato, E.M.C.; Barquinha, P.M.C.; Pimentel, A.C.M.B.G.; Gonçalves, A.M.F.; Marques, A.J.S.; Pereira, L.M.N.; Martins, R.F.P. Fully Transparent ZnO Thin-Film Transistor Produced at Room Temperature. *Adv. Mater.* **2005**, *17*, 590–594. [CrossRef]
28. Pimentel, A.C.; Gonçalves, A.; Marques, A.; Martins, R.; Fortunato, E. Zinc oxide thin films used as an ozone sensor at room temperature. *MRS Proc.* **2011**, *915*, 915-R07-4. [CrossRef]
29. Arya, S.K.; Saha, S.; Ramirez-Vick, J.E.; Gupta, V.; Bhansali, S.; Singh, S.P. Recent advances in ZnO nanostructures and thin films for biosensor applications: Review. *Anal. Chim. Acta* **2012**, *737*, 1–21. [CrossRef] [PubMed]
30. Zhang, Y.; Kang, Z.; Yan, X.; Liao, Q. ZnO nanostructures in enzyme biosensors. *Sci. China Mater.* **2015**, *58*, 60–76. [CrossRef]
31. Pimentel, A.; Fortunato, E.; Gonçalves, A.; Marques, A.; Águas, H.; Pereira, L.; Ferreira, I.; Martins, R. Polycrystalline intrinsic zinc oxide to be used in transparent electronic devices. *Thin Solid Films* **2005**, *487*, 212–215. [CrossRef]
32. Panda, S.K.; Jacob, C. Preparation of transparent ZnO thin films and their application in UV sensor devices. *Solid State Electron.* **2012**, *73*, 44–50. [CrossRef]
33. Chang, H.; Sun, Z.; Ho, K.Y.-F.; Tao, X.; Yan, F.; Kwok, W.-M.; Zheng, Z. A highly sensitive ultraviolet sensor based on a facile in situ solution-grown ZnO nanorod/graphene heterostructure. *Nanoscale* **2011**, *3*, 258–264. [CrossRef] [PubMed]
34. Guo, L.; Zhang, H.; Zhao, D.; Li, B.; Zhang, Z.; Jiang, M.; Shen, D. High responsivity ZnO nanowires based UV detector fabricated by the dielectrophoresis method. *Sens. Actuators B Chem.* **2012**, *166–167*, 12–16. [CrossRef]
35. Chai, G.; Lupan, O.; Chow, L.; Heinrich, H. Crossed zinc oxide nanorods for ultraviolet radiation detection. *Sens. Actuators A Phys.* **2009**, *150*, 184–187. [CrossRef]
36. Yao, I.-C.; Tseng, T.-Y.; Lin, P. ZnO nanorods grown on polymer substrates as UV photodetectors. *Sens. Actuators A Phys.* **2012**, *178*, 26–31. [CrossRef]
37. Ridhuan, N.S.; Razak, K.A.; Lockman, Z.; Abdul Aziz, A. Structural and morphology of ZnO nanorods synthesized using ZnO seeded growth hydrothermal method and its properties as UV sensing. *PLoS ONE* **2012**, *7*, e50405. [CrossRef] [PubMed]
38. Lim, Z.H.; Chia, Z.X.; Kevin, M.; Wong, A.S.W.; Ho, G.W. A facile approach towards ZnO nanorods conductive textile for room temperature multifunctional sensors. *Sens. Actuators B Chem.* **2010**, *151*, 121–126. [CrossRef]
39. Jonoobi, M.; Oladi, R.; Davoudpour, Y.; Oksman, K.; Dufresne, A.; Hamzeh, Y.; Davoodi, R. Different preparation methods and properties of nanostructured cellulose from various natural resources and residues: A review. *Cellulose* **2015**, *22*, 935–969. [CrossRef]
40. Zhao, D.; Chen, K.; Yang, F.; Feng, G.; Sun, Y.; Dai, Y. Thermal degradation kinetics and heat properties of cellulosic cigarette paper: Influence of potassium carboxylate as combustion improver. *Cellulose* **2013**, *20*, 3205–3217. [CrossRef]
41. Ornaghi, H.L.; Poletto, M.; Zattera, A.J.; Amico, S.C. Correlation of the thermal stability and the decomposition kinetics of six different vegetal fibers. *Cellulose* **2014**, *21*, 177–188. [CrossRef]
42. Park, S.; Baker, J.O.; Himmel, M.E.; Parilla, P.A.; Johnson, D.K. Cellulose crystallinity index: Measurement techniques and their impact on interpreting cellulase performance. *Biotechnol. Biofuels* **2010**, *3*, 10. [CrossRef] [PubMed]
43. Ju, X.; Bowden, M.; Brown, E.E.; Zhang, X. An improved X-ray diffraction method for cellulose crystallinity measurement. *Carbohydr. Polym.* **2015**, *123*, 476–481. [CrossRef] [PubMed]

44. Lujun, Y.; Maojun, Z.; Changli, L.; Li, M.; Wenzhong, S. Facile synthesis of superhydrophobic surface of ZnO nanoflakes: Chemical coating and UV-induced wettability conversion. *Nanoscale Res. Lett.* **2012**, *7*, 216. [CrossRef] [PubMed]

45. Hewlett, R.M.; McLachlan, M.A. Surface Structure Modification of ZnO and the Impact on Electronic Properties. *Adv. Mater.* **2016**, *28*, 3893–3921. [CrossRef] [PubMed]

46. Murakami, T.N.; Fukushima, Y.; Hirano, Y.; Tokuoka, Y.; Takahashi, M.; Kawashima, N. Surface modification of polystyrene and poly(methyl methacrylate) by active oxygen treatment. *Colloids Surfaces B Biointerfaces* **2003**, *29*, 171–179. [CrossRef]

47. Xu, S.; Wang, Z.L. One-dimensional ZnO nanostructures: Solution growth and functional properties. *Nano Res.* **2011**, *4*, 1013–1098. [CrossRef]

48. Talebian, N.; Amininezhad, S.M.; Doudi, M. Controllable synthesis of ZnO nanoparticles and their morphology-dependent antibacterial and optical properties. *J. Photochem. Photobiol. B* **2013**, *120*, 66–73. [CrossRef] [PubMed]

49. Zhang, Y.; Ram, M.K.; Stefanakos, E.K.; Goswami, D.Y.; Zhang, Y.; Ram, M.K.; Stefanakos, E.K.; Goswami, D.Y. Synthesis, Characterization, and Applications of ZnO Nanowires. *J. Nanomater.* **2012**, *2012*, 1–22. [CrossRef]

50. Vig, J.R. UV/ozone cleaning of surfaces. *J. Vac. Sci. Technol. A Vac. Surf. Films* **1985**, *3*, 1027–1034. [CrossRef]

51. Baruah, S.; Dutta, J. Hydrothermal growth of ZnO nanostructures. *Sci. Technol. Adv. Mater.* **2009**, *10*, 13001. [CrossRef] [PubMed]

52. Fu, X.; Jiang, F.; Gao, R.; Peng, Z. Microstructure and nonohmic properties of SnO_2-Ta_2O_5-ZnO system doped with ZrO_2. *Sci. World J.* **2014**, *2014*, 754890. [CrossRef] [PubMed]

53. Guo, M.; Diao, P.; Wang, X.; Cai, S. The effect of hydrothermal growth temperature on preparation and photoelectrochemical performance of ZnO nanorod array films. *J. Solid State Chem.* **2005**, *178*, 3210–3215. [CrossRef]

54. Pankove, J.I. *Optical Processes in Semiconductors*; Dover Publications, Inc.: Mineola, NY, USA, 1971.

55. Srikant, V.; Clarke, D.R. On the optical band gap of zinc oxide. *J. Appl. Phys.* **1998**, *83*, 5447. [CrossRef]

56. Kim, Y.-S.; Tai, W.-P.; Shu, S.-J. Effect of preheating temperature on structural and optical properties of ZnO thin films by sol–gel process. *Thin Solid Films* **2005**, *491*, 153–160. [CrossRef]

57. Shinde, S.D.; Patil, G.E.; Kajale, D.D.; Gaikwad, V.B.; Jain, G.H. Synthesis of ZnO nanorods by spray pyrolysis for H_2S gas sensor. *J. Alloys Compd.* **2012**, *528*, 109–114. [CrossRef]

58. Chou, C.-S.; Wu, Y.-C.; Lin, C.-H. Oxygen sensor utilizing ultraviolet irradiation assisted ZnO nanorods under low operation temperature. *RSC Adv.* **2014**, *4*, 52903–52910. [CrossRef]

59. Schmidt-Mende, L.; MacManus-Driscoll, J.L. ZnO—Nanostructures, defects, and devices. *Mater. Today* **2007**, *10*, 40–48. [CrossRef]

60. Zhai, T.; Fang, X.; Liao, M.; Xu, X.; Zeng, H.; Yoshio, B.; Golberg, D. A Comprehensive Review of One-Dimensional Metal-Oxide Nanostructure Photodetectors. *Sensors* **2009**, *9*, 6504–6529. [CrossRef] [PubMed]

61. Mamat, M.H.; Khusaimi, Z.; Zahidi, M.M.; Mahmood, M.R. *Nanorods*; Yaln, O., Ed.; InTech: Rijeka, Croatia, 2012.

62. Kalantar-zadeh, K.; Fry, B. Sensor Characteristics and Physical Effects. In *Nanotechnology-Enabled Sensors*; Springer: Boston, MA, USA, 2008; pp. 13–62.

63. Fryxell, G.E.; Cao, G. *Environmental Applications of Nanomaterials: Synthesis, Sorbents and Sensors*; Imperial College Press: London, UK, 2007.

64. Cullity, B.D. *Elements of X Ray Diffraction*; Addison-Wesley Publisher Companym Inc.: Boston, MA, USA, 1956.

65. Chen, T.-P.; Young, S.-J.; Chang, S.-J.; Hsiao, C.-H.; Hsu, Y.-J. Bending effects of ZnO nanorod metal–semiconductor–metal photodetectors on flexible polyimide substrate. *Nanoscale Res. Lett.* **2012**, *7*, 214. [CrossRef] [PubMed]

66. Schneider, C.A.; Rasband, W.S.; Eliceiri, K.W. NIH Image to ImageJ: 25 years of image analysis. *Nat. Methods* **2012**, *9*, 671–675. [CrossRef] [PubMed]

![materials logo] *materials*

MDPI

Article

In-Doped ZnO Hexagonal Stepped Nanorods and Nanodisks as Potential Scaffold for Highly-Sensitive Phenyl Hydrazine Chemical Sensors

Ahmad Umar [1,2,*], Sang Hoon Kim [1,2], Rajesh Kumar [3], Mohammad S. Al-Assiri [2,4], A. E. Al-Salami [5], Ahmed A. Ibrahim [1,2,6] and Sotirios Baskoutas [6,*]

1   Department of Chemistry, Faculty of Science and Arts, Najran University, P.O. Box 1988, Najran 11001, Saudi Arabia; semikim77@gmail.com (S.H.K.); ahmedragal@yahoo.com (A.A.I.)
2   Promising Centre for Sensors and Electronic Devices (PCSED), Najran University, P.O. Box-1988, Najran 11001, Saudi Arabia; msassiri@gmail.com
3   PG Department of Chemistry, JCDAV College, Dasuya 144205, India; rk.ash2k7@gmail.com
4   Department of Physics, Faculty of Science and Arts, Najran University, P.O. Box 1988, Najran 11001, Saudi Arabia
5   Department of Physics, Faculty of Science, King Khalid University, P.O. Box-9004, Abha 61413, Saudi Arabia; Salami11@gmail.com
6   Department of Materials Science, University of Patras, 26504 Patras, Greece
*   Correspondence: ahmadumar786@gmail.com (A.U.); bask@upatras.gr (S.B.); Tel.: +966-534574597 (A.U.); +30-2610969349 (S.B.)

Received: 26 October 2017; Accepted: 18 November 2017; Published: 21 November 2017

Abstract: Herein, we report the growth of In-doped ZnO (IZO) nanomaterials, i.e., stepped hexagonal nanorods and nanodisks by the thermal evaporation process using metallic zinc and indium powders in the presence of oxygen. The as-grown IZO nanomaterials were investigated by several techniques in order to examine their morphological, structural, compositional and optical properties. The detailed investigations confirmed that the grown nanomaterials, i.e., nanorods and nanodisks possess well-crystallinity with wurtzite hexagonal phase and grown in high density. The room-temperature PL spectra exhibited a suppressed UV emissions with strong green emissions for both In-doped ZnO nanomaterials, i.e., nanorods and nanodisks. From an application point of view, the grown IZO nanomaterials were used as a potential scaffold to fabricate sensitive phenyl hydrazine chemical sensors based on the I–V technique. The observed sensitivities of the fabricated sensors based on IZO nanorods and nanodisks were 70.43 $\mu A \cdot mM^{-1} \cdot cm^{-2}$ and 130.18 $\mu A \cdot mM^{-1} \cdot cm^{-2}$, respectively. For both the fabricated sensors, the experimental detection limit was 0.5 μM, while the linear range was 0.5 μM–5.0 mM. The observed results revealed that the simply grown IZO nanomaterials could efficiently be used to fabricate highly sensitive chemical sensors.

Keywords: In-doped ZnO; stacked nanorods; flower-shaped; sensing; phenyl hydrazine

---

1. Introduction

Phenyl hydrazine is used for the synthesis of agricultural pesticides and insecticides, in pharmaceutical industries, rocket propellant, and explosives. Even a very low concentration of it may prove very fatal and produces drastic environmental and health hazards. The exposure of phenyl hydrazine may lead to dermatitis, skin irritation, liver and kidney damage, and hemolytic anemia [1]. According to World Health Organization and Environmental Protection Agency, phenyl hydrazine is carcinogenic and has been classified as group B2 human carcinogen [2]. Thus, a fast, reliable, highly sensitive, and selective analytical method is desired to detect even traces of phenyl hydrazine.

Modifications of the structural, morphological, optical, chemical, electronic, and physicochemical properties of the ZnO semiconductor nanomaterials either through doping or through the formation of composites have always attracted the researcher to synthesize such materials [3–9]. Various doped ZnO nanomaterials have been recently synthesized and explored for their photocatalytic [10–12], gas sensing [13,14], photonic crystals [15], spintronics [16], electrochemical sensing [17–19], optoelectronics [20,21], dye-sensitized solar cells [22], light emitting diodes [23], field emission transistors [24,25], and many more applications. Methods like sol-gel [24,26], hydrothermal [18,27], ceramics vapor deposition [28], spin coating [29], solution combustion [30], RF sputtering [31], pulse laser deposition [32,33], etc. are reported for the synthesis of doped ZnO nanostructures. However, through the thermal evaporation technique, the directional growth of the nanostructures can be easily controlled by controlling the temperature and source material in contrast with the solution method in which the growth of nanomaterials are depends upon the pH, concentration of the source materials, growth time, template/capping agent/directing agents, etc. Moreover, the use of pure metals as source material grows the nanostructures of high purity and crystallinity.

ZnO is an II–VI, n-type semiconductor with a wide direct band gap of ~3.27 eV, large exciton binding energy (~60 meV), and high electron mobility of 115–155 $cm^2 \cdot V^{-1} \cdot s^{-1}$ at room temperature [34–36]. Doping with higher valence impurities, such as In^{3+}, further enhances the conductivity [37,38]. It has been reported that both the conduction band, as well as the valence band energies of the ZnO, are shifted to lower energy levels while there is no appreciable change in the band gap energy [39]. Additionally, as the ionic radius of In^{3+} (0.094 nm) is greater than that of Zn^{2+} ions (0.074 nm), the incorporation of the In^{3+} ions into the ZnO crystal lattice further produces tensile strains leading to higher surface defects [40]. These properties make In-doped ZnO nanostructures suitable candidates for the fabrication of electrode for the gas, electrochemical sensing, and solar cell applications.

Chava et al. [35] synthesized In-doped ZnO nanoparticles through a cost-effective, low-temperature aqueous solution method, and explored them for the fabrication of photoanode in dye-sensitized solar cells. In-doped ZnO, photoanode exhibited a high short-circuit photocurrent density of 12.58 mA/cm^2 and a power conversion efficiency of 2.7% as compared to the current density of 8.02 mA/cm^2 with 1.8% efficiency for the bare ZnO nanoparticles. Badadhe et al. [41] observed a very high gas response and short response and recovery times of 13,000, ~2 s and ~4 min, respectively, for 1000 ppm H_2S at 250 °C through 3 at. %. In-doped ZnO thin films deposited onto pre-cleaned glass substrates through a conventional spray pyrolysis technique. However, for 50 ppm CO gas, 1 at. % and 2 at. % In-doped ZnO nanostructures prepared by Dhahri et al. [42] showed an excellent gas response with short response and recovery times as compared to pure ZnO at 300 °C. Wang et al. [43] explored the ethanol gas sensing applications of In-doped three-dimensionally ordered macroporous (3DOM) ZnO structures synthesized via a colloidal crystal templating method. High selectivity and sensitivity of the 5 at. % In-doped ZnO structures as compared to pure 3DOM ZnO was attributed to the increased surface area, the increased electron carrier concentration due to the replacement of the Zn^{2+} ions from the crystal lattice by In^{3+} ions, along with the higher rate of O_2 adsorption. Ge et al. [44] observed significant responses towards trace of various nitro-explosive vapors, including trinitrotoluene (TNT), dinitrotoluene (DNT), para-nitrotoluene (PNT), picric acid (PA), and Research Department Explosive (RDX). At room temperature using 5% In-doped ZnO nanoparticles based gas sensors. The use of metal oxide nanomaterials modified electrodes, through electrochemical sensing analysis, is preferred over other traditional methods due to the excellent reliability, high sensitivity, and short response and recovery times [45].

In this paper, a facile thermal evaporation method is reported for the growth of In-doped ZnO (IZO) with two distinct morphologies as a function of reaction conditions on a silicon substrate. The grown nanomaterials were characterized for their morphological, structural, optical, and crystal properties through relevant characterization techniques. Further, the grown IZO nanomaterials were utilized as efficient electron mediators for the fabrication of phenyl hydrazine chemical sensor. Finally, a comparison was made between the In-doped ZnO nanorods and nanodisks for the electrochemical sensing properties towards phenyl hydrazine.

2. Experimental Details

2.1. Growth of In-Doped ZnO Nanostructures

The stepped hexagonal nanorods and disk-shaped IZO nanomaterials were grown on silicon substrate by thermal evaporation of metallic zinc (Zn) and indium (In) powders in the presence of oxygen. For the growth of stepped hexagonal nanorods, metallic Zn and In powders were mixed well in the ratio of 5:2, while for the disk-shaped structures, the Zn and In powders were mixed in the ratio of 10:3. Both of the mixtures were separately transferred to the ceramic boats and were placed at the center of the high-temperature ceramic tube furnace. Two different experiments were performed for the growth of stepped hexagonal nanorods and disk-shaped structures. Prior to the experiments, the Si(100) substrates were cleaned with DI water, ethanol, and acetone, sequentially and were dried by nitrogen gas. For both of the experiments, several pieces of Si(100) were placed adjacent to the source boat and ceramic tube furnace chamber was down to 1 torr using a rotary vacuum pump. For the nanorods growth, the reaction was done at 700 °C, while disks were grown at 850 °C under the continuous flow of high purity nitrogen and oxygen gases with the flow-rates of 40 and 100 sccm, respectively. The reaction was terminated in 1.5–2.5 h. After completing the reaction for the desired time, the furnace was allowed to cool to room temperature and the deposited materials were characterized in terms of their morphological, structural, and optical properties.

2.2. Characterization Techniques

The necessary characterization techniques were utilized for the evaluation of the morphological, structural, optical, compositional, and crystal phases of the as-synthesized IZO nanostructures. Field emission scanning electron microscopy (FESEM; JEOL-JSM-7600F, JEOL, Tokyo, Japan) attached with energy dispersive spectroscopy (EDS) explored the morphological, structural, and compositional characteristics. In-doped ZnO nanostructures were subjected to X-ray diffraction (XRD; JDX-8030W, JEOL, Tokyo, Japan) analysis using Cu-Kα radiation ($\lambda = 1.54$ Å) in the diffraction angle range of 20–65° elaborated the crystal structure, purity, crystal sizes and crystallinity. The optical properties of the deposited materials were examined by room-temperature photoluminescence (PL), measured with the He-Cd (325 nm) laser lines as the exciton source.

2.3. Fabrication of Phenyl Hydrazine Electrochemical Sensor Based on IZO Nanomaterials

To fabricate the phenyl hydrazine chemical sensors, firstly, the grown IZO nanomaterials were transferred from the substrates to the glassy carbon electrodes (GCE) and wetted by phosphate buffer solution (0.1 M PBS) with pH = 7.4 and dried gently with the flow of high purity nitrogen gas.

The PBS (0.1 M, pH = 7.4) solution was prepared by mixing 0.2 M disodium phosphate (Na_2HPO_4) and 0.2 M monosodium phosphate (NaH_2PO_4) solution in 100 mL de-ionized water. Consequently, 5 weight % Nafion solution was dropped onto the modified electrode to form a net-like structure on the electrode, which can hold the functional materials during the sensing measurements. Prior to the experiments, the GCE was polished with alumina slurry and was then sonicated in de-ionized water. The modified electrode, i.e., nafion/In-doped ZnO/GCE was finally dried at 40 °C for 12 h. All of the electrochemical experiments were performed at room temperature by Keithley 6517A-USA (Tektronix, OR, USA) electrometer with computer interfacing with a simple current-voltage (I–V) technique. For the two electrode system, the modified electrodes were used as working electrode, in which the Pt wire was employed as a counter electrode. The sensitivity of the fabricated sensors was determined by plotting a calibration curve of current vs. concentration.

3. Results and Discussion

3.1. Characterization of IZO Nanostructures

The crystallinity and crystal properties of as-grown IZO nanomaterials were examined by X-ray diffraction and the observed results are shown in Figure 1. It was observed that the stepped hexagonal IZO nanorods exhibited several sharp and well-defined diffractions peaks corresponding to the diffraction planes of (100), (002), (101), (102), (110), and (103) at diffraction angles of 31.79°, 34.35°, 36.23°, 47.79°, 56.57°, and 62.81°, respectively. However, the IZO nanodisks show several well-defined diffraction peaks appeared at 2θ = 31.77°, 34.45°, 36.27°, 47.79°, 56.43°, and 62.79°, corresponding to the diffraction planes of (100), (002), (101), (102), (110), and (103), respectively. All of the observed diffraction peaks are well matched with the characteristic peaks of wurtzite hexagonal phase of ZnO. The observed diffraction peaks are well-matched with the JCPDS data card no. 36-1451 and the reported literature [18,45–47]. The incorporation of the In^{3+} ions into the crystal lattice of the ZnO can be confirmed by the presence of additional peaks for In_2O_3 corresponding to diffraction planes (220), (222), (411), and (332) [48–50]. However, the peak for diffraction planes (220) was of very low intensity in case of IZO nanorods, which may be due to the very low concentration of the In^{3+} ions.

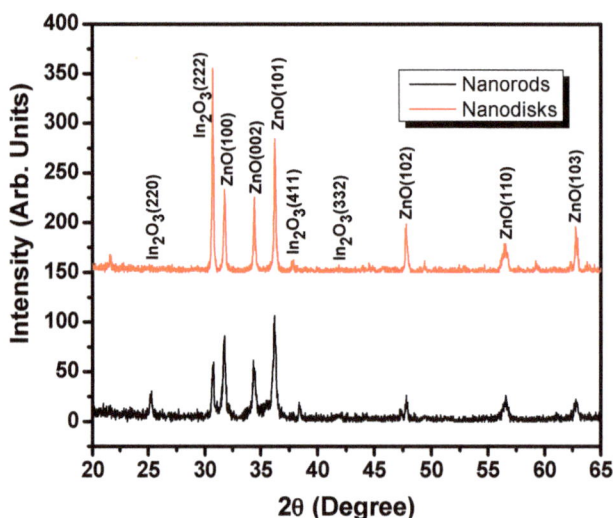

Figure 1. X-ray diffraction (XRD) diffraction patterns for In-doped ZnO (IZO) nanorods and nanodisks.

The crystallite sizes (d) of the thermally deposited IZO nanostructures were evaluated by using Debye–Scherrer equation (Equation (1)) [18,47].

$$d = \frac{0.89\lambda}{\beta.Cos\theta} \qquad (1)$$

where, λ = the wavelength of X-rays used (Cu-Kα radiation with λ = 1.54 Å), θ is the Bragg diffraction angle and β is the peak width at half maximum (FWHM). Three most intense peaks were considered for the calculation of FWHM, and thus the crystallite sizes of the IZO nanorods and nanodisks and the corresponding results are represented in Table 1. The average crystallite sizes of 24.80 nm and 42.24 nm were calculated for the In-doped ZnO nanorods and nanodisks, respectively.

Table 1. The peak width at half maximum (FWHM) values and crystallite sizes of IZO nanostructures.

S.N.	(hkl)	IZO Nanorods			IZO Nanodisks		
		2θ (°)	FWHM (β)	Crystallite Size (nm)	2θ (°)	FWHM (β)	Crystallite Size (nm)
1.	(100)	31.79	0.29564	27.65	31.77	0.19751	41.39
2.	(002)	34.35	0.35795	22.99	34.45	0.1810	45.48
3.	(101)	36.23	0.34833	23.75	36.27	0.20768	39.84

The morphology of the IZO nanostructures thermally synthesized at 700 °C is shown in Figure 2a–d. Stepped hexagonal nanorod shaped morphologies with uniform shape and sizes but with orientations in different directions can be clearly seen from low (Figure 2c,d) as well high (Figure 2a,b) magnification FESEM images. Some of these nanorods seem to be originated from a common base resulting in an urchin shaped structures. The average length of the nanorods was ~1.5 μm, however, the diameter was not uniform. Each IZO nanorod further showed a hexagonal cross-section with a thickness of ~150 nm (Figure 2b). The elemental composition of the IZO nanorods, as examined by the EDS attached with FESEM is shown in Figure 2e. The IZO nanorods are made of Indium, Zinc, and oxygen only. No other peak corresponding to any elemental impurity, prove that thermally synthesized nanorods are of high purity.

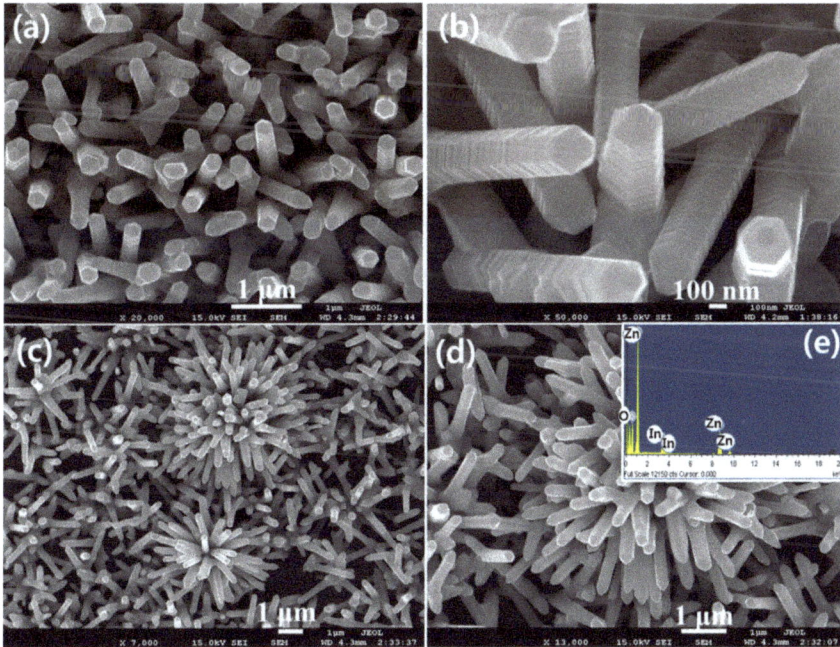

Figure 2. (**a,b**) High magnification (**c,d**) low magnification Field emission scanning electron microscopy (FESEM) images and (**e**) EDS spectrum of stepped hexagonal IZO nanorods synthesized at the 700 °C growth temperature.

The morphology of the IZO nanostructures thermally synthesized at 850 °C is shown in Figure 3a–c. Low magnification FESEM images shown in Figure 3a,b exhibit disk shaped morphologies. However, high magnification FESEM images, as shown in Figure 3c, revealed that the disk-shaped morphologies are further composed of hook-shaped structures that are interlocked with each other. The average diameter of disk shaped morphology was ~250–300 nm. Figure 3d shows the EDS spectrum for the IZO nanodisks, which confirmed the purity of thermally deposited nanodisks as peaks for Indium, Zinc, and oxygen atoms are present.

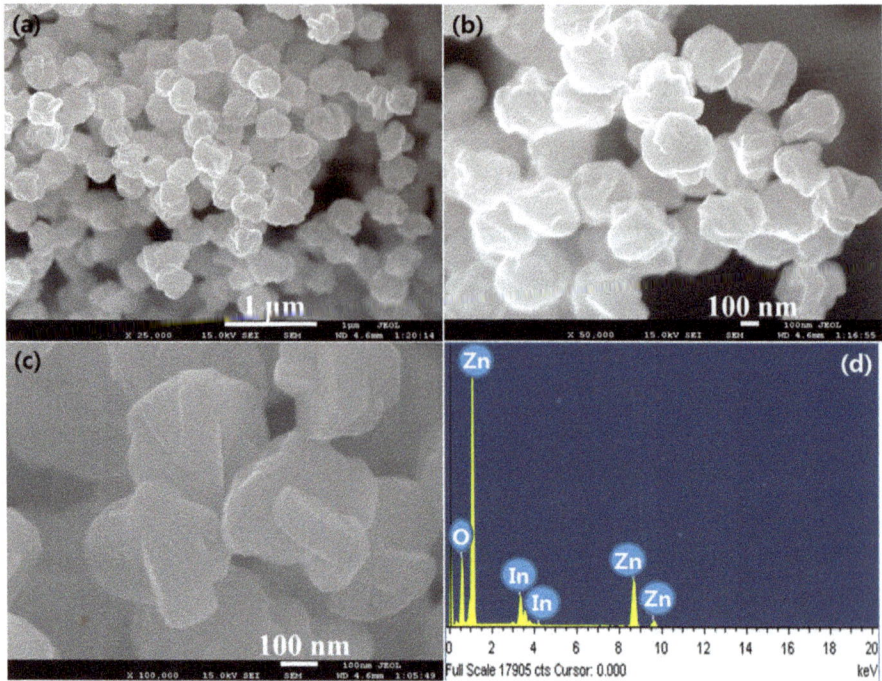

Figure 3. (**a**,**b**) Low magnification (**c**) high magnification FESEM images and (**d**) energy dispersive spectroscopy (EDS) spectrum of IZO nanodisks synthesized at the 850 °C growth temperature.

In order to evaluate the structural and optical property of the thermally deposited In-doped ZnO nanostructures, Photoluminescence spectroscopic analysis was performed at room temperature using a He-Cd source having 325 nm excitation wavelength. The corresponding PL spectra are shown in Figure 4. Strong UV emission peaks centered at 381.6 (3.249 eV) and 381.4 nm (3.251 eV) were observed for In-doped ZnO nanorods and nanodisks, respectively. These peaks may be attributed to the near band edge emission (NBE) resulting due to recombination of free excitons as well as to the transition from 1 longitudinal optical (LO)-phonon replica of two electron satellites (TES) lines of ZnO [51–54]. A slight shift in band edge peak with a significant increase in the UV emission intensity for IZO nanodisks as compared to nanorods may be attributed to the better crystallinity of former, which is also confirmed by the XRD analysis (Figure 1). Additionally, broad but strong bands in the visible region with emission peaks at 547.5 nm (2.264 eV) and 539.4 nm (2.299 eV) for IZO nanorods and nanodisks, respectively, were also observed and can be assigned to is due to the superposition of green, yellow-orange, and red emissions [51,55,56]. The green emission bands for both the morphologies originate due to the radial recombination of a photogenerated positively charged hole (h^+) with a negatively charged electron (e^-) of the singly ionized oxygen (O) vacancies on ZnO surface lattice [57,58].

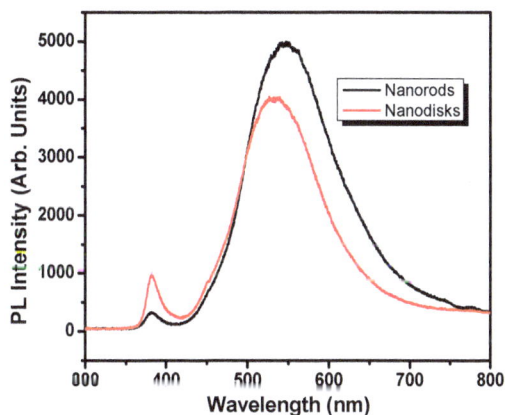

Figure 4. Typical room-temperature photoluminescence (PL) spectra of stepped hexagonal nanorods and flower-shaped IZO nanomaterials.

3.2. Electrochemical Sensing Applications of IZO Nanostructures

IZO nanostructures based electrochemical sensors were tested for the detection 0.05 μM phenyl hydrazine in 0.1 M phosphate buffer solution (PBS) with pH = 7, as compared to blank PBS.

The current responses were measured from 0.0 to +1.5 V. Significant increase in the current response for even very low concentration of 0.05 μM phenyl hydrazine as compared to blank PBS confirms that both the doped ZnO nanostructures could act as efficient electron mediators and electro-catalysts for the electrochemical detection of phenyl hydrazine at room temperature (Figure 5a,b). For both of the sensors, a continuous increase in the current response was observed with an increase in the potential applied. At +1.5 V current responses of 7.268 μM and 6.903 μM were recorded, respectively, for IZO nanorods and nanodisks based sensors.

For the estimation of important sensing parameters like sensitivity, linear dynamic range, detection limit, and correlation coefficient, a series electrochemical sensing experiments were conducted using both the IZO nanostructured modified electrodes using different concentrations. A series of phenyl hydrazine solutions with a concentration range of 0.5 μM–5.0 M in 0.1 M PBS were prepared. Figure 6a,b represent the I–V response curves for IZO nanorods and nanodisks modified GCE, respectively, against various concentrations of phenyl hydrazine in 0.1 M PBS. Expectedly, a continuous increase in current response is seen with the sequential increase in the phenyl hydrazine concentrations from 0.5 μM to 5.0 M in 0.1 M PBS. This may be attributed to the generation of a large number of ions due to ionization of phenyl hydrazine into resulting in large ionic strength at higher concentrations [45,59]. Current responses of 33.56 μA and 51.29 μA were observed for 5.0 M concentrations of phenyl hydrazine in 0.1 M PBS at +1.5 V using GCE modified with IZO nanorods and nanodisks, respectively. Thus, from these results, it can be concluded that IZO nanodisks modified GCE exhibits better sensing performances than stepped hexagonal IZO nanorods modified GCE.

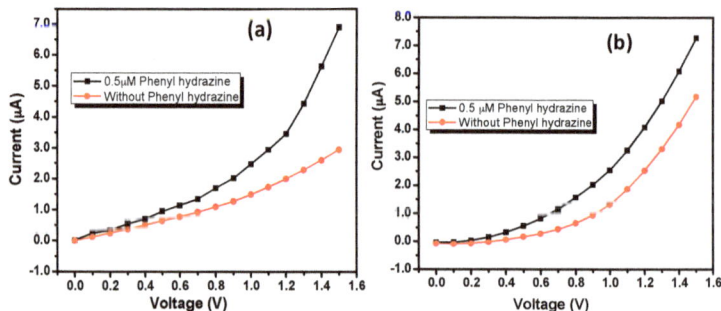

Figure 5. I–V response with 0.5 µM phenyl hydrazine and without phenyl hydrazine using IZO (**a**) nanorods and (**b**) nanodisks modified GCE in 0.1 M PBS solution.

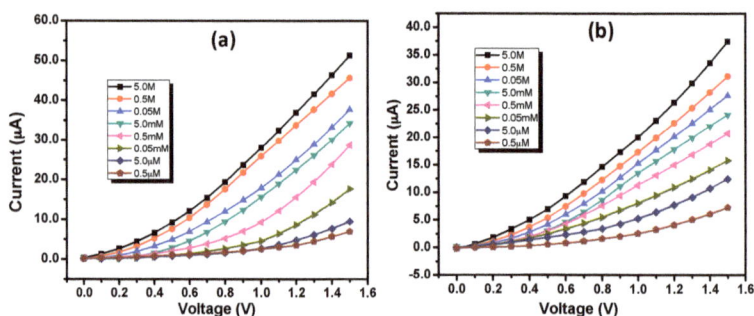

Figure 6. I–V responses for IZO (**a**) nanorods and (**b**) nanodisks against various concentrations of phenyl hydrazine (0.5 µM–5.0 M) in 0.1 M phosphate buffer solution (PBS).

Figure 7 represents the respective calibration plots for IZO nanostructures. Sensitivity, linear dynamic range (LDR) and detection limits were evaluated from these calibration plots. Sensitivity was measured from the ratio of the slope of the calibration plot to the active surface area of the modified GCE [18,45]. For stepped hexagonal IZO nanorods modified GCE, sensitivity was 70.43 µA·mM^{-1}·cm^{-2}, whereas for IZO nanodisks modified GCE it was 130.18 µA·mM^{-1}·cm^{-2}. For both of the modified electrodes the LDR and experimental detection limits were 0.5 µM–5.0 mM and 0.5 µM, respectively. The high sensitivity of IZO nanodisks based sensors are due to the high surface to volume ratio of as-grown nanodisks when compared to the IZO nanorods. The low-dimensionality and reduced dimensions of the IZO nanodisks can be well-understood by the observed SEM images, as shown in Figures 2 and 3. Thus, it can be concluded that nanomaterials dimensions are significantly important for the sensing performance, and hence with reduced dimensions, enhanced sensitivity for the fabricated sensor based on IZO nanodisks can be achieved.

It has been reported that the greater the growth temperature the greater the density of defects on the surface of the nanomaterials due to the diffusion of the oxygen from the crystal lattices creating anion vacancies and enhancing the positive charge density [60–62]. As IZO nanodisks were grown at higher temperature i.e., 850 °C as compared to a 700 °C growth temperature for In-doped ZnO nanorods, the former exhibited superior sensing behavior.

The phenyl hydrazine responses for different sensors reported in the literature are summarized in Table 2. As grown IZO nanodisks based phenyl hydrazine sensor exhibited better sensitivities as compared to other reported sensors. Thus, IZO nanodisks could be efficient electron mediator and electro-catalyst for the fabrication of phenyl hydrazine chemical sensors.

Table 2. Reported phenyl hydrazine sensing parameters of various nanostructures.

Sensing Materials	Sensitivity ($\mu A \cdot mM^{-1} \cdot cm^{-2}$)	LDR (μM–mM)	Detection limit (μM)	Ref.
ZnO nanourchin	42.1	98.0–3.126	78.6	[45]
ZnO–SiO$_2$ nanocomposite	10.80	390.0–50.0	1.42	[63]
ZnO-Fe$_2$O$_3$ microwires	8.33	10^{-3}–10.0	6.7×10^{-4}	[64]
Al- doped ZnO Nanoparticles	1.143	10.0–50.0	1.215 ± 0.02	[65]
CuO hollow spheres	0.578	5×10^3–10.0	2.4×10^3	[66]
CuO flowers	7.145			
Fe$_2$O$_3$ nanoparticles	57.88	97.0–1.56	97	[67]
Cd$_{0.5}$Mg$_{0.5}$Fe$_2$O$_4$ ferrite nanoparticles	7.01	3×10^3–100	3×10^3	[68]
TiO$_2$ nanotubes	40.9	0.25–0.10	0.22	[69]
Ferrocene-modified carbon nanotube	25.3	0.85–0.7	0.6	[70]
IZO nanorods	70.43	0.5–5.0	0.5	*This study*
IZO nanodisks	130.18			

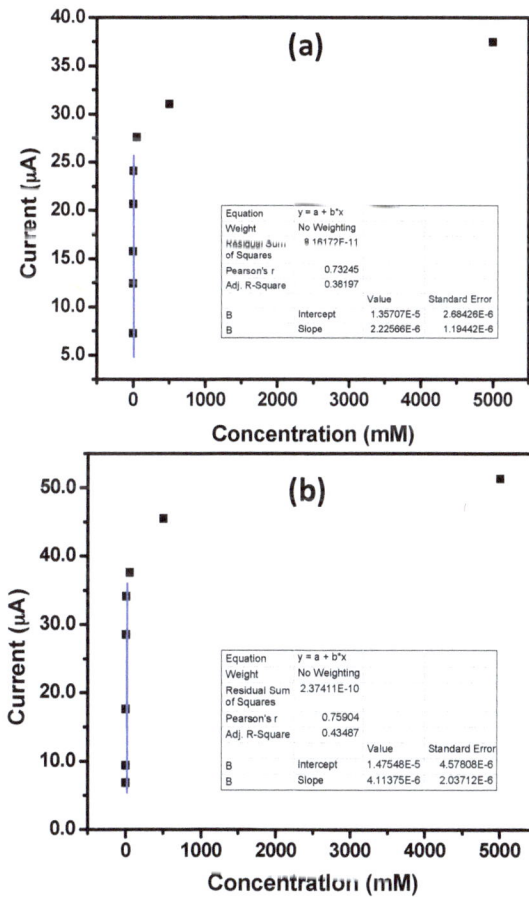

Figure 7. Calibration curves for IZO (**a**) nanorods and (**b**) nanodisks against various concentrations of phenyl hydrazine (0.5 μM–5.0 M) in 0.1 M PBS.

3.3. Proposed Mechanism

Doping of the ZnO nanostructures with In^{3+} ions induces the surface lattice defects along with an active surface area for the effective adsorption of the O_2 molecules from the surrounding air, as well as the phenyl hydrazine molecules from the PBS [71,72]. The presence of electron donor –NH_2 groups and π-electron density of the phenyl rings of phenyl hydrazine molecules further enhance the attractions between the analyte molecules and the active sites of the IZO nanostructures. The adsorbed O_2 molecules are subsequently converted into oxygenated anionic species i.e., O_2^-, O^{2-}, and O^- etc. by extracting the conduction band electrons of the IZO nanostructures [73] (Equations (2)–(4)).

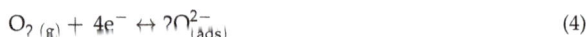

$$O_{2\,(g)} + e^- \leftrightarrow O_{2\,(ads)}^- \tag{2}$$

$$O_{2\,(g)} + 2e^- \leftrightarrow 2O_{(ads)}^- \tag{3}$$

$$O_{2\,(g)} + 4e^- \leftrightarrow 2O_{(ads)}^{2-} \tag{4}$$

These oxygenated chemical species oxidize the phenyl hydrazine molecules into diazenyl benzene (Figure 8). This oxidation process releases the electrons that are transferred back to the conduction band of the IZO nanostructures, thereby increasing the conductivity and hence the response current.

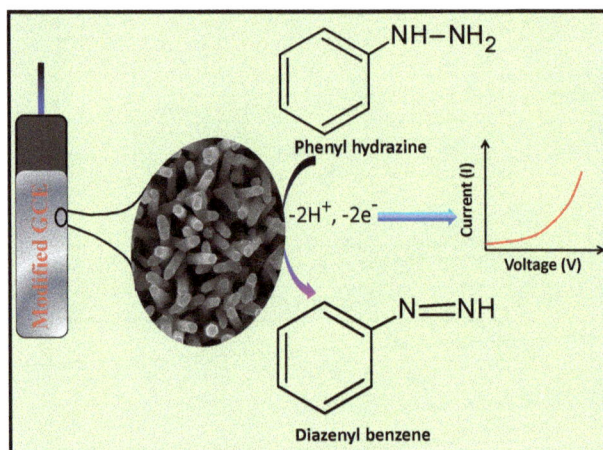

Figure 8. Proposed mechanism for phenyl hydrazine chemical sensing based on IZO nanostructures coated GCE by I–V technique.

4. Conclusions

In summary, In-doped ZnO (IZO) nanomaterials, i.e., stepped hexagonal nanorods and nanodisks were grown on silicon substrate by simple thermal evaporation process and characterized in detail using several techniques. The detailed morphological studies confirmed that both IZO nanomaterials possess well-crystallinity with wurtzite hexagonal phase and grown in high density. A suppressed UV emissions and strong green emissions for both IZO nanomaterials, i.e., nanorods and nanodisks were seen in the room-temperature PL spectra. The fabricated phenyl hydrazine chemical sensors based on as-grown IZO nanomaterials exhibited high sensitivities, i.e., 70.43 $\mu A \cdot mM^{-1} \cdot cm^{-2}$ and 130.18 $\mu A \cdot mM^{-1} \cdot cm^{-2}$, respectively, for nanorods and nanodisks. The experimental detection limits for both of the sensors were 0.5 μM, while the linear ranges were 0.5 μM–5.0 mM.

Acknowledgments: The authors extend their appreciation to the Deanship of Scientific Research at King Khalid University for funding this work through research group program under grant number R.P.G.2/5/38.

Materials **2017**, *10*, 1337

Author Contributions: Ahmad Umar, Sang Hoon Kim and Ahmed A. Ibrahim conceived and designed the experiments; Ahmad Umar, Sang Hoon Kim, Ahmed A. Ibrahim and Mohammad S. Al-Assiri performed the experiments; Ahmad Umar, Rajesh Kumar, Ahmed A. Ibrahim, A. E. Al-Salami and Sotirios Baskoutas analyzed the data; Ahmad Umar wrote the paper. All authors read, reviewed, revised and approved the manuscript.

Conflicts of Interest: The authors declare no conflict of interest.

References

1. Ibrahim, A.A.; Dar, G.N.; Zaidi, S.A.; Umar, A.; Abaker, M.; Bouzid, H.; Baskoutas, S. Growth and properties of Ag-doped ZnO nanoflowers for highly sensitive phenyl hydrazine chemical sensor application. *Talanta* **2012**, *93*, 257–263. [CrossRef] [PubMed]

2. Gholamian, F.; Sheikh-Mohseni, M.A.; Naeimi, H. Simultaneous determination of phenylhydrazine and hydrazine by a nanostructured electrochemical sensor. *Mater. Sci. Eng. C* **2012**, *32*, 2344–2348. [CrossRef]

3. Kennedy, J.; Murmu, P.P.; Leveneur, J.; Markwitz, A.; Futter, J. Controlling preferred orientation and electrical conductivity of zinc oxide thin films by post growth annealing treatment. *Appl. Surf. Sci.* **2016**, *367*, 52–58. [CrossRef]

4. Murmu, P.P.; Kennedy, J.; Williams, G.V.M.; Ruck, B.J.; Granville, S.; Chong, S.V. Observation of magnetism, low resistivity, and magnetoresistance in the near-surface region of Gd implanted ZnO. *Appl. Phys. Lett.* **2012**, *101*. [CrossRef]

5. Kaviyarasu, K.; Geetha, N.; Kanimozhi, K.; Magdalane, C.M.; Sivaranjani, S.; Ayeshamariam, A.; Kennedy, J.; Maaza, M. In vitro cytotoxicity effect and antibacterial performance of human lung epithelial cells A549 activity of Zinc oxide doped TiO₂ nanocrystals: Investigation of bio-medical application by chemical method. *Mater. Sci. Eng. C* **2017**, *74*, 325–333. [CrossRef] [PubMed]

6. Sathyaseelan, B.; Manikandan, E.; Sivakumar, K.; Kennedy, J.; Maaza, M. Enhanced visible photoluminescent and structural properties of ZnO/KIT-6 nanoporous materials for white light emitting diode (w-LED) application. *J. Alloys Compd.* **2015**, *651*, 479–482. [CrossRef]

7. Kennedy, J.; Fang, F.; Futter, J.; Leveneur, J.; Murmu, P.P.; Panin, G.N.; Kang, T.W.; Manikandan, E. Synthesis and enhanced field emission of zinc oxide incorporated carbon nanotubes. *Diam. Relat. Mater.* **2017**, *71*, 79–84. [CrossRef]

8. Kaviyarasu, K.; Magdalane, C.M.; Kanimozhi, K.; Kennedy, J.; Siddhardha, B.; Reddy, E.S.; Rotte, N.K.; Sharma, C.S.; Thema, F.T.; Letsholathebe, D.; et al. Elucidation of photocatalysis, photoluminescence and antibacterial studies of ZnO thin films by spin coating method. *J. Photochem. Photobiol. B Biol.* **2017**, *173*, 466–475. [CrossRef] [PubMed]

9. Kennedy, J.; Murmu, P.P.; Leveneur, J.; Williams, V.M.; Moody, R.L.; Maity, T.; Chong, S.V. Enhanced power factor and increased conductivity of aluminum doped zinc oxide thin films for thermoelectric applications. *J. Nanosci. Nanotechnol.* **2018**, *18*, 1384–1387. [CrossRef]

10. Thi, V.H.T.; Lee, B.K. Effective photocatalytic degradation of paracetamol using La-doped ZnO photocatalyst under visible light irradiation. *Mater. Res. Bull.* **2016**, *96*, 171–182. [CrossRef]

11. Alam, U.; Khan, A.; Raza, W.; Khan, A.; Bahnemann, D.; Muneer, M. Highly efficient Y and V co-doped ZnO photocatalyst with enhanced dye sensitized visible light photocatalytic activity. *Catal. Today* **2017**, *284*, 169–178. [CrossRef]

12. Xing, L.-L.; Wang, Q.; Xue, X.-Y. One-Step Synthesis of Pt–ZnO Nanoflowers and Their Enhanced Photocatalytic Activity. *Sci. Adv. Mater.* **2016**, *8*, 1275–1279. [CrossRef]

13. Hassan, M.M.; Khan, W.; Mishra, P.; Islam, S.S.; Naqvi, A.H. Enhancement in alcohol vapor sensitivity of Cr doped ZnO gas sensor. *Mater. Res. Bull.* **2017**, *93*, 391–400. [CrossRef]

14. Sankar Ganesh, R.; Durgadevi, E.; Navaneethan, M.; Patil, V.L.; Ponnusamy, S.; Muthamizhchelvan, C.; Kawasaki, S.; Patil, P.S.; Hayakawa, Y. Low temperature ammonia gas sensor based on Mn-doped ZnO nanoparticle decorated microspheres. *J. Alloys Compd.* **2017**, *721*, 182–190. [CrossRef]

15. Li, X.; Hu, Z.; Liu, J.; Li, D.; Zhang, X.; Chen, J.; Fang, J. Ga doped ZnO photonic crystals with enhanced photocatalytic activity and its reaction mechanism. *Appl. Catal. B Environ.* **2016**, *195*, 29–38. [CrossRef]

16. Movlarooy, T. Transition metals doped and encapsulated ZnO nanotubes: Good materials for the spintronic applications. *J. Magn. Magn. Mater.* **2017**, *441*, 139–148. [CrossRef]

17. Ibrahim, A.A.; Tiwari, P.; Al-Assiri, M.S.; Al-Salami, A.E.; Umar, A.; Kumar, R.; Kim, S.H.; Ansari, Z.A.; Baskoutas, S. A Highly-Sensitive Picric Acid Chemical Sensor Based on ZnO Nanopeanuts. *Materials* **2017**, *10*, 795. [CrossRef] [PubMed]

18. Ibrahim, A.A.; Umar, A.; Kumar, R.; Kim, S.H.; Bumajdad, A.; Baskoutas, S. Sm_2O_3-doped ZnO beech fern hierarchical structures for nitroaniline chemical sensor. *Ceram. Int.* **2016**, *42*, 16505–16511. [CrossRef]

19. Ngai, K.S.; Tan, W.T.; Zainal, Z.; Zawawi, R.M.; Juan, J.C. Electrochemical Sensor Based on Single-Walled Carbon Nanotube/ZnO Photocatalyst Nanocomposite Modified Electrode for the Determination of Paracetamol. *Sci. Adv. Mater.* **2016**, *8*, 788–796. [CrossRef]

20. Sun, H.; Jen, S.U.; Chiang, H.P.; Chen, S.C.; Lin, M.H.; Chen, J.Y.; Wang, X. Investigation of optoelectronic performance in In, Ga co-doped ZnO thin films with various In and Ga levels. *Thin Solid Films* **2016**, *641*, 12–18. [CrossRef]

21. Zhong, W.; Li, Z.; Zhang, L.; Wang, G.; Liu, Y.; Chen, W. Dependence of Electronic and Optical Properties of Zinc Oxide on Hydrostatic Pressure. *Sci. Adv. Mater.* **2016**, *8*, 1112–1115. [CrossRef]

22. Parthiban, R.; Balamurugan, D.; Jeyaprakash, B.G. Spray deposited ZnO and Ga doped ZnO based DSSC with bromophenol blue dye as sensitizer: Efficiency analysis through DFT approach. *Mater. Sci. Semicond. Process.* **2015**, *31*, 471–477. [CrossRef]

23. Park, N.M.; Oh, M.; Na, Y.B.; Cheong, W.S.; Kim, H. Sputter deposition of Sn-doped ZnO/Ag/Sn-doped ZnO transparent contact layer for GaN LED applications. *Mater. Lett.* **2016**, *180*, 72–76. [CrossRef]

24. Kumar, M.; Jeong, H.; Lee, D. Sol-gel derived Hf- and Mg-doped high-performance ZnO thin film transistors. *J. Alloys Compd.* **2017**, *720*, 230–238. [CrossRef]

25. Wang, Y.; Khan, M.Y.; Lee, S.-K.; Park, Y.-K.; Hahn, Y.-B. Parametric Study of Nozzle-Jet Printing for Directly Drawn ZnO Field-Effect Transistors. *Sci. Adv. Mater.* **2016**, *8*, 148–155. [CrossRef]

26. Liau, L.C.K.; Huang, J.S. Energy-level variations of Cu-doped ZnO fabricated through sol-gel processing. *J. Alloys Compd.* **2017**, *702*, 153–160. [CrossRef]

27. Bernardo, M.S.; Villanueva, P.G.; Jardiel, T.; Calatayud, D.G.; Peiteado, M.; Caballero, A.C. Ga-doped ZnO self-assembled nanostructures obtained by microwave-assisted hydrothermal synthesis: Effect on morphology and optical properties. *J. Alloys Compd.* **2017**, *722*, 920–927. [CrossRef]

28. Nebatti, A.; Pflitsch, C.; Atakan, B. Unusual application of aluminium-doped ZnO thin film developed by metalorganic chemical vapour deposition for surface temperature sensor. *Thin Solid Films* **2017**, *636*, 532–536. [CrossRef]

29. Srinatha, N.; Raghu, P.; Mahesh, H.M.; Angadi, B. Spin-coated Al-doped ZnO thin films for optical applications: Structural, micro-structural, optical and luminescence studies. *J. Alloys Compd.* **2017**, *722*, 888–895. [CrossRef]

30. Wang, Y.; Xu, M.; Li, J.; Ma, J.; Wang, X.; Wei, Z.; Chu, X.; Fang, X.; Jin, F. Sol-combustion synthesis of Al-doped ZnO transparent conductive film at low temperature. *Surf. Coat. Technol.* **2017**, *330*, 255–259. [CrossRef]

31. Bhati, V.S.; Ranwa, S.; Fanetti, M.; Valant, M.; Kumar, M. Efficient hydrogen sensor based on Ni-doped ZnO nanostructures by RF sputtering. *Sens. Actuators B Chem.* **2018**, *255*, 588–597. [CrossRef]

32. Shewale, P.S.; Lee, S.H.; Yu, Y.S. Pulse repetition rate dependent structural, surface morphological and optoelectronic properties of Ga-doped ZnO thin films grown by pulsed laser deposition. *J. Alloys Compd.* **2017**, *725*, 1106–1114. [CrossRef]

33. Heo, J.; Kwon, S.J.; Cho, E.S. Nd: YVO4 Laser Direct Patterning of Aluminum-Doped Zinc Oxide Films Sputtered Under Different Process Conditions. *Sci. Adv. Mater.* **2016**, *8*, 1783–1789. [CrossRef]

34. Kaidashev, E.M.; Lorenz, M.; Von Wenckstern, H.; Rahm, A.; Semmelhack, H.C.; Han, K.H.; Benndorf, G.; Bundesmann, C.; Hochmuth, H.; Grundmann, M. High electron mobility of epitaxial ZnO thin films on c-plane sapphire grown by multistep pulsed-laser deposition. *Appl. Phys. Lett.* **2003**, *82*, 3901–3903. [CrossRef]

35. Chava, R.K.; Kang, M. Improving the photovoltaic conversion efficiency of ZnO based dye sensitized solar cells by indium doping. *J. Alloys Compd.* **2017**, *692*, 67–76. [CrossRef]

36. Chung, C.; Kim, Y.J.; Han, H.H.; Lim, D.; Jung, W.S.; Choi, M.S.; Nam, H.-J.; Son, S.-K.; Sergeevich, A.S.; Park, J.-H.; et al. Synthesis of P-Type ZnO Thin Films with Arsenic Doping and Post Annealing. *Sci. Adv. Mater.* **2016**, *8*, 1857–1860. [CrossRef]

37. Han, N.; Chai, L.; Wang, Q.; Tian, Y.; Deng, P.; Chen, Y. Evaluating the doping effect of Fe, Ti and Sn on gas sensing property of ZnO. *Sens. Actuators B Chem.* **2010**, *147*, 525–530. [CrossRef]

38. Park, J.C.; Kim, D.J.; Lee, H.-N. Characteristics of Amorphous Indium-Zinc-Oxide Thin-Film Transistors Fabricated with a Self-Aligned Coplanar Structure and an NH_3 Plasma Contact Doping Process. *Sci. Adv. Mater.* **2016**, *8*, 295–300. [CrossRef]

39. Liau, C.K.; Huang, J.-S. Effect of indium- and gallium-doped ZnO fabricated through sol-gel processing on energy level variations. *Mater. Res. Bull.* **2018**, *97*, 6–12. [CrossRef]

40. Ghosh, S.; Saha, M.; De, S.K. Tunable surface plasmon resonance and enhanced electrical conductivity of In doped ZnO colloidal nanocrystals. *Nanoscale* **2014**, *6*, 7039–7051. [CrossRef] [PubMed]

41. Badadhe, S.S.; Mulla, I.S. H_2S gas sensitive indium-doped ZnO thin films: Preparation and characterization. *Sens. Actuators B Chem.* **2009**, *143*, 164–170. [CrossRef]

42. Dhahri, R.; Hjiri, M.; El Mir, L.; Alamri, H.; Bonavita, A.; Iannazzo, D.; Leonardi, S.G.; Neri, G. CO sensing characteristics of In-doped ZnO semiconductor nanoparticles. *J. Sci. Adv. Mater. Devices* **2017**, *2*, 34–40. [CrossRef]

43. Wang, Z.; Tian, Z.; Han, D.; Gu, F. Highly Sensitive and Selective Ethanol Sensor Fabricated with In-Doped 3DOM ZnO. *ACS Appl. Mater. Interfaces* **2016**, *8*, 5466–5474. [CrossRef] [PubMed]

44. Ge, Y.; Wei, Z.; Li, Y.; Qu, J.; Zu, B.; Dou, X. Highly sensitive and rapid chemiresistive sensor towards trace nitro-explosive vapors based on oxygen vacancy-rich and defective crystallized In-doped ZnO. *Sens. Actuators B Chem.* **2017**, *244*, 983–991. [CrossRef]

45. Umar, A.; Akhtar, M.S.; Al-Hajry, A.; Al-Assiri, M.S.; Dar, G.N.; Saif Islam, M. Enhanced photocatalytic degradation of harmful dye and phenyl hydrazine chemical sensing using ZnO nanourchins. *Chem. Eng. J.* **2015**, *262*, 588–596. [CrossRef]

46. Kumar, R.; Kumar, G.; Umar, A. ZnO nano-mushrooms for photocatalytic degradation of methyl orange. *Mater. Lett.* **2013**, *97*, 100–103. [CrossRef]

47. Al-Hadeethi, Y.; Umar, A.; Al-Heniti, S.H.; Kumar, R.; Kim, S.H.; Zhang, X.; Raffah, B.M. 2D Sn-doped ZnO ultrathin nanosheet networks for enhanced acetone gas sensing application. *Ceram. Int.* **2017**, *43*, 2418–2423. [CrossRef]

48. Liu, X.; Jiang, L.; Jiang, X.; Tian, X.; Sun, X.; Wang, Y.; He, W.; Hou, P.; Deng, X.; Xu, X. Synthesis of Ce-doped In_2O_3 nanostructure for gas sensor applications. *Appl. Surf. Sci.* **2018**, *428*, 478–484. [CrossRef]

49. Anand, K.; Kaur, J.; Singh, R.C.; Thangaraj, R. Preparation and characterization of Ag-doped In_2O_3 nanoparticles gas sensor. *Chem. Phys. Lett.* **2017**, *682*, 140–146. [CrossRef]

50. Sun, J.; Wang, Q.; Wang, Q.; Zhang, D.-A.; Xing, L.-L.; Xue, X.-Y. High Capacity and Cyclability of SnO_2–In_2O_3/Graphene Nanocomposites as the Anode of Lithium-Ion Battery. *Sci. Adv. Mater.* **2016**, *8*, 1280–1285. [CrossRef]

51. Kennedy, J.; Murmu, P.P.; Manikandan, E.; Lee, S.Y. Investigation of structural and photoluminescence properties of gas and metal ions doped zinc oxide single crystals. *J. Alloys Compd.* **2014**, *616*, 614–617. [CrossRef]

52. Mannam, R.; Kumar, E.S.; Priyadarshini, D.M.; Bellarmine, F.; DasGupta, N.; Ramachandra Rao, M.S. Enhanced photoluminescence and heterojunction characteristics of pulsed laser deposited ZnO nanostructures. *Appl. Surf. Sci.* **2017**, *418*, 335–339. [CrossRef]

53. Lv, Y.; Zhang, Z.; Yan, J.; Zhao, W.; Zhai, C.; Liu, J. Growth mechanism and photoluminescence property of hydrothermal oriented ZnO nanostructures evolving from nanorods to nanoplates. *J. Alloys Compd.* **2017**, *718*, 161–169. [CrossRef]

54. Jiang, X.; Shang, F.; Zhou, Z.; Wang, F.; Liu, C.; Gong, W.; Lv, J.; Zhang, M.; He, G.; Sun, Z. Temperature-Dependent Photoluminescence of ZnO Nanorods Grown by a Copper-Assisted Hydrothermal Method. *Sci. Adv. Mater.* **2015**, *7*, 1800–1803. [CrossRef]

55. Yi, X.Y.; Ma, C.Y.; Yuan, F.; Wang, N.; Qin, F.W.; Hu, B.C.; Zhang, Q.Y. Structural, morphological, photoluminescence and photocatalytic properties of Gd-doped ZnO films. *Thin Solid Films* **2017**, *636*, 339–345. [CrossRef]

56. Kennedy, J.; Carder, D.A.; Markwitz, A.; Reeves, R.J. Properties of nitrogen implanted and electron beam annealed bulk ZnO. *J. Appl. Phys.* **2010**, *107*, 103518. [CrossRef]

57. Kumar, A.S.; Huang, N.M.; Nagaraja, H.S. Influence of Sn doping on photoluminescence and photoelectrochemical properties of ZnO nanorod arrays. *Electron. Mater. Lett.* **2014**, *10*, 753–758. [CrossRef]

58. Nishad, K.K.; Joseph, J.; Tiwari, N.; Kurchania, R.; Pandey, R.K. Investigation on Size Dependent Elemental Binding Energies and Structural Properties of ZnO Nanoparticles and Their Correlation with Observed Photo-Luminescence Behavior. *Sci. Adv. Mater.* **2015**, *7*, 1368–1378. [CrossRef]

59. Umar, A.; Akhtar, M.S.; Dar, G.N.; Baskoutas, S. Low-temperature synthesis of α-Fe$_2$O$_3$ hexagonal nanoparticles for environmental remediation and smart sensor applications. *Talanta* **2013**, *116*, 1060–1066. [CrossRef] [PubMed]

60. Chen, C.-Y.; Liu, Y.-R.; Lin, S.-S.; Hsu, L.-J.; Tsai, S.-L. Role of Annealing Temperature on the Formation of Aligned Zinc Oxide Nanorod Arrays for Efficient Photocatalysts and Photodetectors. *Sci. Adv. Mater.* **2016**, *8*, 2197–2203. [CrossRef]

61. Umar, A.; Kumar, R.; Kumar, G.; Algarni, H.; Kim, S.H. Effect of annealing temperature on the properties and photocatalytic efficiencies of ZnO nanoparticles. *J. Alloys Compd.* **2015**, *648*, 46–52. [CrossRef]

62. Ren, H.; Xiang, G.; Gu, G.; Zhang, X. Enhancement of ferromagnetism of ZnO:Co nanocrystals by post-annealing treatment: The role of oxygen interstitials and zinc vacancies. *Mater. Lett.* **2014**, *122*, 256–260. [CrossRef]

63. Ali, A.M.; Harraz, F.A.; Ismail, A.A.; Al Ouyail, S.A.; Algarni, H.; Al-Sehemi, A.G. Synthesis of amorphous ZnO-SiO2 nanocomposite with enhanced chemical sensing properties. *Thin Solid Films* **2016**, *605*, 277–282. [CrossRef]

64. Rahman, M.M.; Gruner, G.; Al-Ghamdi, M.S.; Daous, M.A.; Khan, S.B.; Asiri, A.M. Fabrication of highly sensitive phenyl hydrazine chemical sensor based on as-grown ZnO-Fe$_2$O$_3$ microwires. *Int. J. Electrochem. Sci.* **2013**, *8*, 520–534.

65. Rahman, M.M.; Khan, S.B.; Jamal, A.; Faisal, M.; Asiri, A.M. Fabrication of phenyl-hydrazine chemical sensor based on Al-doped ZnO Nanoparticles. *Sens. Transducers* **2011**, *134*, 32–44.

66. Khan, S.B.; Faisal, M.; Rahman, M.M.; Abdel-Latif, I.A.; Ismail, A.A.; Akhtar, K.; Al-Hajry, A.; Asiri, A.M.; Alamry, K.A. Highly sensitive and stable phenyl hydrazine chemical sensors based on CuO flower shapes and hollow spheres. *New J. Chem.* **2013**, *37*, 1098–1104. [CrossRef]

67. Hwang, S.W.; Umar, A.; Dar, G.N.; Kim, S.H.; Badran, R.I. Synthesis and characterization of iron oxide nanoparticles for phenyl hydrazine sensor applications. *Sens. Lett.* **2014**, *12*, 97–101. [CrossRef]

68. Al-Heniti, S.H.; Umar, A.; Zaki, H.M.; Dar, G.N.; Al-Ghamdi, A.A.; Kim, S.H. Synthesis and Characterizations of Ferrite Nanomaterials for Phenyl Hydrazine Chemical Sensor Applications. *J. Nanosci. Nanotechnol.* **2014**, *14*, 3765–3770. [CrossRef] [PubMed]

69. Ameen, S.; Shaheer Akhtar, M.; Seo, H.K.; Shin, H.S. TiO$_2$ nanotube arrays via electrochemical anodic oxidation: Prospective electrode for sensing phenyl hydrazine. *Appl. Phys. Lett.* **2013**, *103*. [CrossRef]

70. Afzali, D.; Karimi-Maleh, H.; Khalilzadeh, M.A. Sensitive and selective determination of phenylhydrazine in the presence of hydrazine at a ferrocene-modified carbon nanotube paste electrode. *Environ. Chem. Lett.* **2011**, *9*, 375–381. [CrossRef]

71. Xiang, D.; Lin, H.; Ma, J.; Chen, X.; Jiang, J.; Qu, F. A Facile Method for the Synthesis of Porous ZnO Hollow Microspheres and Their Gas Sensing Properties for Acetone. *Sci. Adv. Mater.* **2015**, *7*, 1319–1325. [CrossRef]

72. Khan, M.Y.; Ahmad, R.; Lee, G.H.; Suh, E.-K.; Hahn, Y.-B. Effect of Annealing Atmosphere on the Optical and Electrical Properties of Al-Doped ZnO Films and ZnO Nanorods Grown by Solution Process. *Sci. Adv. Mater.* **2016**, *8*, 1523–1529. [CrossRef]

73. Ibrahim, A.A.; Hwang, S.W.; Dar, G.N.; Kim, S.H.; Abaker, M.; Ansari, S.G. Synthesis and Characterization of Gd-Doped ZnO Nanopencils for Acetone Sensing Application. *Sci. Adv. Mater.* **2015**, *7*, 1241–1246. [CrossRef]

materials

MDPI

Article

Study of Perfluorophosphonic Acid Surface Modifications on Zinc Oxide Nanoparticles

Rosalynn Quiñones [1,*], **Deben Shoup** [1], **Grayce Behnke** [1], **Cynthia Peck** [1], **Sushant Agarwal** [2], **Rakesh K. Gupta** [2], **Jonathan W. Fagan** [3], **Karl T. Mueller** [3,4], **Robbie J. Iuliucci** [5] and **Qiang Wang** [6,7]

[1]   Department of Chemistry, Marshall University, Huntington, WV 25755, USA; shoup2@marshall.edu (D.S.); behnke3@marshall.edu (G.B.); peck24@marshall.edu (C.P.)

[2]   Department of Chemical & Biomedical Engineering, West Virginia University, Morgantown, WV 26506, USA; Sushant.Agarwal@mail.wvu.edu (S.A.); Rakesh.Gupta@mail.wvu.edu (R.K.G.)

[3]   Department of Chemistry, Pennsylvania State University, State College, PA 16802, USA; jwf188@psu.edu (J.W.F.); Karl.Mueller@pnnl.gov (K.T.M.)

[4]   Physical and Computational Sciences Directorate, Pacific Northwest National Laboratory, Richland, WA 99352, USA

[5]   Chemistry Department, Washington and Jefferson College, Washington, PA 15391, USA; riuliucci@washjeff.edu

[6]   Department of Physics and Astronomy, West Virginia University, Morgantown, WV 25606, USA; qiang.wang@mail.wvu.edu

[7]   Shared Research Facilities, West Virginia University, Morgantown, WV 25606, USA

*   Correspondence: quinonesr@marshall.edu; Tel.: +1-304-696-6731; Fax: +1-304-696-3243

Received: 27 October 2017; Accepted: 22 November 2017; Published: 28 November 2017

Abstract: In this study, perfluorinated phosphonic acid modifications were utilized to modify zinc oxide (ZnO) nanoparticles because they create a more stable surface due to the electronegativity of the perfluoro head group. Specifically, 12-pentafluorophenoxydodecylphosphonic acid, 2,3,4,5,6-pentafluorobenzylphosphonic acid, and (1H,1H,2H,2H-perfluorododecyl)phosphonic acid have been used to form thin films on the nanoparticle surfaces. The modified nanoparticles were then characterized using infrared spectroscopy, X-ray photoelectron spectroscopy, and solid-state nuclear magnetic resonance spectroscopy. Dynamic light scattering and scanning electron microscopy-energy dispersive X-ray spectroscopy were utilized to determine the particle size of the nanoparticles before and after modification, and to analyze the film coverage on the ZnO surfaces, respectively. Zeta potential measurements were obtained to determine the stability of the ZnO nanoparticles. It was shown that the surface charge increased as the alkyl chain length increases. This study shows that modifying the ZnO nanoparticles with perfluorinated groups increases the stability of the phosphonic acids adsorbed on the surfaces. Thermogravimetric analysis was used to distinguish between chemically and physically bound films on the modified nanoparticles. The higher weight loss for 12-pentafluorophenoxydodecylphosphonic acid and (1H,1H,2H,2H-perfluorododecyl)phosphonic acid modifications corresponds to a higher surface concentration of the modifications, and, ideally, higher surface coverage. While previous studies have shown how phosphonic acids interact with the surfaces of ZnO, the aim of this study was to understand how the perfluorinated groups can tune the surface properties of the nanoparticles.

Keywords: self-assembly films; zinc oxide; perfluorophosphonic acid; solid-state NMR; zeta potential

1. Introduction

Zinc oxide (ZnO) nanoparticles have distinct properties that allow for a wide variety of applications. For example, it is an n-type semiconducting nanomaterial, which has allowed for its use as a biosensor and as a layer in light emitting diodes [1,2]. Due to its abundance and semiconductive properties, ZnO has been used as an ideal electron transfer layer in inverted solar cells [3–9]. The implementation of ZnO into electronic and solar devices is largely due to its unique properties, such as a wide band gap (3.37 meV), stable wurtzite crystal structure, and high exciton binding energy (60 meV) [10–14]. Much attention has been focused on modifying the surfaces of ZnO to make it more suitable for applications, such as in heterojunction solar cells and organic light emitting diodes [15,16]. Modifying the surface via organic acid thin films, specifically phosphonic acids, has already shown promise as a means of tailoring and enhancing the aforementioned properties of ZnO [17].

Organic acids are chemically adsorbed onto a surface by forming self-assembled monolayers (SAMs). SAMs have been used to modify surfaces for a variety of purposes, such as forming thiols on the surface of gold for use as a biosensor, modifying metal alloys for use in biomaterials, and serving as a platform for surface initiated polymerization [18–20]. Head groups bind to the surface of a substrate via chemisorption, and commonly consist of thiols, phosphonates, and carboxylic acids [21–23]. Phosphonic acid SAMs have previously been shown to strongly bind to the surface of zinc oxide and other metal oxides, so they have been utilized to produce favorable changes, such as work function, and hydrophobicity [17,24–26]. Interface modifiers with self-assembling properties have been used to improve the charge transfer between organic layers and metal oxides through covalently bonding the modifiers onto the surface of the metal oxide [4]. The modifier can serve multiple purposes, including passivation of the surface charge traps to improve forward charge transfer, tuning of the energy level offset between semiconductors and organic layers, and affecting the upper organic layer morphology [4,27]. The surfaces of the ZnO nanostructures have the potential to be improved via SAMs to make them less corrosive, more stable, and electronically favorable.

There has been interest in using perfluorophosphonic acids to modify ZnO single crystal surfaces in order to fine tune work function and contribute to other favorable characteristics [28]. For example, SAMs of organic phosphonic acids that contain perfluoro groups have been utilized due to the high electronegativity of fluorine and low surface tension, and because phosphonic acids that have tail groups containing fluorine typically have stronger surface adhesion than SAMs with carboxylic acid tail groups [29]. Perfluorophosphonic acids SAMs have also been shown to exhibit a better control of electronic properties than alkyl phosphonic acid SAMs. For instance, they have been used to modify the surface of aluminum and have successfully controlled voltage of thin film transistors [30]. Another ideal property of the perfluorophosphonic acids is that when a fluorinated compound is introduced to a surface, a fluorination effect occurs, which allows for changes in energy levels without steric hindrance [31]. Aromatic perfluorophosphonic acids have previously been shown to have a more densely packed monolayer with a higher fraction of bidentate binding than alkyl chain phosphonic acids, which has contributed to the ability to favorably tune the work function [28,32].

In this study, 12-pentafluorophenoxydodecylphosphonic acid (PFPDPA), 2,3,4,5,6-pentafluorobenzylphosphonic acid (5FBPA), and (1H,1H,2H,2H-Perfluorododecyl)phosphonic acid (F$_{21}$DDPA) were used to form thin films on the surface of ZnO nanoparticles (Figure 1).

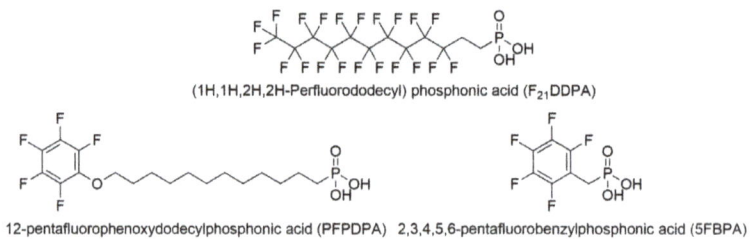

(1H,1H,2H,2H-Perfluorododecyl) phosphonic acid (F$_{21}$DDPA)

12-pentafluorophenoxydodecylphosphonic acid (PFPDPA)   2,3,4,5,6-pentafluorobenzylphosphonic acid (5FBPA)

Figure 1. Molecular structures of the perfluorophosphonic acids used to modify the Zinc oxide (ZnO) nanoparticles.

The nanoparticles were characterized using X-ray photoelectron spectroscopy (XPS) (Physical Electronics Inc., Chanhassen, MN, USA) to identify surface composition and surface ratio, and a scanning electron microscope with energy dispersive X-ray spectroscopy (SEM/EDS) (JEOL, Peabody, MA, USA) to observe the morphology and particle sizes of the nanoparticles. Dynamic light scattering (DLS) (Brookhaven Instrument Corporation, Long Island, NY, USA), in addition to the particle size data that is obtained from the SEM images, and zeta potential were used to determine the size and surface charge of the ZnO nanoparticles. Attenuated total reflectance infrared spectroscopy (ATR-IR) (Thermo Fisher Scientific, Waltham, MA, USA) was used to analyze the presence and ordering of the acids that were bonded to the surfaces of the nanoparticles. Solid-state nuclear magnetic resonance (SS-NMR) (Bruker Corporation, Billerica, MA, USA) identified the mode(s) of attachment of the phosphorous on the surface of the ZnO. Thermogravimetric analysis (TGA) (TA Instruments, New Castle, DE, USA) was used to measure the mass changes of the modified samples.

2. Results and Discussion

2.1. Attenuated Total Reflectance Infrared Spectroscopy (ATR-IR)

ATR–IR was used to distinguish chemisorbed from physisorbed organic thin films on the surfaces of the nanoparticles. In addition, changes on phosphonic head group IR frequencies indicate the manner by which the self-assembled thin films are bonded to the surface. The ZnO nanoparticles were modified via surface chemical adsorption of three organic acids, with phosphonic functional groups (PFPDPA, 5FBPA, and F_{21}DDPA). These reactions led to the formation of self-assembled organic thin films on the surface of the nanoparticles. Since ZnO by itself cannot result in detection of organic IR stretches, spectra readily reveal the thin film formation. Rinsing nanoparticles in THF, followed by sonication and vacuum centrifugation is an effective means to remove weakly bound, physisorbed films [19,33,34]. The presence of organic stretches that exist after sonication, as shown in Figure 2, confirm that the attachments were both stable and chemically bonded.

Typically, the C–H stretches (3000 to 2800 cm^{-1}) of SAMs are used as reference bands for SAM organization and the presence of the film on the modified surface [17,35]. However, with the phosphonic acids being used to modify the substrates, only the PFPDPA substrates have these recognizable reference bands, as observed in Figure 2a. There are two characteristic vibrations that are associated with this C–H region: An asymmetrical stretch at ~2918 cm^{-1} and a symmetric stretch at ~2850 cm^{-1}. However, these stretches are shifted depending on the conformation of the alkyl chain. Both C–H bands will shift to lower wavenumbers when the alkyl chain is organized on the surface by forming mostly *trans* conformations [36]. After the samples were rinsed and sonicated, the values of $\bar{v}_{CH2}$ for PFPDPA were 2918 cm^{-1} for $\bar{v}_{CH2\,asym}$ and 2848 cm^{-1} for $\bar{v}_{CH2\,sym}$. The values show that the attachment of the organic acids form strongly-bound and ordered thin films.

ATR-IR was used to identify the vibrational fingerprints of the molecules that were bonded to ZnO. This allows for the verification of attachment using the P–O vibrations and the analysis of the C–F vibrations to confirm the expected surface functionalization. Adsorbed films exhibit bands in the region from 1400 to 900 cm^{-1} related to C–F and P–O vibrations. In the PFPDPA on ZnO surface IR absorption spectrum (Figure 2b), pronounced bands at 1154, 1046, and 961 cm^{-1} are displayed, which are assigned to the C–F stretching vibration and the P=O and P–O stretching modes of adsorbed phosphonate, respectively. The 5FBPA absorption spectrum (Figure 2c) shows characteristic bands at 1128, 1079, 1023, and 980 cm^{-1}, which are assigned to C–F stretching modes of the pentafluorobenzyl group and the formation of phosphonate bonding on the surface. Furthermore, ZnO modified with F_{21}DDPA displays fewer bands at 1208, 1150, and 1071 cm^{-1} when compared to the control, as observed in Figure 2d. These bands are assigned to C–F stretching modes of the perfluoroalkyl chain and the formation of a bidentate binding motif for the phosphonate on the ZnO surface, rather than a monodentate or tridentate binding, as previously reported [28,32]. Some reports have claimed that ZnO prefers a bidentate binding motif [17,37]. The relative intensities and placements of the bands (particularly the broad P=O absorption for PFPDPA) and the shift of P=O vibration when compared to the control in the IR spectra suggest mainly bidentate binding of the phosphonic acid for F_{21}PPA modifications. Since small signals from free P–O–H vibrations (around 961 and 981 cm^{-1}) for 5FBPA and PFPDPA are observed, the possibility that a fraction of the molecules are binding through monodentate attachment and/or a bidentate attachment with a free P–O–H group cannot be ignored.

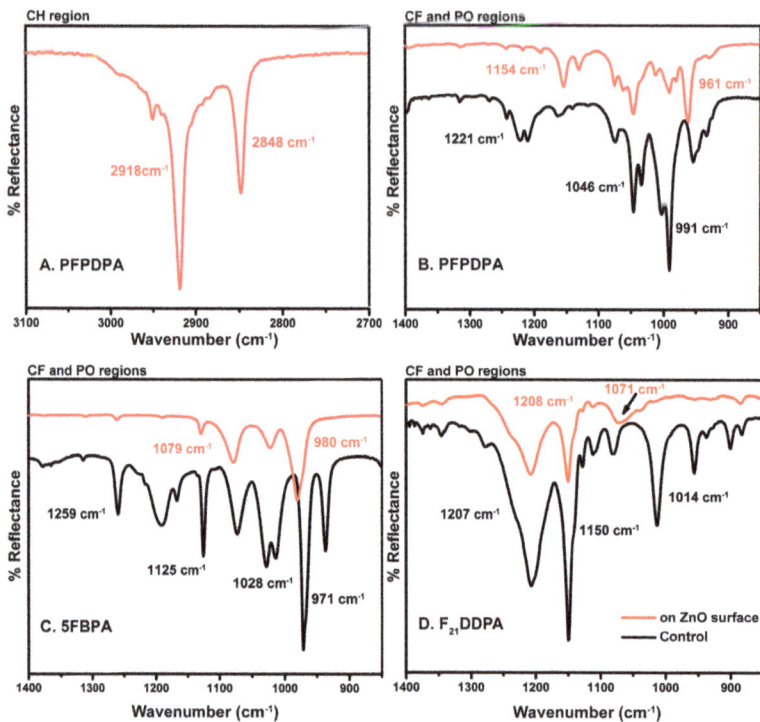

Figure 2. Infrared spectra of ZnO nanoparticles modified with (**A**) PFPDPA C–H region; (**B**) PFPDPA C–F and P–O regions; (**C**) 5FBPA C–F and P–O regions; and (**D**) F_{21}DDPA C–F and P–O regions are displayed. The spectral regions occurred after rinsing and sonication (red spectra) compared to each control (black spectra).

2.2. X-ray Photoelectron Spectroscopy (XPS)

XPS was used to further analyze the bonding motif, the chemical states, and composition of elements of the perfluorophosphonic acid modifications on ZnO nanoparticles. Compositional survey scans were acquired using a pass energy of 117.4 eV and 0.5 eV scan step and high-resolution spectra of Zn2p, O1s, C1s, F1s, and P2p were acquired for both the modified and unmodified ZnO samples using a pass energy of 23.5 eV and 0.05 eV scan step. The representative survey spectrum of unmodified ZnO nanoparticles, presented in Figure S1, in the supporting material, exhibits intense Zn and O peaks, and clearly shows the absence of fluorine and phosphorus, as expected. The insets of Figure S1 show the high-resolution spectra of Zn_{2p} and O_{1s} peaks from the unmodified ZnO control sample.

For the samples that were modified with perfluorinated phosphonic acids, the survey spectra (Figure S1) show the presence of fluorine, carbon, and phosphorus. The high-resolution spectra for each element (Zn, C, O, F, and P) modifications are shown in Figure 3. In addition, Figure 3 shows all of the fitted spectra for each element (Zn, C, O, F, and P) after modification. The binding energy and assigned peaks that were obtained from the fitting are presented in Table S1 in the supporting material.

Common to all of the samples, Zn2p at 1022 eV (2p3/2) and 1045 eV (2p1/2) was attributed to the ZnO nanoparticles and O1s at 530.5 eV was attributed to be ZnO. Furthermore, when the samples are modified, new peaks are observed. C1s peaks at 284.8 eV and 287.8 eV were attributed to –CH$_2$ and –CF functional groups, respectively. F1s at 688 eV was attributed to C–F functional group. The P2p peak for all of the modifications appears at the similar binding energy around 134 eV, indicating the formation of stable covalent bonds between the oxide surface and the deprotonated phosphonic acid headgroup [38]. As for O1s, a peak at around 532 eV was present and is attributed to oxygen of the phosphonic groups anchored to the ZnO surface [38,39]. When ZnO was modified with F$_{21}$DDPA, two new peaks for C1s at 291 and 293 eV were observed and attributed to –CF$_2$–CF$_2$ and –CF$_2$–CF$_3$ groups, respectively [40]. Furthermore, the carbon-to-phosphorus (C/P) and fluorine-to-phosphorus (F/P) ratios were calculated from the relative intensity of the elements and are summarized in Table 1. These numbers are consistent with the molecules that bonded on the surface creating a thin film, and it is a direct measurement of the molecular coverage [24,41]. These values approach the theoretical values shown in Table 1 with the deviation attributed to the possibility of multi-films on the modified surfaces. The agreement between the stoichiometric compositions and the measured C/P and F/P ratios indicate that the films are uniform and homogeneous. Fluorine signal was noticed to increase as the number of fluorine atoms was changed from the PFPDPA and 5FBPA modifications, which had five fluorine atoms, to twenty-one fluorine atoms in the F$_{21}$DDPA modification. This resulted in a corresponding increase in the atomic % concentration and F/P ratios as observed in Table 1. In summary, the XPS results confirm that perfluorophosphonic acid films have modified the surface of the ZnO nanoparticles.

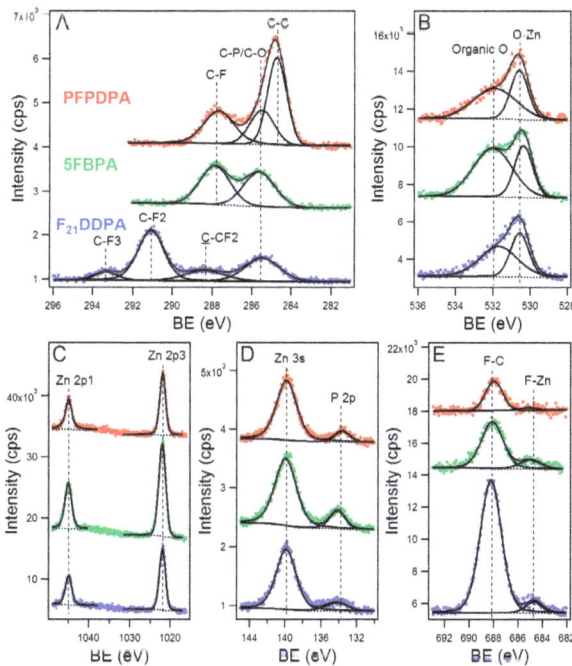

Figure 3. High-resolution X-ray photoelectron spectroscopy (XPS) spectra and fits (**A**) C1s (**B**) O1s (**C**) Zn2p (**D**) P2p, and (**E**) F1s core level spectra of PFPDPA on ZnO (red lines), 5FBPA on ZnO (green lines), and F21DDPA on ZnO nanoparticles (blue lines). Curves are shifted along the Y-axis to enable better comparison.

Table 1. Atomic percentages and C/P and F/P ratios determined by XPS and compared with theoretical values calculated from the chemical formula (numbers in parenthesis).

Sample	Zn2p %	O1s %	C1s %	F1s %	P2p %	C/P	F/P
PFPDPA	14.9	28.3	44.9	10.1	1.8	24.8 (18)	5.5 (5)
5FBPA	25.0	31.2	23.5	17.6	2.7	8.7 (7)	6.5 (5)
F_{21}DDPA	13.5	19.5	25.9	39.1	2.1	12.4 (12)	18.7 (21)
ZnO Control	47.5	52.5	-	-	-	-	-

2.3. Solid State Nuclear Magnetic Resonance Spectroscopy (SS-NMR)

Solid-state NMR spectroscopy provides a convenient examination of molecular motions and bonding motifs. Here, ^{31}P SS-NMR experiments were used to characterize the surface modifications of perfluorophosphonic acid on ZnO nanoparticles.

^{31}P NMR exhibits only one signal centered at 27.6 ppm for 5FBPA control and at 29.8 ppm for 5FBPA adsorbed on ZnO surfaces, as observed in Figure 4A. In this study, only a single peak is observed, suggesting a preferred hydrogen bonding affinity onto the ZnO surface. This conclusion is due to the fact to the sharpest and least shifted of the peaks is associated with the physisorbed films of the 5FBPA on the ZnO surface, and, therefore, this will not allow for precise conclusion regarding the type of surface bonding [42].

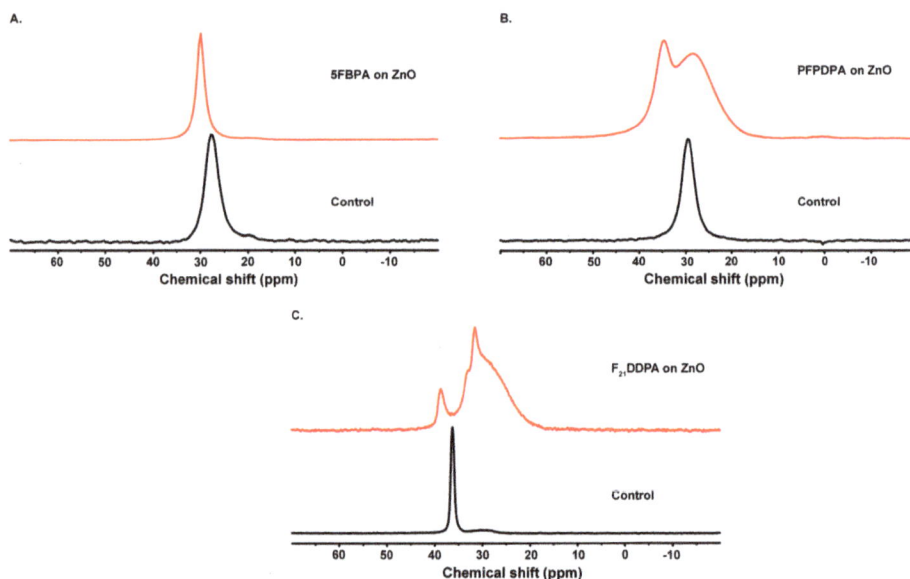

Figure 4. Solid-state ^{31}P CP-MAS NMR spectra of (**A**) 5FBPA; (**B**) PFPDPA; and (**C**) F_{21}DDPA modifications and bulk controls.

The ZnO samples modified with PFPDPA show two ^{31}P resonances between 28.3 and 34.6 ppm, reflecting the option of various bonding motifs while the control has a single peak at 29.4 ppm (Figure 4B). These resonances are broader when compared to the other two modifications and the control, confirming the presence of phosphonate adsorbed to ZnO surface. The peak at 34.6 ppm is likely due to a physisorbed species, while the remaining broad peak arises from a variety of chemisorbed bonding configurations. The broadening effect has been attributed to chemical shift heterogeneity due to a distribution in sites at the nanoparticle interface [43]. Furthermore,

ZnO modified with PFPDPA was re-analyzed two years after being modified and the ^{31}P SS-NMR spectrum was recorded. This spectrum is shown in Figure 4B and demonstrates that the film remained intact on the surface after ambient storage over a long period of time.

^{31}P CP/MAS-NMR spectra of the ZnO modified nanoparticles with F_{21}DDPA (Figure 4C) are compared to the spectra of the SS-NMR ^{31}P spectra of the bulk F_{21}DDPA control. The F_{21}DDPA control ^{31}P chemical shift changes from a single peak at 36.3 ppm with a small peak at 29.9 ppm. As the ZnO surface was modified with F_{21}DDPA, two peaks ranging from 38.6 to 31.5 ppm with the resonance broadened were observed in Figure 4C. The broadening of the peaks was previously noticed due to the possibility of multi-films or multiple bonding sites, as observed by the above XPS data [17,44,45]. Therefore, it is reasonable to associate the three peaks that were observed at lower shift values to the phosphate involved in the surface bonding via monodentate, bidentate, and/or tri-dentate motifs. The peak positioned at 38.6 ppm can be associated to a multilayer stack due to strong hydrogen bonds or fluorine with the OH groups on the ZnO surface, resulting in some chain torsions and chain motions that may improve the interactions with the ZnO surface [46].

2.4. Scanning Electron Microscopy-Energy Dispersive X-ray Spectroscopy (SEM/EDS)

SEM allowed the visualization of the morphology of the ZnO nanoparticles and EDS assisted in the identification of the elemental composition of the modified ZnO nanoparticles. The SEM images show that both unmodified and modified materials are composed of uniform oblong nanoparticles with typical particle diameters <200 nm. However, a distribution of sizes occurred. There was no visible change in the morphology of the nanoparticles after the modifications of SAMs when compared to the unmodified ZnO nanoparticles (Figure 5). The diameter of the modified nanoparticles is slightly larger than the diameter of the unmodified nanoparticles, which is expected due to the adsorbed perfluorophosphonic acid layer. These measurements are summarized in Table 2. The particle agglomeration that causes the particle sizes to appear to be greater than the value indicated from the manufacturer (>100 nm) is likely due to an increase of interaction between the particles and the charge on the surface, as observed using zeta potential analysis (see below). Aromatic rings are known to have highly delocalized electron density and structural rigidity, and these properties have potential applications in charge transfer and electronic functionality [23,47]. Here, PFPDPA films created more agglomeration when compared to the other two modifications. Nanoparticles that were modified with long alkyl chains, as observed for PFPDPA, have been shown to increase the agglomeration as the order of the films increases [48]. The SEM images confirm that the morphologies of the nanoparticles do not change but the sizes of the particles increase after the surfaces are modified with the phosphonic acids.

EDS elemental analysis obtained in conjunction with the SEM images revealed the presence of zinc, oxygen, fluorine, and phosphorus atoms in all of the examined sections of ZnO modified with 5FBPA, PFPDPA, and F_{21}DDPA, whereas the unmodified sample revealed zinc and oxygen atoms. EDS spectra for all of the perfluoro modifications indicates a high relative concentration of fluorine on the surface (Figure S2 in supporting material). Furthermore, EDS mapping (Figure 5D) indicates that fluorine (F_k) is present homogenously across the sample image, indicating that modification is occurring uniformly, as previously reported [49].

Table 2. Particle size and distribution obtained from scanning electron microscope (SEM).

Modifications	Average Particle Size (nm)	Particle Distribution ($\pm$nm)
ZnO	139.8	18.6
5FBPA	167.0	27.9
PFPDPA	194.1	40.8
F_{21}DDPA	166.0	32.7

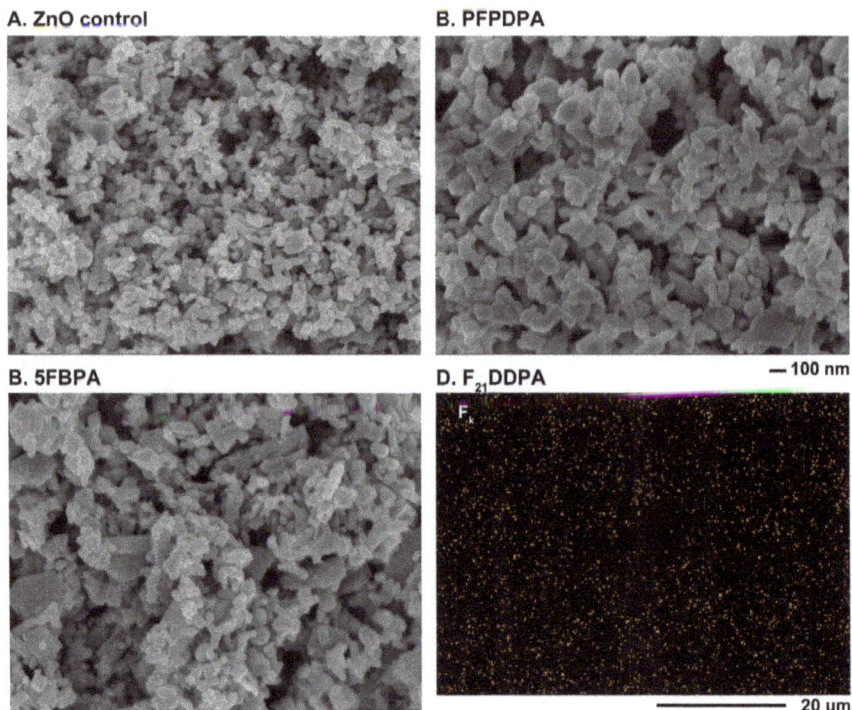

Figure 5. SEM survey spectra of (**A**) ZnO nanoparticles unmodified (control); (**B**) PFPDPA on ZnO; (**C**) 5FBPA on ZnO; and (**D**) energy dispersive X-ray spectroscopy (EDS) fluorine (F_k) elemental mapping pattern for F_{21}DDPA on ZnO nanoparticles.

2.5. Dynamic Light Scattering (DLS) and Zeta Potential Measurements

DLS was used to measure the particle sizes of the unmodified and modified ZnO nanoparticles in water and THF. The SEM measurements indicate that the particles are significantly smaller (139 to 166 nm) than the results from DLS (217 to 497 nm). DLS measures the hydrodynamic diameter, which is the diameter of the particle and surface associated ligands, ions, or molecules that travel along with the particle in colloidal solution, increasing the average particle size [50]. Therefore, a discrepancy occurs between the two different measurement methods because DLS accounts for this hydrodynamic diameter and Brownian particle displacement of particles in solution, increasing the apparent particle size while SEM does not [48,51]. In SEM, which measures solid surfaces, the counter ion effects and/or electrostatic interactions between the oppositely charged ions are limited when compared to DLS analysis.

Zeta potential is correlated to the surface charge of the particle and the nature and composition of the surrounding medium, in which the particle is dispersed [52]. After surface modification with perfluorinated phosphonic acids, zeta potential values varied significantly. The zeta potential of ZnO was −11.48 mV in THF, and became more negative following modification, with the most negative being −89.12 mV for ZnO nanoparticles that were modified with F_{21}DDPA (Table 3). The perfluorinated phosphonic acid modifications were negatively charged, imparting a negative charge to the dispersed nanoparticles, as previously reported [53,54]. This negative charge led to electrostatic repulsion between molecules, stabilizing the nanoparticles. Consequently, the higher absolute value of the zeta potential means an increased stability of the suspended particles against agglomeration [54,55]. All of the perfluorinated phosphonic acid surface modifications that were included in this study led to a significant increase in surface stability when compared to the ZnO control.

Table 3. Particle size and zeta potential values of the hydrodynamic diameters of ZnO and surface modified nanoparticles obtained using dynamic light scattering (DLS).

Modification	Water		THF	
	Particle Size (nm)	Zeta Potential (mV)	Particle Size (nm)	Zeta Potential (mV)
ZnO	497.2 ± 12.2	-11.09 ± 0.42	413.2 ± 18.0	11.48 ± 4.40
5FBPA	247.1 ± 8.9	-20.31 ± 0.66	279.1 ± 3.5	-48.04 ± 2.68
PFPDPA	275.3 ± 9.5	-22.53 ± 0.39	243.5 ± 0.7	-52.29 ± 3.02
F_{21}DDPA	217.5 ± 8.4	-18.64 ± 0.56	244.1 ± 2.2	-89.12 ± 1.96

Two different solvents with different polarities were used to analyze the stability of the modified surfaces using zeta potential. As shown in Table 3, these observations suggest that the interactions between the ZnO nanoparticles that were modified with perfluorophosphonic acids and the solvent, either water or THF, are relatively significant. While the zeta potential was solvent dependent for the surface modified samples, the zeta potential of unmodified ZnO did not change significantly with a change in solvent. By tailoring the solvent selection process to the desired application, the adsorption capacity of various surface modifications on ZnO nanoparticles could be improved. Additionally, the surface charge is dependent on the solvent viscosity. Usually, viscosity increases as particle size decreases, as shown in Table 3 [56,57]. While water evaporates more slowly than other solvents, allowing for the nanoparticles to remain in solution and preventing the formation of films, which makes it an ideal solvent for nanofluids [57]. THF evaporates more quickly, due to its higher viscosity, and increases the surface stability, as seen in the zeta potential results, making it ideal for use in electronic devices and solar cells applications. Additionally, the surfaces are covered uniformly, as has already been reported for the obtained EDS mapping results. The observed variations in zeta potential may be due to the different solvents used due to their effect on surface chemical composition, surface polarity, and swelling behavior [58].

The surface charges of unmodified and modified ZnO nanoparticles were assessed by zeta potential measurements over various pH ranges (Figure 6). Based on prior electrophoresis experiments, it is known that solid oxides in aqueous suspension are generally electrically charged [59]. The isoelectric point (IEP) represents the pH value at which the zeta potential value is equal to zero [60]. Unmodified ZnO nanoparticles reach their isoelectric point at a pH of approximately 10, which agrees with the value that was obtained from a previous study (Figure 6) [61]. The F_{21}DDPA and 5FBPA modified ZnO reached their isoelectric point at pHs of 9 and 8.5, respectively. PFPDPA never reaches its isoelectric point and remains negative over the pHs tested. A lack of particle surface charges leads to the absence of inter-particle repulsive forces, causing the colloidal system to be the least stable at the IEP [62]. However, ZnO modified with PFPDPA is not very soluble in aqueous solution, and exhibited a negative surface charge by zeta potential measurements. In neutral solution, ZnO, F_{21}DDPA, and 5FBPA are slightly positively charged, and PFPDPA is slightly negatively charged. As previously reported, the zeta potential becomes more negative as the alkyl chain length increases [63].

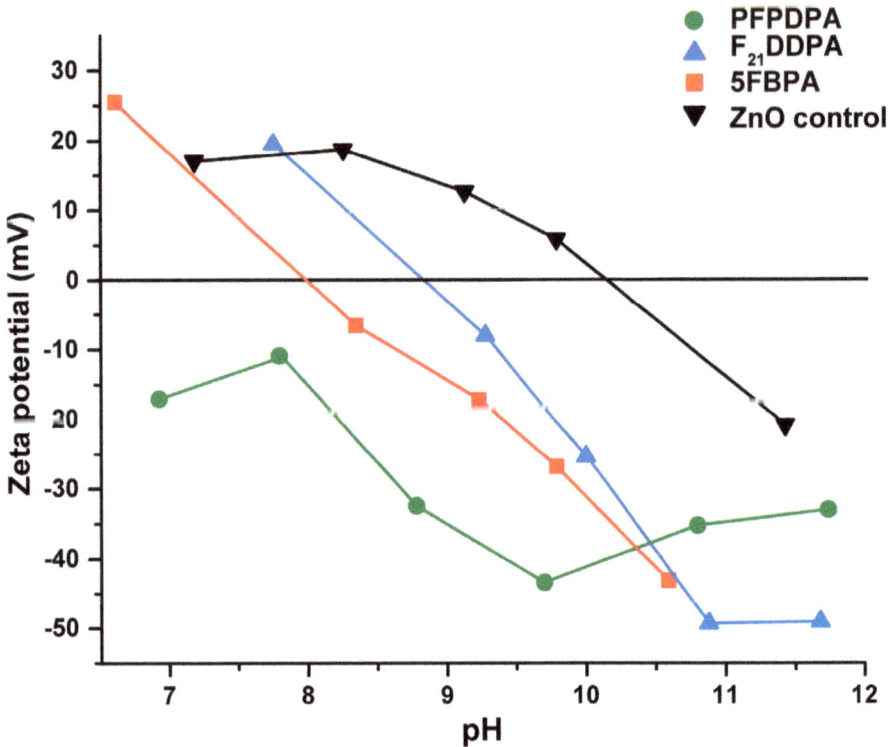

Figure 6. pH dependence of the zeta potentials of surface functionalized ZnO films and unmodified ZnO nanoparticles.

2.6. Thermogravimetric Analysis (TGA)

TGA analysis of all the perfluorophosphonic acid modification on ZnO showed a significant weight loss between 350–500 °C, as shown in Figure 7. The overall weight loss is assigned to the decomposition of the surface modifications bonded onto ZnO surfaces [64]. The weight loss in the initial stage is the films that physically absorbed on ZnO surface [65]. The decomposition of 5FBPA had the lowest weight loss % when compared to F_{21}DDPA, which has a 14% weight loss, and PFPDPA, which has a 12% weight loss. In addition to a phosphonic acid group, 5FBPA and PFPDPA have a complex benzene ring with resonance possibilities that have C–F functional groups. Therefore, comparing 5FBPA and PFPDPA, PFPDPA has less steric hindrance effect due to the long C–F chain and has well-organized and stronger films on the surface as compared to 5FBPA films [65]. Therefore, 5FBPA has more physically adsorbed films when compared to strongly chemically bonded films for PFPDPA and F_{21}DDPA modifications on a ZnO surface. It is concluded that the higher weight loss PFPDPA and F_{21}DDPA modifications correspond to a higher surface concentration of the modifications, and, ideally, higher surface coverage [66].

Materials **2017**, *10*, 1363

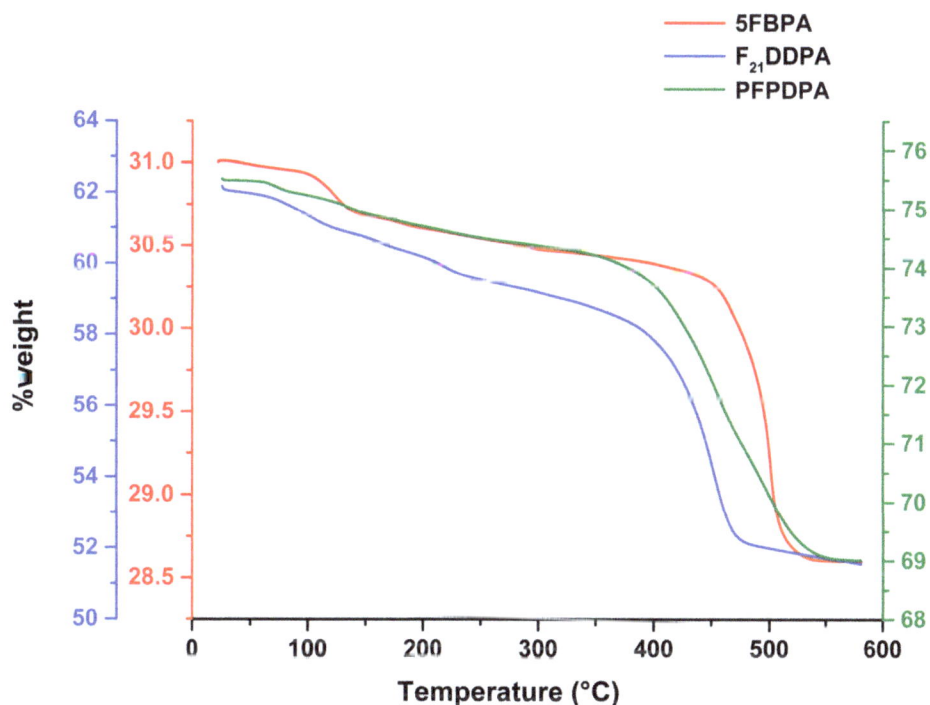

Figure 7. Thermogravimetric analysis (TGA) data analysis of surface modifications with the 5FBPA (red line and *y*-axis), F_{21}DDPA (blue line and *y*-axis), and PFPDPA (green line and *y*-axis) modifications on ZnO nanoparticles.

3. Materials and Methods

3.1. Materials

ZnO nanopowder, PFPDPA, (99.0%), and 5FBPA, (97.0%) were purchased from Sigma Aldrich (St. Louis, MO, USA). F_{21}DDPA was purchased from Synquest Laboratories (Alachua, FL, USA). For ZnO nanopowders, the manufacturer reported average particle sizes below 100 nm with a Brunauer-Emmett-Teller surface area of 15 to 25 m^2/g. Tetrahydrofuran (THF, Optima grade), was purchased from Fisher Scientific (Waltham, MA, USA). All of the chemicals and reagents were used without further purification.

3.2. Preparation of the Samples

For preparation of the adsorbed molecules, 0.35 g of ZnO nanoparticles were dispersed in 30 mL of THF by sonication at 33 ± 2 °C for 15 min. Then, 26.6 mM of each organic acid was added to 6 mL THF and sonicated for 30 min to dissolve. The 30 mL ZnO solutions were combined with the 6 mL acid solutions and sonicated together for 15 min. After sonication, the mixtures were left stirring for 48 h and were allowed to evaporate at room temperature after 24 h. The dry samples were dispersed again in 15 mL of THF and further sonicated for 15 min. The modified nanoparticles were recovered using a vacuum centrifuge (<20 mbar, 1400 rpm for 25 min) and the particles were again left under a fume hood overnight to dry.

3.3. Characterization of the Films

3.3.1. ATR–IR

ATR–IR was performed using a Thermo Scientific Nicolet iS50 FT-IR and was used to analyze the alkyl chain ordering and bonding motif of the molecules to the surface. The unmodified ZnO nanoparticles were used to collect a background spectrum for analysis purposes. Typically, 256 scans were collected with a resolution of 2 cm^{-1}.

3.3.2. XPS

XPS measurements were performed with a PHI 5000 VersaProbe ESCA Microprobe system (ULVAC-PHI). XPS measurements were performed using a focused Al K-Alpha X-ray source at 1486 eV energy and 25 W, with an X-ray spot size of 100 μm. The take-off angle of the photoelectron was set at 45°. An analyzer pass energy of 117.4 eV was used for a survey scan, and high-resolution scans for fluorine, oxygen, phosphorus, and carbon elements were carried out at an analyzer pass energy of 23.5 eV. The XPS spectra were referenced to the C1s peak at a binding energy of 284.8 eV.

3.3.3. SS-NMR

Solid state NMR spectra were acquired with a Bruker Avance 300 spectrometer and 7 T Bruker magnet. Samples were packed into 4 mm zirconia rotors. A Bruker double resonance MAS probe was tuned to a ^{1}H frequency of 300.405 MHz and a ^{31}P frequency of 121.606 MHz. The ^{31}P direct-polarization pulse sequence used a 3.50 μs 90° pulse on the ^{31}P channel and 30 kHz of proton decoupling during acquisition. The free induction decay (FID) was Fourier transformed with 16 Hz of line broadening after zero filling to a total of 32 k points. The ^{31}P spectra were referenced externally by measuring the resonance frequency of an aqueous solution of 85% phosphoric acid (set to 0.0 ppm).

3.3.4. SEM/EDS

SEM/EDS was performed using a JEOL JSM-7600F field emission SEM. The EDS was collected with an Oxford INCA EDS system and data were analyzed using the Oxford Aztec Energy Analyzer software. The chamber of the SEM was held under high vacuum conditions. The accelerating voltage for the EDS ranged from 3–10 kV. Samples were prepared individually in pin stubs and were sputtered with a 10-nm thin coat of gold/palladium. SEM/EDS was used to analyze the surface composition of the nanoparticles and obtain information about particle size and elemental composition.

3.3.5. DLS/Zeta Potential

A Brookhaven ZetaPlus Potential Analyzer (90Plus PALS) was used to perform DLS and zeta potential measurements of the unmodified and modified ZnO nanoparticles. The measurements were performed at 25 °C in water and THF. At least three measurements were made for each sample, and the collected values were averaged. The Zeta Potential Analyzer was employed to determine the direction of particles under the influence of an electric field, allowing for the estimation of the zeta potentials of the nanoparticles.

3.3.6. TGA

The thermogravimetric analysis was acquired with a TA Q500 TGA at a heating rate of 10 °C/min and a flow rate of high purity nitrogen of 100 mL/min. Approximately 20–30 mg of samples was measured in alumina pans.

4. Conclusions

In this study, ZnO nanoparticles were modified by self-assembly of perfluorinated phosphonic acids. The IR and XPS data showed that the modifications attached at the phosphonic head group and formed strong and ordered thin films on the surfaces of the nanoparticles. The IR and SS-NMR spectra also indicated that these films remained strongly bonded on the surfaces after sonication and after two years.

SEM images indicate that the morphologies of the nanoparticles do not change with the modifications from their initial oblong shape, and EDS mapping data shows that the modifications are homogenously distributed throughout the samples. As expected, particle sizes from SEM indicate that the sizes of the nanoparticles increase with the addition of the perfluorinated phosphonic acids. However, DLS particle sizing shows a decrease in particle sizes after modification, which correlates to the zeta potential measurements that show the surface stability of the modified nanoparticles increases when compared to the unmodified ZnO. Over a range of pH values, the ZnO control IEP corresponds with literature values, while the modified samples IEP decreased (5FBPA and F_{21}DDPA) or remained negative, and never reached an isoelectric point (PFPDPA modification). Thermogravimetric analysis indicated that higher weight loss corresponds to less steric hindrance and chemically bonded films, rather than physically bonded films.

These results indicate that modifying the ZnO nanoparticles with perfluorinated phosphonic acids increases the stability of the phosphonic acids that are adsorbed on the surfaces, as revealed by zeta potential measurements, even though multi-layer or physisorbed films (as shown by XPS and SS-NMR analysis) are formed that show sufficient stability. Here, both XPS and SS-NMR analysis indicate a heterogeneous population on the ZnO surface, involving various bonding motifs that are presumably bound at different types of adsorption sites. While it is already understood how phosphonic acids and the surfaces of ZnO interact, this study is a crucial step in understanding how perfluorinated groups can tune the surface properties of the nanoparticles. These modified nanoparticles could be incorporated into systems where a stable surface is necessary.

Supplementary Materials: The following are available online at www.mdpi.com/1996-1944/10/12/1363/s1, Table S1: Binding energies determined by XPS, Figure S1: XPS survey spectra, Figure S2: EDS spectra for modified ZnO nanoparticles.

Acknowledgments: This research was partially supported by a grant from NASA WV Space Grant Consortium (NASA Grant #NNX15AI01H) and internal funding from Marshall University. We would like to thank Marcela Redigolo from West Virginia University (WVU) Shared Research Facilities for the use of the SEM/EDS. KTM and JWF acknowledge the support of the National Science Foundation under Grant CHE-1411687.

Author Contributions: R.Q. wrote the paper, conceived, and designed the project. G.B. and D.S. performed the ATR-IR and DLS with zeta potential measurements experiments. D.S. and G.B. analyzed the SEM data. C.P. performed the ATR-IR experiments. R.K.G. and S.A. performed the TGA experiment. J.W.F., K.T.M. and R.I.J. performed the SS-NMR experiments. Q.W. performed and analyzed the XPS experiments.

Conflicts of Interest: The authors declare no conflict of interest.

References

1. Uthirakumar, P.; Hong, C.-H.; Suh, E.-K.; Lee, Y.-S. Novel fluorescent polymer/zinc oxide hybrid particles: Synthesis and application as a luminescence converter for white light-emitting diodes. *Chem. Mater.* **2006**, *18*, 4990–4992. [CrossRef]
2. Fathil, M.; Arshad, M.M.; Ruslinda, A.; Gopinath, S.C.; Adzhri, R.; Hashim, U.; Lam, H. Substrate-gate coupling in ZnO-FET biosensor for cardiac troponin I detection. *Sens. Actuators B Chem.* **2017**, *242*, 1142–1154. [CrossRef]
3. Hau, S.K.; Yip, H.-L.; Baek, N.S.; Zou, J.; O'Malley, K.; Jen, A.K.-Y. Air-stable inverted flexible polymer solar cells using zinc oxide nanoparticles as an electron selective layer. *Appl. Phys. Lett.* **2008**, *92*, 225. [CrossRef]
4. Yip, H.L.; Hau, S.K.; Baek, N.S.; Ma, H.; Jen, A.K.Y. Polymer Solar Cells That Use Self-Assembled-Monolayer-Modified ZnO/Metals as Cathodes. *Adv. Mater.* **2008**, *20*, 2376–2382. [CrossRef]
5. Matsui, M.; Mase, H.; Jin, J.-Y.; Funabiki, K.; Yoshida, T.; Minoura, H. Application of semisquaric acids as sensitizers for zinc oxide solar cell. *Dyes Pigment.* **2006**, *70*, 48–53. [CrossRef]
6. Bekci, D.R.; Karsli, A.; Cakir, A.C.; Sarica, H.; Guloglu, A.; Gunes, S.; Erten-Ela, S. Comparison of ZnO interlayers in inverted bulk heterojunction solar cells. *Appl. Energy* **2012**, *96*, 417–421. [CrossRef]

7. Das, J.; Khushalani, D. Nonhydrolytic Route for Synthesis of ZnO and Its Use as a Recyclable Photocatalyst. *J. Phys. Chem. C* **2010**, *114*, 2544–2550. [CrossRef]
8. Lai, M.-H.; Tubtimtae, A.; Lee, M.-W.; Wang, G.-J. ZnO-Nanorod Dye-Sensitized Solar Cells: New Structure without a Transparent Conducting Oxide Layer. *Int. J. Photoenergy* **2010**, *2010*, 497095. [CrossRef]
9. Hong, R.; Pan, T.; Qian, J.; Li, H. Synthesis and surface modification of ZnO nanoparticles. *Chem. Eng. J.* **2006**, *119*, 71–81. [CrossRef]
10. Wang, X.; Kong, X.; Yu, Y.; Zhang, H. Synthesis and characterization of water-soluble and bifunctional ZnO-Au nanocomposites. *J. Phys. Chem. C* **2007**, *111*, 3836–3841. [CrossRef]
11. Senthilkumar, O.; Yamauchi, K.; Senthilkumar, K.; Yamamae, T.; Fujita, Y.; Nishimoto, N. UV-Blue Light Emission from ZnO Nanoparticles. *J. Korean Phys. Soc.* **2008**, *53*, 46–49.
12. Srikant, V.; Clarke, D.R. On the optical band gap of zinc oxide. *J. Appl. Phys.* **1998**, *83*, 5447–5451. [CrossRef]
13. Lin, H.-F.; Liao, S.-C.; Hung, S.-W. The dc thermal plasma synthesis of ZnO nanoparticles for visible-light photocatalyst. *J. Photochem. Photobiol. A Chem.* **2005**, *174*, 82–87. [CrossRef]
14. Mohamed, R.M.; McKinney, D.; Kadi, M.W.; Mkhalid, I.A.; Sigmund, W. Platinum/zinc oxide nanoparticles: Enhanced photocatalysts degrade malachite green dye under visible light conditions. *Ceram. Int.* **2016**, *42*, 9375–9381. [CrossRef]
15. Maity, S.; Bhunia, C.; Sahu, P. Improvement in optical and structural properties of ZnO thin film through hexagonal nanopillar formation to improve the efficiency of a Si–ZnO heterojunction solar cell. *J. Phys. D Appl. Phys.* **2016**, *49*, 205104. [CrossRef]
16. Brine, H.; Sánchez-Royo, J.F.; Bolink, H.J. Ionic liquid modified zinc oxide injection layer for inverted organic light-emitting diodes. *Organ. Electron.* **2013**, *14*, 164–168. [CrossRef]
17. Quiñones, R.; Rodriguez, K.; Iuliucci, R.J. Investigation of phosphonic acid surface modifications on zinc oxide nanoparticles under ambient conditions. *Thin Solid Films* **2014**, *565*, 155–164. [CrossRef]
18. Bain, C.D.; Evall, J.; Whitesides, G.M. Formation of monolayers by the coadsorption of thiols on gold: Variation in the head group, tail group, and solvent. *J. Am. Chem. Soc.* **1989**, *111*, 7155–7164. [CrossRef]
19. Quiñones, R.; Gawalt, E.S. Study of the Formation of Self-Assembled Monolayers on Nitinol. *Langmuir* **2007**, *23*, 10123–10130. [CrossRef] [PubMed]
20. Quiñones, R.; Gawalt, E.S. Polystyrene Formation on Monolayer-Modified Nitinol Effectively Controls Corrosion. *Langmuir* **2008**, *24*, 10858–10864. [CrossRef] [PubMed]
21. Raman, A.; Dubey, M.; Gouzman, I.; Gawalt, E.S. Formation of Self-Assembled Monolayers of Alkylphosphonic Acid on the Native Oxide Surface of SS316L. *Langmuir* **2006**, *22*, 6469–6472. [CrossRef] [PubMed]
22. Jadhav, S.A. Self-assembled monolayers (SAMs) of carboxylic acids: An overview. *Cent. Eur. J. Chem.* **2011**, *9*, 369–378. [CrossRef]
23. Reese, S.; Fox, M.A. Self-Assembled Monolayers on Gold of Thiols Incorporating Conjugated Terminal Groups. *J. Phys. Chem. B* **1998**, *102*, 9820–9824. [CrossRef]
24. Ford, W.E.; Abraham, F.; Scholz, F.; Nelles, G.; Sandford, G.; Wrochem, F.V. Spectroscopic Characterization of Fluorinated Benzylphosphonic Acid Monolayers on AlO x/Al Surfaces. *J. Phys. Chem. C* **2017**, *121*, 1690–1703. [CrossRef]
25. Kim, D.H.; Park, J.-H.; Lee, T.I.; Myoung, J.-M. Superhydrophobic Al-doped ZnO nanorods-based electrically conductive and self-cleanable antireflecting window layer for thin film solar cell. *Sol. Energy Mater. Sol. Cells* **2016**, *150*, 65–70. [CrossRef]
26. Jouet, R.J.; Warren, A.D.; Rosenberg, D.M.; Bellitto, V.J.; Park, K.; Zachariah, M.R. Surface Passivation of Bare Aluminum Nanoparticles Using Perfluoroalkyl Carboxylic Acids. *Chem. Mater.* **2005**, *17*, 2987–2996. [CrossRef]
27. Colvin, V.; Goldstein, A.; Alivisatos, A. Semiconductor nanocrystals covalently bound to metal surfaces with self-assembled monolayers. *J. Am. Chem. Soc.* **1992**, *114*, 5221–5230. [CrossRef]
28. Lange, I.; Reiter, S.; Pätzel, M.; Zykov, A.; Nefedov, A.; Hildebrandt, J.; Hecht, S.; Kowarik, S.; Wöll, C.; Heimel, G.; et al. Tuning the Work Function of Polar Zinc Oxide Surfaces using Modified Phosphonic Acid Self-Assembled Monolayers. *Adv. Funct. Mater.* **2014**, *24*, 7014–7024. [CrossRef]
29. Boardman, L.D.; Pellerite, M.J. Fluorinated Phosphonic Acids. Google Patents US6824882 B2, 30 November 2004.

30. Kraft, U.; Zschieschang, U.; Ante, F.; Kalblein, D.; Kamella, C.; Amsharov, K.; Jansen, M.; Kern, K.; Weber, E.; Klauk, H. Fluoroalkylphosphonic acid self-assembled monolayer gate dielectrics for threshold-voltage control in low-voltage organic thin-film transistors. *J. Mater. Chem.* **2010**, *20*, 6416–6418. [CrossRef]

31. Xu, T.; Yu, L. How to design low bandgap polymers for highly efficient organic solar cells. *Mater. Today* **2014**, *17*, 11–15. [CrossRef]

32. Timpel, M.; Nardi, M.V.; Krause, S.; Ligorio, G.; Christodoulou, C.; Pasquali, L.; Giglia, A.; Frisch, J.; Wegner, B.; Moras, P. Surface Modification of ZnO (0001)-Zn with Phosphonate-Based Self-Assembled Monolayers: Binding Modes, Orientation, and Work Function. *Chem. Mater.* **2014**, *26*, 5042–5050. [CrossRef]

33. Raman, A.; Quiñones, R.; Barriger, L.; Eastman, R.; Parsi, A.; Gawalt, E.S. Understanding Organic Film Behavior on Alloy and Metal Oxides. *Langmuir* **2010**, *26*, 1747–1754. [CrossRef] [PubMed]

34. Quiñones, R.; Raman, A.; Gawalt, E.S. An approach to differentiating between multi- and monolayers using MALDI-TOF MS. *Surf. Interface Anal.* **2007**, *39*, 593–600. [CrossRef]

35. Whitesides, G.M.; Laibinis, P.E. Wet chemical approaches to the characterization of organic surfaces: Self-assembled monolayers, wetting, and the physical-organic chemistry of the solid-liquid interface. *Langmuir* **1990**, *6*, 87–96. [CrossRef]

36. Snyder, R.G.; Struss, H.L.; Elliger, C.A. Carbon-hydrogen atretching modes and structure of n-alkyl chains. 1. Long, disordered Chains. *J. Phys. Chem.* **1982**, *86*, 5145–5150. [CrossRef]

37. Wood, C.; Li, H.; Winget, P.; Brédas, J.-L. Binding Modes of Fluorinated Benzylphosphonic Acids on the Polar ZnO Surface and Impact on Work Function. *J. Phys. Chem. C* **2012**, *116*, 19125–19133. [CrossRef]

38. Rechmann, J.; Sarfraz, A.; Götzinger, A.C.; Dirksen, E.; Müller, T.J.J.; Erbe, A. Surface Functionalization of Oxide-Covered Zinc and Iron with Phosphonated Phenylethynyl Phenothiazine. *Langmuir* **2015**, *31*, 7306–7316. [CrossRef] [PubMed]

39. Di Mauro, A.; Smecca, E.; D'Urso, A.; Condorelli, G.G.; Fragalà, M.E. Tetra-anionic porphyrin loading onto ZnO nanoneedles: A hybrid covalent/non covalent approach. *Mater. Chem. Phys.* **2014**, *143*, 977–982. [CrossRef]

40. Ishizaki, T.; Teshima, K.; Masuda, Y.; Sakamoto, M. Liquid phase formation of alkyl- and perfluoro-phosphonic acid derived monolayers on magnesium alloy AZ31 and their chemical properties. *J. Colloid Interface Sci.* **2011**, *360*, 280–288. [CrossRef] [PubMed]

41. Wei, Q.; Tajima, K.; Tong, Y.; Ye, S.; Hashimoto, K. Surface-Segregated Monolayers: A New Type of Ordered Monolayer for Surface Modification of Organic Semiconductors. *J. Am. Chem. Soc.* **2009**, *131*, 17597–17604. [CrossRef] [PubMed]

42. Davidowski, S.K.; Holland, G.P. Solid-State NMR Characterization of Mixed Phosphonic Acid Ligand Binding and Organization on Silica Nanoparticles. *Langmuir* **2016**, *32*, 3253–3261. [CrossRef] [PubMed]

43. Holland, G.P.; Sharma, R.; Agola, J.O.; Amin, S.; Solomon, V.C.; Singh, P.; Buttry, D.A.; Yarger, J.L. NMR Characterization of Phosphonic Acid Capped SnO_2 Nanoparticles. *Chem. Mater.* **2007**, *19*, 2519–2526. [CrossRef]

44. Yah, W.O.; Takahara, A.; Lvov, Y.M. Selective Modification of Halloysite Lumen with Octadecylphosphonic Acid: New Inorganic Tubular Micelle. *J. Am. Chem. Soc.* **2011**, *134*, 1853–1859. [CrossRef] [PubMed]

45. Cao, G.; Lee, H.; Lynch, V.M.; Mallouk, T.E. Synthesis and structural characterization of a homologous series of divalent-metal phosphonates, $M^{II}(O_3PR) \cdot H_2O$ and $M^{II}(HO_3PR)_2$. *Inorg. Chem.* **1988**, *27*, 2781–2785. [CrossRef]

46. Hotchkiss, P.J.; Malicki, M.; Giordano, A.J.; Armstrong, N.R.; Marder, S.R. Characterization of phosphonic acid binding to zinc oxide. *J. Mater. Chem.* **2011**, *21*, 3107–3112. [CrossRef]

47. Shervedani, R.K.; Hatefi-Mehrjardi, A.; Babadi, M.K. Comparative electrochemical study of self-assembled monolayers of 2-mercaptobenzoxazole, 2-mercaptobenzothiazole, and 2-mercaptobenzimidazole formed on polycrystalline gold electrode. *Electrochim. Acta* **2007**, *52*, 7051–7060. [CrossRef]

48. Feichtenschlager, B.; Pabisch, S.; Peterlik, H.; Kickelbick, G. Nanoparticle Assemblies as Probes for Self-Assembled Monolayer Characterization: Correlation between Surface Functionalization and Agglomeration Behavior. *Langmuir* **2012**, *28*, 741–750. [CrossRef] [PubMed]

49. Quiñones, R.; Garretson, S.; Behnke, G.; Fagan, J.W.; Mueller, K.T.; Agarwal, S.; Gupta, R.K. Fabrication of phosphonic acid films on nitinol nanoparticles by dynamic covalent assembly. *Thin Solid Films* **2017**, *642*, 195–206. [CrossRef]

50. Anuradha, K.; Bangal, P.; Madhavendra, S.S. Macromolecular Arabinogalactan Polysaccharide Mediated Synthesis of Silver Nanoparticles, Characterization and Evaluation. *Macromol. Res.* **2015**, *24*, 152–162. [CrossRef]

51. Ling, X.Y.; Reinhoudt, D.N.; Huskens, J. Ferrocenyl-Functionalized Silica Nanoparticles: Preparation, Characterization, and Molecular Recognition at Interfaces. *Langmuir* **2006**, *22*, 8777–8783. [CrossRef] [PubMed]

52. Sizovs, A.; Song, X.; Waxham, M.N.; Jia, Y.; Feng, F.; Chen, J.; Wicker, A.C.; Xu, J., Yu, Y.; Wang, J. Precisely Tunable Engineering of Sub-30 nm Monodisperse Oligonucleotide Nanoparticles. *J. Am. Chem. Soc.* **2014**, *136*, 234–240. [CrossRef] [PubMed]

53. Schmitt Pauly, C.; Genix, A.-C.; Alauzun, J.G.; Guerrero, G.; Appavou, M.-S.; Pérez, J.; Oberdisse, J.; Mutin, P.H. Simultaneous Phase Transfer and Surface Modification of TiO$_2$ Nanoparticles Using Alkylphosphonic Acids: Optimization and Structure of the Organosols. *Langmuir* **2015**, *31*, 10966–10974. [CrossRef] [PubMed]

54. Oueiny, C.; Berlioz, S.; Patout, L.; Perrin, F.X. Aqueous dispersion of multiwall carbon nanotubes with phosphonic acid derivatives. *Colloids Surf. A* **2016**, *493*, 41–51. [CrossRef]

55. Mikolajczyk, A.; Gajewicz, A.; Rasulev, B.; Schaeublin, N.; Maurer-Gardner, E.; Hussain, S.; Leszczynski, J.; Puzyn, T. Zeta Potential for Metal Oxide Nanoparticles: A Predictive Model Developed by a Nano-Quantitative Structure–Property Relationship Approach. *Chem. Mater.* **2015**, *27*, 2400–2407. [CrossRef]

56. Fedele, L.; Colla, L.; Bobbo, S. Viscosity and thermal conductivity measurements of water-based nanofluids containing titanium oxide nanoparticles. *Int. J. Refrig.* **2012**, *35*, 1359–1366. [CrossRef]

57. Borlaf, M.; Colomer, M.T.; Cabello, F.; Serna, R.; Moreno, R. Electrophoretic Deposition of TiO$_2$/Er^{3+} Nanoparticulate Sols. *J. Phys. Chem. B* **2013**, *117*, 1556–1562. [CrossRef] [PubMed]

58. Raman, A.; Gawalt, E.S. Self-Assembled Monolayers of Alkanoic Acids on the Native Oxide Surface of SS316L by Solution Deposition. *Langmuir* **2007**, *23*, 2284–2288. [CrossRef] [PubMed]

59. Parks, G.A. The Isoelectric Points of Solid Oxides, Solid Hydroxides, and Aqueous Hydroxo Complex Systems. *Chem. Rev.* **1965**, *65*, 177–198. [CrossRef]

60. Lopes, K.P.; Cavalcante, L.S.; Simões, A.Z.; Gonçalves, R.F.; Escote, M.T.; Varela, J.A.; Longo, E.; Leite, E.R. NiTiO$_3$ nanoparticles encapsulated with SiO$_2$ prepared by sol–gel method. *J. Sol-Gel Sci. Technol.* **2008**, *45*, 151–155. [CrossRef]

61. Marsalek, R. Particle size and zeta potential of ZnO. *APCBEE Procedia* **2014**, *9*, 13–17. [CrossRef]

62. Clement, S.; Deng, W.; Drozdowicz-Tomsia, K.; Liu, D.; Zachreson, C.; Goldys, E.M. Bright, water-soluble CeF$_3$ photo-, cathodo-, and X-ray luminescent nanoparticles. *J. Nanopart. Res.* **2015**, *17*, 7. [CrossRef]

63. Ivanov, M.R.; Haes, A.J. Anionic Functionalized Gold Nanoparticle Continuous Full Filling Separations: Importance of Sample Concentration. *Anal. Chem.* **2012**, *84*, 1320–1326. [CrossRef] [PubMed]

64. Helmy, R.; Fadeev, A.Y. Self-Assembled Monolayers supported on TiO$_2$: Comparison of C$_{18}$H$_{37}$SiX$_3$ (X=H, Cl, OCH$_3$), C$_{18}$H$_{37}$Si(CH$_3$)$_2$Cl and C$_{18}$H$_{37}$PO(OH)$_2$. *Langmuir* **2002**, *18*, 8924–8928. [CrossRef]

65. Ye, H.-J.; Shao, W.-Z.; Zhen, L. Tetradecylphosphonic acid modified BaTiO3 nanoparticles and its nanocomposite. *Colloids Surf. A* **2013**, *427* (Suppl. C), 19–25. [CrossRef]

66. Ruiterkamp, G.J.; Hempenius, M.A.; Wormeester, H.; Vancso, G.J. Surface functionalization of titanium dioxide nanoparticles with alkanephosphonic acids for transparent nanocomposites. *J. Nanopart. Res.* **2011**, *13*, 2779–2790. [CrossRef]

materials

MDPI

Review

Molecular Mechanisms of Zinc Oxide Nanoparticle-Induced Genotoxicity Short Running Title: Genotoxicity of ZnO NPs

Agmal Scherzad, Till Meyer, Norbert Kleinsasser and Stephan Hackenberg *

Department of Oto-Rhino Laryngology, Plastic, Aesthetic and Reconstructive Head and Neck Surgery, University of Wuerzburg, 97080 Wuerzburg, Germany; scherzad_a@ukw.de (A.S.); meyer_t2@ukw.de (T.M.); kleinsasser_n@ukw.de (N.K.)

* Correspondence: hackenberg_s@ukw.de; Tel.: +49-931-201-21323; Fax: +49-931-201-21321

Received: 16 November 2017; Accepted: 9 December 2017; Published: 14 December 2017

Abstract: Background: Zinc oxide nanoparticles (ZnO NPs) are among the most frequently applied nanomaterials in consumer products. Evidence exists regarding the cytotoxic effects of ZnO NPs in mammalian cells; however, knowledge about the potential genotoxicity of ZnO NPs is rare, and results presented in the current literature are inconsistent. Objectives: The aim of this review is to summarize the existing data regarding the DNA damage that ZnO NPs induce, and focus on the possible molecular mechanisms underlying genotoxic events. Methods: Electronic literature databases were systematically searched for studies that report on the genotoxicity of ZnO NPs. Results: Several methods and different endpoints demonstrate the genotoxic potential of ZnO NPs. Most publications describe in vitro assessments of the oxidative DNA damage triggered by dissolved Zn^{2+} ions. Most genotoxicological investigations of ZnO NPs address acute exposure situations. Conclusion: Existing evidence indicates that ZnO NPs possibly have the potential to damage DNA. However, there is a lack of long-term exposure experiments that clarify the intracellular bioaccumulation of ZnO NPs and the possible mechanisms of DNA repair and cell survival.

Keywords: zinc oxide nanoparticles; genotoxicity; DNA damage; ROS; autophagy

1. Introduction

Over the past 15 years, nanotechnology has increasingly gained in importance in industry, biomedicine, and research. According to the current definition of the European Union (EU), nanomaterials are natural, incidental, or manufactured materials that contain particles in an unbound state, either as aggregates or as agglomerates. At least 50% of these particles must exhibit one or more external dimension within the size range of 1–100 nm [1]. Surface properties become more important as a function of the size reduction of a material. Thus, nanoparticles (NPs) have completely different mechanical, optical, electrical, magnetic, and catalytic properties compared with larger particles of the same composition. Hence, the bioactivity of NPs significantly differs from that of their fine-size analogues [2]. Zinc oxide (ZnO) NPs are among the most commonly used nanomaterials in industrial applications. Despite their increasing usage in consumer products, the safety aspects of ZnO NPs remain uncertain. In particular, information regarding the possible genotoxicity of ZnO NPs is rare, and partially contradictory. The aim of this review is to summarize the literature published between 2009 and 2017 that covers the genotoxicity of ZnO NPs in mammalian and non-mammalian in vitro and in vivo systems, and to estimate the current risk of using ZnO NPs in consumer products. Furthermore, information on the molecular mechanisms of ZnO NP-induced DNA damage will also be outlined and discussed.

2. Application of ZnO NPs

ZnO forms a whitish powder and has quite a broad spectrum of applications. ZnO formulations are particularly important in the production of rubber, as an additive in cement, and as a main ingredient in pigments and paints. They are also used as catalysts in the chemical industry, and as standard materials in both the pharmaceutical and cosmetic industries. Numerous electronic components contain ZnO due to its favorable semiconductor properties [3]. A further eminent characteristic is its ability to reflect UV irradiation, which makes ZnO an important physical UV filter in sunscreens. Nanoparticulate ZnO has a very high UV-protective value, and is not as occlusive and whitish as bulky ZnO powder. Thus, ZnO NPs are preferentially applied in cosmetic products compared with larger particles. For consumers, skin exposure is the most likely way to come into contact with ZnO NPs, whereas for industrial workers, airway exposure is more relevant [4].

Currently, approximately 1800 industrial products are available that contain nanomaterials [5]. According to article 16 of the Cosmetic Regulation from 2013, cosmetic products containing nanomaterials have to be notified. Prior to 2013, there was no legal requirement for the declaration of NPs in consumer products, and the number could only be estimated. The EU is currently discussing the introduction of such a regulation in order to facilitate the information flow to the public and research institutions. According to consumer product inventories, there are approximately 40 products available on the United States (US) market containing ZnO NPs.

3. Exposure Routes

For the toxicological evaluation of NPs, knowledge regarding the routes of intake is essential. Knowledge regarding its bioavailability and resorption is also important. Possible intake routes of NPs in humans are the gastrointestinal tract, the skin, and the airways. For consumers, dermal exposure is the most likely way to come into contact with ZnO NPs due to the high number of cosmetic products containing ZnO NPs. The stratum corneum, known as the upper layer of the skin, seems to be a sufficient barrier against ZnO NP penetration into the epidermis, as shown by several authors [6,7]. It was clearly demonstrated that ZnO NPs were not able to penetrate healthy and intact human or porcine skin. Although NPs may be retained in the hair follicle ostium or skin folds, they are usually sufficiently eliminated by sebum flow [8]. However, skin damages, for example after excessive sun bathing, may harm this protection layer, and lead to possible toxicological effects from NPs. Cytotoxic or genotoxic effects only seem to be relevant in proliferating cells, which can be found in the basal layers of the epidermis. This is why the application of ZnO NPs to injured or defective skin is discussed as being potentially dangerous. The ingestion of ZnO NPs and contact with intestinal mucosa must be evaluated equally. In particular, chronic intestinal illness may lead to a defect in the mucosa barrier, which consequently may lead to an enhanced toxicity. Further studies are needed to evaluate the correlation between the grade of skin damage and the hazard of ZnO NPs.

Airway exposure via inhalation is the predominant means of contact for workers in the chemical, cosmetic, or paint industries [4]. Nanosized particles are able to reach the peripheral airway sites, such as the bronchiolar and alveolar regions. If not carried away by mucociliary transport mechanisms, NPs may affect alveolar cells and cause toxic, genotoxic, or inflammatory effects [4]. Inhalation exposure to ZnO NPs seems to be an important hazard, and risk assessment is urgently needed within this context [9]. Indeed, the airway exposure of NPs seems to be very important in the toxicological circumstances. According to Vermylen et al., the intake of superfine structures via inhalation has profound negative local and systemic side effects, such as an enhanced risk of cardiovascular diseases [10]. These very small particles are able to penetrate the tracheobronchial tree. In particular, ultrafine particles, which have a diameter less than 100 nm, are able to pass directly into the blood stream [10,11]. Some studies hypothesize that NPs might be able to reach the brain along peripheral nerves [12,13]. This may offer a therapeutic option as well. However, toxicological evaluations are warranted.

4. Genotoxicity of ZnO NPs

The difference between the volume and surface of NPs enables their variety of chemical, physical, and biological properties [14]. Due to their small size, large surface area, and physicochemical characteristics, NPs may exhibit unpredictable genotoxic properties. The biological properties depend on the manufacturing procedure, agglomeration and aggregation tendencies, and surface coating. During the manufacturing processes, the particle diameters are not homogeneous. Due to their surface, NPs tend to aggregate, which implicates the need for dispersions. Surface coating is a suitable method for preventing the aggregation of NPs. These above-mentioned circumstances significantly influence the toxicity of NPs. Kwon et al. showed that small NPs cross the cellular membranes more easily, which leads to an increased potency of DNA damage. Accumulated NPs might be internalized into the cell mainly during the mitosis process. According to Liu et al., a crucial determinant of toxicity is the solubility of ZnO NPs, which is influenced by various factors, including the pH of the environment in tissues, cells, and organelles [15]. ZnO NPs and other particles such as silver are soluble, and may release ions. Unlike silver, Zn is an important component of several enzymes and transcription factors in the human body. After incorporation, ZnO NPs may dissolve into Zn^{2+} and trigger several signaling pathways and cascades, which might lead to an enhanced influx of calcium, gene upregulation, or the release of pro-inflammatory markers [16]. The solubility of NPs such as silver (Ag), copper (Cu), or ZnO is one of the main contributors to their toxicity. Ag, Cu, and ZnO NPs have some commonalities. Their elemental composition is metallic, they fight the growth of microorganisms, they have a negative surface charge, and most importantly, all of them are soluble [17]. Nevertheless, there are also differences between these metallic particles. According to Bondarenko, although their particle size is similar, their toxicity is likely different. Cu ions may be involved in electron-transfer processes, in contrast to Ag and Zn [17].

According to Golbamaki, the genotoxic effects of NPs may be classified as either "primary genotoxicity" or "secondary genotoxicity". Reactive oxygen species (ROS) generation during particle-induced inflammation is the cause of secondary genotoxicity, whereas primary genotoxicity refers to genotoxic effects without inflammation [18]. There are studies that point to the correlation between particle size and toxicity [9]. However, information concerning the size dependency of NP-induced toxicity is contradictory. Warheit et al. did not observe any variation in the toxicity levels of large and small TiO2 NPs [19]. However, Golbamaki and Karlsson detected significantly increased DNA damage after cell exposure to larger micrometer-sized particles compared with smaller NPs [18,20]. Due to these inhomogeneous statements, the size dependency of nanotoxicity and nanogenotoxicity needs to be clarified. NP size must always be characterized exactly in order to provide comparable data in the context of the current literature.

Over the past 10 years, studies focusing on the nanotoxicity of ZnO have been continuously published. However, most of these studies primarily address the cytotoxic aspects of ZnO NPs. Dose–response correlations between ZnO NP concentration and cellular viability are investigated in most studies. However, DNA damage occurs at significantly lower concentrations compared with cytotoxic effects. Hence, genotoxicological evaluations of NPs must be performed at non-cytotoxic doses. Although ZnO NPs are frequently applied in industry and research, data on the genotoxic potential of this material is quite limited [21].

4.1. Molecular Mechanisms of Genotoxicity and Evaluation of Oxidative DNA Damage

It is crucial to understand the molecular mechanisms of genotoxicity caused by ZnO NPs in order to provide a valid risk assessment. Although several groups have contributed data towards elucidating these pathways, the associated mechanisms and correlations still remain unclear. The role of Zn ions cannot definitely be ruled out at this stage. Auffan et al. showed that chemically stable metallic nanoparticles have no significant cellular toxicity, whereas nanoparticles that are able to be oxidized, reduced, or dissolved are cytotoxic and genotoxic for cellular organisms [22]. Results from the Wuerzburg group suggest a correlation between ion concentration and genotoxic effects [23], but other groups could not confirm these results in several test systems (micronucleus test, comet assay, and γ H2AX) in a human neural cell line [24].

Autophagy is a lysosome-dependent degradation process that is usually activated in stress situations. Roy et al. identified autophagy activation as a major modulator of ZnO NP-induced cellular toxicity [25]. The detection of increased autophagosome formation and several autophagy marker proteins was reported. ROS generation was identified to be a major trigger for the induction of autophagy. Antioxidant enzymes inhibited cell death and reduced autophagy marker protein expression. The important role of autophagy in ZnO NP-induced toxicity was demonstrated by our group as well. Similar to the results reported by Roy et al., cellular damage could be reduced by counteracting oxidative stress and autophagy [26]. The correlation between autophagosome formation and apoptosis is controversially discussed in the literature. According to Vessoni et al., autophagy is a reaction to DNA damage, and plays an ambiguous role in regulating cell fate [27]. On the one hand, autophagy may promote cell protection, e.g., by degrading pro-apoptotic proteins or by supporting DNA repair. On the other hand, autophagy may also lead to cytotoxic events through the degradation of anti-apoptotic and DNA repair-related proteins [28]. In fact, ZnO NP-induced oxidative DNA damage stimulates autophagy pathways, and thus may influence the balance between cell survival and cytotoxicity. Pati et al. demonstrated an inhibition of DNA repair mechanisms. The reduction in the macrophage cell viability was due to the arrest of the cell cycle at the G0/G1 phase, the inhibition of superoxide dismutase, catalase, and eventually ROS [29].

Kononenko et al. demonstrated a concentration-dependent genotoxicity and cytotoxicity. DNA and chromosomal damage was accompanied by a reduction of glutathione S-transferase and catalase activity [30].

The amount of DNA damage does not only depend on the tested NP itself, but also on the exposed target cell, and the cell's genetic and proteomic properties in particular. ZnO NPs were shown to activate the p53 pathway by several groups [25,31,32]. ZnO NP-induced DNA damage should usually be forced by p53-associated apoptosis. Ng et al. examined a p53 knockdown fibroblast cell line exposed to ZnO NPs, and found a resistance to ZnO NP-mediated apoptosis, as well as a progressive cellular proliferation, indicating a possible first step to carcinogenesis.

The photogenotoxicity of ZnO NPs is a very important topic. UV irradiation was shown to enhance the cytotoxic properties of ZnO NPs in the A549 cell line by Yang and Ma [33]. Wang et al. reported on the oxidative DNA damage induced by ZnO NPs during UVA (ultraviolet) and visible light irradiation in a dose-dependent manner in HaCaT human skin keratinocytes [34]. The authors proclaimed a photogenotoxic potential of ZnO NPs in combination with UV light. These findings must be discussed critically, especially with respect to the use of ZnO NPs in sunscreen products. Contrary results were published by Demir et al., who demonstrated ZnO NP-related DNA damage in human and mouse cell lines using the micronucleus test and comet assay [35]. Furthermore, they observed anchorage-independent cell growth after NP exposure, which can be interpreted as an initial step towards malignant cell transformation. However, UVB exposure antagonized these effects. Future research projects can be expected to illuminate the interactions between UV light and ZnO NPs regarding DNA damage or DNA protection.

Certainly, a detailed characterization of the physicochemical properties of ZnO NPs is crucial in order to understand the partially divergent statements in the literature. Bhattacharya et al. underscored the important role of the physical properties of NPs. They showed that rod-shaped ZnO NPs induced significantly more DNA damage in peripheral blood mononuclear cells compared with spherical NPs [36]. Coatings may also influence the genotoxic potential of ZnO NPs, as shown by Yin et al., who demonstrated the extended DNA damage of NPs coated with poly (methacrylic acid) (PMAA) compared with uncoated particles [37]. The surface activity and large surface area of NPs lead to a high sorption capability, and thus induce further toxic effects. NPs can function as carriers of absorbed toxic substances, and thus enhance their bioavailability [38].

The majority of the current data regarding the genotoxic effects of nanoparticulate ZnO are based on in vitro investigations. In cells, NPs induce inflammation, genotoxic effects, and damages via the generation of reactive oxygen species (ROS). Sharma et al. published several studies on the genotoxicity of ZnO NPs in a variety of cell systems. They observed DNA damage using the single cell microgel electrophoresis (comet) assay in the HepG2 human liver cell line and the A-431 human epidermal cell line. Cells were exposed to ZnO NPs for 6 h [39,40]. The generation of ROS was demonstrated and discussed as a possible trigger of in vitro genotoxicity in both studies. Patel et al. found the generation of ROS in the A-431 cell line following the application of ZnO NPs. In this publication, ZnO NPs induced cell death, as well as a cell cycle arrest in the S and G2/M phase [41]. Tyrosine phosphorylation was shown to be another promoter of DNA damage in HepG2 cells [42]. Transmission and scanning electron microscopy are the usual tools for the investigation of cellular NP uptake, although these methods are quite time-consuming and technically challenging. Condello et al. demonstrated the entrance of ZnO NPs into human colon carcinoma cells, either by passive diffusion, endocytosis, or both. The entrance mode was dependent on the agglomeration state of the nanomaterial [43]. Toduka et al. used side-scattered light in flow cytometry as an indicator of NP uptake into mammalian cells [44]. Several nanomaterials were tested, including ZnO NPs in Chinese hamster ovary (CHO)-K1 cells using this method. Particles were internalized into the cells, and thus induced a high ROS production, which was directly correlated with the genotoxic events shown by the generation of the phosphorylated histone γ-H2AX. The co-cultivation with the antioxidant N-acetylcysteine (NAC) counteracted DNA damage. Kermanizadeh et al. also showed the important role of oxidative stress through demonstrating a suppression of the toxic potential of ZnO NPs by the antioxidant Trolox in a hepatocyte cell line [45]. DNA damage and pro-inflammatory IL-8 production were induced by oxidative stress and ROS generation. Other groups also published similar results demonstrating the positive correlation between oxidative stress and DNA damage [46,47]. The generation of ROS was mainly assessed by the dichloro-dihydro-fluorescein diacetate (DCFH-DA) assay. Various markers for oxidative stress were evaluated, e.g., glutathione (GSH) reduction, elevated gluthatione, malondialdehyde, superoxide dismutase, and catalase. The photogenotoxicity of ZnO NPs, including a high cellular uptake, was shown in Allium cepa [48]. Other groups also demonstrated the connection between DNA damage and ROS production [43,49].

Most of the studies on nanogenotoxicity were performed using cell lines instead of primary cells. Due to high interindividual variation and the difficulty of standardizing cellular harvest, repetitive experiments with large numbers of patients are necessary in order to assess representative data on primary cells. However, primary cells are neither immortalized nor transformed. Thus, the similarity to cells within the origin tissue is usually higher compared with transformed cell lines. This is why studies with primary cells are supplementary to those using standardized cell lines, and can contribute to common knowledge on nanogenotoxicology. Sharma et al. presented a study using primary human epidermal keratinocytes, a relevant target cell for ZnO NPs, which are mainly used in cosmetics applied to the human skin. ZnO NPs were internalized by the cells, as shown by transmission electron microscopy, where they induced a DNA fragmentation after 6 h of exposure at a concentration of 8 µg/mL [50]. Our own group used primary human nasal mucosa to evaluate the genotoxicity of ZnO NPs. Nasal mucosa belongs to the most important primary contact regions of humans with volatile xenobiotics. Cells of the nasal mucosa are representative of the entire human upper aerodigestive tract. Distinct three-dimensional cell culture systems serve to imitate the in vivo situation quite closely [51]. The genotoxic potential of ZnO NPs was proven in human nasal mucosa cells in an air–liquid interface cell culture, as well as by the extended secretion of IL-816. Primary human adipose tissue derived mesenchymal stem cells showed DNA damage and pro-inflammatory cytokine production after ZnO NP exposure as well. The stem cell migration capacity was impaired significantly after NP exposure. Interestingly, ZnO NPs were internalized into the cells in high amounts after 24 h, and remained in the cytoplasm for over three weeks, indicating bioaccumulation of the particles. Future studies should illuminate cellular uptake dynamics and exclusion mechanisms. The intracellular persistence of NPs could be a severe problem, since even low exposure doses can lead to critical intracellular concentrations after repetitive contact [52]. The repetitive exposure of nasal mucosa mini organ cultures induced an enhanced genotoxic effect, and 24 h after exposure the DNA damage even increased, probably due to persisting NPs in the cells and the ongoing production of ROS [53]. Ghosh et al. investigated the genotoxic effects of ZnO NPs on human peripheral blood mononuclear cells. The in vitro tests revealed weak genotoxic effects. A significant decrease of mitochondrial membrane potential was also detected [54]. Branica et al. demonstrated a significant increase of DNA damage in human lymphocytes after exposure to ZnO NPs [55].

In contrast to the series of publications stating the possible genotoxicity of ZnO NPs, there are other studies showing no evidence of DNA damage. Nam et al. classified ZnO NPs as well as Zn ions as non-genotoxic in the so-called SOS chromotest [56]. In addition, Kwon et al. did not find any genotoxic effects in several in vitro and in vivo assays that used differently sized and differently charged particles [57]. In a study conducted by Alaraby et al., no toxicity or oxidative stress induction was observed in vivo. Furthermore, no significant changes in the frequency of mutant clones or percentage of DNA in tail (comet assay) were measured, although significant changes in Hsp70 and p53 gene expression were detected [58].

Sahu et al. demonstrated the cytotoxic effects and inflammatory potential of ZnO NPs in a human monocyte cell line, but did not observe any DNA damage [59]. Bayat et al. critically discussed the test systems that are routinely used for genotoxicity assessments. They stated that in vitro genotoxicity testing is probably unreliable because different test systems produce inconsistent results [60].

4.2. In Vivo Studies

Only a few studies can be found that evaluate the genotoxicity of ZnO NPs in vivo. Pati et al. investigated the toxicity of ZnO NPs in mice. In this publication, ZnO NPs were dispersed in water by vortex mixing. Afterwards, the animals were fed with water containing NPs in order to demonstrate oral exposure. ZnO NP-treated animals showed signs of toxicity, which was associated with severe DNA damage in peripheral blood and bone marrow cells. Moreover, DNA repair mechanisms were inhibited and enhanced organ inflammation was detected, as well as a disturbance of wound healing [29]. Sharma et al. used a mouse model for subacute oral exposure to ZnO NPs for two weeks. NPs accumulated in the liver and induced DNA damage in liver cells. The authors used an Fpg-modified comet assay to prove that oxidative stress induced DNA damage [32]. Ali et al. found a reduction in glutathione, glutathione-S-transferase, and glutathione peroxidase, as well as an increase in malondialdehyde and catalase in Lymnaea luteola freshwater snails after ZnO NP exposure for 24 and 96 h. Genotoxic effects were found in the digestive gland cells treated with ZnO NPs [61]. Li et al. used a mouse model to prove the biodistribution and genotoxicity of orally administered and intraperitoneally injected ZnO NPs [62]. Baky et al. examined the cardiotoxic effects of ZnO NPs in rats [63], and found that alpha-lipoic acid and vitamin E reduced the DNA damage in cardiac cells. Zhao et al. found DNA damage in embryo-larval zebrafish [64]. The authors compared the toxic effects of Zn ions and ZnO NPs, and demonstrated that ions only partially contribute to the toxic effects. In contrast, triethoxycaprylylsilane-coated ZnO NPs did not induce DNA damage in lung cells from rats after inhalation exposure [65]. Ghosh et al. showed a reduced mitochondrial membrane potential in bone marrow cells in vivo. Furthermore, an enhanced oxidative stress, a G0/G1 cell cycle arrest, and chromosomal aberration with micronuclei formation were measured [54]. In the study conducted by Ng et al., a significant toxicity was observed in melanogaster F1 progenies upon ingestion of ZnO NPs. The egg-to-adult viability of the flies was significantly reduced, which was closely associated with ROS induction by ZnO NPs. Nuclear factor E2-related factor 2 was identified to play a role in ZnO NP-mediated ROS production [49]. Anand et al. investigated the effects of ZnO NPs in Drosophila melanogaster. Food containing 0.1 mM to 10 mM of ZnO NPs induced distinctive phenotypic changes, such as a deformed segmented thorax and a single or deformed wing, which were transmitted to offspring in subsequent generations [66]. Manzo et al. investigated the effects of ZnO NPs in sea urchins. ZnO NPs provoked damages to immune cells in adult echinoids and transmissible effects to offspring [67].

5. Summary

Although evaluations of the genotoxicity of ZnO NPs are not consistent, there seems to be reliable evidence supporting the potential for them to damage the DNA in human cells. Genotoxic events were demonstrated using several methods and different endpoints. Besides the comet assay, the micronucleus test, the chromosomal aberration assay, and the γ H2AX method were used. The correlation between oxidative stress and DNA damage can be easily proved by the Fpg-modified comet assay and by the interaction with antioxidants such as N-acetylcysteine. Research has shown the internalization of ZnO NPs into the cells via endocytosis or several other mechanisms such as macropinocytosis. Intracellular distribution was observed by transmission electron microscopy as well as by alternative methods such as side scatter flow cytometry. While there is still some controversy surrounding the possible transfer of ZnO NPs into the nucleus, a distribution into cell organelles can definitely be observed. The inclusion into lysosomes seems to be of major importance, since due to the low pH milieu of lysosomes, ZnO dissolves and Zn^{2+} ions are released. Ion release from ZnO NPs may already occur in the cellular expansion medium. Research studies also discuss both intracellular and extracellular Zn^{2+} release as main triggers for DNA damage. Even if ZnO NPs are not able to enter the nucleus, Zn^{2+} ions affect DNA integrity in a dose-dependent manner. Lysosomes release Zn^{2+} ions into the cytoplasm, which is then a trigger for ROS generation. Several research groups have proven this phenomenon by using the DCFH-DA assay. Markers for oxidative stress such as GSH reduction, elevated glutathione, malondialdehyde, superoxide dismutase, and catalase were analyzed after ZnO NP exposure. As a reaction to disrupted DNA integrity, lysosomes develop into autophagosomes, which can be detected by transmission electron microscopy or indirectly by several protein markers such as LC3 II or beclin-1. The role of autophagy on apoptosis or cell survival is still unclear, and only a few studies address the topic of DNA repair capacity after NP exposure. There is evidence indicating the insufficient repair of DNA disintegrity after ZnO NP exposure, which can be explained by trapped NPs in intracellular departments, and an ongoing trigger for ROS-induced DNA damage. Figure 1 shows a hypothetical model of ZnO NP-induced genotoxicity.

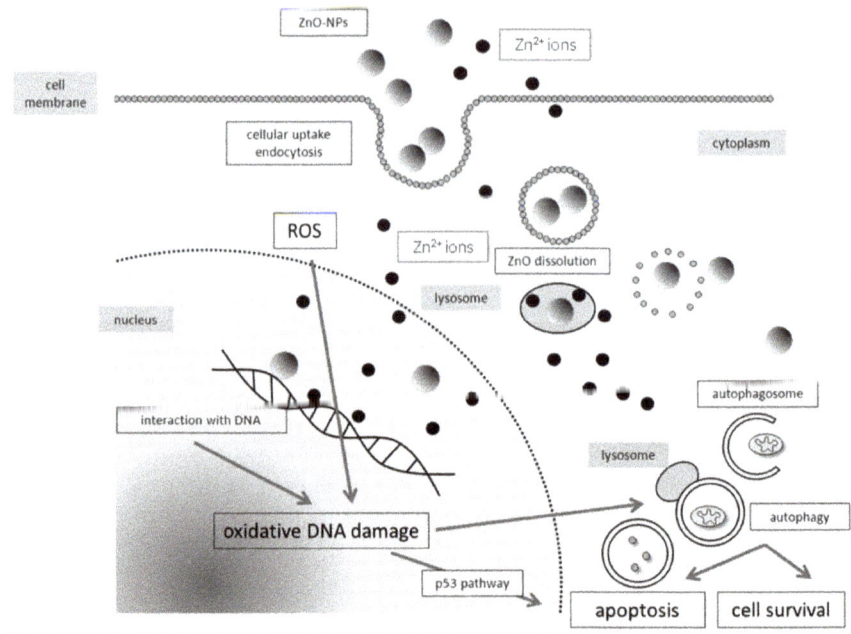

Figure 1. A hypothetical model of Zinc oxide nanoparticle (ZnO NP)-induced genotoxicity.

6. Conclusions and Recommendations for Future Research

At present, there is still limited information regarding the genotoxic potential of ZnO NPs. Due to inconsistencies in the data available, it is nearly impossible to give recommendations or properly assess the risk of ZnO NP application. Most studies on the hazardous effects of ZnO NPs focus on cytotoxicity. However, ZnO NPs seem to belong to a group of nanomaterials that are able to cause DNA damage. Thus, further genotoxicological evaluation is needed. A strictly detailed and standardized physicochemical characterization of the tested NPs is obligatory in order to produce comparable and informative genotoxicological data. The authors refer to the recommendations of Landsiedel et al. (2010) [65] regarding nanotoxicological study design. Most genotoxicological investigations on ZnO NPs address acute exposure situations. That is why our knowledge of bioaccumulation and long-term exposure effects is only fragmentary. Hence, test systems need to be established in order to clarify these questions, and the biological mechanisms responsible for DNA damage must be analyzed continuously. ZnO NPs are very promising and highly effective materials, and a proper characterization of the genotoxic issues is mandatory in order to apply them reasonably and safely.

Table 1 summarizes relevant publications on ZnO NP-associated genotoxicity mechanisms. The order of listed NPs in Table 1 was sorted according to the particle size, beginning with the smallest. We did not observe any tendency that the results regarding genotoxic potency varied within the two groups of particles smaller or larger than 100 nm. Although the group of larger particles did not exactly fit the definition of NPs, they seem to be still small enough to exhibit comparable toxic properties as compared with NPs <100 nm.

Table 1. Current literature review of the genotoxic effects of ZnO nanoparticles.

Characteristics of Nanomaterial(s)	In Vivo	Exposure	Methods	Results	Reference
ZnO NPs: average-size 10–20 nm	Earthworm Eisenia fetida (Savigny, 1826)	0.1, 0.5, 1.0, 5.0 g/kg for 7 days	Comet assay	DNA damages were observed at dosages greater than 1.0 g/kg	[66]
ZnO NPs: average size 12 ± 3 nm	Cells of bronchoalveolar lavage fluid, day 1 and 3 after ZnO exposure, in female wild-type C57BL/6JBomTac (C57) mice	Intratracheal instillation of 2, 6, 18 µg ZnO NPs	Comet assay	DNA damage was dose dependent. However, three days post-exposure genotoxicity decreased	[67]
ZnO NPs: average size 22 nm	Freshwater snail Lymnaea luteola (L. luteola)	10, 21.33, and 32 µg/mL for 96 h	Comet assay	Comet assay revealed DNA damage after treatment with ZnO NPs	[61]
ZnO NPs: average size 28 ± 5 nm Zeta potential −22 mV	Drosophila melanogaster	Food containing 0.1 mM, 1 mM, and 10 mM of ZnO NPs throughout the entire life cycle from egg to egg stage	TUNEL (TdT-mediated dUTP-biotin nick end labeling) assay ROS detection assay	ZnO NPs exposure induced a increase of DNA fragmentation and phenotypic changes, which were transmitted to the offspring	[68]
ZnO NPs: average size 30 nm	Cells of liver and kidney of mice after oral exposure	50 and 300 mg/kg of ZnO for 14 days	Comet assay	The Comet assay revealed a significant increase in the Fpg-specific DNA lesions in liver and kidney cells	[69]
ZnO-NPs: average size: ~70 nm Zeta potential +8.8 mV	MRC5 human lung fibroblasts, Drosophila melanogaster	0, 1, 10, 25, 50, 75, and 100 µg/mL for 24, 48 and 72 h	Comet assay ROS detection assay	Significant genotoxicity was induced by ZnO NPs	[49]
ZnO NPs: average size <100 nm	Human peripheral blood mononuclear cells (PBMCs) and Swiss albino male mice	Cell treatment: 0, 25, 50, and 100 µg/mL for 3 h Animal treatment: 25, 50, and 100 mg/kg body weight 18 h before sacrifice	Comet assay Chromosome aberration assay Micronucleus assay	Apoptosis mediated by ROS generation, reduced mitochondrial membrane potential (MMP) in bone marrow cells, a G0/G1 cell cycle arrest, and chromosomal aberration with micronuclei formation	[54]
ZnO NPs: average size 100 nm Surface area: 15–25 m²/g 14 nm Surface area: 30 ± 5 m²/g	Sea urchin	1 mg/kg food for three weeks	Comet assay	ZnO NPs 100 nm provoked in adult echinoids damages to immune cells and transmissible effects to offspring, ZnO NPs 14 nm provoked nucleus damages in immune cells and malformed larvae	[70]
ZnO NPs: average size 72 ± 46 nm Zeta potential −13.3 ± 2.3 mV ZnO microparticles particles (MPs)	Madin–Darby canine kidney (MDCK) cells	1, 5, 10, 15, 30, and 60 µg/mL ZnO for 24 h	Comet assay Cytokinesis-block micronucleus assay	ZnO NPs significantly elevated DNA and chromosomal damage, whereas equimolar concentrations of ZnO MPs did not	[30]
ZnO NPs: average size <100 nm	Broodstock zebrafish larvae, Danio rerio	0.2, 1, 2, 4, 6 mg/L for 96 h	Comet assay	Comet assay revealed significant DNA damage after ZnO NP exposure	[71]

Table 1. *Cont.*

Characteristics of Nanomaterial(s)	In Vivo	Exposure	Methods	Results	Reference
ZnO NPs: average size 20 nm (+) charge: 35 ± 5, 20 nm (−) charge: 28 ± 8, 70 nm (+) charge: 70 ± 19, 70 nm (−) charge: 72 ± 11 nm; Hydrodynamic size of ZnO nanoparticles: 20 nm (+) charge: 200 to 400 nm, 20 nm (−) charge: 180–300, 70 nm (+) charge: 300–900 nm, 70 nm (−) charge: 200–500 nm; zeta potential: 20 nm (+) charge: +25.9 mV, 20 nm (−) charge: −38.5 mV, 70 nm (+) charge: +25.9 mV, 70 nm (−) charge: −40.6 mV	SD rat: liver and stomach cells	500, 1000, and 2000 mg/kg body weights, three times by gavage at 0, 24, and 45 h	Bacterial mutagenicity assay in vitro chromosomal aberration test in vivo comet assay in vivo micronucleus test	Surface modified ZnO NPs did not induce genotoxicity in vitro and in vivo	[57]
ZnO NPs: average size 104.17 ± 66.77 nm	Mouse embryonic fibroblast (MEF Ogg1+/+) and mouse embryonic fibroblast knockout (MEF Ogg1−/−) cell lines	Sub-toxic dose (1 µg/mL) for 12 weeks, Short-term exposure (0.3125 to 40 µg/mL) for 48 h	Comet assay	Short-term ZnO NPs exposure induce ROS, genotoxicity, and oxidative DNA damage. No effects after long-term exposure	[72]
ZnO NPs: average size 106.55 ± 64.79 nm Zeta potential: −21.00 ± 0.80 mV ZnO NPs bulk: average size 4.2 µm	Haemolymph cells from Drosophila melanogaster	6, 12, 24, mM for 24 h	Wing-spot test Comet assay	No increases in the frequency of mutant spots was detected Significant increase in DNA damage was observed	[73]
ZnO NPs: average size 200–250 nm Zeta potential −0.56 mV	Mice and cells isolated from mice	0–500 µg/mL for 24 h Mice were treated with 200 and 500 mg/kg bodyweight of ZnO NPs	Comet assay Micronucleus Assay	The comet assay revealed severe DNA damage in peripheral blood and bone marrow cells. Moreover, DNA repair mechanism were inhibited	[29]
ZnO NPs: average size 291.66 ± 6.59 nm Zeta potential −11.40 ± 0.26 mV	Drosophila melanogaster	0.02, 0.1, 0.2, 1 and 2 mg/g of food media	The wing-spot assay Comet assay	ZnO NPs were not genotoxic	[58]
ZnO NPs: average size 470 ± 45 nm Zeta potential: −10.35 ± 0.83 mV ZnO NPs: average size 1040 ± 70 nm Zeta potential: −10.51 ± 1.43 mV	Dunaliella tertiolecta	0.1, 2, 5, 10, 25, 50 mg/L for 24 and 72 h	Comet assay	Genotoxic action was evident only starting from 5 mg/L	[74]
ZnO NPs: average size 15–18 nm	Cell line (A549)	0.1, 10, 100 µg/mL	γH2AX immunofluorescence assay	Foci analyses showed the induction of DNA double strand breaks by ZnO NPs. Reduction of DNA damage was achieved by the treatment with the ROS scavenger N-acetyl-L-cysteine	[75]
ZnO NPs: average size 15–25 nm	Human neuroblastoma SHSY5Y cell line	20, 30, 40 µg/mL for 3 h and 6 h	Micronuclei evaluation by flow cytometry γH2AX assay Comet assay Oxidative DNA damage	Micronuclei were induced by ZnO NPs, H2AX phosphorylation and DNA damage were observed in all cases	[24]

Table 1. *Cont.*

Characteristics of Nanomaterial(s)	In Vivo	Exposure	Methods	Results	Reference
ZnO NPs: average size 17 nm Zeta potential: −14.0 mV	Human malignant melanoma skin (A375) cell line	5, 10, 20 μg/mL for 24 and 48 h	Comet assay	ZnO NPs induced DNA damage. A gradual nonlinear increase in cell DNA damage was observed as concentration and duration of ZnO nanoparticle exposure increased	[76]
ZnO NPs: average size 10–50 nm	Rat kidney epithelial cell line (NRK-52E)	25.0–100.0 mg/mL for cytotoxicity assays and 12.5–50.0 mg/mL for genotoxicity assay	Comet assay	ZnO NPs caused statistically significant DNA damage	[77]
ZnO NPs: average size 20 nm	Chinese hamster lung fibroblasts (V79 cells)	30.0, 60.0, 120.0 μM for 3 h	Cytokinesis-block micronucleus Assay somatic mutation and Recombination test micronucleus assay	ZnO NPs increase the frequency of micronuclei, results were not dose related	[78]
ZnO NPs: average size 19.5 ± 5.8 nm	Primary mouse embryo fibroblasts (PMEF)	5 and 10 μg/mL for 24 h	Comet assay	ZnO NPs caused statistically significant DNA damage	[79]
ZnO NPs: average size 25.8 ± 8.9 nm Zeta potential: +17.4 mV	Human intestinal carcinoma epithelial cell lines, SW480 and DLD-1 and the normal human intestinal mucosa epithelial cell line, NCM460	Cell exposure concentrations 62.5, 250, and 1000 μM for 12 or 24 h	Oxidative stress measurement Cell cycle analysis	The elevated ROS levels induce significant damage to the DNA of the cells, resulting in cell-cycle arrest and subsequently cell death	[80]
ZnO NPs: average size 25.12 ± 9.2 nm	Cell line from gill tissue of Wallago attu (WAG)	0, 12.5, 25, 50 mg/L for 24 h	Comet assay Micronucleus assay	ZnO NPs induced DNA damage in a dose dependent manner	[81]
ZnO-S ZnO NPs-S: average size 26 ± 9 nm Zeta potential: +19.2 ± 0.3 mV ZnO NPs-M average size 78 ± 25 nm Zeta potential: +200 ± 0.6 mV ZnO NPs-L: average size 147 ± 53 nm Zeta potential: +21.1 ± 0.4 mV	Human lymphoblastoid (WIL2-NS) cells	10 mg/L for 24 h	Genotoxicity-cytokinesis-block micronucleus (CBMN) Cytome Assay	Genotoxicity was significantly enhanced in the presence of the medium-sized and large-sized particles	[82]
ZnO NPs: average size 30 nm Zeta potential: −13.4 mV	Human monocytic cell line (THP-1)	0.5, 1, 5, 10, 15, 20 μg/mL for 3 h	Comet assay micronucleus assays	ZnO NP induced an enhanced DNA damage and micronucleated cells	[83]
ZnO NPs: average size 30 nm Zeta potential: −26 mV	Human epidermal cell line (A431)	0.008–20 mg/mL for 3, 6, 24, 48 h	Comet assay	ZnO NPs induced an enhanced DNA damage	[39]
ZnO NPs: average size 29 ± 10 nm	WIL2-NS human lymphoblastoid cells	10 μg/mL for 24 h	Comet assay	PMA-coated ZnO had significant genotoxicity compared to uncoated ZnO	[37]
ZnO NPs: average size <35 nm	Human lymphocyte	1.0, 2.5, 5, and 7.5 μg/mL over 2 weeks	Comet assay Comet-FISH	ZnO NPs induced DNA damage	[55]

Table 1. Cont.

Characteristics of Nanomaterial(s)	In Vivo	Exposure	Methods	Results	Reference
ZnO NPs: average size ≤35 nm; Zeta potential: +46.2 mV; ZnO NPs: average size 50–80 nm; Zeta potential: −23 mV	Human embryonic kidney (HEK293) and mouse embryonic fibroblast (NIH/3T3) cells	10, 100, 1000 μg/mL for 1 h	Comet assay; Micronucleus assay	ZnO NPs induced a significant of DNA damage with and without enzymes. The frequency of micronuclei was enhanced as well	[35]
ZnO NPs: average size ≤35 nm; Zeta potential: +46.2 mV; ZnO NPs: average size 50–80 nm; Zeta potential: −23 mV	Allium cepa root meristem cells	10, 100, 1000 μg/mL for 1 h	Comet assay	ZnO NPs were genotoxic in a dose-dependent manner	[84]
ZnO NPs: average size (given by producer) nanosized (30–35 nm) fine (150–300 nm)	human bronchial epithelial BEAS-2B cells	0.5–3.0 μg/cm² for 48 h; Comet assay 3 h to 6 h	Comet assay	ZnO NPs exposure induced DNA damage, fine ZnO did not induce DNA damages	[85]
ZnO NPs: average size 40–70 nm	human peripheral lymphocytes, human sperm cells	11.5, 46.2, 69.4, 93.2 μg/mL for 30 min, simultaneous or pre-irradiation with UV light	Comet assay	ZnO NPs are capable of inducing genotoxic effects on human sperm and lymphocytes. The effect is enhanced by UV	[86]
ZnO NPs: average size 50–70 nm	human colon carcinoma cells (LoVo)	Treatment concentration and duration was not unique e.g., cell death assay: 5 μg/cm² ZnO NPs for 2, 4, and 6 h; Zr^{2+} ions release: 5 and 10 μg/cm² for 30 min, 1 h, 2 h, 4 h, 6 h, 24 h	DNA damage assessment by 8-oxodG steady-state levels and γ-H2AX histone phosphorylation	ZnO NPs entered LoVo cells. The simultaneous presence of ZnO NPs and Zn^{2+} ions in the LoVo cells induced severe DNA damage	[43]
ZnO NPs: average size 75 ± 5 nm	Human lymphocyte cells	0, 125, 500, 1000 μg/mL for 3 h	Comet assay	1000 μg/mL ZnO NPs induced significant genotoxic effects	[87]
ZnO NPs: average size 86 ± 41 nm; mean lateral diameter: 42 ± 21 nm	Primary human nasal mucosa cells	0.01, 0.1, 5, 10, 50 μg/mL for 24 h	Comet assay	ZnO NPs induced DNA damage in a dose dependent manner	[23]
ZnO NPs: average size <100 nm; Zeta potential −33.8 ± 10	Saccharomyces cerevisiae cells	0, 0.125, 0.25, 0.5, 1, 2, 4, and 8 μg/mL for 24, 48, and 72 h; Colony-forming assay: cells were treated for 10 days.	GreenScreen assay; Comet assay	GreenScreen assay: No genotoxic effects could be measured. Comet assay: ZnO NPs were genotoxic	[56]
ZnO NPs: average size <100 nm (given by producer)	lung fibroblast (MRC5) cell line		Immunochemical assay DNA methyltransferase activity; Quantification of the 5-mC content in genomic DNA	dose-related decrease in global DNA methylation and DNA methyltransferase activity direct correlation between the concentration of NPs, global methylation levels, and expression levels of Dnmt1, 3A, and 3B genes upon exposure	[88]
ZnO NPs: average 20–200 nm, Zeta potential: 26.9 mV	A549 cells	1, 20, 40 μg/cm (=2, 40, 80 μg/mL) for 4 h; fpg-sensitive sites: 20 and 40 μg/cm after 4 h	Comet assay	ZnO NPs induced DNA damage	[89]

Table 1. *Cont.*

Characteristics of Nanomaterial(s)	In Vivo	Exposure	Methods	Results	Reference
ZnO NPs: average size NM-110: 70–100 nm; NM-111: 58–93 nm Zeta potential: NM-110: −11.5 mV; NM-111: −11.4 mV	HK2-cells	Ten concentrations between 0.16 and 80 µg/cm for 4 h	Comet assay	Increase of tail DNA following nanomaterials exposure	[90]
ZnO NPs: average size 45–170 nm Zeta potential: −15.6 ± 2.4 mV	Human colon carcinoma (Caco-2) cells	CBMN assay: 6.4, 12.8, 22.4, 64.0 µg/mL for 6 or 24 h Comet assay: 6.4, 16.0 µg mL⁻¹ for 24 h	CBMN assay Comet assay	ZnO NPs induced DNA damage	[91]
ZnO NPs: average size 120 ± 2.6 nm	Root cells of *Allium cepa*	25, 50, 75, 100 µg/mL for 4 h	Analysis of mitotic index, micronuclei index and chromosomal aberration index	Dose dependent depression of mitotic index, an increase of pyknotic cells, an increase of micronuclei index and chromosomal aberration index	[48]
ZnO NPs: average size NM-110: 20–250/50–350 nm; NM-111: 20–200/10–450 nm	Human hepatoblastoma C3A cells, in vitro	NM concentrations between 0.16 µg cm⁻² and 80 µg/cm for 4 h	Comet assay	significant increase in percentage tail DNA	[45]
ZnO NPs: average size 64–510 nm Zeta potential: −25.30 mV	human peripheral blood lymphocytes	50–1000 µg/mL for 24 h (cytotoxicity) 25, 50 and 100 µg/mL for 4 h (genotoxicity)	Comet assay	The smaller NPs are more genotoxic, treatment with vitamin C or quercetin significantly reduces the genotoxicity	[92]
	human peripheral blood lymphocytes	0.01–10 mM for 4, 8, 24 h	Comet assay	ZnO NPs induced DNA damage in a dose dependent manner	[93]
ZnO NPs: average size 250–970 nm Zeta potential 20 mV	human bronchial cells (3D model)	30 µL of a 1.06 mg/mL suspension with a dosage of 50 µg/cm² for 24 to 72 h	Comet assay	ZnO NPs were genotoxic in a dose-dependent manner	[94]
ZnO NPs (50 wt %) were purchased From Sigma-Aldrich (St. Louis, MO, USA). No data about particle size	human promyelocytic leukemia (HL-60) cells, and peripheral blood mononuclear cells (PBMC)	0, 0.05, 5, 10, 15, and 20 mg/L for 24 h	Comet assay	ZnO NPs were genotoxic in a dose-dependent manner	[95]

Author Contributions: Agmal Scherzad and Stephan Hackenberg wrote part of the MS, Till Meyer created the table, Norbert Kleinsasser designed figures and edited paper.

Conflicts of Interest: The authors declare no conflict of interest.

References

1. Official Journal of the European Union. COMMISSION RECOMMENDATION of 18 October 2011 on the Definition of Nanomaterial (Text with EEA Relevance) (2011/696/EU). Available online: https://ec.europa.eu/research/industrial_technologies/pdf/policy/commission-recommendation-on-the-definition-of-nanomater-18102011_en.pdf (accessed on 11 December 2017).
2. Shi, H.B.; Magaye, R.; Castranova, V.; Zhao, J.S. Titanium dioxide nanoparticles: A review of current toxicological data. *Part. Fibre Toxicol.* **2013**. [CrossRef] [PubMed]
3. Klingshirn, C. ZnO: Material, physics and applications. *Chemphyschem* **2007**, *8*, 782–803. [CrossRef] [PubMed]
4. Osmond, M.J.; Mccall, M.J. Zinc oxide nanoparticles in modern sunscreens: An analysis of potential exposure and hazard. *Nanotoxicology* **2010**, *4*, 15–41. [CrossRef] [PubMed]
5. Vance, M.E.; Kulheu, T.; Vejerano, E.P.; McGinnis, S.P.; Hochella, M.F., Jr.; Rejeski, D.; Hull, M.S. Nanotechnology in the real world: Redeveloping the nanomaterial consumer products inventory. *Beilstein J. Nanotechnol.* **2015**, *6*, 1769–1780. [CrossRef] [PubMed]
6. Gamer, A.O.; Leibold, E.; van Ravenzwaay, B. The in vitro absorption of microfine zinc oxide and titanium dioxide through porcine skin. *Toxicol. In Vitro* **2006**, *20*, 301–307. [CrossRef] [PubMed]
7. Cross, S.E.; Innes, B.; Roberts, M.S.; Tsuzuki, T.; Robertson, T.A.; McCormick, P. Human skin penetration of sunscreen nanoparticles: In Vitro assessment of a novel micronized zinc oxide formulation. *Skin Pharmacol. Phys.* **2007**, *20*, 148–154. [CrossRef] [PubMed]
8. Lademann, J.; Otberg, N.; Richter, H.; Weigmann, H.J.; Lindemann, U.; Schaefer, H.; Sterry, W. Investigation of follicular penetration of topically applied substances. *Skin Pharmacol. Appl.* **2001**, *14*, 17–22. [CrossRef] [PubMed]
9. Vandebriel, R.J.; De Jong, W.H. A review of mammalian toxicity of ZnO nanoparticles. *Nanotechnol. Sci Appl.* **2012**, *5*, 61–71. [CrossRef] [PubMed]
10. Vermylen, J.; Nemmar, A.; Nemery, B.; Hoylaerts, M.F. Ambient air pollution and acute myocardial infarction. *J. Thromb. Haemost.* **2005**, *3*, 1955–1961. [CrossRef] [PubMed]
11. Nemmar, A.; Vanbilloen, H.; Hoylaerts, M.F.; Hoet, P.H.M.; Verbruggen, A.; Nemery, B. Passage of intratracheally instilled ultrafine particles from the lung into the systemic circulation in hamster. *Am. J. Respir. Crit. Care* **2001**, *164*, 1665–1668. [CrossRef] [PubMed]
12. Kim, J.S.; Yoon, T.J.; Kim, B.G.; Park, S.J.; Kim, H.W.; Lee, K.H.; Park, S.B.; Lee, J.K.; Cho, M.H. Toxicity and tissue distribution of magnetic nanoparticles in mice. *Toxicol. Sci.* **2006**, *89*, 338–347. [CrossRef] [PubMed]
13. Oberdorster, G.; Sharp, Z.; Atudorei, V.; Elder, A.; Gelein, R.; Kreyling, W.; Cox, C. Translocation of inhaled ultrafine particles to the brain. *Inhal. Toxicol.* **2004**, *16*, 437–445. [CrossRef] [PubMed]
14. Kwon, J.Y.; Koedrith, P.; Seo, Y.R. Current investigations into the genotoxicity of zinc oxide and silica nanoparticles in mammalian models in vitro and in vivo: Carcinogenic/genotoxic potential, relevant mechanisms and biomarkers, artifacts, and limitations. *Int. J. Nanomed.* **2014**, *9*, 271–286.
15. Liu, J.; Feng, X.L.; Wei, L.M.; Chen, L.J.; Song, B.; Shao, L.Q. The toxicology of ion-shedding zinc oxide nanoparticles. *Crit. Rev. Toxicol.* **2016**, *46*, 348–384. [CrossRef] [PubMed]
16. Saptarshi, S.R.; Duschl, A.; Lopata, A.L. Biological reactivity of zinc oxide nanoparticles with mammalian test systems: An overview. *Nanomedicine* **2015**, *10*, 2075–2092. [CrossRef] [PubMed]
17. Bondarenko, O.; Juganson, K.; Ivask, A.; Kasemets, K.; Mortimer, M.; Kahru, A. Toxicity of Ag, CuO and ZnO nanoparticles to selected environmentally relevant test organisms and mammalian cells in vitro: A critical review. *Arch. Toxicol.* **2013**, *87*, 1181–1200. [CrossRef] [PubMed]
18. Golbamaki, N.; Rasulev, B.; Cassano, A.; Robinson, R.L.M.; Benfenati, E.; Leszczynski, J.; Cronin, M.T.D. Genotoxicity of metal oxide nanomaterials: Review of recent data and discussion of possible mechanisms. *Nanoscale* **2015**, *7*, 2154–6398. [CrossRef] [PubMed]
19. Warheit, D.B.; Webb, T.R.; Sayes, C.M.; Colvin, V.L.; Reed, K.L. Pulmonary instillation studies with nanoscale TiO_2 rods and dots in rats: Toxicity is not dependent upon particle size and surface area. *Toxicol. Sci.* **2006**, *91*, 227–236. [CrossRef] [PubMed]

20. Karlsson, H.L.; Gustafsson, J.; Cronholm, P.; Moller, L. Size-dependent toxicity of metal oxide particles—A comparison between nano- and micrometer size. *Toxicol. Lett.* **2009**, *188*, 112–118. [CrossRef] [PubMed]

21. Singh, N.; Manshian, B.; Jenkins, G.J.; Griffiths, S.M.; Williams, P.M.; Maffeis, T.G.; Wright, C.J.; Doak, S.H. NanoGenotoxicology: The DNA damaging potential of engineered nanomaterials. *Biomaterials* **2009**, *30*, 3891–3914. [CrossRef] [PubMed]

22. Auffan, M.; Rose, J.; Wiesner, M.R.; Bottero, J.Y. Chemical stability of metallic nanoparticles: A parameter controlling their potential cellular toxicity in vitro. *Environ. Pollut.* **2009**, *157*, 1127–1133. [CrossRef] [PubMed]

23. Hackenberg, S.; Scherzed, A.; Technau, A.; Kessler, M.; Froelich, K.; Ginzkey, C.; Koehler, C.; Burghartz, M.; Hagen, R.; Kleinsasser, N. Cytotoxic, genotoxic and pro-inflammatory effects of zinc oxide nanoparticles in human nasal mucosa cells in vitro. *Toxicol. In Vitro* **2011**, *25*, 657–663. [CrossRef] [PubMed]

24. Valdiglesias, V.; Costa, C.; Kilic, G.; Costa, S.; Pasaro, E.; Laffon, B.; Teixeira, J.P. Neuronal cytotoxicity and genotoxicity induced by zinc oxide nanoparticles. *Environ. Int.* **2013**, *55*, 92–100. [CrossRef] [PubMed]

25. Roy, R.; Singh, S.K.; Chauhan, L.K.; Das, M.; Tripathi, A.; Dwivedi, P.D. Zinc oxide nanoparticles induce apoptosis by enhancement of autophagy via PI3K/Akt/mTOR inhibition. *Toxicol. Lett.* **2014**, *227*, 29–40. [CrossRef] [PubMed]

26. Hackenberg, S.; Scherzed, A.; Gohla, A.; Technau, A.; Froelich, K.; Ginzkey, C.; Koehler, C.; Burghartz, M.; Hagen, R.; Kleinsasser, N. Nanoparticle-induced photocatalytic head and neck squamous cell carcinoma cell death is associated with autophagy. *Nanomedicine* **2014**, *9*, 21–33. [CrossRef] [PubMed]

27. Vessoni, A.T.; Filippi-Chiela, E.C.; Menck, C.F.M.; Lenz, G. Autophagy and genomic integrity. *Cell Death Differ.* **2013**, *20*, 1444–1454. [CrossRef] [PubMed]

28. Mizushima, N.; Levine, B.; Cuervo, A.M.; Klionsky, D.J. Autophagy fights disease through cellular self-digestion. *Nature* **2008**, *451*, 1069–1075. [CrossRef] [PubMed]

29. Pati, R.; Das, I.; Mehta, R.K.; Sahu, R.; Sonawane, A. Zinc-Oxide Nanoparticles Exhibit Genotoxic, Clastogenic, Cytotoxic and Actin Depolymerization Effects by Inducing Oxidative Stress Responses in Macrophages and Adult Mice. *Toxicol. Sci.* **2016**, *150*, 454–472. [CrossRef] [PubMed]

30. Kononenko, V.; Repar, N.; Marusic, N.; Drasler, B.; Romih, T.; Hocevar, S.; Drobne, D. Comparative in vitro genotoxicity study of ZnO nanoparticles, ZnO macroparticles and $ZnCl_2$ to MDCK kidney cells: Size matters. *Toxicol. In Vitro* **2017**, *40*, 256–263. [CrossRef] [PubMed]

31. Ng, K.W.; Khoo, S.P.K.; Heng, B.C.; Setyawati, M.I.; Tan, E.C.; Zhao, X.X.; Xiong, S.J.; Fang, W.R.; Leong, D.T.; Loo, J.S.C. The role of the tumor suppressor p53 pathway in the cellular DNA damage response to zinc oxide nanoparticles. *Biomaterials* **2011**, *32*, 8218–8225. [CrossRef] [PubMed]

32. Sharma, V.; Anderson, D.; Dhawan, A. Zinc oxide nanoparticles induce oxidative DNA damage and ROS-triggered mitochondria mediated apoptosis in human liver cells (HepG2). *Apoptosis* **2012**, *17*, 852–870. [CrossRef] [PubMed]

33. Yang, Q.B.; Ma, Y.F. Irradiation-Enhanced Cytotoxicity of Zinc Oxide Nanoparticles. *Int. J. Toxicol.* **2014**, *33*, 187–203. [CrossRef] [PubMed]

34. Wang, C.C.; Wang, S.G.; Xia, Q.S.; He, W.W.; Yin, J.J.; Fu, P.P.; Li, J.H. Phototoxicity of Zinc Oxide Nanoparticles in HaCaT Keratinocytes-Generation of Oxidative DNA Damage During UVA and Visible Light Irradiation. *J. Nanosci. Nanotechnol.* **2013**, *13*, 3880–3888. [CrossRef] [PubMed]

35. Demir, E.; Akca, H.; Kaya, B.; Burgucu, D.; Tokgun, O.; Turna, F.; Aksakal, S.; Vales, G.; Creus, A.; Marcos, R. Zinc oxide nanoparticles: Genotoxicity, interactions with UV-light and cell-transforming potential. *J. Hazard. Mater.* **2014**, *264*, 420–429. [CrossRef] [PubMed]

36. Bhattacharya, D.; Santra, C.R.; Ghosh, A.N.; Karmakar, P. Differential Toxicity of Rod and Spherical Zinc Oxide Nanoparticles on Human Peripheral Blood Mononuclear Cells. *J. Biomed. Nanotechnol.* **2014**, *10*, 707–716. [CrossRef] [PubMed]

37. Yin, H.; Casey, P.S.; McCall, M.J.; Fenech, M. Effects of Surface Chemistry on Cytotoxicity, Genotoxicity, and the Generation of Reactive Oxygen Species Induced by ZnO Nanoparticles. *Langmuir* **2010**, *26*, 15399–15408. [CrossRef] [PubMed]

38. Yang, K.; Zhu, L.Z.; Xing, B.S. Sorption of phenanthrene by nanosized alumina coated with sequentially extracted humic acids. *Environ. Sci. Pollut. Res.* **2010**, *17*, 410–419. [CrossRef] [PubMed]

39. Sharma, V.; Shukla, R.K.; Saxena, N.; Parmar, D.; Das, M.; Dhawan, A. DNA damaging potential of zinc oxide nanoparticles in human epidermal cells. *Toxicol. Lett.* **2009**, *185*, 211–218. [CrossRef] [PubMed]

40. Sharma, V.; Anderson, D.; Dhawan, A. Zinc oxide nanoparticles induce oxidative stress and genotoxicity in human liver cells (HepG2). *J. Biomed. Nanotechnol.* **2011**, *7*, 98–99. [CrossRef] [PubMed]

41. Patel, P.; Kansara, K.; Senapati, V.A.; Shanker, R.; Dhawan, A.; Kumar, A. Cell cycle dependent cellular uptake of zinc oxide nanoparticles in human epidermal cells. *Mutagenesis* **2016**, *31*, 481–490. [CrossRef] [PubMed]

42. Osman, I.F.; Baumgartner, A.; Cemeli, E.; Fletcher, J.N.; Anderson, D. Genotoxicity and cytotoxicity of zinc oxide and titanium dioxide in HEp-2 cells. *Nanomedicine* **2010**, *5*, 1193–1203. [CrossRef] [PubMed]

43. Condello, M.; De Berardis, B.; Ammendolia, M.G.; Barone, F.; Condello, G.; Degan, P.; Meschini, S. ZnO nanoparticle tracking from uptake to genotoxic damage in human colon carcinoma cells. *Toxicol. In Vitro* **2016**, *35*, 169–179. [CrossRef] [PubMed]

44. Toduka, Y.; Toyooka, T.; Ibuki, Y. Flow cytometric evaluation of nanoparticles using side-scattered light and reactive oxygen species-mediated fluorescence-correlation with genotoxicity. *Environ. Sci. Technol.* **2012**, *46*, 7629–7636. [CrossRef] [PubMed]

45. Kermanizadeh, A.; Gaiser, B.K.; Hutchison, G.R.; Stone, V. An in vitro liver model—Assessing oxidative stress and genotoxicity following exposure of hepatocytes to a panel of engineered nanomaterials. *Part. Fibre Toxicol.* **2012**, *9*, 28. [CrossRef] [PubMed]

46. Kumar, A.; Pandey, A.K.; Singh, S.S.; Shanker, R.; Dhawan, A. Engineered ZnO and TiO(2) nanoparticles induce oxidative stress and DNA damage leading to reduced viability of Escherichia coli. *Free Radic. Biol. Med.* **2011**, *51*, 1872–1881. [CrossRef] [PubMed]

47. Guan, R.; Kang, T.; Lu, F.; Zhang, Z.; Shen, H.; Liu, M. Cytotoxicity, oxidative stress, and genotoxicity in human hepatocyte and embryonic kidney cells exposed to ZnO nanoparticles. *Nanoscale Res. Lett.* **2012**, *7*, 602. [CrossRef] [PubMed]

48. Kumari, M.; Khan, S.S.; Pakrashi, S.; Mukherjee, A.; Chandrasekaran, N. Cytogenetic and genotoxic effects of zinc oxide nanoparticles on root cells of Allium cepa. *J. Hazard. Mater.* **2011**, *190*, 613–621. [CrossRef] [PubMed]

49. Ng, C.T.; Yong, L.Q.; Hande, M.P.; Ong, C.N.; Yu, L.E.; Bay, B.H.; Baeg, G.H. Zinc oxide nanoparticles exhibit cytotoxicity and genotoxicity through oxidative stress responses in human lung fibroblasts and *Drosophila melanogaster*. *Int. J. Nanomed.* **2017**, *12*, 1621–1637. [CrossRef] [PubMed]

50. Sharma, V.; Singh, S.K.; Anderson, D.; Tobin, D.J.; Dhawan, A. Zinc oxide nanoparticle induced genotoxicity in primary human epidermal keratinocytes. *J. Nanosci. Nanotechnol.* **2011**, *11*, 3782–3788. [CrossRef] [PubMed]

51. Kleinsasser, N. Toxicological evaluation of inhalation noxae: Test methods, assessment of toxic action and hazard potential, threshold limit values. *Laryngo-Rhino-Otologie* **2004**, *83*, S36–S53. [PubMed]

52. Hackenberg, S.; Scherzed, A.; Technau, A.; Froelich, K.; Hagen, R.; Kleinsasser, N. Functional responses of human adipose tissue-derived mesenchymal stem cells to metal oxide nanoparticles in vitro. *J. Biomed. Nanotechnol.* **2013**, *9*, 86–95. [CrossRef] [PubMed]

53. Hackenberg, S.; Zimmermann, F.Z.; Scherzed, A.; Friehs, G.; Froelich, K.; Ginzkey, C.; Koehler, C.; Burghartz, M.; Hagen, R.; Kleinsasser, N. Repetitive exposure to zinc oxide nanoparticles induces dna damage in human nasal mucosa mini organ cultures. *Environ. Mol. Mutagen.* **2011**, *52*, 582–589. [CrossRef] [PubMed]

54. Ghosh, M.; Sinha, S.; Jothiramajayam, M.; Jana, A.; Nag, A.; Mukherjee, A. Cyto-genotoxicity and oxidative stress induced by zinc oxide nanoparticle in human lymphocyte cells in vitro and Swiss albino male mice in vivo. *Food Chem. Toxicol.* **2016**, *97*, 286–296. [CrossRef] [PubMed]

55. Branica, G.; Mladinic, M.; Omanovic, D.; Zeljezic, D. An alternative approach to studying the effects of ZnO nanoparticles in cultured human lymphocytes: Combining electrochemistry and genotoxicity tests. *Arhiv za higijenu rada i toksikologiju* **2016**, *67*, 277–288. [CrossRef] [PubMed]

56. Nam, S.H.; Kim, S.W.; An, Y.J. No evidence of the genotoxic potential of gold, silver, zinc oxide and titanium dioxide nanoparticles in the SOS chromotest. *J. Appl. Toxicol.* **2013**, *33*, 1061–1069. [CrossRef] [PubMed]

57. Kwon, J.Y.; Lee, S.Y.; Koedrith, P.; Lee, J.Y.; Kim, K.M.; Oh, J.M.; Yang, S.I.; Kim, M.K.; Lee, J.K.; Jeong, J.; et al. Lack of genotoxic potential of ZnO nanoparticles in in vitro and in vivo tests. *Mutat. Res. Genet. Toxicol. Environ. Mutagen.* **2014**, *761*, 1–9. [CrossRef] [PubMed]

58. Alaraby, M.; Annangi, B.; Hernandez, A.; Creus, A.; Marcos, R. A comprehensive study of the harmful effects of ZnO nanoparticles using *Drosophila melanogaster* as an in vivo model. *J. Hazard. Mater.* **2015**, *296*, 166–174. [CrossRef] [PubMed]

59. Sahu, D.; Kannan, G.M.; Vijayaraghavan, R. Size-Dependent Effect of Zinc Oxide on Toxicity and Inflammatory Potential of Human Monocytes. *J. Toxicol. Environ. Health* **2014**, *77*, 177–191. [CrossRef] [PubMed]

60. Bayat, N.; Rajapakse, K.; Marinsek-Logar, R.; Drobne, D.; Cristobal, S. The effects of engineered nanoparticles on the cellular structure and growth of Saccharomyces cerevisiae. *Nanotoxicology* **2014**, *8*, 363–373. [CrossRef] [PubMed]

61. Ali, D.; Alarifi, S.; Kumar, S.; Ahamed, M.; Siddiqui, M.A. Oxidative stress and genotoxic effect of zinc oxide nanoparticles in freshwater snail *Lymnaea luteola* L. *Aquat. Toxicol.* **2012**, *124*, 83–90. [CrossRef] [PubMed]

62. Li, C.H.; Shen, C.C.; Cheng, Y.W.; Huang, S.H.; Wu, C.C.; Kao, C.C.; Liao, J.W.; Kang, J.J. Organ biodistribution, clearance, and genotoxicity of orally administered zinc oxide nanoparticles in mice. *Nanotoxicology* **2012**, *6*, 746–756. [CrossRef] [PubMed]

63. Baky, N.A.; Faddah, L.M.; Al-Rasheed, N.M.; Fatani, A.J. Induction of inflammation, DNA damage and apoptosis in rat heart after oral exposure to zinc oxide nanoparticles and the cardioprotective role of alpha-lipoic acid and vitamin E. *Drug Res.* **2013**, *63*, 228–236. [CrossRef] [PubMed]

64. Zhao, X.; Wang, S.; Wu, Y.; You, H.; Lv, L. Acute ZnO nanoparticles exposure induces developmental toxicity, oxidative stress and DNA damage in embryo-larval zebrafish. *Aquat. Toxicol.* **2013**, *136–137*, 49–59. [CrossRef] [PubMed]

65. Landsiedel, R.; Ma-Hock, L.; Van Ravenzwaay, B.; Schulz, M.; Wiench, K.; Champ, S.; Schulte, S.; Wohlleben, W.; Oesch, F. Gene toxicity studies on titanium dioxide and zinc oxide nanomaterials used for UV-protection in cosmetic formulations. *Nanotoxicology* **2010**, *4*, 364–381. [CrossRef] [PubMed]

66. Hu, C.W.; Li, M.; Cui, Y.B.; Li, D.S.; Chen, J.; Yang, L.Y. Toxicological effects of TiO_2 and ZnO nanoparticles in soil on earthworm Eisenia fetida. *Soil Biol. Biochem.* **2010**, *42*, 586–591. [CrossRef]

67. Jacobsen, N.R.; Stoeger, T.; van den Brule, S.; Saber, A.T.; Beyerle, A.; Vietti, G.; Mortensen, A.; Szarek, J.; Budtz, H.C.; Kermanizadeh, A.; et al. Acute and subacute pulmonary toxicity and mortality in mice after intratracheal instillation of ZnO nanoparticles in three laboratories. *Food Chem. Toxicol.* **2015**, *85*, 84–95. [CrossRef] [PubMed]

68. Anand, A.S.; Prasad, D.N.; Singh, S.B.; Kohli, E. Chronic exposure of zinc oxide nanoparticles causes deviant phenotype in *Drosophila melanogaster*. *J. Hazard. Mater.* **2017**, *327*, 180–186. [CrossRef] [PubMed]

69. Sharma, V.; Singh, P.; Pandey, A.K.; Dhawan, A. Induction of oxidative stress, DNA damage and apoptosis in mouse liver after sub-acute oral exposure to zinc oxide nanoparticles. *Mutat. Res.* **2012**, *745*, 84–91. [CrossRef] [PubMed]

70. Manzo, S.; Schiavo, S.; Oliviero, M.; Toscano, A.; Ciaravolo, M.; Cirino, P. Immune and reproductive system impairment in adult sea urchin exposed to nanosized ZnO via food. *Sci. Total Environ.* **2017**, *599*, 9–13. [CrossRef] [PubMed]

71. Boran, H.; Ulutas, G. Genotoxic effects and gene expression changes in larval zebrafish after exposure to ZnCl2 and ZnO nanoparticles. *Dis. Aquat. Org.* **2016**, *117*, 205–214. [CrossRef] [PubMed]

72. Annangi, B.; Rubio, L.; Alaraby, M.; Bach, J.; Marcos, R.; Hernandez, A. Acute and long-term in vitro effects of zinc oxide nanoparticles. *Arch. Toxicol.* **2016**, *90*, 2201–2213. [CrossRef] [PubMed]

73. Carmona, E.R.; Inostroza-Blancheteau, C.; Rubio, L.; Marcos, R. Genotoxic and oxidative stress potential of nanosized and bulk zinc oxide particles in *Drosophila melanogaster*. *Toxicol. Ind. Health* **2016**, *32*, 1987–2001. [CrossRef] [PubMed]

74. Schiavo, S.; Oliviero, M.; Miglietta, M.; Rametta, G.; Manzo, S. Genotoxic and cytotoxic effects of ZnO nanoparticles for *Dunaliella tertiolecta* and comparison with SiO_2 and TiO_2 effects at population growth inhibition levels. *Sci. Total Environ.* **2016**, *550*, 619–627. [CrossRef] [PubMed]

75. Heim, J.; Felder, E.; Tahir, M.N.; Kaltbeitzel, A.; Heinrich, U.R.; Brochhausen, C.; Mailander, V.; Tremel, W.; Brieger, J. Genotoxic effects of zinc oxide nanoparticles. *Nanoscale* **2015**, *7*, 8931–8938. [CrossRef] [PubMed]

76. Alarifi, S.; Ali, D.; Alkahtani, S.; Verma, A.; Ahamed, M.; Ahmed, M.; Alhadlaq, H.A. Induction of oxidative stress, DNA damage, and apoptosis in a malignant human skin melanoma cell line after exposure to zinc oxide nanoparticles. *Int. J. Nanomed.* **2013**, *8*, 983–993.

77. Uzar, N.K.; Abudayyak, M.; Akcay, N.; Algun, G.; Ozhan, G. Zinc oxide nanoparticles induced cyto- and genotoxicity in kidney epithelial cells. *Toxicol. Mech. Methods* **2015**, *25*, 334–339. [CrossRef] [PubMed]

78. Reis, E.D.; de Rezende, A.A.A.; Santos, D.V.; de Oliveria, P.F.; Nicolella, H.D.; Tavares, D.C.; Silva, A.C.A.; Dantas, N.O.; Spano, M.A. Assessment of the genotoxic potential of two zinc oxide sources (amorphous and nanoparticles) using the in vitro micronucleus test and the in vivo wing somatic mutation and recombination test. *Food Chem. Toxicol.* **2015**, *84*, 55–63. [CrossRef] [PubMed]

79. Yang, H.; Liu, C.; Yang, D.; Zhang, H.; Xi, Z. Comparative study of cytotoxicity, oxidative stress and genotoxicity induced by four typical nanomaterials: The role of particle size, shape and composition. *J. Appl. Toxicol.* **2009**, *29*, 69–78. [CrossRef] [PubMed]

80. Setyawati, M.I.; Tay, C.Y.; Leong, D.T. Mechanistic Investigation of the Biological Effects of SiO$_2$, TiO$_2$, and ZnO Nanoparticles on Intestinal Cells. *Small* **2015**, *11*, 3458–3468. [CrossRef] [PubMed]

81. Dubey, A.; Goswami, M.; Yadav, K.; Chaudhary, D. Oxidative Stress and Nano-Toxicity Induced by TiO$_2$ and ZnO on WAG Cell Line. *PLoS ONE* **2015**, *10*, e0127493. [CrossRef] [PubMed]

82. Yin, H.; Casey, P.S.; McCall, M.J.; Fenech, M. Size-dependent cytotoxicity and genotoxicity of ZnO particles to human lymphoblastoid (WIL2-NS) cells. *Environ. Mol. Mutagen.* **2015**, *56*, 767–776. [CrossRef] [PubMed]

83. Senapati, V.A.; Kumar, A.; Gupta, G.S.; Pandey, A.K.; Dhawan, A. ZnO nanoparticles induced inflammatory response and genotoxicity in human blood cells: A mechanistic approach. *Food Chem. Toxicol.* **2015**, *85*, 61–70. [CrossRef] [PubMed]

84. Demir, E.; Kaya, N.; Kaya, B. Genotoxic effects of zinc oxide and titanium dioxide nanoparticles on root meristem cells of Allium cepa by comet assay. *Turk. J. Biol.* **2014**, *38*, 31–39. [CrossRef]

85. Roszak, J.; Catalan, J.; Jarventaus, H.; Lindberg, H.K.; Suhonen, S.; Vippola, M.; Stepnik, M.; Norppa, H. Effect of particle size and dispersion status on cytotoxicity and genotoxicity of zinc oxide in human bronchial epithelial cells. *Mutat. Res. Genet. Toxicol. Environ.* **2016**, *805*, 7–18. [CrossRef] [PubMed]

86. Gopalan, R.C.; Osman, I.F.; Amani, A.; De Matas, M.; Anderson, D. The effect of zinc oxide and titanium dioxide nanoparticles in the Comet assay with UVA photoactivation of human sperm and lymphocytes. *Nanotoxicology* **2009**, *3*, 33–39. [CrossRef]

87. Sarkar, J.; Ghosh, M.; Mukherjee, A.; Chattopadhyay, D.; Acharya, K. Biosynthesis and safety evaluation of ZnO nanoparticles. *Bioprocess Biosyst. Eng.* **2014**, *37*, 165–171. [CrossRef] [PubMed]

88. Patil, N.A.; Gade, W.N.; Deobagkar, D.D. Epigenetic modulation upon exposure of lung fibroblasts to TiO$_2$ and ZnO nanoparticles: Alterations in DNA methylation. *Int. J. Nanomed.* **2016**, *11*, 4509–4519.

89. Karlsson, H.L.; Cronholm, P.; Gustafsson, J.; Moller, L. Copper oxide nanoparticles are highly toxic: A comparison between metal oxide nanoparticles and carbon nanotubes. *Chem. Res. Toxicol.* **2008**, *21*, 1726–1732. [CrossRef] [PubMed]

90. Kermanizadeh, A.; Vranic, S.; Boland, S.; Moreau, K.; Baeza-Squiban, A.; Gaiser, B.K.; Andrzejczuk, L.A.; Stone, V. An in vitro assessment of panel of engineered nanomaterials using a human renal cell line: Cytotoxicity, pro-inflammatory response, oxidative stress and genotoxicity. *BMC Nephrol.* **2013**, *14*. [CrossRef] [PubMed]

91. Zijno, A.; De Angelis, I.; De Berardis, B.; Andreoli, C.; Russo, M.T.; Pietraforte, D.; Scorza, G.; Degan, P.; Ponti, J.; Rossi, F.; et al. Different mechanisms are involved in oxidative DNA damage and genotoxicity induction by ZnO and TiO$_2$ nanoparticles in human colon carcinoma cells. *Toxicol. In Vitro* **2015**, *29*, 1503–1512. [CrossRef] [PubMed]

92. Shalini, D.; Senthilkumar, S.; Rajaguru, P. Effect of size and shape on toxicity of zinc oxide (ZnO) nanomaterials in human peripheral blood lymphocytes. *Toxicol. Mech. Methods* **2017**, 1–8. [CrossRef] [PubMed]

93. Sliwinska, A.; Kwiatkowski, D.; Czarny, P.; Milczarek, J.; Toma, M.; Korycinska, A.; Szemraj, J.; Sliwinski, T. Genotoxicity and cytotoxicity of ZnO and Al$_2$O$_3$ nanoparticles. *Toxicol. Mech. Methods* **2015**, *25*, 176–183. [CrossRef] [PubMed]

94. Haase, A.; Dommershausen, N.; Schulz, M.; Landsiedel, R.; Reichardt, P.; Krause, B.C.; Tentschert, J.; Luch, A. Genotoxicity testing of different surface-functionalized SiO_2, ZrO_2 and silver nanomaterials in 3D human bronchial models. *Arch. Toxicol.* **2017**. [CrossRef] [PubMed]

95. Soni, D.; Gandhi, D.; Tarale, P.; Bafana, A.; Pandey, R.A.; Sivanesan, S. Oxidative Stress and Genotoxicity of Zinc Oxide Nanoparticles to Pseudomonas Species, Human Promyelocytic Leukemic (HL-60), and Blood Cells. *Biol. Trace Elem. Res.* **2017**, *178*, 218–227. [CrossRef] [PubMed]

MDPI

Brief Report

Two-Dimensional Fluorescence Difference Spectroscopy of ZnO and Mg Composites in the Detection of Physiological Protein and RNA Interactions

Amanda Hoffman [1], Xiaotong Wu [1], Jianjie Wang [2], Amanda Brodeur [2], Rintu Thomas [2], Ravindra Thakkar [1], Halena Hadi [3], Garry P. Glaspell [4], Molly Duszynski [5], Adam Wanekaya [5] and Robert K. DeLong [1,*]

[1] Department of Anatomy and Physiology, Nanotechnology Innovation Center of Kansas State (NICKS), Kansas State University, Manhattan, KS 66506, USA; arhoffman@ksu.edu (A.H.); maggie.wu1992@vet.k-state.edu (X.W.); ravithakkar@vet.k-state.edu (R.T.)
[2] Department of Biomedical Science, Missouri State University, Springfield, MO 65897, USA; JWang@MissouriState.edu (J.W.); ABrodeur@MissouriState.edu (A.B.); rtthomas3@uh.edu (R.T.)
[3] University of Notre Dame, Notre Dame, IN 46556, USA; hhadi@nd.edu
[4] Department of Chemistry, Virginia Commonwealth University, Richmond, VA 23284, USA; Garry.P.Glaspell@erdc.dren.mil
[5] Department of Chemistry, Missouri State University, Springfield, MO 65897, USA; Duszynski314@live.missouristate.edu (M.D.); Wanekaya@MissouriState.edu (A.W.)
* Correspondence: robertdelong@vet.k-state.edu; Tel.: +1-785-532-6313

Received: 31 October 2017; Accepted: 8 December 2017; Published: 15 December 2017

Abstract: Two-dimensional fluorescence difference spectroscopy (2-D FDS) was used to determine the unique spectral signatures of zinc oxide (ZnO), magnesium oxide (MgO), and 5% magnesium zinc oxide nanocomposite (5% Mg/ZnO) and was then used to demonstrate the change in spectral signature that occurs when physiologically important proteins, such as angiotensin-converting enzyme (ACE) and ribonuclease A (RNase A), interact with ZnO nanoparticles (NPs). When RNase A is bound to 5% Mg/ZnO, the intensity is quenched, while the intensity is magnified and a significant shift is seen when torula yeast RNA (TYRNA) is bound to RNase A and 5% Mg/ZnO. The intensity of 5% Mg/ZnO is quenched also when thrombin and thrombin aptamer are bound to the nanocomposite. These data indicate that RNA–protein interaction can occur unimpeded on the surface of NPs, which was confirmed by gel electrophoresis, and importantly that the change in fluorescence excitation, emission, and intensity shown by 2-D FDS may indicate specificity of biomolecular interactions.

Keywords: zinc oxide; nanocomposites; aptamer; thrombin; angiotensin-converting enzyme; ribonuclease A; RNA; two-dimensional fluorescence difference spectroscopy

1. Introduction

Metamaterials or composites combine the advantages of multiple elements in the nanoscale and have unique physico-chemical properties [1–3]. There is currently a great deal of interest in nanobio sensors where many of these applications involve zinc oxide (ZnO) nanoparticles (NPs) or doped derivatives [4–14]. A variety of different target molecules, including but not limited to riboflavin, mucin-1, bisphenol A, ATP, acetamiprid, micro-RNA, thrombin, and different types of cancer cells (Hela, SK-BR-3, K562), have been detected [4–14]. ZnO has been fabricated into a variety of different structures with various other components including carbon quantum dots [4], platinum [5], AlGaN [6], Au (gold) [7,8,11], Co (cobalt) [9], and graphene [12]. In many cases, detection by these nanocomposite sensors is based on electrochemistry or photoelectrochemistry [5,7,9,11,12,14]. However, in some cases, detection has been based on electrochemiluminescence [4], field effect transistors (FET) [13], or 3-D quantitative fluorescence imaging [6].

Our group has recently reported that two-dimensional fluorescence difference spectroscopy (2-D FDS) can be used as a new characterization technique for nanomaterials and can be used to probe the nanobio interface [15]. This method has an advantage over the other methods of detection listed above because it does not rely on a dye or fluorophore to detect the presence of nanomaterials and their interactions, and it is a simple technique that does not require complex analysis [15]. We noted that this technique is sensitive to the nanoparticle synthetic method and composition and that a spectral shift in two dimensions occurs upon either RNA or protein interactions at the surface [15,16]. Here, a fluorescence quenching or shift occurs via energy transfer or via overlap between the biomolecule and the nanomaterial. We demonstrate that 2-D FDS can provide unique spectral signature of composite NPs and, further, can be utilized to determine RNA and protein interaction specificity.

2. Results

2.1. Iteration of 2-D FDS in the Analysis of Nanoparticle Composition

Synthesis of the 5% Mg/ZnO composite was compared to the pure parent material (Figure 1).

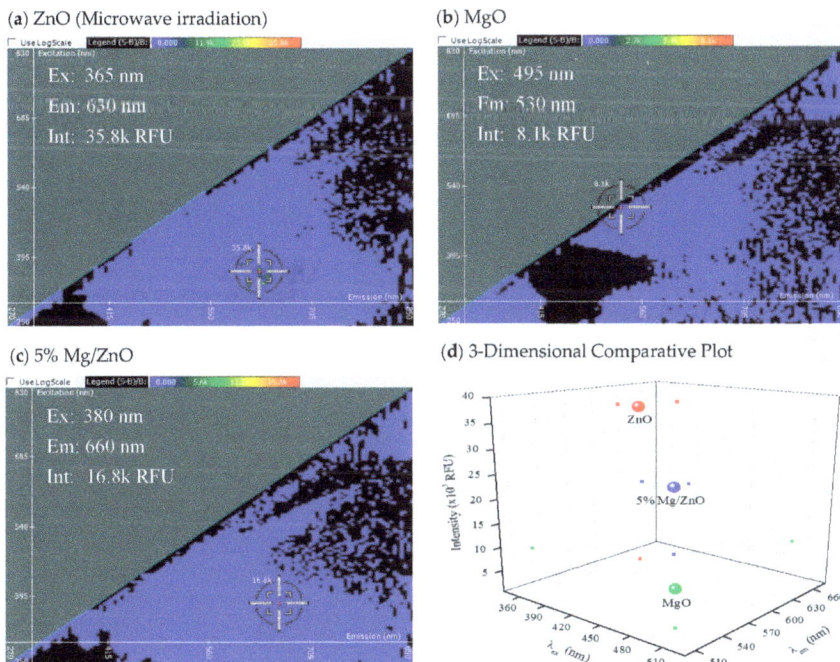

Figure 1. 2-D FDS of (**a**) zinc oxide nanoparticles (NPs) synthesized by microwave irradiation; (**b**) magnesium oxide nanoparticles; and (**c**) 5% magnesium with 95% zinc oxide; (**d**) Three-dimensional comparative plot of the three nanocomposites.

In Figure 1, the spectral signatures of three different nanomaterials are compared. ZnO, MgO, or the composite excitation and emission spectrum is shown in 2-D in Panels a, b, and c, respectively. The *y*-axis represents the excitation wavelength in nm, the *x*-axis represents the emission wavelength in nm, and the scale above the 2-D FDS plot represents the intensity in relative fluorescence units

(RFU). Each spectral signature is unique to the nanomaterial, as shown by the 3-D graph where the intensity of the composite was midway between ZnO (highest) and MgO (lowest), as shown in Panel d. The *x*-axis denotes excitation wavelength in nm, the *y*-axis denotes emission wavelength in nm, and the *z*-axis denotes intensity in RFU. The large spheres in Panel d denote the intersection of the excitation, emission, and intensity values of each nanomaterial as labeled. The small dots show the individual values for excitation, emission, and intensity for simple reading of the 3-D graph, and these values are applied to the following figures (Figures 1–4). The 2-D FDS method can be used to distinguish between different nanomaterials. This also demonstrates the nature of nanocomposites, where certain properties of each nanoparticle are combined to form new properties. The fluorescent intensity of ZnO is 35.8 k RFU, while the fluorescent intensity of MgO is 8.1 k RFU. When they are combined to form 5% Mg/ZnO, the fluorescent intensity is in between the two nanoparticles at 16.8 k RFU.

2.2. Iteration of 2-D FDS in the Analysis of Nanoparticle Interaction to Protein

The binding events to angiotensin-converting enzyme (ACE) or ribonuclease A (RNase A), two physiologically relevant proteins, were characterized by 2-D FDS (Figure 2).

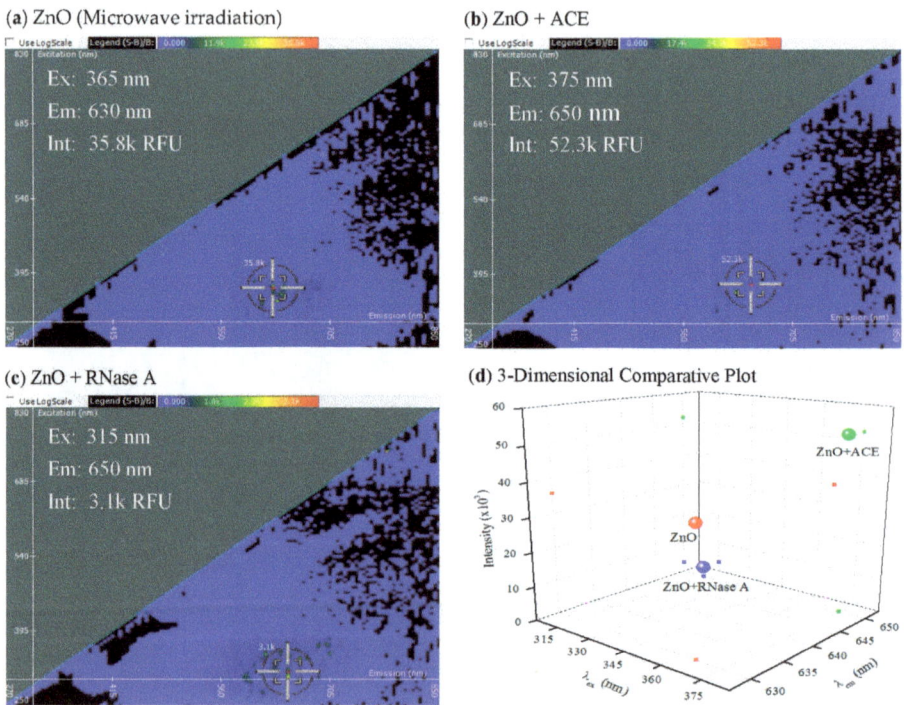

Figure 2. 2-D FDS of (**a**) ZnO nanoparticles synthesized by microwave irradiation; (**b**) ACE bound to microwave irradiation ZnO nanoparticles; and (**c**) ribonuclease A (RNase A) bound to microwave irradiation ZnO nanoparticles; The (**d**) 3-dimensional comparative plot of the three ZnO nanoparticle interactions.

In Figure 2, the spectral signature of ZnO is compared to the spectral signatures of ZnO with either ACE or RNase A bound to it, and it is seen that the interaction of protein with ZnO causes three shifts, in the excitation wavelength, emission wavelength, and the fluorescence intensity. The same shift occurs in the emission when either ACE or RNase A is bound to ZnO. There is a shift in the excitation as well, but there is a significant difference between the ACE and RNase A excitation wavelengths. There is also a significant change in the intensities when ACE or RNase A are bound to ZnO. The intensity of ZnO is 35.8 k RFU, and this value is quenched to 3.1 k RFU when RNase A is bound to ZnO. Binding RNase A to ZnO has the opposite effect and magnifies the intensity to 52.3 k RFU. This is likely due to different orientations of the proteins at the surface of the NPs and whether interaction exposes fluorescently active amino acids (e.g., tryptophan), hence resulting in fluorescent quenching or activation (unpublished observations).

2.3. Iteration of 2-D FDS in the Analysis of Protein–RNA Interaction to the Nanoparticle

In cells and tissues, magnesium (Mg) is well-known to mediate protein and especially RNA structure and stability. Hence, we hypothesized that the incorporation of Mg into ZnO NPs may allow for interaction and that, for an RNA binding protein (RNase A), the addition of RNA (from torula yeast) may exchange RNA at the surface as a function of binding to protein and that this signal might be detected by 2-D FDS. These data are shown next (Figure 3).

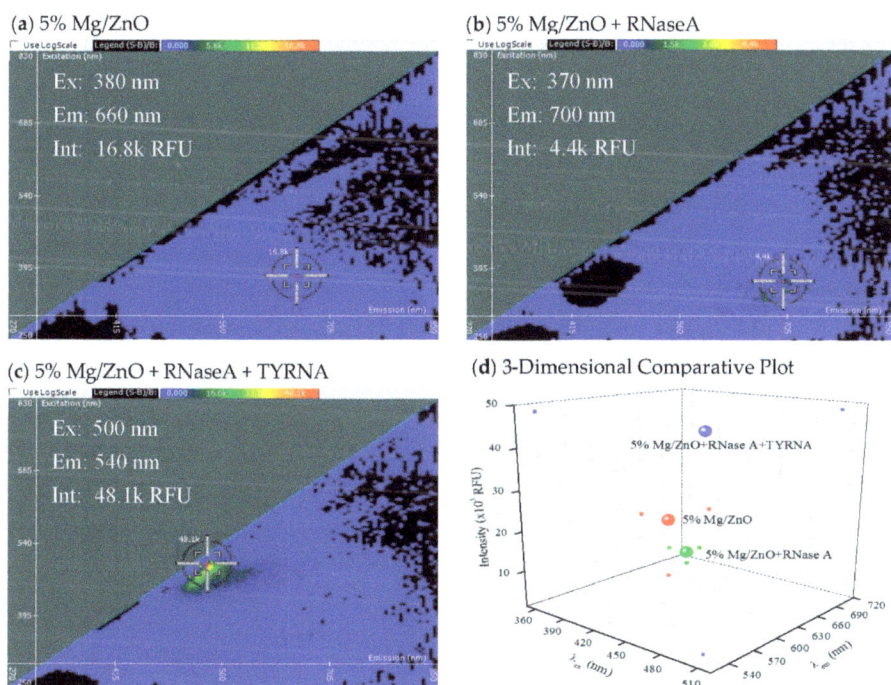

Figure 3. 2-D FDS of (**a**) 5% magnesium with 95% zinc oxide; (**b**) RNase A bound to 5% Mg/ZnO; and (**c**) torula yeast RNA and RNase A bound to 5% Mg/ZnO; (**d**) Three-dimensional comparative plot of the three 5% Mg/ZnO RNase A interactions.

In Figure 3, the spectral signature of 5% Mg/ZnO is compared to the spectral signatures of RNase A, as well as both TYRNA and RNase A, bound to 5% Mg/ZnO. As expected, the data reveal a shift in the emission during both interactions, but in opposite directions. There is a slight shift in the excitation when RNase A only is bound to 5% Mg/ZnO, and a significant excitation shift when RNase A is bound followed by the addition of TYRNA. The fluorescent intensity of 5% Mg/ZnO is quenched when RNase A is bound to it, changing from a value of 16.8 to 4.4 k RFU. The fluorescent intensity of 5% Mg/ZnO is significantly magnified when TYRNA and RNase A are bound to it, with an intensity of 48.1 k RFU. As mentioned previously, this is likely explained by the addition of RNA to RNase, causing conformational change allowing exposure of the protein's own fluorescence moieties and some energy transfer between it, the RNA molecule, and NP surface.

2.4. Iteration of 2-D FDS in the Detection of Specific Aptamer–Protein Interaction

To detect aptamer target interaction, the classic and perhaps most well characterized thrombin RNA aptamer system [17,18] was employed in conjunction with the 5% Mg/ZnO NP composite (Figure 4).

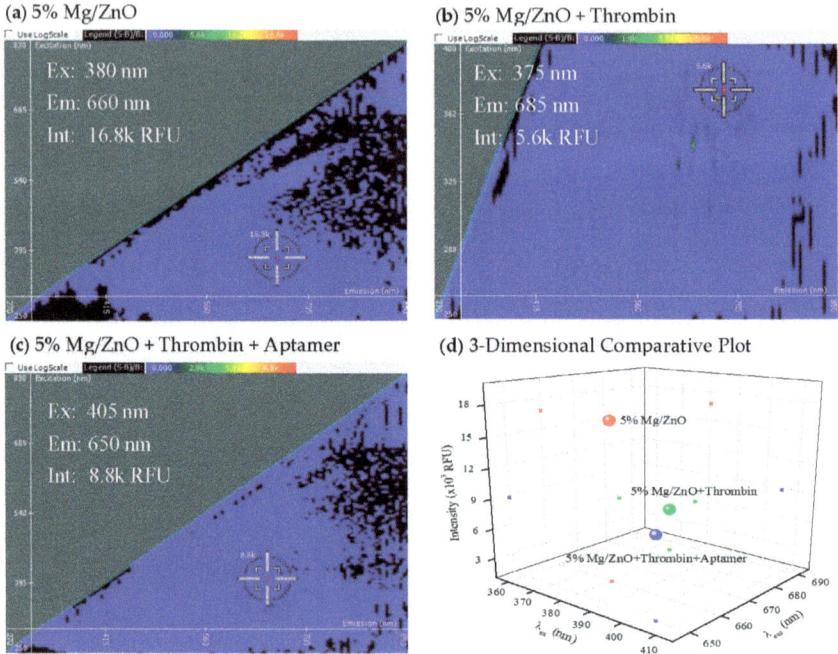

Figure 4. 2-D FDS of (**a**) 5% magnesium with 95% zinc oxide; (**b**) thrombin bound to 5% Mg/ZnO; and (**c**) thrombin and thrombin aptamer bound to 5% Mg/ZnO; (**d**) Three-dimensional comparative plot of the three 5% Mg/ZnO thrombin interactions.

In Figure 4, the spectral signature of 5% Mg/ZnO is compared with the spectral signatures of 5% Mg/ZnO bound to either thrombin or both thrombin aptamer and thrombin. When thrombin is bound to 5% Mg/ZnO, there is a shift in emission and a slight change in excitation. A quenching effect is also created, with the fluorescent intensity shifting from 16.8 to 5.6 k RFU. Binding thrombin aptamer and thrombin to 5% Mg/ZnO creates a shift in the emission and a larger shift in excitation. The fluorescent intensity is quenched to a value of 8.8 k RFU as well. These results are interpreted as a strong association of the aptamer/target protein at the surface quenching fluorescence.

2.5. NPs Do Not Affect Specific Aptamer–Protein Interaction by Gel Mobility Shift

To confirm that the NPs do not interfere with aptamer–target interactions, we employed a classical electrophoretic mobility shift assay (EMSA). In this experiment, an increasing amount of thrombin was added to the interaction tube and was titrated with or without the addition of NPs. NP–protein interactions have been well documented by researchers in our lab and by others [19,20], making it possible to show that the introduction of the thrombin aptamer after the thrombin protein was bound resulted in total fluorescence quenching, as shown previously above, and that the results of the electrophoretic mobility shift assay (EMSA) shows this by the decreasing fluorescent intensity of the bands in Figure 5a as the thrombin concentration increases.

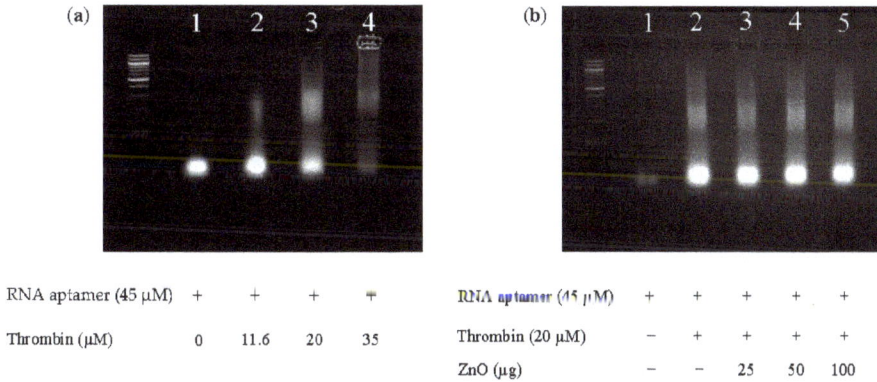

| RNA aptamer (45 µM) | + | + | + | + |
| Thrombin (µM) | 0 | 11.6 | 20 | 35 |

RNA aptamer (45 µM)	+	+	+	+	+
Thrombin (20 µM)	−	+	+	+	+
ZnO (µg)	−	−	25	50	100

Figure 5. Electrophoretic mobility shift assay (EMSA) of (**a**) RNA aptamer with varying concentrations of thrombin and (**b**) RNA aptamer with or without thrombin and with varying concentrations of ZnO.

In Figure 5, binding of thrombin to RNA aptamer is detected by EMSA and is concentration-dependent (Panel a). The ladder fragments of the 25 nucleotide Toggle-25t RNA aptamer phosphothioate is denoted by "25 NT" under the ladder. In Figure 5b, the band intensity in Lane 1 is lower than the band intensity in Lane 1 of Figure 5a. This is caused by gel-to-gel staining variability, and shown in the figure due to the gel imager automatically correcting fluorescence, leading to different band intensities. Introduction of the NP, ZnO, was used in this case because of its protein binding affinity [19,20] and was shown not to disrupt the specific biomolecular interaction, if anything leading to fewer non-specific aggregates (staining in the well) and stabilizing the complex at all concentrations tested.

3. Discussion

In summary, 2-D FDS was used to assess the fluorescence of composite materials (5% Mg/ZnO) and their interaction with RNA and protein. Fluorescence excitation occurs when electrons absorb energy to an excited state; and, when it relaxes to the ground state, fluorescent emission results. Fluorescent quenching occurs when electron clouds of nanomaterials interact with each other at the van der Waals radii, and energy is dissipated as heat when the excited electron moves back to the ground state [21]. We postulate that the binding of a biomolecule, such as RNA or protein, onto the surface of a nanomaterial with similar electronic fluorescence properties will yield a quenching or fluorescent shift effect. This technique can detect the presence of physiologically relevant proteins such as angiotensin-converting enzyme (ACE), ribonuclease A (RNase A), and thrombin, as well as RNA such as that derived from torula yeast (TYRNA) or RNA aptamer. Using the thrombin–aptamer system, we have shown that 2-D FDS can be used to detect specific RNA–protein interaction, where the changes in fluorescence are unique in comparison to the TYRNA–RNase system. These findings imply that 2-D FDS in conjunction with nanomaterial composites may have ramifications in the detection of protein, RNA, and their interactions and may be a useful detection platform for analytical assays more generally, aiding and supporting classical methods such as EMSA.

4. Materials and Methods

4.1. Materials

Mg/ZnO with a 5% concentration was provided by Dr. A. Wanekaya (Missouri State University, Springfield, MO, USA) and synthesized via hydrothermal methods. ZnO nanoparticles synthesized by microwave irradiation methods were obtained from Dr. G. Glaspell (Virginia Commonwealth University, Richmond, VA, USA). MgO was purchased from Sigma-Aldrich (St. Louis, MO, USA). HPLC grade water was purchased from Acros Organics (Belgium, WI, USA). Angiotensin-converting enzyme from rabbit lung (ACE), ribonucleic acid from torula yeast (TYRNA), and thrombin were obtained from Sigma-Aldrich. Thrombin aptamer was obtained from Trilink Biotechnology (San Diego, CA, USA). RNase A was purchased from Thermo Fisher Scientific (Waltham, MA, USA). A black 96-well microplate was purchased from Midsci Corp. (Valley Park, MO, USA).

4.2. Stock Preparation

Using the Mettler Toledo Excellence XS Analytical Balance (Mettler-Toledo, LLC., Columbus, OH, USA), nanomaterials were weighed out at 3 mg and suspended in 1 mL of HPLC grade water. The nanomaterial solution is then dispersed for 5 min with 20 s pulses at 50% amplitude using Sonics Vibracell VCX 130 probe ultra-sonicator (Sonics & Materials, Inc., Newtown, CT, USA).

4.3. Nanomaterial Alone

For the 2-D FDS of each nanomaterial on its own, 66.7 µL of the dispersed nanomaterial solution was added to a sterile microcentrifuge tube with 133.3 µL of HPLC grade water and then re-suspended, giving a final nanomaterial concentration of 1 mg/mL. This solution was then added to a well in a black 96-well microplate and read by the Molecular Devices Spectramax i3x spectrophotometer (Molecular Devices, LLC., Sunnyvale, CA, USA).

4.4. Protein Interactions

For the protein interaction tests, 1 mg/mL of 5% Mg/ZnO was incubated on ice with 0.1 mg/mL thrombin or RNase A for 30 min on the orbital shaker at 150 rpm, then transferred to a well of a black 96-well microplate and read by the Molecular Device Spectramax i3x spectrophotometer. For the specificity tests, the procedure above was followed, and 0.1 mg/mL thrombin aptamer or TYRNA were added to their respective solutions, incubated on ice again under the same conditions, and then transferred to a microplate and read by the spectrophotometer. The procedure above was followed for the ZnO tests with 0.038 mg/mL ACE or RNase A added instead.

4.5. Molecular Device Settings

The Molecular Devices Spectramax i3x spectrophotometer utilizes the Spectral Optimization Wizard that is included in the Softmax Pro 6.4.2 accompaniment software (Sunnyvale, CA, USA) to scan the black 96-well microplate without the lid. The device reads the fluorescence endpoints of unknown wavelengths. The photomultiplier (PMT) gain was high, flashes per read was six, and wavelength increment was 5 nm. The microplate was shaken in a linear mode at medium intensity before the first read and was read from the top at a height of 1 mm. The range of excitation and emission wavelengths was 250–830 nm and 270–850 nm, respectively. The previously observed range in emission values of metal oxides is 25 nm [15].

4.6. RNA Aptamer-Thrombin and Gel Mobility Shift

Binding specificity of Toggle-25t RNA aptamer phosphothioate (5′-GGG AAC AAA GCU GAA GUA CUU ACC C-3′) (Integrated DNA Technology, Reference number: 118200374, Coralville, IA, USA) to varying concentrations of human alpha thrombin (Enzyme Research Laboratories, Catalog number:

HT 1002a, Pittsburgh, PA, USA) was observed using agarose gel electrophoresis. The lyophilized RNA aptamer was suspended in diethyl pyrocarbonate (DEPC) water at a final concentration of 163 μM. Human alpha thrombin at a concentration of 3.41 mg/mL was obtained in buffer solution containing 50 mM sodium citrate, 0.2 M NaCl, and 0.1% polyethylene glycol (PEG) (pH 6.5). All reaction samples in 20 μL was composed of 45 μM RNA aptamer in HBS buffer containing 150 mM NaCl, 2 mM CaCl$_2$, 20 mM HEPES (pH = 7.35 ± 0.1), and 6.25 mM MgCl$_2$. With the exception of control (RNA aptamer alone), the rest of the reaction mixtures were titrated with protein alpha thrombin with increasing concentrations (11.61, 20, and 35 μM) to achieve aptamer/protein molar ratios as indicated: 1:0.25, 1:0.50, and 1:0.7. ZnO nanoparticles were washed with water three times and then with ethanol with concentrations of 50% (v/v), 70% (v/v), and 95% (v/v) three times. Colloidal suspension of ZnO nanoparticles was prepared by ultrasonication (Fischer Scientific, Model number: FS20, Pittsburgh, PA, USA) for 45 min at a concentration of 1 mg/mL in water. Reaction samples of RNA aptamer–thrombin–ZnO nanoparticles in 20 μL contained 6.25 mM MgCl$_2$, 45 μM RNA aptamer, 20 μM alpha-thrombin, and ZnO colloidal suspension at desired concentrations in HBS buffer The samples were incubated at room temperature for 30 min and then at 37 °C for 10 min prior to gel electrophoresis analysis. Following incubation, the reaction mixture was loaded on to a 3% (w/v) agarose gel containing 0.005% (v/v) ethidium bromide. The agarose gel was electrophoresed at room temperature (approximately 25 °C) in 1 X TAE containing 40 mM Tris (pH = 7.6), 20 mM acetic acid, and 1 mM EDTA for 45 min at 110 V (Bulletin 6205). Agarose gels were imaged using EL Logic 200 imaging system (Eastman Kodak Company, Rochester, NY, USA) and quantified using Kodak Image Pro software. Each of the experiments were replicated three times and the quantification results were imported into Excel, where the mean, standard deviation, and error for the data sets were calculated.

5. Conclusions

We used two-dimensional fluorescence difference spectroscopy to study ZnO and Mg composites and their interactions with physiological proteins and RNA.

Acknowledgments: This work was supported by an NIH grant NIH 7R15CA139390-03 Anti-Cancer RNA Nanoconjugates.

Author Contributions: Amanda Brodeur and Jianjie Wang co-mentored Rintu Thomas who provided gel shift data shown in Figure 5. Robert K. DeLong mentored Amanda Hoffman and Xiaotong Wu who completed the remaining experiments. Ravindra Thakkar analyzed the nanoparticles for chemical composition synthesized by Garry P. Glaspell and Adam Wanekaya and student Molly Duszynski. Halena Hadi researched the background and literature and wrote the summary used for the introduction.

Conflicts of Interest: The authors declare no conflict of interest. The founding sponsors had no role in the design of the study; in the collection, analyses, or interpretation of data; in the writing of the manuscript; or in the decision to publish the results.

References

1. Linden, S.; Enkrich, C.; Wegener, M.; Zhou, J.; Koschny, T.; Soukoulis, C. Magnetic Response of Metamaterials at 100 Terahertz. *Science* **2004**, *306*, 1351–1353. [CrossRef] [PubMed]
2. Chen, P.C.; Liu, X.; Hedrick, J.L.; Xie, Z.; Wang, S.; Lin, Q.Y.; Hersam, M.C.; Dravid, V.P.; Mirkin, C.A. Polyelemental nanoparticle libraries. *Science* **2016**, *352*, 1565–1569. [CrossRef] [PubMed]
3. Smith, D.R.; Pendry, J.B.; Wiltshire, M.C. Metamaterials and Negative Refractive Index. *Science* **2004**, *305*, 788–792. [CrossRef] [PubMed]
4. Zhang, M.; Liu, H.; Chen, L.; Yan, M.; Ge, L.; Ge, S.; Yu, J. A disposable electrochemiluminescence device for ultrasensitive monitoring of K562 leukemia cells based on aptamers and ZnO@carbon quantum dots. *Biosens. Bioelectron.* **2013**, *49*, 79–85. [CrossRef] [PubMed]
5. Lv, J.J.; Yang, Z.H.; Zhuo, Y.; Yuan, R.; Chai, Y.Q. A novel aptasensor for thrombin detection based on alkaline phosphatase decorated ZnO/Pt nanoflowers as signal amplifiers. *Analyst* **2015**, *140*, 8088–8091. [CrossRef] [PubMed]

6. Shrivastava, S.; Triel, N.M.; Son, Y.; Lee, W.; Lee, N. Seesawed fluorescence nano-aptasensor based on highly vertical ZnO nanorods and three-dimensional quantitative fluorescence imaging for enhanced detection accuracy of ATP. *Biosens. Bioelectron.* **2017**, *90*, 450–458. [CrossRef] [PubMed]

7. Qiao, Y.; Li, J.; Li, H.; Fang, H.; Fan, D.; Wang, W. A label-free photoelectrochemical aptasensor for bisphenol A based on surface plasmon resonance of gold nanoparticle-sensitized ZnO nanopencils. *Biosens. Bioelectron.* **2016**, *86*, 315–320. [CrossRef] [PubMed]

8. He, Z.; Zhang, P.; Li, X.; Zhang, J.; Zhu, J. A Targeted DNAzyme-Nanocomposite Probe Equipped with Built-in Zn^{2+} Arsenal for Combined Treatment of Gene Regulation and Drug Delivery. *Sci. Rep.* **2016**, *6*, 22737. [CrossRef] [PubMed]

9. Li, H.; Qiao, Y.; Li, J.; Fang, H.; Fan, D.; Weng, W. A sensitive and label-free photoelectrochemical aptasensor using Co-doped ZnO diluted magnetic semiconductor nanoparticles. *Biosens. Bioelectron.* **2016**, *77*, 378–384. [CrossRef] [PubMed]

10. Han, Z.; Wang, X.; Heng, C.; Han, Q.; Cai, S.; Li, J.; Qi, C.; Liang, W.; Yang, R.; Wang, C. Synergistically enhanced photocatalytic and chemotherapeutic effects of aptamer-functionalized ZnO nanoparticles towards cancer cells. *Phys. Chem. Chem. Phys.* **2015**, *17*, 21576–21582. [CrossRef] [PubMed]

11. Yang, Z.; Zhuo, Y.; Yuan, R.; Chai, Y. Amplified Thrombin Aptasensor Based on Alkaline Phosphatase and Hemin/G-Quadruplex-Catalyzed Oxidation of 1-Naphthol. *ACS Appl. Mater. Interfaces* **2015**, *7*, 10308–10315. [CrossRef] [PubMed]

12. Liu, F.; Zhang, Y.; Yu, J.; Wang, S.; Ge, S.; Song, X. Application of ZnO/graphene and S6 aptamers for sensitive photoelectrochemical detection of SK-BR-3 breast cancer cells based on a disposable indium tin oxide device. *Biosens. Bioelectron.* **2014**, *51*, 413–420. [CrossRef] [PubMed]

13. Hagen, J.A.; Kim, S.N.; Bayraktaroglu, B.; Leedy, K.; Chavez, J.L.; Kelley-Loughnane, N.; Naik, R.R.; Stone, M.O. Biofunctionalized Zinc Oxide Field Effect Transistors for Selective Sensing of Riboflavin with Current Modulation. *Sensors* **2011**, *11*, 6645–6655. [CrossRef] [PubMed]

14. Pang, X.; Qi, J.; Zhang, Y.; Ren, Y.; Su, M.; Jia, B.; Wang, Y.; Wei, Q.; Du, B. Ultrasensitive photoelectrochemical aptasensing of miR-155 using efficient and stable $CH_3NH_3PbI_3$ quantum dots sensitized ZnO nanosheets as light harvester. *Biosens. Bioelectron.* **2016**, *85*, 142–150. [CrossRef] [PubMed]

15. Hurst, M.N.; DeLong, R.K. Two-Dimensional Fluorescence Difference Spectroscopy to Characterize Nanoparticles and their Interactions. *Sci. Rep.* **2016**, *6*, 33287. [CrossRef] [PubMed]

16. DeLong, R.K.; Hurst, M.N.; Aryal, S.; Inchun, N.K. Unique Boron Carbide Nanoparticle Nanobio Interface: Effects on Protein-RNA Interactions and 3-D Spheroid Metastatic Phenotype. *Anticancer Res.* **2016**, *36*, 2097–2103. [PubMed]

17. Long, S.B.; Long, M.B.; White, R.R.; Sullenger, B.A. Crystal structure of an RNA aptamer bound to thrombin. *RNA* **2008**, *14*, 2504–2512. [CrossRef] [PubMed]

18. Jeter, M.L.; Ly, L.V.; Fortenberry, Y.M.; Whinna, H.C.; White, R.R.; Rusconi, C.P.; Sullenger, B.A.; Church, F.C. RNA aptamer to thrombin binds anion-binding exosite-2 and alters protease inhibition by heparin-binding serpins. *FEBS Lett.* **2004**, *568*, 10–14. [CrossRef] [PubMed]

19. Bhogale, A.; Patel, N.; Sarpotdar, P.; Mariam, J.; Dongre, P.M.; Miotello, A.; Kothari, D.C. Systematic investigation on the interaction of bovine serum albumin with ZnO nanoparticles using fluorescence spectroscopy. *Colloids Surf. B* **2013**, *102*, 257–264. [CrossRef] [PubMed]

20. Gann, H.; Glaspell, G.; Garrad, R.; Wanekaya, A.; Ghosh, K.; Cillessen, L.; Scholz, A.; Parker, B.; Warner, M.; DeLong, R.K. Interaction of MnO and ZnO Nanomaterials with Biomedically Important Proteins and Cells. *J. Biomed. Nanotechnol.* **2010**, *6*, 37–42. [CrossRef] [PubMed]

21. Lakowicz, J.R. *Principles of Fluorescence Spectroscopy*, 3rd ed.; Mechanisms and Dynamics of Fluorescence Quenching; Springer: Boston, MA, USA, 2006; pp. 331–351; ISBN 978-0-387-31278-1.

materials

MDPI

Article

NH$_4$OH Treatment for an Optimum Morphological Trade-off to Hydrothermal Ga-Doped n-ZnO/p-Si Heterostructure Characteristics

Abu ul Hassan Sarwar Rana and Hyun-Seok Kim *

Division of Electronics and Electrical Engineering, Dongguk University-Seoul, Seoul 04620, Korea;
a.hassan.rana@gmail.com
* Correspondence: hyunseokk@dongguk.edu; Tel.: +82-2-2260-3996; Fax: +82-2-2277-8735

Received: 25 October 2017; Accepted: 23 December 2017; Published: 27 December 2017

Abstract: Previous studies on Ga-doped ZnO nanorods (GZRs) have failed to address the change in GZR morphology with increased doping concentration. The morphology-change affects the GZR surface-to-volume ratio and the real essence of doping is not exploited for heterostructure optoelectronic characteristics. We present NH$_4$OH treatment to provide an optimum morphological trade-off to n-GZR/p-Si heterostructure characteristics. The GZRs were grown via one of the most eminent and facile hydrothermal method with an increase in Ga concentration from 1% to 5%. The supplementary OH$^-$ ion concentration was effectively controlled by the addition of an optimum amount of NH$_4$OH to synchronize GZR aspect and surface-to-volume ratio. Hence, the probed results show only the effects of Ga-doping, rather than the changed morphology, on the optoelectronic characteristics of n-GZR/p-Si heterostructures. The doped nanostructures were characterized by scanning electron microscopy, energy dispersive X-ray spectroscopy, X-ray diffraction, photoluminescence, Hall-effect measurement, and Keithley 2410 measurement systems. GZRs had identical morphology and dimensions with a typical wurtzite phase. As the GZR carrier concentration increased, the PL response showed a blue shift because of Burstein-Moss effect. Also, the heterostructure current levels increased linearly with doping concentration. We believe that the presented GZRs with optimized morphology have great potential for field-effect transistors, light-emitting diodes, ultraviolet sensors, and laser diodes.

Keywords: ZnO; nanorod; Ga; doping; heterostructure; optoelectronics; hydrothermal

---

1. Introduction

Because of its direct bandgap of 3.37 eV and high exciton binding energy of 60 meV at room temperature, ZnO has become one of the most important semiconductors in recent decades. The ease of fabrication processes has allowed the researchers to fabricate plenty of one-dimensional ZnO nanoscale shapes such as nanorods (NRs), nanowires, nanotubes, nanoflowers, nanoparticles, nanobelts, and many more [1–4]. Due to its enticing properties and structure, it has shown great potential in the realm of optoelectronic devices such as solar cells, field effect transistors, sensors, light emitting diodes, UV sensors, and laser diodes [5–9]. Furthermore, its chemical properties, such as biocompatibility, non-toxicity, and chemical stability, are useful for applications in cosmetics, medicine, and catalysis [10–12].

It is well known that ZnO is n-type because of the presence of many intrinsic donor defects [13]. Notwithstanding, it is important to control the intrinsic carrier concentration for optoelectronic device applications. It is believed that the highly doped ZnO, with least resistivity, may replace indium tin oxide, which is on the verge of extinction, as a transparent electrode [14]. Hence, ZnO doping is inevitable to control the majority carrier density for optoelectronic device applications. For this reason, group III elements, such as In (M$_W$ 114.82), Ga (M$_W$ 69.73), and Al (M$_W$ 26.98), have been considered as the most suitable candidates because of the presence of an extra electron in their outermost shell [15–17]. Ga, being highly soluble in ZnO and a similar atomic radii with Zn, is one of the finest elements to dope ZnO without compromising its optoelectronic structure. Methods used to dope ZnO with Ga include radio frequency magnetron sputtering, molecular-beam epitaxy, arc-discharge, sol-gel, thermal evaporation, spray pyrolysis, pulsed laser deposition, metal-organic chemical vapor deposition, and hydrothermal method [18–26]. Nonetheless, the optoelectronic character of the fabricated devices with all the sophisticated methods may ensure better results, but we preferred hydrothermal method because of its simplicity, low cost, and ease of use [27].

Although, Ga-doping has already been used to influence the ZnO electronic and optical structure [28,29]. But, instead of mere speculations, it was difficult to cite the real reason of change in gallium-doped ZnO nanorods (GZR) optical and electrical characteristics because of changed ZnO morphology. For example, Wang et al. reported a redshift in photo-luminescent (PL) high-intensity UV peak which was ascribed to the combined effect of GZR decreased diameter and increased doping concentration [28]. On the contrary, Park et al. witnessed an increase in GZR diameter and a blue shift of high-intenslty PL UV peak with an increase in doping concentration [29]. Furthermore, not only the morphology but the growth mechanisms were antithetical to each other and the reasons were ought to be addressed. In this study, we introduce NH_4OH treatment for an optimum trade-off to hydrothermal Ga-doped n-ZnO/p-Si heterostructure characteristics. The goal of the study is to synchronize the NR morphology and dimensions so as the change in NR optical and electrical characteristics be conceived because of doping rather than changed morphology. In this context, the properties of undoped ZnO nanorods (UZRs) were compared and contrasted with GZRs grown via NH_4OH treatment and with the GZR properties reported in the previous studies [28,29]. The GZR morphology was optimized by effectively controlling OH^- ion provision to the solution via NH_4OH decomposition. Hence, despite morphology-induced change in surface-to-volume ratio, only the effects of Ga-doping were realized for GZR optoelectronic devices. The GZRs were characterized for morphological, structural, optical, elemental, and electrical characteristics.

2. Results and Discussion

2.1. GZR Morphology Dependence on Doped and Undoped Seeds

Figure 1 shows the plan-view scanning electron microscope (SEM) images of doped and undoped ZnO seeds and the UZR growth dependence upon seeds. It is already established that ZnO morphology and diameter depend upon seed particle size [30]. The doped and undoped seeds were used to monitor if there was any change in particle size of doped ZnO seeds. Instead of GZRs, only UZRs were grown on seeds to confirm the synchronized morphology change because of seeds and not because of Ga content in GZR growth solution. It is seen in Figure 1a,b that the particle size is shrunk in Ga-doped seeds. We believe that the shrunk morphology is because of the formation of Ga-Zn or Ga-OH clusters in seed solution. Similarly, the grown NRs on small diameter doped seeds have smaller dimensions than NRs grown on undoped seeds, as shown in Figure 1c,d. Furthermore, the vertical NR alignment confirms that the preferred orientation provided by Ga-doped seeds is along 0001 rather than $2\bar{1}\bar{1}0$ or $11\bar{2}0$ directions. Hence, throughout the experiments, we used Ga-doped ZnO seeds for GZR growth to minimize surface free energy between GZRs and Si substrates and to provide a smooth basic growth units to GZRs. Similarly, it is also substantiated that Ga-doped ZnO thin films can also be fabricated for thin-film-based solar cell applications.

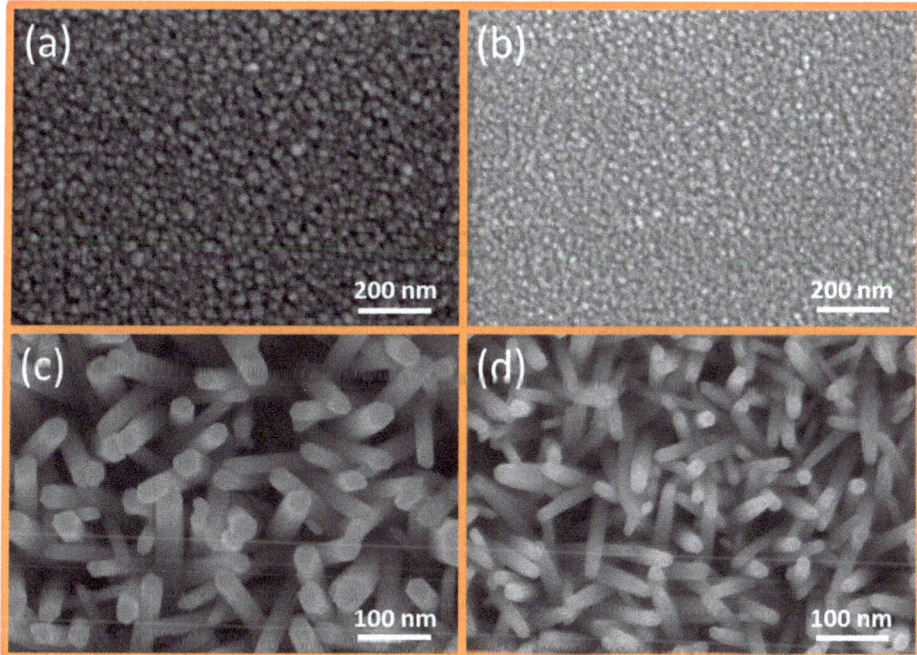

Figure 1. SEM images of: (**a**) undoped ZnO seeds; (**b**) Ga-doped ZnO seeds; (**c**) UZRs on undoped ZnO seeds; and (**d**) UZRs on Ga-doped ZnO seeds.

2.2. GZR Morphology without NH₄OH Treatment

To probe into the effects of NH_4OH treatment, different doping concentration GZRs were first grown without the use of any surfactants or NH_4OH. Figure 2a–d show the plan-view SEM images of UZRs, 1%, 2%, and 5% GZRs, respectively. In contrast to the findings of Park et al. our results support experimental results of Wang et al. [28,29]. In the absence of any additives, the Ga^{3+} reacts with OH^- ions in the solution provided by the decomposition of methenamine and supports homogeneous nucleation of reactants in the solution against heterogeneous nucleation on seeds. The regular OH^- ions supply is quite vital for NR growth and their shortage may result in morphological changes in general or decrease in NR diameter in particular. The UZRs have the largest diameter which keeps on decreasing as the doping level increases from 1% to 5%, which supports high homogeneous nucleation rates in the solution. Only a small change is seen in the diameters of UZRs and 1% and 2% GZRs, but a gross change in diameter is seen between UZRs and 5% GZRs. However, instead of the large diameter GZR lateral growth, the high concentration 5% GZRs are also oriented well along 0001 direction, which is in contrast to the findings of Wang et al. [28]. We believe that the axial growth of even a highly doped sample is because of Ga-doped seeds which support one of the highest growth rates in 0001 direction and render an additional benefit to GZR growth [31].

Figure 2. SEM images of: (**a**) UZRs; (**b**) 1% GZRs; (**c**) 2% GZRs; and (**d**) 5% GZRs.

2.3. NH$_4$OH Treatment for Optimum Morphological Trade-off

The longstanding controversy regarding a doping-centric change in ZNR morphology and its reasons and solutions are addressed via NH$_4$OH treatment. The GZR morphology control is important because it affects the GZR surface-to-volume ratio and ultimately influences their optical and electrical properties [32]. Previously, it was difficult to substantiate either the change in optical and electrical properties was doping-centric or because of the morphology-induced change in GZR surface-to-volume ratio. Hence, we exploited NH$_4$OH treatment to address the problem by optimizing GZR morphology, specifically GZR diameter, for different doping concentrations, and the phenomenon is called as GZR morphological trade-off. Figure 3 shows the SEM images of GZRs grown with NH$_4$OH treatment. The idea was to control the morphology by an optimum provision of OH$^-$ ions via NH$_4$OH decomposition in the solution. The additional OH$^-$ ions impart a trade-off for OH$^-$ ions that were wasted in Ga-OH complex formation and homogeneous nucleation. Hence, the average diameter of 1%, 2%, and 5% GZRs remained fixed at ~60 nm, as shown in Figure 3a–c, respectively. For best results, the amount of NH$_4$OH ought to be controlled judiciously because a slight increase or decrease may alter results. In this study, we used 5, 7, and 10 mL NH$_4$OH for 1%, 2%, and 5% GZRs, respectively.

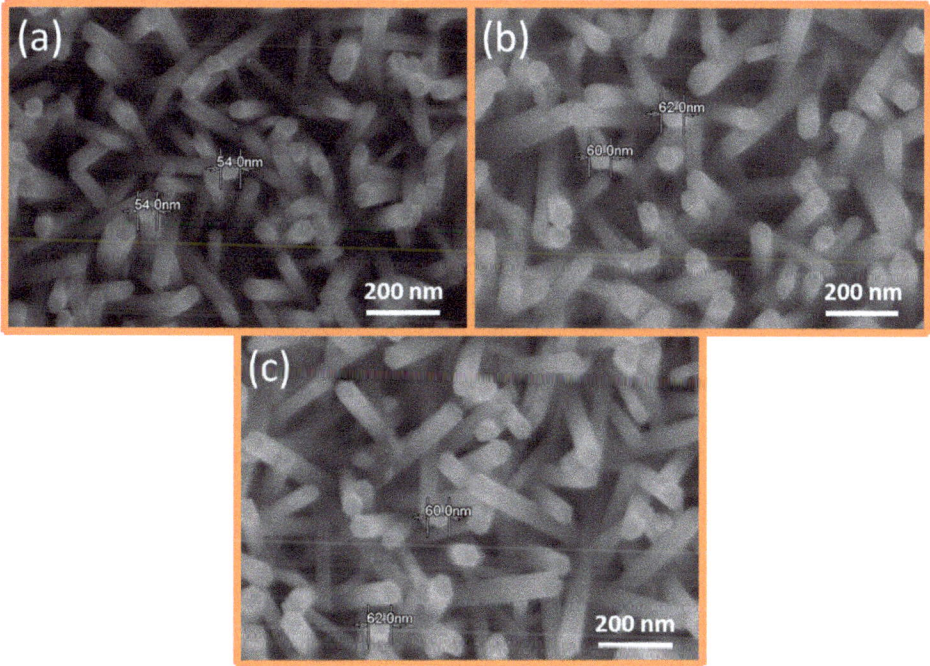

Figure 3. SEM images of: (**a**) 1%; (**b**) 2%; and (**c**) 5% GZRs grown with NH₄OH treatment.

2.4. GZR Isoelectric Point-Dependent Growth Mechanism

The GZR growth mechanism is better explained by isoelectric point-centric surface charge reversal phenomenon, as shown in Figure 4. The ZnO isoelectric point, without the addition of any surfactant, ranges from 7.4 to 8.2 [33]. ZnO nucleation and growth depend upon pH-centric surface charge. Initially, the solution pH was 7 and no nucleation was promoted. With methenamine decomposition, the solution pH was naturally raised to 9 and 0001 surface charge was reversed from positive to negative, as shown in Figure 4a. As soon as the surface charge was reversed, the Zn^{2+} and Ga^{3+} ions were deposited on the negative 0001 and the O^- ions were deposited on the $000\bar{1}$ positive surfaces. The heterogeneous nucleation process kept working similarly until pH was reduced to 7.5 because of OH^- ion extinction in the solution and the 0001 surface charge was reversed to positive. This particular point is called as growth stoppage point beyond which the growth is not promoted.

The said phenomenon was also substantiated with the help of cross-sectional SEM images of NRs on corresponding pH values in Figure 4. The NRs kept growing as the pH value decreased from its zenith at 9 and back to 7.5. Any rise in temperature or time beyond this point did not support nucleation, where NR length remained the same, as shown in Figure 4c,d. In the presence of Ga, the solution isoelectric point was shortly reached because of an utter wastage of OH^- ions via homogeneous nucleation in the form of Ga-OH complexes. Hence, GZR diameter was reduced because of an early surface charge reversal as the Ga concentration was increased in the solution. With the addition of NH₄OH, the surface charge reversal duration and nucleation process were extended by raising the pH to the maximum value of 10.5 for 5% GZRs.

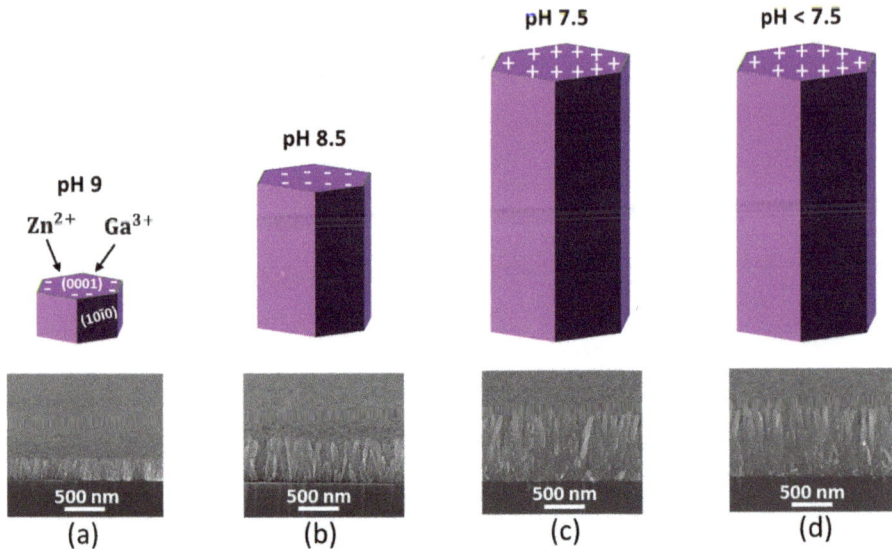

Figure 4. GZR isoelectric point-dependent growth mechanism with their corresponding cross-sectional SEM images at (**a**) 30 min; (**b**) 2 h; (**c**) 4 h; and (**d**) 6 h.

2.5. Elemental Characteristics of Ga-Doped Seeds and GZRs

The elemental characteristics of Ga-doped ZnO seeds and GZRs grown with NH_4OH treatment were measured with energy dispersive X-ray spectroscopy (EDS), as shown in Figure 5. The insets explain the detailed atomic and weight percentages of the found elements in a nanostructure. The found elements in all the samples were Zn, O, Si, and Ga. Figure 5a confirms the incorporation of Ga into ZnO seeds. The large Si peak is from the substrate because of a very thin layer of twice coated ZnO seeds. Similarly, all the GZR samples in Figure 5b–d have Ga peaks with an increased intensity as the doping level increases. The point to ponder is an increase in the Ga atomic percentage as the doping level increases from 1% to 5%. The highest number of Ga atoms are found in 5% GZRs, which confirms the effective incorporation of Ga even in high doping concentrations via our method. The sole purpose of EDS was to confirm the incorporation of Ga into ZnO lattice and the effectiveness of the doping method. With the presented EDS results, the presence of Ga ions in the host can be determined. However, in order to measure the proper stoichiometry of elements, the silicon substrate must not be taken into account in the calculation, which is present in all the EDS samples in Figure 5. Perhaps, we tried a lower EDS acceleration voltage to avoid excitations from the Si substrate but all went in vain and we found Si peaks in all quantifications. Hence, it is impossible to drive conclusions from the current EDS quantification regarding the stoichiometry and proper ratio of elements.

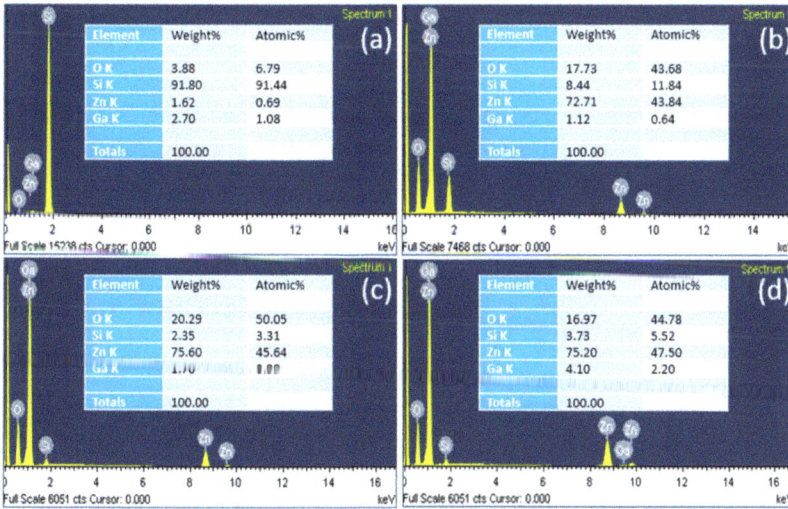

Figure 5. EDS of (**a**) Ga-doped ZnO seeds; and (**b**) 1%; (**c**) 2%; and (**d**) 5% GZRs grown with NH₄OH treatment.

2.6. Structural Characteristics of GZRs

GZR structural characteristics were found with X-ray diffraction (XRD) crystallography, as shown in Figure 6. All the samples show a typical hexagonal wurtzite ZnO phase and none of the secondary diffraction phases, such as $ZnGa_2O_4$ and Ga_2O_3, are seen in the XRD response of all the samples [34]. Hence, despite Ga incorporation into ZnO crystal lattice, it is inferred that Ga-doping does not influence ZnO structural phase. The presence of multiple peaks along 100, 002, 101, 102, 110, 103, and 112 affirms the GZR polycrystalline nature in all samples. Despite different intensity XRD peaks, the highest peaks in all the samples are along 002 direction. The low intensity peaks along 100 and 101 certified a slight *a*-axis orientation in all the samples because of partly inclined GZRs. However, the 100 and 101 peak intensities are smaller and impotent as compared to 002 peak intensity, which confirms the *c*-axis orientation of GZRs in all the samples. Since the ionic radii of Ga^{3+} (0.06 nm) and Zn^{2+} (0.07 nm) are almost identical, large ZnO lattice distortions were already not expected, which was certified by an increase in 002 peak intensity with an increase in doping concentration [35]. The detailed lattice parameters are provided in Table 1. The lattice constant was calculated with Bragg's law [36]. It is certified that the 2θ position of 002 peaks remains almost fixed for 1% and 2% GZRs except for heavily-doped 5% GZRs. Furthermore, a decrease in full-width of half maximum (FWHM) of 002 peak affirms the GZR improved crystallinity. Expecting a large stress/strain in GZR direct growth on bare Si substrate because of large lattice mismatch, GZRs were grown on a buffer layer of Ga-doped ZnO seeds. Hence, the extrinsic factors for stress can be eliminated and the intrinsic stress/stain along the GZR *c*-axis and in thin GZR film were calculated via XRD data analysis. The *c*-axis strain (ε_c) is

$$\varepsilon_c = \frac{c - c_o}{c_o} \times 100,$$ (1)

where c_o and c are the lattice constants of stress free bulk ZnO (0.5205 nm) and GZRs, respectively. Although Ga adds minimum lattice vibrations to GZRs, yet the heavily-doped sample shows the highest strain along *c*-axis (Table 1). The stress (σ) in GZRs can also be calculated via biaxial stress model [37],

$$\sigma = \frac{2(C_{13})^2 - C_{33}(C_{11} + C_{12})}{2C_{13}} \varepsilon,$$ (2)

where C_{11}, C_{12}, C_{13}, and C_{33} are the bulk ZnO elastic stiffness constants with value 208.8, 119.7, 104.2, and 213.8 GPa, repectively. Hence,

$$\sigma = -233\ \varepsilon. \tag{3}$$

The calculated stress/stain values are reported in Table 1, where the negative sign for strain shows that the strain is compressive for all the doped samples.

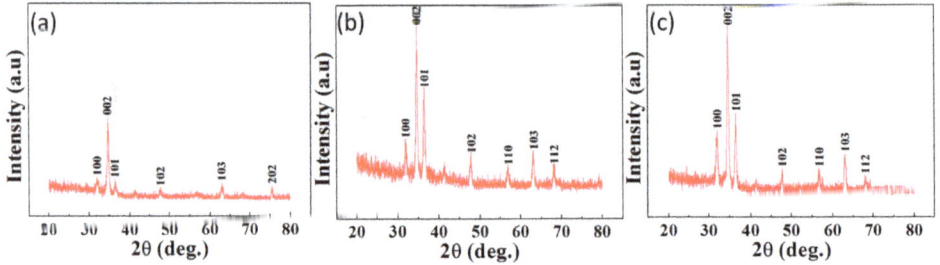

Figure 6. XRD of: (**a**) 1%; (**b**) 2%; and (**c**) 5% GZRs.

Table 1. Lattice parameters of GZRs.

Ga-Doping (%)	002 Position (Deg.)	a (Å)	c (Å)	c/a	FWHM (Rad.)	ε_c (%)	σ (GPa)
1	34.38	3.25	5.20	1.60	0.005	−0.07	0.17
2	34.39	3.25	5.20	1.60	0.004	−0.03	0.08
5	34.60	3.24	5.18	1.59	0.002	−0.48	0.94

2.7. Optical Properties of GZRs

Figure 7 depicts the room temperature PL spectra of GZRs in the range of 300 nm to 800 nm. The PL spectra show three distinct peaks in UV, visible, and near IR regions. The sharp and highest peak in UV region is a direct response of free exciton recombination via a near band edge exciton-exciton collision process [38]. The UV peak positions on X and Y scales are highlighted in Figure 7 panels. It can be seen in the Figure 7 that the UV peak intensity increases with Ga-doping concentration which verifies the better optical characteristics of heavily doped 5% GZRs. Another point to ponder in this particular backdrop is a blue shift of UV peaks with an increase in doping concentration. There are four conflicting theories regarding GZR-assisted UV peak position shift which are of paramount importance. First, a redshift in UV peak position is reported with an increase in Ga concentration because of doping-induced band gap renormalization (BGR) effect [39]. An increase in carrier concentration results in free carrier screening via many-body interactions which influence BGR effect. Second, extrinsic and intrinsic stress and lattice distortions may also influence peak shift position [40]. Third, a morphology induced change in surface-to-volume ratio can have a grave impact upon a change in the peak position [41]. Forth, the peaks are blue shifted because of doping-induced Burstein-Moss (BM) effect [42]. Previously, it was difficult to trace the real reason of a change in optical characteristics because of changed morphology and stress centers in GZRs. In this study, we support the description given my BM effects because all the samples were blue shifted. Furthermore, the reduced dimeter-induced change in surface-to-volume ratio cannot be considered as the GZR diameter was synchronized to capture the real reason for this particular shift. It has also been reported in the structural characteristics that the grown GZRs were stress-free and no lattice distortions were monitored because of GZR growth on Ga-doped seeds. Hence, the samples were blue shifted because of BM effects, where the Fermi level was moved towards the conduction band as the doping concentration increased from 1% to 5%. The exciton recombination from an increased energy bandgap emits in lower wavelength regions.

Similarly, the GZR optical properties differ in the wavelength range of 500 to 700 nm. Generally, ZnO shows a broad and high peak in the visible band is believed to be a direct response to many Zn interstitial and oxygen vacancy defects [35]. Herein, the 1% GZRs show more or less a similar trend with UZRs but the peak moves toward a flat band condition with an increased Ga concentration. It is inferred that Ga atoms tend to replace the defect centers which respond in the visible region. An increase in UV peak intensity and a decrease in visible peak intensity with an increased doping concentration show that Ga doping improves the ZnO optical properties. Hence, Ga-doping is an important way to fix the naturally occurring defects and to improve the ZnO crystalline structure. Furthermore, it is also found that Ga-doping creates some defects which respond to near IR region as shown by the peaks around 750 nm.

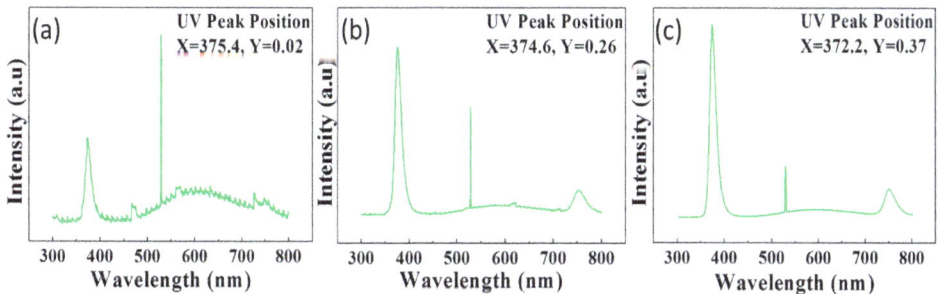

Figure 7. PL spectra of (a) 1%; (b) 2%; and (c) 5% GZRs.

2.8. Carrier Concentration and Electrical Characteristics of GZRs

Despite bandgap engineering and exciton energy, the optoelectronic device characteristics largely depends upon the GZR electronic character such as majority carrier concentration and carrier mobility. The GZR carrier concentration was found with a four-probe Hall-effect measurement system under dark conditions. For a smooth carrier transport during Hall-measurement, GZRs were grown on an insulating glass substrate and an ohmic-In contacts were made around the four corners of the grown GZR film. For better results, the GZR dimensions on glass substrate were synchronized with the dimensions on p-Si substrate. The detailed experimental findings are illustrated in inset table in Figure 8. It is found that Ga-doping increases majority carrier concentration. The hydrothermally grown UZRs already have high intrinsic carrier density because of the incorporation of many donor defects during growth. Because of having an extra electron in the outermost shell ($4S^2$ $4P^1$), Ga-doing is an effective way to increase the average carrier concentration from 10^{16} in UZRs to 10^{21} in 5% GZRs.

The current-voltage (I-V) response of UZRs and GZRs-based n-ZnO/p-Si heterostructures were found with Keithley 2410 in a probe station under dark conditions. The I-V response is shown in Figure 8 and the schematics of the heterostructure device with probe station contacts are shown in Figure 9g. It can be seen that all the samples show a diode-like I-V behavior forming an ohmic contact with In. The current density increases with an average increase in majority carrier concentration and the highest current levels are achieved in the highest doped 5% GZRs. The electrical characteristics of our samples are comparable with the electrical characteristics of sol-gel spin-coated Al-doped ZnO films [43]. It is affirmed that the electrical behavior is a carrier concentration-dependent direct response of GZRs rather than an increase or decrease in morphology-induced surface-to-volume ratio.

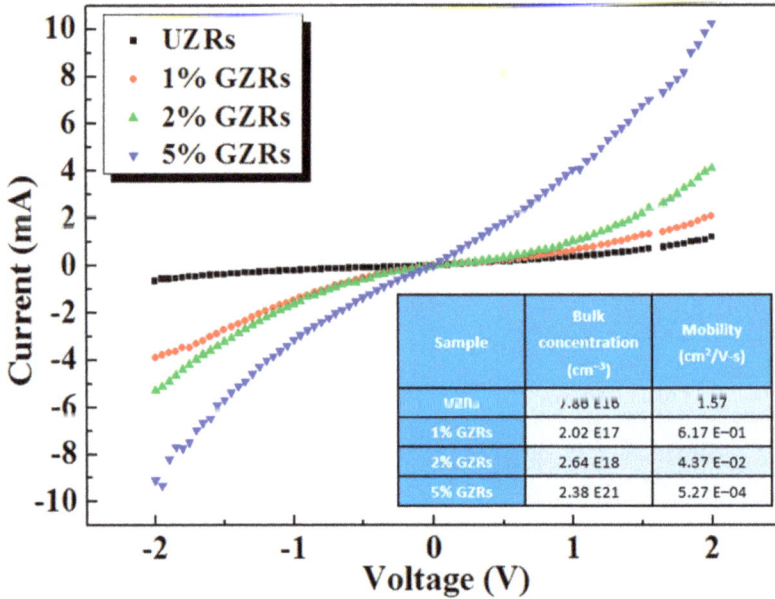

Figure 8. Electrical characteristics of UZRs and GZRs. Bottom inset table reveals the Hall-effect measurements.

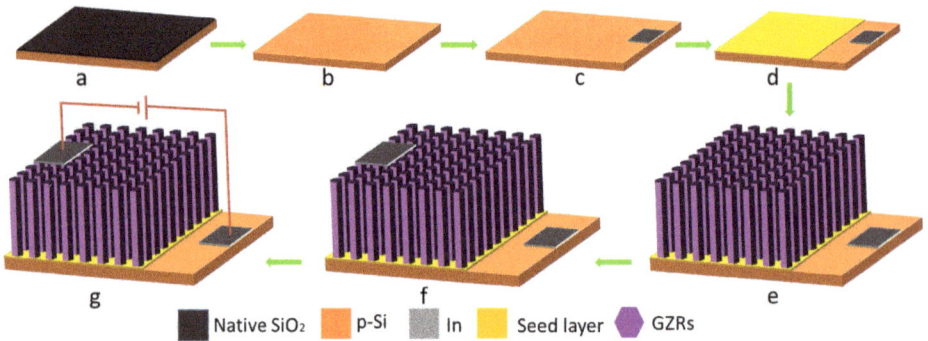

Figure 9. Device fabrication process flow diagram.

3. Materials and Methods

3.1. Substrate Cleaning, Bottom Electrode Deposition, and Seed Coating

To conduct all the experiments, commercially available chemicals were purchased from Sigma Aldrich. To make a p-n heterojunction, n-GZRs were grown on p-Si (100, 1–10 Ω·cm) substrates. Figure 9 shows the process flow diagram of the experimental procedures for device fabrication. Si reacts with oxygen in ambient conditions to form an insulating SiO_2 layer on the surface, as shown in Figure 9a. Hence, Si substrates were immersed in buffered oxide etchant (BOE) (6:1) to remove the native SiO_2 layer for a smooth working of a p-n junction device. After two min, the substrates were removed from BOE and cleaned with deionized (DI) water (18 MΩ) and N_2 gas and a clean substrate was ready for device fabrication, as shown in Figure 9b. Next step was bottom In electrode deposition, which was done via photolithography and metal evaporation, as depicted in Figure 9c. Doped ZnO seeds were made by mixing gallium nitrate hydrate [$Ga(NO_3)_3 \cdot xH_2O$] (M_W 255.74 anhydrous basis) and zinc acetate dihydrate [$Zn(CH_3COO)_2 \cdot 2H_2O$] ($M_W$ 219.51 g/M) in n-propanol [C_3H_8O] (M_W 60.10 g/M). The chemicals were sonicated for 30–40 min for saturated seed solution. The Ga-doped seeds were spun twice on the substrate surface at 3000 RPM for 2 min. To provide seed stability, the spin-coated seeds were annealed at 100 and 300 °C on a hotplate for successive coatings. Finally, a thin layer of Ga-doped ZnO seeds was formed to assist vertical GZR growth, as shown in Figure 9d.

3.2. GZR Growth and Heterostructure Device Fabrication

The growth solution was made by mixing 25 mM each of zinc nitride hexahydrate [$Zn(NO_3)_2 \cdot 6H_2O$] (M_W 297.48 g/mol) and methenamine [$C_6H_{12}N_4$] (M_W 140.186 g/mol) in DI water. For doping, three different growth solutions were prepared by adding 0.252, 0.510, and 1.315 mM gallium nitrate hydrate into the above solutions to fix the Ga-doping to 1%, 2%, and 5%, respectively. The doping molarity to percentage conversion was calculated by the relation: Ga% = $M_{Ga}/(M_{Ga} + M_{Zn}) \times 100\%$, where M_{Zn} and M_{Ga} are the respective molar concentrations of Zn and Ga. Also, different amounts of ammonium hydroxide [NH_4OH] (M_W 35.05 g/M with 28% NH_3 in H_2O) were added into the solutions after intermittent intervals to provide GZR morphology trade-off. The growth solutions were subjected to a magnetic stirring for 1 h without heating. After 1 h, the seeded substrates were immersed upside down into solution autoclaves and was heated at 90 °C. After 4 h, the substrates were removed from the autoclaves, rinsed in DI water, and dried with N_2 and vertical GZRs were grown on the seeded substrates, as shown in Figure 9e. The top In electrode was deposited directly upon GZRs with photolithography and metal evaporation in the opposite direction to the bottom In electrode, as depicted in Figure 9f. The probe station contacts for electrical measurements were made in a way portrayed in Figure 9g.

3.3. Material Characterizations

The photolithography and In contacts were made with Karl Suss MA-6 (Dongguk University, Seoul, Korea) and E-beam metal evaporator (Dongguk University, Seoul, Korea). The GZR morphology was seen with SEM (Hitachi S-4800, Suwon, Gyeonggi-do, Korea) operating at 25 KeV. Ga incorporation into ZnO and elemental proportions were confirmed with EDS attached to SEM equipment. The GZR structural characteristics were measured with XRD (Rigaku Ultima IV, Dongguk University, Seoul, Korea) from 2θ values of 20 to 80 degrees at λ = 1.5418 Å. The optical properties were monitored with PL spectroscopy (Accent RPM 2000, Suwon, Gyeonggi-do, Korea) from 300 nm to 800 nm at room temperature and pressure. The I-V diode characteristics were measured with Keithley 2410 in a probe station under the dark condition and the majority carrier concentration was found using Hall-effect measurement system (ECOPIA AHT55T5, Dongguk University, Seoul, Korea). The in-situ solution temperature and pH were monitored with a digital thermometer and pH meter, respectively.

4. Conclusions

In this study, we present NH_4OH treatment for an optimum morphological trade-off to hydrothermal n-GZR/p-Si heterostructures. NH_4OH treatment was necessary to provide an additional OH^- ion supplement to synchronize GZR morphology for high doping levels. For best results, the supplementary OH^- ions were optimally tuned for different doping concentrations. Hence, rather than changed morphology induced characteristic transition, it was easy to probe the real essence of GZR optoelectronic characteristics, which was not the case in previous studies. In this context, the GZR morphological, elemental, structural, optical, and electrical properties were analyzed. The morphology of all the doped samples was tuned to an average diameter of ~60 nm. EDS results showed that Ga was incorporated into ZnO lattice and Ga atomic concentration increased with doping. The decreased lattice constant and increased 002 XRD peak intensity depicted that Ga-doping improved the crystalline ZnO structure without any lattice distortions. PL results showed that increased Ga concentration shifted the UV peak towards the lower wavelengths because of carrier concentration-induced BM effects, and Ga tend to fix the intrinsic ZnO defects and improve GZR optical characteristics. Furthermore, the GZR electrical characteristics were also improved by Ga-doping. The carrier concentration reached as high as 10^{21} cm^{-3} for 5% GZRs. All the samples showed a diode like I-V behavior with a uniform increase in current intensity with doping concentration. The propounded results may provide impetus to the study on optoelectronic characteristics of ZnO-based heterojunction devices.

Acknowledgments: This work was supported by the Korea Institute of Energy Technology Evaluation and Planning (KETEP) and the Ministry of Trade, Industry & Energy (MOTIE) of the Republic of Korea (No. 20174030201520) and the Basic Science Research Program through the National Research Foundation of Korea (NRF) funded by the Ministry of Education (No. 2017R1D1A1A09000823).

Author Contributions: A.u.H.S.R. fabricated the heterostructure devices, designed and conducted all the experiments, and analyzed the data. H.S.K. planned and supervised the project. Both authors contributed to discussing the results and writing the manuscript.

Conflicts of Interest: The authors declare no conflict of interest.

References

1. Shi, Y.; Bao, S.; Shi, R.; Huang, C.; Amini, A.; Wu, Z.; Zhang, L.; Wang, N.; Cheng, C. Y-shaped ZnO Nanobelts Driven from Twinned Dislocations. *Sci. Rep.* **2016**, *6*, 22494. [PubMed]
2. Yang, Z.; Wang, M.; Shukla, S.; Zhu, Y.; Deng, J.; Ge, H.; Wang, X.; Xiong, Q. Developing Seedless Growth of ZnO Micro/Nanowire Arrays towards ZnO/FeS$_2$/CuI P-I-N Photodiode Application. *Sci. Rep.* **2015**, *5*, 11377. [CrossRef] [PubMed]
3. Rana, A.S.; Kang, M.; Jeong, E.S.; Kim, H.S. Transition between ZnO Nanorods and ZnO Nanotubes with Their Antithetical Properties. *J. Nanosci. Nanotechnol.* **2016**, *16*, 10772–10776. [CrossRef]
4. Tachikawa, S.; Noguchi, A.; Tsuge, T.; Hara, M.; Odawara, O.; Wada, H. Optical Properties of ZnO Nanoparticles Capped with Polymers. *Materials* **2011**, *4*, 1132–1143. [CrossRef] [PubMed]
5. Alenezi, M.R.; Henley, S.J.; Silva, S.R.P. On-chip Fabrication of High Performance Nanostructured ZnO UV Detectors. *Sci. Rep.* **2015**, *5*, 8516. [CrossRef] [PubMed]
6. Lee, C.T. Fabrication Methods and Luminescent Properties of ZnO Materials for Light-Emitting Diodes. *Materials* **2010**, *3*, 2218–2259. [CrossRef]
7. Kim, K.H.; Utashiro, K.; Abe, Y.; Kawamura, M. Structural Properties of Zinc Oxide Nanorods Grown on Al-Doped Zinc Oxide Seed Layer and Their Applications in Dye-Sensitized Solar Cells. *Materials* **2014**, *7*, 2522–2533. [CrossRef] [PubMed]
8. Kang, M.; Rana, A.S.; Jeong, E.S.; Kim, H.S. Direct Observation of Thermally Generated Electron-Hole Pairs in ZnO Nanorods with Surface Acoustic Wave. *J. Nanosci. Nanotechnol.* **2017**, *17*, 4141–4144. [CrossRef]
9. Zong, X.; Zhu, R. Zinc oxide nanorod field effect transistor for long-time cellular force measurement. *Sci. Rep.* **2017**, *7*, 43661. [CrossRef] [PubMed]
10. Pineda-Hernandez, G.; Escobedo-Morales, A.; Pal, U.; Chigo-Anota, E. Morphology evolution of hydrothermally grown ZnO nanostructures on gallium doping and their defect structures. *Mater. Chem. Phys.* **2012**, *135*, 810–817. [CrossRef]
11. Choi, Y.E.; Kwak, J.W.; Park, J.W. Nanotechnology for Early Cancer Detection. *Sensors* **2010**, *10*, 428–455. [CrossRef] [PubMed]
12. Chu, D.; Masuda, Y.; Ohji, T.; Kato, K. Formation and Photocatalytic Application of ZnO Nanotubes Using Aqueous Solution. *Langmuir* **2010**, *26*, 2811–2815. [CrossRef] [PubMed]
13. Wang, Q.; Yan, Y.; Zeng, Y.; Lu, Y.; Chen, L.; Jiang, Y. Free-Standing Undoped ZnO Microtubes with Rich and Stable Shallow Acceptors. *Sci. Rep.* **2016**, *6*, 27341. [CrossRef] [PubMed]
14. Sun, D.; Tan, C.; Tian, X.; Huang, Y. Comparative Study on ZnO Monolayer Doped with Al, Ga and In Atoms as Transparent Electrodes. *Materials* **2017**, *10*, 703. [CrossRef] [PubMed]
15. Koida, T.; Kaneko, T.; Shibata, H. Carrier Compensation Induced by Thermal Annealing in Al-Doped ZnO Films. *Materials* **2017**, *10*, 141. [CrossRef] [PubMed]
16. Babar, A.R.; Deshamukh, P.R.; Daekate, R.J.; Haranath, D.; Bhosale, C.H.; Rajpure, K.Y. Gallium doping in transparent conductive ZnO thin films prepared by chemical spray pyrolysis. *J. Phys. D Appl. Phys.* **2008**, *41*, 135404–135410. [CrossRef]

Materials **2018**, *11*, 37

17. Biswal, R.; Maldonado, A.; Vega-Perez, J.; Acosta, D.R.; Olvera, D.L.L. Indium Doped Zinc Oxide Thin Films Deposited by Ultrasonic Chemical Spray Technique, Starting from Zinc Acetylacetonate and Indium Chloride. *Materials* **2014**, *7*, 5038–5046. [CrossRef] [PubMed]

18. Tseng, J.Y.; Chen, Y.T.; Yang, M.Y.; Wang, C.Y.; Li, P.C.; Yu, W.C.; Hsu, Y.F.; Wang, S.F. Deposition of low-resistivity gallium-doped zinc oxide films by low-temperature radio-frequency magnetron sputtering. *Thin Solid Films* **2009**, *517*, 6310–6314. [CrossRef]

19. Yang, Z.; Look, D.C.; Liu, J.L. Ga-related photoluminescence lines in Ga-doped ZnO grown by plasma-assisted molecular-beam epitaxy. *Appl. Phys. Lett.* **2009**, *94*, 072101. [CrossRef]

20. Park, G.S.; Choi, W.B.; Kim, J.M.; Choi, Y.C.; Lee, Y.H.; Lim, C.D. Structural investigation of gallium oxide (β-Ga₂O₃) nanowires grown by arc-discharge. *J. Cryst. Growth* **2000**, *220*, 494–500. [CrossRef]

21. Cheong, K.Y.; Muti, N.; Ramanan, S.R. Electrical and optical studies of ZnO:Ga thin films fabricated via the sol-gel technique. *Thin Solid Films* **2002**, *410*, 142–146. [CrossRef]

22. Chang, L.W.; Yeh, J.W.; Cheng, C.L.; Shieu, F.S.; Shih, H.C. Field emission and optical properties of Ga-doped ZnO nanowires synthesized via thermal evaporation. *Appl. Surf. Sci.* **2011**, *257*, 3145–3151. [CrossRef]

23. Rao, T.P.; Kumar, M.C.S. Physical properties of Ga-doped ZnO thin films by spray pyrolysis. *J. Alloys Compd.* **2010**, *506*, 788–793.

24. Park, S.M.; Ikegami, T.; Ebihara, K. Effects of substrate temperature on the properties of Ga-doped ZnO by pulsed laser deposition. *Thin Solid Films* **2006**, *513*, 90–94. [CrossRef]

25. Chen, H.; Du Pasquier, A.; Saraf, G.; Zhong, J.; Lu, Y. Dye-sensitized solar cells using ZnO nanotips and Ga-doped ZnO films. *Semicond. Sci. Technol.* **2008**, *23*, 045004–045010. [CrossRef]

26. Le, H.Q.; Lim, S.K.; Goh, G.K.L.; Chua, S.J.; Ong, J.X. Optical and Electrical Properties of Ga-Doped ZnO Single Crystalline Films Grown on MgAl₂O₄ by Low Temperature Hydrothermal Synthesis. *J. Electrochem. Soc.* **2010**, *157*, H796–H800. [CrossRef]

27. Rana, A.S.; Ko, K.; Hong, S.; Kang, M.; Kim, H.S. Fabrication and Characterization of ZnO Nanorods on Multiple Substrates. *J. Nanosci. Nanotechnol.* **2015**, *15*, 8375–8380. [CrossRef] [PubMed]

28. Wang, H.; Baek, S.; Song, J.; Lee, J.; Lim, S. Microstructural and optical characteristics of solution-grown Ga-doped ZnO nanorod arrays. *Nanotechnology* **2008**, *19*, 075607–075613. [CrossRef] [PubMed]

29. Park, G.C.; Hwang, S.M.; Lim, J.H.; Joo, J. Growth behavior and electrical performance of Ga-doped ZnO nanorod/p-Si heterojunction diodes prepared using a hydrothermal method. *Nanoscale* **2014**, *6*, 1840–1847. [CrossRef] [PubMed]

30. Rana, A.S.; Chang, S.B.; Chae, H.U.; Kim, H.S. Structural, optical, electrical and morphological properties of different concentration sol-gel ZnO seeds and consanguineous ZnO nanostructured growth dependence on seeds. *J. Alloys Compd.* **2017**, *729*, 571–582. [CrossRef]

31. Song, J.; Lim, S. Effect of Seed Layer on the Growth of ZnO Nanorods. *J. Phys. Chem. C* **2007**, *111*, 596–600. [CrossRef]

32. Barnard, A.S.; Russo, S.P.; Snook, I.K. Electronic band gaps of diamond nanowires. *Phys. Rev. B* **2003**, *68*, 235407–235412. [CrossRef]

33. Degen, A.; Kosec, M. Effect of pH and impurities on the surface charge of zinc oxide in aqueous solution. *J. Eur. Ceram. Soc.* **2000**, *20*, 667–673. [CrossRef]

34. Amin, M.; Shah, N.A.; Bhatti, A.S.; Malik, M.A. Effects of Mg doping on optical and CO gas sensing properties of sensitive ZnO nanobelts. *CrystEngComm* **2014**, *16*, 6080–6088. [CrossRef]

35. Shanon, R.D. Revised effective ionic radii and systematic studies of interatomic distances in halides and chalcogenides. *Acta Crystallogr. Sect. A* **1976**, *32*, 751–767. [CrossRef]

36. Rana, A.S.; Lee, J.Y.; Shahid, A.; Kim, H.S. Growth Method-Dependent and Defect Density-Oriented Structural, Optical, Conductive, and Physical Properties of Solution-Grown ZnO Nanostructures. *Nanomaterials* **2017**, *7*, 266. [CrossRef] [PubMed]

37. Wang, Y.G.; Lau, S.P.; Lee, H.W.; Yu, S.F.; Tay, B.K.; Zhang, X.H.; Tse, K.Y.; Hng, H.H. Comprehensive study of ZnO films prepared by filtered cathodic vacuum arc at room temperature. *J. Appl. Phys.* **2003**, *94*, 1597–1604. [CrossRef]

38. Rana, A.S.; Kang, M.; Kim, H.S. Microwave-assisted Facile and Ultrafast Growth of ZnO Nanostructures and Proposition of Alternative Microwave-assisted Methods to Address Growth Stoppage. *Sci. Rep.* **2016**, *6*, 24870. [CrossRef] [PubMed]

39. Reynolds, D.C.; Look, D.C.; Jogai, B. Combined effects of screening and band gap renormalization on the energy of optical transitions in ZnO and GaN. *J. Appl. Phys.* **2000**, *88*, 5760–5763. [CrossRef]

40. Fair, R.B. The effect of strain-induced band-gap narrowing on high concentration phosphorus diffusion in silicon. *J. Appl. Phys.* **1979**, *50*, 860–868. [CrossRef]

41. Chen, C.W.; Chen, K.H.; Shen, C.H.; Ganguly, A.; Chen, L.C.; Wu, J.J.; Wen, H.I.; Pong, W.F. Anomalous blueshift in emission spectra of ZnO nanorods with sizes beyond quantum confinement regime. *Appl. Phys. Lett.* **2006**, *88*, 241905–241907. [CrossRef]

42. Burstein, E. Anomalous Optical Absorption Limit in InSb. *Phys. Rev.* **1954**, *93*, 632–633. [CrossRef]

43. Kumar, K.D.A.; Valanarasu, S.; Kathalingam, A.; Ganesh, V.; Shkir, M.; AlFaify, S. Effect of solvents on sol-gel spin-coated nanostructured Al-doped ZnO thin films: A film for key optoelectronic applications. *Appl. Phys. A* **2017**, *123*, 801. [CrossRef]

![materials logo] *materials*

MDPI

Article

Study on Zinc Oxide-Based Electrolytes in Low-Temperature Solid Oxide Fuel Cells

Chen Xia [1,3], Zheng Qiao [1,2], Chu Feng [1], Jung-Sik Kim [4], Baoyuan Wang [1,*] and Bin Zhu [1,3,*]

1   Hubei Collaborative Innovation Center for Advanced Organic Chemical Materials, Faculty of Physics and Electronic Science, Hubei University, Wuhan 430062, China; cxia@kth.se (C.X.); qiaozheng@hgnu.edu.cn (Z.Q.); musia0803@163.com (C.F.)
2   College of Mechanical and Electrical Engineering, Huanggang Normal University, Huanggang 430062, China
3   Department of Energy Technology, KTH Royal Institute of Technology, 10044 Stockholm, Sweden
4   Department of Aeronautical & Automotive Engineering, Loughborough University, Loughborough LE11 3TU, UK; J.Kim@lboro.ac.uk
*   Correspondence: baoyuanw@163.com (B.W.); binzhu@kth.se or zhubin@hubu.edu.cn (B.Z.)

Received: 29 November 2017; Accepted: 26 December 2017; Published: 28 December 2017

Abstract: Semiconducting-ionic conductors have been recently described as excellent electrolyte membranes for low-temperature operation solid oxide fuel cells (LT-SOFCs). In the present work, two new functional materials based on zinc oxide (ZnO)—a legacy material in semiconductors but exceptionally novel to solid state ionics—are developed as membranes in SOFCs for the first time. The proposed ZnO and ZnO-LCP (La/Pr doped CeO_2) electrolytes are respectively sandwiched between two $Ni_{0.8}Co_{0.15}Al_{0.05}$Li-oxide (NCAL) electrodes to construct fuel cell devices. The assembled ZnO fuel cell demonstrates encouraging power outputs of 158–482 mW cm^{-2} and high open circuit voltages (OCVs) of 1–1.06 V at 450–550 °C, while the ZnO-LCP cell delivers significantly enhanced performance with maximum power density of 864 mW cm^{-2} and OCV of 1.07 V at 550 °C. The conductive properties of the materials are investigated. As a consequence, the ZnO electrolyte and ZnO-LCP composite exhibit extraordinary ionic conductivities of 0.09 and 0.156 S cm^{-1} at 550 °C, respectively, and the proton conductive behavior of ZnO is verified. Furthermore, performance enhancement of the ZnO-LCP cell is studied by electrochemical impedance spectroscopy (EIS), which is found to be as a result of the significantly reduced grain boundary and electrode polarization resistances. These findings indicate that ZnO is a highly promising alternative semiconducting-ionic membrane to replace the electrolyte materials for advanced LT-SOFCs, which in turn provides a new strategic pathway for the future development of electrolytes.

Keywords: semiconducting-ionic conductor; solid oxide fuel cells; zinc oxide; composite electrolyte; proton conduction

---

1. Introduction

In the preceding decades, fuel cells (FC) technologies have attracted enormous attention for power generation due to the imperious demand of humankind for sustainable energy resources [1,2]. As a typical category of FC technologies, solid oxide fuel cells (SOFCs) are currently receiving ever-increasing research interest because of their distinguishing advantages of high energy conversion efficiency, low greenhouse gas emissions and excellent fuel flexibility [3–5]. Unfortunately, current high-temperature SOFCs suffer from high manufacturing costs and technological complexities, due to the fact that yttria-stabilized zirconia (YSZ) electrolyte requires high temperatures (800–1000 °C) or precisely controlled thin film quality by advanced technologies to reach a sufficient ionic conductivity [6,7]. On the other hand, intermediate-temperature (600–800 °C) SOFCs are subject to an issue regarding the reduction reaction of Samarium-doped ceria (SDC) electrolyte in hydrogen atmosphere, which introduces additional electronic conduction and thus results in serious power loss to the cell [8]. Therefore, to realize the widespread application of SOFCs, it is highly critical to overcome the barriers of high-temperature operation and material degradation to develop advanced low-temperature (300–600 °C) SOFCs (LT-SOFCs). Since the electrolyte layer is well known as the heart of a fuel cell device in determining the operational temperatures and durability as well as the ultimate energy conversion efficiency, new strategies for excavating alternative electrolytes with high and stable ionic conductivity at reduced temperatures are strongly desired.

To address this challenge, an efficacious approach based on semiconducting ionic conductors has been proposed very recently to replace the conventional electrolyte YSZ and SDC [9–15]. The developed materials have exhibited extraordinarily high ionic conductivity superior to those of YSZ and SDC, showing tremendous potential as membrane layer in LT-SOFCs [14,15]. For instance, a breakthrough study on $SmNiO_3$ reported a high protonic conductivity in such perovskite semiconductor that compare favorably with those of best-performing solid electrolytes. The corresponding SOFC with Pt/$SmNiO_3$/Pt geometry demonstrated dramatic power output of 225 mW cm^{-2} at 500 °C [16]. Tao et al. also demonstrated that good proton conduction (0.1 S cm^{-1} at 500 °C) can be obtained in semiconductor $Li_xAl_{0.5}Co_{0.5}O_2$ [17]. Our previous work also detected high ionic transport in a natural hematite (α-Fe_2O_3) and applied the semiconducting hematite electrolyte into SOFC, observing an impressive power density of 467 mW cm^{-2} at 600 °C [18,19]. In addition to these single phase semiconductors, high ionic conduction is also found in hetero-structured materials. Garcia-Barriocanal et al. reported a colossal ionic conduction at the interfaces of ionic conductor/semiconductor hetero-structure YSZ/$SrTiO_3$, indicating that substantial ionic conductivity can be achieved even close to room temperature [20]. A series of composite materials consisting of semiconductors and ionic conductors such as $Li_{0.15}Ni_{0.45}Zn_{0.4}O_x$/SDC and SDC/$Na_2CO_3$-$Sr_2Fe_{1.5}Mo_{0.5}O_{6-\delta}$ were also applied as membranes in SOFCs, revealing significantly enhanced ionic conductivity as compared to single phase ionic conductors [21–24]. A new fuel cell technology, named as electrolyte-layer free fuel cell (EFFC) or semiconductor-ion membrane fuel cell (SIMFC) designed by energy band alignment and perovskite solar cell principle has been proposed to realize better integration and functionality of these materials [11,14]. Such type of cell device is assembled using $Ni_{0.8}Co_{0.15}Al_{0.05}$Li-oxide (NCAL) as electrodes into a typical configuration similar to perovskite solar cell: NCAL (ETL)/semiconducting ionic conductor (function layer)/NCAL (HTL) (ETL and HTL mean electron transport layer and hole transport layer), managing to achieve better fuel cell performances in a simpler device architecture [22].

The semiconductor ZnO has gained substantial interest in the research community and industrial applications because of its peculiar properties, such as excellent thermal stability, good oxidation resistance, considerable optoelectronic properties, and band gap in the near ultraviolet [25]. It is well known not only as a versatile semiconductor but also as a probable oxygen-ion conductor due to the enrichment of oxygen vacancies at high temperature [26]. It has been reported that the oxygen vacancy is a deep donor in ZnO with a (2+/0) transition level at ~1.0 eV below the bottom of the conduction band [27]. Liu et al. observed that addition of 0.5 wt % ZnO increased the ionic conductivity of YSZ by as much as 120% at 800 °C [28]. Furthermore, protons also may exist in ZnO and doped ZnO due to the fact that hydrogen is easily ionized to protons in oxide lattice. As reported, it is detected the concentration of protons in ZnO increases with elevating temperature [29]. Economically, ZnO is a cost-effective material in practical applications, for that it is able to be synthesized by remarkably simple crystal-growth technologies. Therefore, taking advantage of the properties of ZnO and following the above strategy, this work accesses the utility of ZnO-based materials for electrolytes in LT-SOFC. Two types of fuel cells are fabricated using pure ZnO and ZnO-LCP (La/Pr-doped CeO_2) composite as membrane layer, respectively, sandwiched between two NCAL electrodes. The structure, morphology and electrical properties of the materials are investigated. The performances of the cells are evaluated within a low temperature range of 450–550 °C.

Materials **2018**, *11*, 40

2. Experimental Section

ZnO powders were obtained through a simple pre-sintering of commercial ZnO at 650 °C for 2 h. The sintered powders were ground thoroughly for electrolyte uses and further composite preparation. LCP ($La_{0.33}Ce_{0.62}Pr_{0.05}O_{2-\delta}$) powder was synthesized through an 800 °C heat treatment of LaCePr-carbonates, which is a mixture of lanthanum, cerium, and praseodymium carbonates. Afterwards, ZnO-LCP composite was prepared by blending the sintered ZnO with LCP in a mass composition of 1:1. The resultant mixture was heated again at 800 °C for 2 h and ground completely to obtain the eventual homogeneous ZnO LCP composite material. The commercial ZnO was purchased from Sigma Aldrich, Shanghai, China, and the raw material LaCePr-carbonate was obtained from a rare-earth company in Baotou, China. Additionally, the electrode material NCAL was processed in a slurry form by mixing NCAL powder with terpineol solvent. The resultant slurry was pasted onto Ni-foam and desiccated at 120 °C for use as an electrode and current collector. The commercial NCAL was purchased from Tianjin Damo Science and Technology Joint Stock Ltd., Tianjin, China.

Regarding the fabrication of fuel cells, two fuel cell devices were assembled based on ZnO electrolyte and ZnO-LCP composite, respectively, with two pieces of Ni-foam pasted by NCAL on both sides in each case; subsequently the three layers were pressed uniaxially under a load of 200–250 MPa into one tablet. For comparison purpose, a device based on LCP electrolyte was also fabricated in the same configuration. The resulting fuel cell devices, NCAL/ZnO/NCAL, NCAL/ZnO-LCP/NCAL, and NCAL/LCP/NCAL are roughly 2 mm in thickness and 13 mm in diameter (active area of 0.64 cm^2). All devices, were pre-treated using an in situ heating step at 600 °C for 1 h after being mounted into the testing setup, before performance measurements at 450–550 °C.

The crystal structures of samples were studied using a Bruker D8 Advanced X-ray diffractometer (XRD, Bruker Corporation, Karlsruhe, Germany) with Cu Kα (λ = 1.54060 Å) as the source, with tube voltage at 45 kV and current of 40 mA. The particle morphology of powder samples, cross section and elemental composition of fuel cell device were investigated using a JEOL JSM7100F field emission scanning electron microscope (FE-SEM, Carl Zeiss, Oberkochen, Germany) under an accelerating voltage of 200 kV, and the equipped energy dispersive spectrometer (EDS, Carl Zeiss, Oberkochen, Germany) that operated at 15 kV.

The electrochemical impedance spectra (EIS) of fuel cells were measured in H$_2$/air atmosphere using an electrochemical work station (Gamry Reference 3000, Gamry Instruments, Warminster, PA, USA). The measurement was performed under open circuit voltage (OCV) conditions, and the applied frequency range was 0.1–10^6 Hz with a AC signal voltage of 10 mV in amplitude. The performance measurements for fuel cells were carried out on a programmable electronic load instrument (IT8511, ITECH Electrical Co., Ltd., Nanjing, China) at 450–550 °C with humidified hydrogen as fuel (120–140 mL min^{-1}) and air as the oxidant (120–150 mL min^{-1}).

3. Results and Discussion

3.1. Crystalline Structure and Morphology

The XRD patterns of the prepared ZnO, LCP and ZnO-LCP composite are presented in Figure 1. The XRD of ZnO displays a series of characteristic diffraction peaks that correspond to the (100), (002), (101), (102), (110), (103) and (112) planes in JCPDS File No. 36-1451, which can be well indexed to the typical hexagonal wurtzite structure of zinc oxide [30]. The XRD diagram of LCP is characteristic of a cubic fluorite structure, with a slight shift to lower angle compared with the standard pattern of ceria (JCPDS File No. 34-0394), echoing the fact that LCP is a La/Pr co-doped CeO$_2$ as reported previously [31]. In the diffractogram of ZnO-LCP composite, all diffraction peaks emerged can be assigned to wurtzite phase of ZnO and cubic fluorite phase of LCP, no extra phases and peak shift could be identified, which confirms that no chemical interaction occurred between the two materials. Compared to the XRD patterns of ZnO and LCP, the composite sample shows less intense peaks, revealing a larger full width at half maximum (FWHM) of the characteristic peak. According to the Scherrer equation $D = \frac{K\gamma}{B\cos\theta}$, it can be calculated that the average grain size (D) of the material decreased from 26 (ZnO) and 15 (LCP) to 12 nm (ZnO-LCP). Moreover, the interplanar spacing values and lattice parameters of ZnO and LCP from the XRD analysis are given in Table 1.

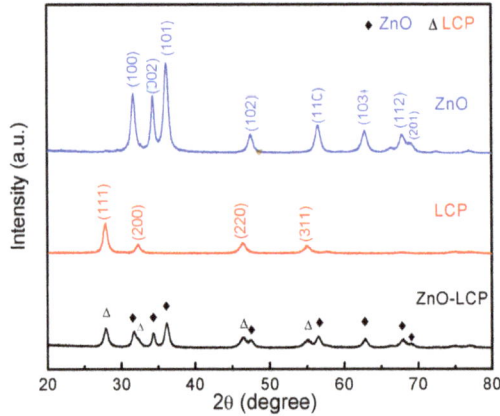

Figure 1. XRD patterns of the prepared ZnO, LCP and ZnO-LCP composite.

Table 1. The lattice parameters of ZnO and LCP.

Sample	d Spacing (nm)	Lattice Constant (nm)
ZnO	0.2485 (101) plane 0.2612 (002) plane	a = b = 0.3243 c = 0.5205
LCP	0.3203 (111) plane	a = b = c = 0.5470

The micro-structure of the resultant materials and fuel cell device were investigated by SEM. Figure 2a,b shows the recorded morphology of ZnO particles and ZnO-LCP particles. As can be seen, both ZnO and ZnO-LCP exhibit nano-sized particles and irregular shape particles, which is owing to the utilization of commercial and industrial-grade materials without elaborate control of nano-structure, while the composite material appears to be made up of more condensed particles with better distributions. Moreover, the average grain/particle size of ZnO-LCP were found to be smaller than that of ZnO, which is in good agreement with the XRD result. The observed size values are larger than the calculated grain size results according to the Scherrer equation, indicating that small grains aggregated in the materials and formed larger particles.

Figure 2c illustrates a cross-sectional SEM image of the NCAL/ZnO-LCP/NCAL cell after operation, clearly displaying three individual layers consisting of a membrane layer with thickness of 550 µm and two porous NCAL-Ni electrode layers. The ZnO-LCP membrane layer adheres well with the NCAL-Ni layer without any delamination trace even after scissoring treatment for SEM characterization, as an indication of satisfactory mechanical strength of the device. Three high-magnification images of the NCAL-Ni layer, NCAL particles and the intermediate layer are further presented in Figure 2d,e, respectively. Figure 2d shows a clear view of the NCAL particles located in the three dimension woven structure of Ni-foam, forming a porous structure, which is ulteriorly confirmed in Figure 2e. The particle size of NCAL ranges from 50 to 200 nm. Additionally, as the core component of the fuel cell, ZnO-LCP layer exhibits a gas-tight structure in Figure 2f while no distinct sign of cracking can be observed. Apart from blocking gas leakage/crossover during operation, this dense layer can also ensure fast ion transport and thus aid in reducing the internal resistance of the single cell.

Figure 2. SEM images of (**a**) the resultant ZnO power and (**b**) ZnO-LCP composite; (**c**) Cross-sectional morphology of ZnO-LCP cell; (**d**) NCAL-Ni electrode; (**e**) NCAL particles and (**f**) the intermediate ZnO-LCP layer after operation.

In order to study the compatibility of the cathode and electrolyte membrane materials, Figure 3a presents a SEM image of the NCAL/ZnO-LCP/NCAL cell focusing on the membrane/cathode interface region after operation, and Figure 3b gives a detailed morphology on the basis of Figure 3a, which verifies the dense layer of ZnO-LCP membrane and porous structure of cathode again. The elemental mappings for Zn (elements only from membrane) and Ni (element only from electrode) in Figure 3c,d clearly distinguish the interface between membrane and cathode. These results indicate that ZnO-LCP membrane layers were well bonded with the NCAL cathode layer during operations without any interlaminar separation or fissure, revealing a good thermal compatibility between the membrane materials and cathode materials. Figure 3e–h further give the EDS analysis results based on the cross-section of the ZnO-LCP cell after operation. The borders between membrane layer and electrode are clear and uniform, confirmed by the elemental mappings of Zn, Ce and Ni. It reflects that there was no obvious elemental interdiffusion or segregation occurred at the interfaces during operation, which excludes the possibility of any undesired secondary reaction, and thus confirms the good chemical compatibility of the electrode and electrolyte materials.

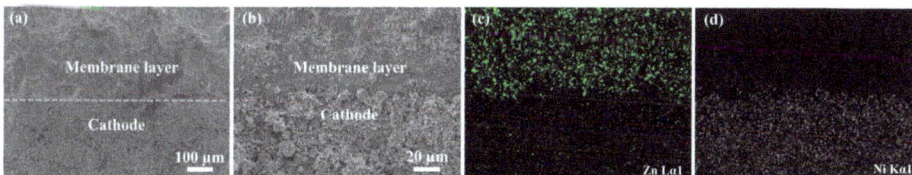

Figure 3. *Cont.*

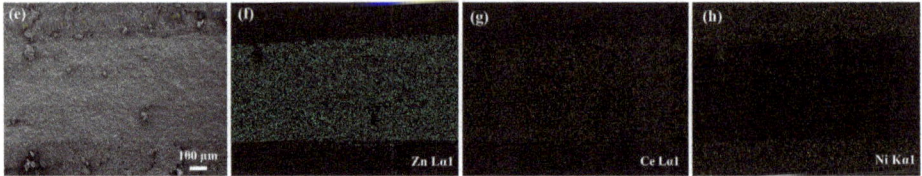

Figure 3. (**a**) SEM image and (**b**) detailed morphology of the membrane/cathode interface for ZnO-LCP cell after operation, and the corresponding elemental mappings for (**c**) Zn and (**d**) Ni; (**e**) Electron image of the cross-section for ZnO-LCP cell; and the corresponding elemental mappings for (**f**) Zn; (**g**) Ce and (**h**) Ni after operation.

3.2. Electrochemical Performance

Figure 4 shows the typical current voltage (I-V) and current-power (I-P) characteristics for the three fuel cells based on ZnO, LCP, and ZnO-LCP electrolytes, respectively at 550 °C. From Figure 3a, it can be observed that ZnO cell delivers an OCV of 1.06 V and a maximum power density of 482 mW cm^{-2}, slightly lower than that of LCP cell with 540 mW cm^{-2} in peak power density. This is the first demonstration of ZnO in fuel cell device that shows considerable performance at low temperature. It suggests the tremendous potential of ZnO from scientific and technological as well as applied perspectives for electrolyte uses. We also note that the performance is comparable to that of a newly reported thin-film SOFC based on YSZ/GDC (Gd-doped CeO$_2$) bi-layer electrolyte [32], and even superior to some other SOFCs using ceria-based electrolytes [33,34]. In the case of the ZnO-LCP cell, a higher OCV of 1.07 V and significantly enhanced power density of 864 mW cm^{-2} were attained at 550 °C compared to other two cells. The power output manifested an almost 2-fold increment over that of ZnO cell. This sharp enhancement is most likely explained by the enhanced ionic conductivity in ZnO-LCP composite through interfacial conduction effect, as confirmed formerly in a number of semiconducting ionic conductors [15,19,35,36].

Based upon the above investigations, the ZnO-based cells were further assessed at reduced temperatures from 450 to 525 °C, with a 30-min dwelling time at each testing point to stabilize the cell. As shown in Figure 4b, the ZnO cell exhibits boosted power output from 158 to 380 mW cm^{-2} at 475 to 525 °C along with a mildly raised OCV from 1 to 1.03 V. Within the same temperature range, the ZnO-LCP cell demonstrates appreciable power outputs, reaching 390, 625 and 794 mW cm^{-2} at 475, 500 and 525 °C, respectively, while the OCV fluctuates in the 1.08~1.1 V window, as shown in the inset of Figure 4b. The enhancements of power density are mainly due to the thermally activated ion transport in Zno and ZnO-LCP with the rise in temperature. The achieved high OCVs can be ascribed to the excellent catalytic activity of NCAL, which has been reported as an efficient catalyst for both anode and cathode with superior triple O^{2-}/H$^+$/e$^-$ conduction [37], and the junction effect of the device [38]. It needs to be emphasized that the junction effect is based on a Schottky junction formed between the Co/Ni alloy layer, which was originated from the anodic NCAL via reduction reaction, and the intermediate ZnO or ZnO-LCP semiconductor layer. The Schottky barrier field in the junction points from alloy layer to semiconductor layer, playing a crucial role in preventing electrons in H$_2$ supply side from passing through the device [11,14]. Consequently, though there is a significant electronic conduction in the ZnO-based electrolytes, high OCVs can be still obtained by the cells. In addition, the two cell devices were further operated at lower temperature, observed is that both cells failed to reach a sufficient OCV at 450 °C, which is chiefly due to the poor catalytic activity of the NCAL electrodes at too low temperatures [39]. However, it still can be concluded from current initial results that both ZnO and ZnO-LCP composite can function well as electrolyte membrane layer in LT-SOFCs.

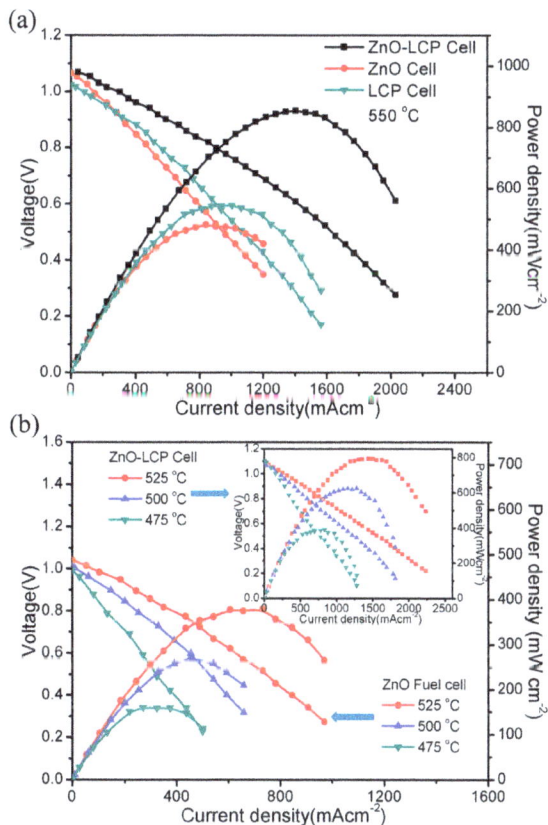

Figure 4. (**a**) Electrochemical performance for fuel cells based on ZnO, LCP and ZnO-LCP at 550 °C for comparison (**b**) Low-temperature performance of fuel cells with ZnO and ZnO-LCP at various temperatures.

3.3. Electrical Conductivity

To understand the excellent electrochemical performances of ZnO-based fuel cells, the conductivity of the used ZnO-based materials were studied. As reported, the linear part in the central region of I-V characteristic curve mainly reflects the ohmic loss of electrolyte in a SOFC [40,41], thus ohmic resistances of the ZnO and ZnO-LCP layer can be estimated from the slope of I-V curves, from which the ionic conductivity (σ) that contributes to cell performance can be calculated according to $\sigma = \frac{L}{R \times S}$, where L is the thickness of the electrolyte layer, S denotes the effective area, and R represents the resistance. Figure 5a shows the estimated ionic conductivities for the used ZnO electrolyte and ZnO-LCP composite as a function of temperatures. The ionic conductivity of ZnO is 0.037 S cm^{-1} at 475 °C and increases to 0.09 S cm^{-1} at 550 °C. This result is slightly ahead of those of well-known oxygen ion electrolytes YSZ, GDC, Mg-doped LaGaO$_3$ (LSGM), and typical proton electrolytes BaCe$_{0.5}$Y$_{0.5}$O$_{3-\delta}$ (BCY) and BaZr$_{0.1}$Ce$_{0.7}$Y$_{0.2}$O$_{3-\delta}$ (BZCY) in previous reports [42,43]. ZnO-LCP reveals a significantly promoted ionic conductivity of 0.082 S cm^{-1} at 475 °C, which then reached as high as 0.156 S cm^{-1} at 550 °C. The corresponding activation energy (E$_a$) for ionic conduction can be obtained based on Arrhenius relationship $\sigma = \frac{A}{T} \exp\left(-\frac{E_a}{kT}\right)$, in which T is the absolute temperature, A is a pre-exponential factor, and k represents the Boltzmann constant. As presented in Figure 5b, the E$_a$ for ionic conduction of ZnO and ZnO-LCP at 475–550 °C are 0.7 and 0.51 eV, respectively, showing smaller values than those of pure O^{2-} conductors YSZ and LSGM. Particularly, the F_a for ZnO is close to the reported activation energies of pure proton conductors BCY and BZCY, which are in the scope of 0.66–0.78 eV [44]. Consider that protons in solid proton electrolytes generally require much lower activation energy to motivate their transport than oxygen ions, it is speculated that the ZnO-based electrolytes possess hybrid proton and oxygen ion conduction, thus resulting in a coupling lower E_a.

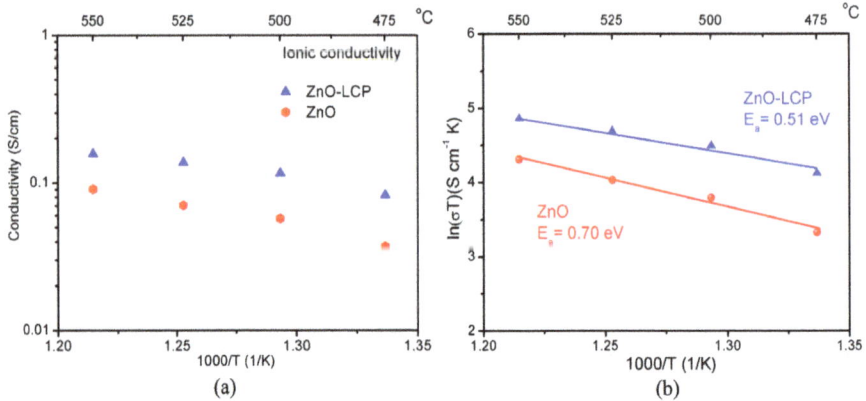

Figure 5. (**a**) Ionic conductivities of ZnO electrolyte and ZnO-LCP composite estimated from I-V curve result; (**b**) The corresponding activation energy of the ionic conductivities for ZnO and ZnO-LCP composite.

To verify the speculation, an additional experiment was undertaken to test the proton conductive behavior of ZnO electrolyte by using a O^{2-} blocking fuel cell in configuration of NCAL/BZCY/ZnO/BZCY/NCAL. Since BZCY is a state-of-the-art proton conductor with major proton conduction and negligible oxygen ion and electron conduction [45,46], the trilayer BZCY/ZnO/BZCY membrane would primarily transport protons from the anode side to cathode side while filtering out oxygen ions and electrons. By this means, the proton conductive property of ZnO can be confirmed from the performance of this multilayer cell device. This method has been reported for testing proton-related properties and conductivities of specific materials [47,48]. Figure 6a illustrates the cross-section of the device characterized by SEM after performance measurements and the corresponding elemental mappings from EDS test. As can be seen, five layers of the NCAL/BZCY/ZnO/BZCY/NCAL architecture can be clearly distinguished in SEM. This is ulteriorly identified by the elemental mappings of Ni, Zn and Ba in Figure 5a, which are exclusive from the NCAL, ZnO and BZCY layer, respectively. Few cracks are detected in the two BZCY layer membranes, probably resulting from scissoring the cross-section for SEM measurement. Figure 6b shows the cell electrochemical performance at 550 °C, exhibiting a maximum power density of 235 mW cm^{-2} with an OCV of 1.05 V. This result reflects only proton transport contribution to the power output. Therefore, from the I-V curve the proton conductivity of ZnO was calculated to be 0.05 S cm^{-1} at 550 °C. Compared to the total ionic conductivity of ZnO (0.09 S cm^{-1} at 550 °C), this value (0.05 S cm^{-1} at 550 °C) indicates that the used ZnO electrolyte might be a hybrid O^{2-}/H^+ conductor, where proton conduction dominates the total ionic conduction. However, with the multilayer configuration, the additional layers and interfaces would induce more power losses, which means the calculated value for the proton conductivity is smaller than the actual value. Therefore, it is more likely that the used ZnO electrolyte is a pure proton conductor rather than mixed proton and oxygen-ion conductor, as reported in previous study [29]. Such conductive behavior could account for the low activation energies of the materials. Our study thus confirms the proton conductive property of ZnO.

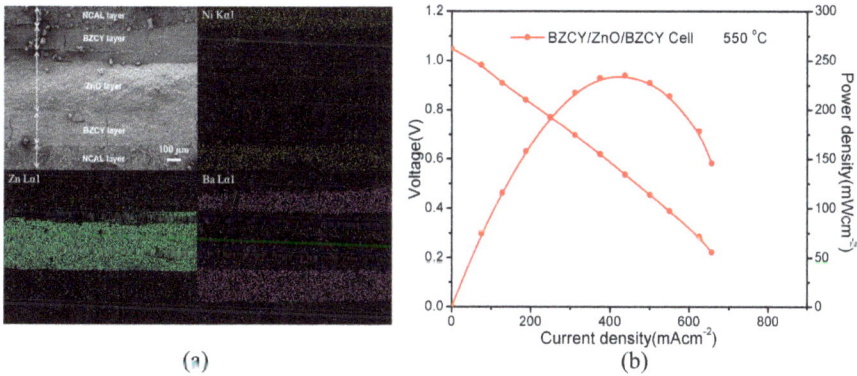

(a)   (b)

Figure 6. (**a**) A cross-sectional SEM image of the prepared NCAL/BZCY/ZnO/BZCY/NCAL fuel cell after operation and corresponding elemental mapping results for Ni, Zn and Ba; (**b**) Electrochemical performance of the NCAL/BZCY/ZnO/BZCY/NCAL fuel cells tested at 550 °C.

3.4. Impedance Spectroscopy Analysis

Notably, the ZnO-based fuel cell exhibited significantly enhanced performance by incorporating ionic conducting LCP to form a composite. For comparative study of the electrochemical processes between our ZnO-based composite fuel cell and conventional doped-CeO_2-based fuel cell, impedance spectroscopy analysis was carried out for the two types of cells. Figure 7 presents the EIS results of LCP cell and ZnO-LCP cell acquired in H_2/air at 525 and 550 °C. In each impedance spectrum, the intercept with the real axis at high frequencies reflects the bulk resistance, the semicircle at intermediate frequencies represents the grain-boundary process, while the semicircle at low frequencies region corresponds to the electrode polarization behavior [41,49]. An intuitive comparison from the curves indicates that the EIS for ZnO-LCP cell have smaller high-frequency intercepts and smaller semicircles than LCP cell. We employed an empirical equivalent circuit model of $LR_b(R_{gb}Q_{gb})(R_eQ_e)$ to fit the EIS data to get internal resistances information, in which L is inductance of the instrument leads and current collectors, R_b, R_{gb} and R_e stand for bulk resistance, grain boundary resistance and electrode polarization resistance respectively, and Q is the constant phase element (CPE) representing a non-ideal capacitor. Thereby, $R_{gb}Q_{gb}$ and R_eQ_e denote the semicircles of grain boundary conduction and electrode polarization process, respectively.

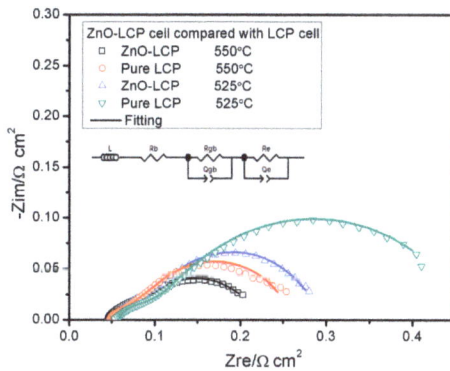

Figure 7. Impedance spectra for ZnO-LCP fuel cell and LCP fuel cell measured in H_2/air at two different temperatures and the corresponding fitting lines. The inset is equivalent circuit adopted for fitting the EIS data.

The simulated parameters extracted from the fitting results are summarized in Table 2. It can be discerned that the R_b of ZnO-LCP cell shows slightly smaller R_b than that of LCP cell at both 525 and 550 °C. With respect to the R_{gb}, ZnO-LCP exhibits evidently reduced values of 0.034 Ω cm^2 (550 °C) and 0.045 Ω cm^2 (525 °C) as compared to LCP. This partly manifests that the migration of ions at the grain boundary is less resistive in the composite. We attributed this phenomenon to the heterophasic interfacial conduction effect at semiconductor oxide/ionic conductor oxide interface regions in hetero-structured composite material [19,50]. Such behavior has been reported in many semiconducting ionic systems, such as La$_{0.6}$Sr$_{0.4}$Co$_{0.2}$Fe$_{0.8}$O$_{3-\delta}$-Ca$_{0.04}$Ce$_{0.8}$Sm$_{0.16}$O$_{2-\delta}$ (LSCF-SCDC), YSZ-SrTiO$_3$, and CoFe$_2$O$_4$-GDC [15,20,51]. Furthermore, greater difference is observed between the R_e of the two devices, whereby ZnO-LCP possesses smaller R_e with values of 0.144 Ω cm^2 at 550 °C and 0.197 Ω cm^2 at 525 °C. Since these electrolyte/electrode interfacial polarization resistances often cause significant power losses in SOFCs, the smaller R_e of ZnO-LCP would help in attaining higher power output of the cell. The above results regarding internal resistances commendably interpret the promoted performances in ZnO-LCP composite fuel cell.

Table 2. The equivalent circuit analysis results of ZnO-LCP and LCP samples at 525 and 550 °C, the R and Q have a unit of Ω cm^2 and S Sec^n cm^{-2}, respectively.

Sample	T	R_b	R_{gb}	Q_{gb}	n	R_e	Q_e	n	Chi Squared
ZnO-LCP	550 °C	0.046	0.034	0.610	0.6362	0.144	2.810	0.6307	1.675×10^{-4}
LCP		0.048	0.042	0.820	0.506	0.173	1.650	0.7092	6.063×10^{-4}
ZnO-LCP	525 °C	0.052	0.045	0.472	0.6312	0.197	1.325	0.7296	1.696×10^{-4}
LCP		0.054	0.054	0.277	0.5322	0.370	1.164	0.6196	8.338×10^{-4}

Figure 8 displays the EIS of ZnO-LCP measured in H$_2$/air at various temperatures. The EIS curves present in a form of flat-shaped arc or semicircle, because of the mixed electron and ion conductive behavior in the composite. With the testing temperature increases from 475 to 550 °C, the bulk resistance drops slightly from 0.072 to 0.046 Ω cm^2, while the polarization resistance which is reflected by the intercept of arc or semicircle on the real axis reveals a dramatic shrunken tendency. For instance, at 475 °C the polarization resistance of the cell is about 1.2 Ω cm^2 whereas the bulk resistance is smaller by one order of magnitude. It is also noted that the polarization resistance at 475 °C are far greater than those at temperatures over 500 °C. This should arise from the fact that both catalytic activity of electrode and ionic conduction of electrolyte require a sufficient thermal condition to realize their functions, suggesting that the currently designed ZnO-based cells are more applicable to operate at over 500 °C. Clearly, on the one hand, the above results prove the operational feasibility of ZnO-based electrolyte fuel cells at LT. On the other hand, it signifies that the 300–475 °C LT operation of the cell remains a huge challenge, which requires more scientific and technological studies on the materials.

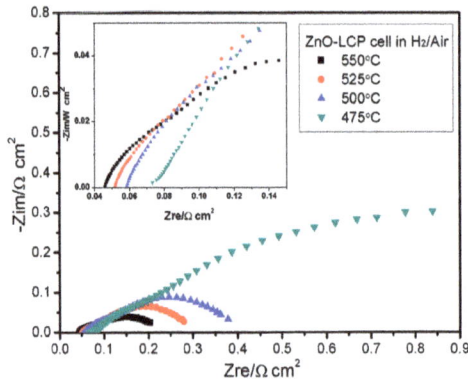

Figure 8. Impedance spectra of ZnO-LCP fuel cell acquired in H$_2$/air at various temperatures.

4. Conclusions

In summary, two zinc oxide-based electrolyte materials, pure ZnO and ZnO-LCP composite, have been developed for LT-SOFC applications for the first time. The two types of fuel cells based on pure ZnO and ZnO-LCP composite exhibited excellent power outputs of 482 and 864 mW cm^{-2} at low temperature of 550 °C, respectively. On this basis, our investigation found that ZnO electrolyte possessed decent ionic conductivity of 0.09 S cm^{-1} at 550 °C along with activation energy of 0.70 eV, while ZnO-LCP composite exhibited promoted ionic conductivity of 0.156 S cm^{-1} at the same temperature with low activation energies of 0.51 eV. These results are ahead of some standard electrolytes in previous reports. More profoundly, the proton conductive property of ZnO was detected using an oxygen-ion blocking fuel cell, showing a considerable proton conductivity of 0.05 S cm^{-1} at 550 °C. Besides, the improved performance and electrochemical processes of the ZnO-LCP cell were investigated through impedance spectra measurements. The improvements are discovered to be majorly owing to the reduced grain boundary and electrode polarization resistances. These findings suggest that zinc oxide-based semiconductors and composites are attractive materials for developing new electrolyte membranes for LT-SOFCs. It deserves more investigation into the synthesis methods and electrochemical properties regarding the electron/ion coupling effect of the materials as well as device working principle.

Acknowledgments: C.X. and Z.Q. contributed equally to this work. This work was supported by the National Natural Science Foundation of China (Grant No. 51502084 and 51372075), the Natural Science Foundation of Hubei Province (Grant No. 2015CFA120), the Swedish Research Council (Grant No. 621-2011-4983), and the European Commission FP7 TriSOFC-project (Grant No. 303454).

Author Contributions: C.X. and B.Z. conceived and designed the experiments; Z.Q., C.X., C.F. and B.W. performed the experiments; C.X. and Z.Q. analyzed the data; B.W. and B.Z. contributed the used materials and analysis tools; C.X., Z.Q. and J.K. wrote the paper.

Conflicts of Interest: The authors declare no conflict of interest.

References

1. Ormerod, R.M. Solid oxide fuel cells. *Chem. Soc. Rev.* **2003**, *32*, 17–28. [CrossRef] [PubMed]
2. Dyer, C.K. Replacing the battery in portable electronics. *Sci. Am.* **1999**, *281*, 88–93. [CrossRef]
3. Wang, W.; Su, C.; Wu, Y.; Ran, R.; Shao, Z. Progress in solid oxide fuel cells with nickel-based anodes operating on methane and related fuels. *Chem. Rev.* **2013**, *113*, 8104–8151. [CrossRef] [PubMed]
4. Chen, Y.; Zhou, W.; Ding, D.; Liu, M.; Ciucci, F.; Tade, M.; Shao, Z. Advances in cathode materials for solid oxide fuel cells: Complex oxides without alkaline earth metal elements. *Adv. Energy Mater.* **2015**, *5*, 1500537. [CrossRef]
5. Atkinson, A.; Barnett, S.; Gorte, R.J.; Irvine, J.T.S.; McEvoy, A.J.; Mogensen, M.; Singhal, S.C.; Vohs, J. Advanced anodes for high-temperature fuel cells. *Nat. Mater.* **2004**, *3*, 17–27. [CrossRef] [PubMed]
6. Steele, B.C.H. Appraisal of Ce$_{1-y}$Gd$_y$O$_{2-y/2}$ Electrolytes for ITSOFC Operation at 500 °C. *Solid State Ion.* **2000**, *129*, 95–110. [CrossRef]
7. Singhal, S.C. Advances in solid oxide fuel cell technology. *Solid State Ion.* **2000**, *135*, 305–313. [CrossRef]
8. Andersson, D.A.; Simak, S.I.; Skorodumova, N.V.; Abrikosov, I.A.; Johansson, B. Optimization of ionic conductivity in doped ceria. *Proc. Natl. Acad. Sci. USA* **2006**, *103*, 3518–3521. [CrossRef] [PubMed]
9. Zhu, B.; Raza, R.; Abbas, G.; Singh, M. An electrolyte-free fuel cell constructed from one homogenous layer with mixed conductivity. *Adv. Funct. Mater.* **2011**, *21*, 2465–2469. [CrossRef]
10. Zhu, B.; Raza, R.; Qin, H.; Liu, Q.; Fan, L. Fuel cells based on electrolyte and non-electrolyte separators. *Energy Environ. Sci.* **2011**, *4*, 2986–2992. [CrossRef]
11. Wang, B.; Cai, Y.; Xia, C.; Kim, J.S.; Liu, Y.; Dong, W.; Wang, H.; Afzal, M.; Li, J.; Raza, R.; et al. Semiconductor-ionic membrane of lasrcofe-oxide-doped ceria solid oxide fuel cells. *Electrochim. Acta* **2017**, *248*, 496–504. [CrossRef]
12. Zagórski, K.; Wachowski, S.; Szymczewska, D.; Mielewczyk-Gryń, A.; Jasiński, P.; Gazda, M. Performance of a single layer fuel cell based on a mixed proton-electron conducting composite. *J. Power Sources* **2017**, *353*, 230–236. [CrossRef]

13. Xia, C.; Wang, B.; Ma, Y.; Cai, Y.; Afzal, M.; Liu, Y.; He, Y.; Zhang, W.; Dong, W.; Li, J.; et al. Industrial-grade rare-earth and perovskite oxide for high-performance electrolyte layer-free fuel cell. *J. Power Sources* **2016**, *307*, 270–279. [CrossRef]

14. Zhu, B.; Wang, B.; Wang, Y.; Raza, R.; Tan, W.; Kim, J.S.; van Aken, P.A.; Lund, P. Charge separation and transport in $La_{0.6}Sr_{0.4}Co_{0.2}Fe_{0.8}O_{3-\delta}$ and ion-doping ceria heterostructure material for new generation fuel cell. *Nano Energy* **2017**, *37*, 195–202. [CrossRef]

15. Wang, B.; Wang, Y.; Fan, L.; Cai, Y.; Xia, C.; Liu, Y.; Raza, R.; van Aken, P.A.; Wang, H. Preparation and characterization of Sm and Ca co-doped ceria-$La_{0.6}Sr_{0.4}Co_{0.2}Fe_{0.8}O_{3-\delta}$ semiconductor-ionic composites for electrolyte-layer-free fuel cells. *J. Mater. Chem. A* **2016**, *4*, 15426–15436. [CrossRef]

16. Zhou, Y.; Guan, X.; Zhou, H.; Ramadoss, K.; Adam, S.; Liu, H.; Lee, S.; Shi, J.; Tsuchiya, M.; Fong, D.D.; et al. Strongly correlated perovskite fuel cells. *Nature* **2016**, *534*, 231–234. [CrossRef] [PubMed]

17. Lan, R.; Tao, S. Novel proton conductors in the layered oxide material $Li_xlAl_{0.5}Co_{0.5}O_2$. *Adv. Energy Mater.* **2014**, *4*, 1301683. [CrossRef]

18. Wu, Y.; Xia, C.; Zhang, W.; Yang, X.; Bao, Z.Y.; Li, J.J.; Zhu, B. Natural Hematite for Next-Generation Solid Oxide Fuel Cells. *Adv. Funct. Mater.* **2016**, *26*, 930–942. [CrossRef]

19. Xia, C.; Cai, Y.; Ma, Y.; Wang, B.; Zhang, W.; Karlsson, M.; Wu, Y.; Zhu, B. Natural mineral-based solid oxide fuel cell with heterogeneous nanocomposite derived from hematite and rare-earth minerals. *ACS Appl. Mater. Interfaces* **2016**, *8*, 20748–20755. [CrossRef] [PubMed]

20. Garcia-Barriocanal, J.; Rivera-Calzada, A.; Varela, M.; Sefrioui, Z.; Iborra, E.; Leon, C.; Pennycook, S.J.; Santamaria, J. Colossal ionic conductivity at interfaces of epitaxial ZrO_2:Y_2O_3/$SrTiO_3$ heterostructures. *Science* **2008**, *321*, 676–680. [CrossRef] [PubMed]

21. Zhu, B.; Fan, L.; Zhao, Y.; Tan, W.; Xiong, D.; Wang, H. Functional semiconductor-ionic composite GDC-KZnAl/LiNiCuZnO$_x$ for single-component fuel cell. *RSC Adv.* **2014**, *4*, 9920–9925. [CrossRef]

22. Zhu, B.; Huang, Y.; Fan, L.; Ma, Y.; Wang, B.; Xia, C.; Afzal, M.; Zhang, B.; Dong, W.; Wang, H.; et al. Novel fuel cell with nanocomposite functional layer designed by perovskite solar cell principle. *Nano Energy* **2016**, *19*, 156–164. [CrossRef]

23. Li, P.; Yu, B.; Li, J.; Yao, X.; Zhao, Y.; Li, Y. A single layer solid oxide fuel cell composed of La_2NiO_4 and doped ceria-carbonate with H_2 and methanol as fuels. *Int. J. Hydrogen Energy* **2016**, *41*, 9059–9065. [CrossRef]

24. Dong, X.; Tian, L.; Li, J.; Zhao, Y.; Tian, Y.; Li, Y. Single layer fuel cell based on a composite of $Ce_{0.8}Sm_{0.2}O_{2-\delta}$-$Na_2CO_3$ and a mixed ionic and electronic conductor $Sr_2Fe_{1.5}Mo_{0.5}O_{6-\delta}$. *J. Power Sources* **2014**, *249*, 270–276. [CrossRef]

25. Chang, S.J.; Duan, B.G.; Hsiao, C.H.; Liu, C.W.; Young, S.J. UV enhanced emission performance of low temperature grown Ga-doped ZnO nanorods. *IEEE Photonics Technol. Lett.* **2014**, *26*, 66–69. [CrossRef]

26. Carrasco, J.; Lopez, N.; Illas, F. First principles analysis of the stability and diffusion of oxygen vacancies in metal oxides. *Phys. Rev. Lett.* **2004**, *93*, 225502. [CrossRef] [PubMed]

27. Janotti, A.; Van de Walle, C.G. Oxygen vacancies in ZnO. *Appl. Phys. Lett.* **2005**, *87*, 122102. [CrossRef]

28. Liu, Y.; Lao, L.E. Structural and electrical properties of ZnO-doped 8 mol% yttria-stabilized zirconia. *Solid State Ion.* **2006**, *177*, 159–163. [CrossRef]

29. Norbya, T. Proton conduction in oxides. *Solid State Ion.* **1990**, *40*, 857–862. [CrossRef]

30. Suwanboon, S.; Amornpitoksuk, P.; Haidoux, A.; Tedenac, J.C. Structural and optical properties of undoped and aluminium doped zinc oxide nanoparticles via precipitation method at low temperature. *J. Alloys Compd.* **2008**, *462*, 335–339. [CrossRef]

31. Xia, C.; Wang, B.; Cai, Y.; Zhang, W.; Afzal, M.; Zhu, B. Electrochemical properties of LaCePr-oxide/K_2WO_4 composite electrolyte for low-temperature SOFCs. *Electrochem. Commun.* **2017**, *77*, 44–48. [CrossRef]

32. Kim, H.J.; Kim, M.; Neoh, K.C.; Han, G.D.; Bae, K.; Shin, J.M.; Kim, G.T.; Shim, J.H. Slurry spin coating of thin film yttria stabilized zirconia/gadolinia doped ceria bi-layer electrolytes for solid oxide fuel cells. *J. Power Sources* **2016**, *327*, 401–407. [CrossRef]

33. Liu, Y.H.; Yin, C.Q.; Wang, L.H.; Li, D.B.; Lian, J.S.; Hu, J.D.; Guo, Z.X. Properties of a ceria-based ($C_6S_2G_2$) solid oxide electrolyte sintered with Al_2O_3 additive. *Sci. Sinter.* **2008**, *40*, 13–20. [CrossRef]

34. Yamaguchi, T.; Shimizu, S.; Suzuki, T.; Fujishiro, Y.; Awano, M. Evaluation of micro LSM-supported GDC/ScSZ bilayer electrolyte with LSM–GDC activation layer for intermediate temperature-SOFCs. *J. Electrochem. Soc.* **2008**, *155*, B423–B426. [CrossRef]

Materials **2018**, *11*, 40

35. Fan, L.; Ma, Y.; Wang, X.; Singh, M.; Zhu, B. Understanding the electrochemical mechanism of the core-shell ceria-LiZnO nanocomposite in a low temperature solid oxide fuel cell. *J. Mater. Chem. A* **2014**, *2*, 5399–5407. [CrossRef]

36. Shiratori, Y.; Tietz, F.; Buchkremer, H.P.; Stöver, D. YSZ-MgO composite electrolyte with adjusted thermal expansion coefficient to other SOFC components. *Solid State Ion.* **2003**, *164*, 27–33. [CrossRef]

37. Zhu, B.; Lund, P.D.; Raza, R.; Ma, Y.; Fan, L.; Afzal, M.; Patakangas, J.; He, Y.; Zhao, Y.; Tan, W.; et al. Schottky junction effect on high performance fuel cells based on nanocomposite materials. *Adv. Energy Mater.* **2015**, *5*, 1401895. [CrossRef]

38. Fan, L.; Su, P.C. Layer-structured LiNi$_{0.8}$Co$_{0.2}$O$_2$: A new triple (H$^+$/O^2 /e$^-$) conducting cathode for low temperature proton conducting solid oxide fuel cells. *J. Power Sources* **2016**, *306*, 369–377. [CrossRef]

39. Xia, C.; Cai, Y.; Wang, B.; Afzal, M.; Zhang, W.; Soltaninazarlou, A.; Zhu, B. Strategy towards cost-effective low-temperature solid oxide fuel cells: A mixed-conductive membrane comprised of natural minerals and perovskite oxide. *J. Power Sources* **2017**, *342*, 779–786. [CrossRef]

40. Zhu, B. Using a fuel cell to study fluoride-based electrolytes. *Electrochem. Commun.* **1999**, *1*, 242–246. [CrossRef]

41. Fan, L.; Wang, C.; Chen, M.; Di, J.; Zheng, J.; Zhu, B. Potential low-temperature application and hybrid-ionic conducting property of ceria-carbonate composite electrolytes for solid oxide fuel cells. *Int. J. Hydrogen Energy* **2011**, *36*, 9987–9993. [CrossRef]

42. Mahato, N.; Banerjee, A.; Gupta, A.; Omar, S.; Balani, K. Progress in material selection for solid oxide fuel cell technology: A review. *Prog. Mater. Sci.* **2015**, *72*, 141–337. [CrossRef]

43. Zuo, C.; Zha, S.; Liu, M.; Hatano, M.; Uchiyama, M. Ba (Zr$_{0.1}$Ce$_{0.7}$Y$_{0.2}$)O$_{3-\delta}$ as an electrolyte for low-temperature solid-oxide fuel cells. *Adv. Mater.* **2006**, *18*, 3318–3320. [CrossRef]

44. Sawant, P.; Varma, S.; Wani, B.N.; Bharadwaj, S.R. Synthesis, stability and conductivity of BaCe$_{0.8-x}$Zr$_x$Y$_{0.2}$O$_{3-\delta}$ as electrolyte for proton conducting SOFC. *Int. J. Hydrogen Energy* **2012**, *37*, 3848–3856. [CrossRef]

45. Qian, J.; Tao, Z.; Xiao, J.; Jiang, G.; Liu, W. Performance improvement of ceria-based solid oxide fuel cells with yttria-stabilized zirconia as an electronic blocking layer by pulsed laser deposition. *Int. J. Hydrogen Energy* **2013**, *38*, 2407–2412. [CrossRef]

46. Sun, W.; Shi, Z.; Wang, Z.; Liu, W. Bilayered BaZr$_{0.1}$Ce$_{0.7}$Y$_{0.2}$O$_{3-\delta}$/Ce$_{0.8}$Sm$_{0.2}$O$_{2-\delta}$ electrolyte membranes for solid oxide fuel cells with high open circuit voltages. *J. Membr. Sci.* **2015**, *476*, 394–398. [CrossRef]

47. Cai, Y.; Xia, C.; Wang, B.; Zhang, W.; Wang, Y.; Zhu, B. Bio-derived calcite as a novel electrolyte for solid oxide fuel cells: A strategy toward utilization of waste shells. *ACS Sustain. Chem. Eng.* **2017**, *5*, 10387–10395. [CrossRef]

48. Wang, B.; Cai, Y.; Xia, C.; Liu, Y.; Muhammad, A.; Wang, H.; Zhu, B. CoFeZrAl-oxide based composite for advanced solid oxide fuel cells. *Electrochem. Commun.* **2016**, *73*, 15–19. [CrossRef]

49. Ortiz-Vitoriano, N.; De Larramendi, I.R.; De Muro, I.G.; De Larramendi, J.R.; Rojo, T. Nanoparticles of La$_{0.8}$Ca$_{0.2}$Fe$_{0.8}$Ni$_{0.2}$O$_{3-\delta}$ perovskite for solid oxide fuel cell application. *Mater. Res. Bull.* **2010**, *45*, 1513–1519. [CrossRef]

50. Kant, K.M.; Esposito, V.; Pryds, N. Enhanced conductivity in pulsed laser deposited Ce$_{0.9}$Gd$_{0.1}$O$_{2-\delta}$/SrTiO$_3$ heterostructures. *Appl. Phys. Lett.* **2010**, *97*, 143110. [CrossRef]

51. Lin, Y.; Fang, S.; Su, D.; Brinkman, K.S.; Chen, F. Enhancing grain boundary ionic conductivity in mixed ionic-electronic conductors. *Nat. Commun.* **2015**, *6*. [CrossRef] [PubMed]

materials

MDPI

Article

Structural and Electrochemical Characterization of Zn$_{1-x}$Fe$_x$O—Effect of Aliovalent Doping on the Li$^+$ Storage Mechanism

Gabriele Giuli [1,*], Tobias Eisenmann [2,3], Dominic Bresser [2,3,*], Angela Trapananti [4], Jakob Asenbauer [2,3], Franziska Mueller [2,3] and Stefano Passerini [2,3]

[1] School of Science and Technology-Geology Division, University of Camerino, Via gentile III da Varano, 62032 Camerino, Italy
[2] Helmholtz Institute Ulm (HIU), Helmholtzstrasse 11, 89081 Ulm, Germany; tobias.eisenmann@kit.edu (T.E.); jakob.asenbauer@kit.edu (J.A.); franziska-mueller1@gmx.de (F.M.); stefano.passerini@kit.edu (S.P.)
[3] Karlsruhe Institute of Technology (KIT), P.O. Box 3640, 76021 Karlsruhe, Germany
[4] School of Science and Technology-Physics Division, University of Camerino, Via Madonna delle Carceri, 62032 Camerino, Italy; angela.trapananti@unicam.it
* Correspondence: gabriele.giuli@unicam.it (G.G.); dominic.bresser@kit.edu (D.B.);
   Tel.: +39-0737-402606 (G.G.); +49-731-50-34117 (D.B.)

Received: 5 December 2017; Accepted: 27 December 2017; Published: 29 December 2017

Abstract: In order to further improve the energy and power density of state-of-the-art lithium-ion batteries (LIBs), new cell chemistries and, therefore, new active materials with alternative storage mechanisms are needed. Herein, we report on the structural and electrochemical characterization of Fe-doped ZnO samples with varying dopant concentrations, potentially serving as anode for LIBs (Rechargeable lithium-ion batteries). The wurtzite structure of the Zn$_{1-x}$Fe$_x$O samples (with x ranging from 0 to 0.12) has been refined via the Rietveld method. Cell parameters change only slightly with the Fe content, whereas the crystallinity is strongly affected, presumably due to the presence of defects induced by the Fe^{3+} substitution for Zn^{2+}. XANES (X-ray absorption near edge structure) data recorded ex situ for Zn$_{0.9}$Fe$_{0.1}$O electrodes at different states of charge indicated that Fe, dominantly trivalent in the pristine anode, partially reduces to Fe^{2+} upon discharge. This finding was supported by a detailed galvanostatic and potentiodynamic investigation of Zn$_{1-x}$Fe$_x$O-based electrodes, confirming such an initial reduction of Fe^{3+} to Fe^{2+} at potentials higher than 1.2 V (vs. Li$^+$/Li) upon the initial lithiation, i.e., discharge. Both structural and electrochemical data strongly suggest the presence of cationic vacancies at the tetrahedral sites, induced by the presence of Fe^{3+} (i.e., one cationic vacancy for every two Fe^{3+} present in the sample), allowing for the initial Li$^+$ insertion into the ZnO lattice prior to the subsequent conversion and alloying reaction.

Keywords: lithium-ion battery; anode; crystal chemistry; electrochemistry

---

1. Introduction

Rechargeable lithium-ion batteries (LIBs) are now the technology of choice for electrochemical energy storage, and are consequently employed in a wide range of devices ranging from portable electronics to electric vehicles and stationary energy storage [1–4]. However, in order to eventually achieve the highly challenging goal of fully replacing gasoline-powered vehicles by electric ones, both the energy density and power density require further improvement [5]. With respect to the anode side, the state-of-the-art active material graphite has essentially reached its limits in terms of specific capacity and charge rate capability [6,7]. Therefore, new cell chemistries and, accordingly, new anode materials, presumably based on alternative lithium-ion storage mechanisms, are being presently investigated. The two most prominent ones in this context are conversion- and alloying-type materials [6,8–10].

Both mechanisms, however, suffer from substantial intrinsic drawbacks, such as large volume changes upon (de-)lithiation in the case of alloying materials (leading to short cycle life) [8,10] or high voltage hysteresis in the case of conversion materials [9]. As an example, pure ZnO, a material widely investigated for semiconductors [11], suffers from essentially irreversible Li_2O formation upon initial lithiation when applied as an anode material in LIBs. Hence, it typically offers low specific capacities of around 330 mAh g^{-1}, solely originating from the reversible Li-Zn de-/alloying reaction (theoretical specific capacity of 988 mAh g^{-1}, if fully reversible) [12]. When Zn is partially substituted by transition metals (e.g., Fe, Co) within the wurtzite structure, however, the reversibility of the Li_2O formation is substantially enhanced [12–14]. Consequently, transition metal-doped zinc oxides combine both conversion and alloying mechanisms in one active material and are accordingly classified as conversion-alloying materials (CAMs) [14]. Several studies have already revealed the outstanding electrochemical performance of, for instance, carbon-coated Fe-doped ZnO, when applied as negative electrode materials in LIBs, delivering reversible specific capacities of almost 1000 mAh g^{-1} [12,15]. Nevertheless, the exact role of the transition metal for the lithiation mechanism in CAMs, enabling the reversible formation and decomposition of Li_2O, has still not been fully unraveled.

Herein, we report the detailed investigation of the structural and electrochemical properties of $Zn_{1-x}Fe_xO$ (with x = 0, 0.02, 0.04, 0.06, 0.08, 0.10, and 0.12) by means of X-ray diffraction (XRD), X-ray absorption spectroscopy (XAS), galvanostatic cycling, and cyclic voltammetry. The results suggest that an increasing amount of cationic vacancies as a result of the substitution of Zn^{2+} by Fe^{3+} leads to a substantial lithium ion insertion into the wurtzite lattice, accompanied by the reduction of Fe^{3+} to Fe^{2+}, prior to the subsequent—in such case facilitated—conversion/alloying reaction.

2. Results and Discussion

2.1. Synthesis and Basic Characterization

The synthesis of the $Zn_{1-x}Fe_xO$ sample series was conducted following a synthesis route developed earlier [12]. In excellent agreement with a previous study [15], the color of the samples varied, becoming increasingly orange for increasing iron concentration. The particle size was well in the range of a few nm (10–20 nm—depending on the Fe content; see Appendix A Figure A1). The Zn:Fe ratio was determined by means of inductively coupled plasma optical emission spectroscopy (ICP-OES), confirming the targeted Fe concentrations of x = 0.02, 0.04, 0.06, 0.08, 0.10, and 0.12 within an experimental error of around 1%.

2.2. Powder XRD Characterization and Structural Refinement

The X-ray diffraction patterns of the as synthesized samples are shown in Figure 1, arranged from bottom to top for increasing Fe content. All the diffraction reflections could be indexed according to the hexagonal ZnO structure ($P6_3mc$ space group), whereas the absence of other reflections allowed the exclusion of the presence of crystalline impurities such as other zinc and/or iron oxide phases. The increase of the diffraction reflections' Full Width at Half Maximum (FWHM) was evident (from bottom to top), revealing that the ZnO crystallinity and/or crystallite size was remarkably affected by the Fe concentration.

Rietveld refinement provided an accurate determination of the unit cell parameters and atomic positions for all the samples. Table 1 summarizes the disagreement indexes, the most relevant structural parameters, and the average crystallites sizes (calculated using the Williamson-Hall (W.-H.) method [16]), whereas in Figure 2, a typical comparison between observed and calculated XRD patterns is displayed.

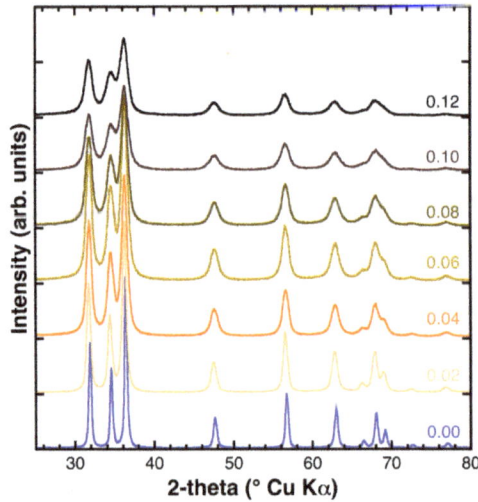

Figure 1. Powder X-ray diffraction (XRD) patterns of the as synthesized samples. The Fe fraction is indicated at the right side above each diffractogram.

Table 1. Structural parameters, average crystallite size, and disagreement indexes of the Rietveld refinements performed using isotropic temperature factors and isotropic broadening of the diffraction reflections. The errors of the Unit cell parameters and <T–O> distances are provided in brackets.

	ZnO	$Zn_{0.98}Fe_{0.02}O$	$Zn_{0.96}Fe_{0.04}O$	$Zn_{0.94}Fe_{0.06}O$	$Zn_{0.92}Fe_{0.08}O$	$Zn_{0.9}Fe_{0.1}O$	$Zn_{0.88}Fe_{0.12}O$
a_0 (Å)	3.2511 (3)	3.2524 (1)	3.2549 (1)	3.2553 (1)	3.2549 (2)	3.2552 (2)	3.2573 (2)
c_0 (Å)	5.2098 (1)	5.2099 (1)	5.2081 (3)	5.2075 (3)	5.2079 (5)	5.2043 (4)	5.2096 (4)
V_0 (Å³)	47.687 (1)	47.728 (1)	47.784 (3)	47.791 (3)	47.781 (5)	47.760 (5)	47.868 (4)
<T-O>	1.9795 (50)	1.9800 (50)	1.9802 (50)	1.9815 (50)	1.9818 (50)	1.9815 (50)	1.9830 (50)
wRp	8.36	7.11	7.44	7.88	7.88	7.60	8.7.44
Rp	6.60	5.64	5.79	6.26	6.07	5.87	5.92
R_F^2	3.49	3.43	4.51	5.09	3.78	4.19	4.26
R_F	1.92	1.79	2.59	3.05	2.02	2.30	2.17
a_0/c_0	0.6240	0.6188	0.6250	0.6251	0.6250	0.6255	0.6252
W.-H. intercept [1]	0.0033	0.0048	0.0069	0.0104	0.0063	0.011	0.0088
W.-H. slope [1]	0.0003	0.0013	0.0023	0.0006	0.0045	0.0031	0.0046
Crystallite size (nm)	42	29	20	13	22	13	16

[1] Fitted intercepts and slope of the W.-H. plots, obtained using the refined peak shape parameters.

The absence of any relevant amounts (i.e., at the 0.5 wt % level) of additional Fe-bearing phases in the XRD patterns of the doped samples suggests that Fe was successfully inserted as dopant into the ZnO lattice in the whole Fe concentration range analyzed here (x ranging between 0 and 0.12). Similar to previously published data [17], the substitution of Zn by Fe resulted in a small but appreciable variation of the unit cell volume (Figure 3), while the variation of the <T–O> distances was within the experimental error (Figure A2 in the Appendix A). In this regard, it is interesting to notice that the unit cell volume of the doped samples increased—albeit only slightly—as a function of the Fe content despite the fact that Fe^{3+} in tetrahedral coordination had a smaller ionic radius than Zn^{2+} (i.e., 0.49 and 0.60 Å, respectively [18]). In particular, the increase of the a_0 parameter could be fitted by a linear trend $a_0 = 3.2518(6) + 0.043(9)X_{Fe}$ (R = 91.4%), for which the slope of the linear function is about four times its standard deviation and thus, significant. Contrarily, the c_0 parameter could also be fitted by a linear trend $c_0 = 5.2094(13) - 0.021(18)X_{Fe}$ (R = 47.1%), but the slope of this linear function had a value comparable to its standard deviation and was therefore not significant. In sum, the unit cell volume displayed a slight increase with the iron content according to the following function:

$V_0 = 47.706(24) + 1.08(33)X_{Fe}$ (R = 82.6%). As a conclusion, we assigned this anisotropic variation of the unit cell volume to two driving forces: the Fe doping and the presence of structural defects as a result of the aliovalent substitution of Zn^{2+} by Fe^{3+} [17].

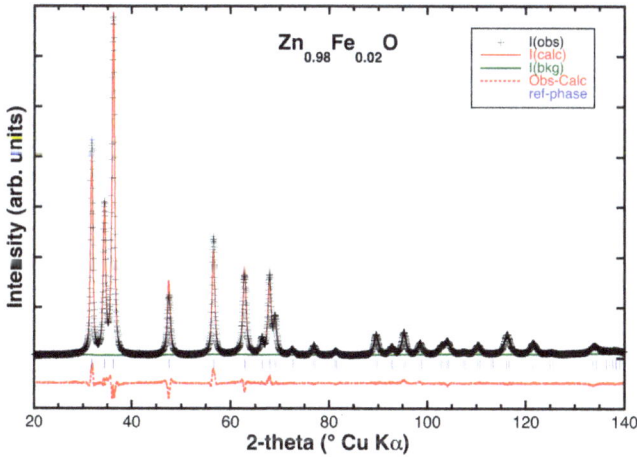

Figure 2. Typical Rietveld refinement of a Fe-doped sample ($Zn_{0.98}Fe_{0.02}O$): Black crosses mark the experimental data; the solid red and green lines represent the theoretical pattern and background function, respectively, while the dotted red line is the residual. Vertical blue lines mark the angular positions of the ZnO reflections.

The FWHM of the XRD reflections increases progressively with the increase of the Fe content, x from 0 to 0.12, meaning that the crystallinity of the samples decreases in the same order. The reflection shape parameters obtained from the Rietveld refinement were used to determine the FWHM as a function of 2-theta and to build the corresponding W.-H. plots. The crystallite size, calculated by means of the Scherrer formula based on the intercepts of the W.-H. plots, decreased from 42 to 13 nm (Figure 4), whereas the strain displayed a more scattered trend. In absence of a careful calibration of the instrumental parameters of the utilized diffractometer, these values have to be taken solely as indicative and should not be considered as an absolute measure of the crystallite size. However, we noted that these values were of the same order of magnitude as those obtained earlier by means of transmission electron microscopy (TEM) and small-angle X-ray scattering (SAXS) for the same sample series [15].

These results obtained for the series of zinc oxide samples with varying Fe dopant concentrations, ranging from 0 to 0.12, indicated that the doping remarkably affected the crystallinity. This effect was particularly pronounced for Fe contents as small as x = 0.02 and 0.04, while it was less pronounced for higher Fe contents. A similar effect of the Fe content on the crystallite size was also observed by Kumar et al. for Fe-doped ZnO samples (with x ranging from 0 to 0.06), prepared by a sol-gel synthesis method [19], and Reddy et al. for example (with x ranging from 0 to 0.05) synthesized by a low-temperature combustion route [20].

It is worth mentioning once more that the cell parameter variations observed here were not solely related to the ionic radius of the substituting cation (0.49 vs. 0.60 for tetrahedrally coordinated Fe^{3+} vs. Zn^{2+}, respectively [18]). In fact, besides displaying compositional variations, the samples studied here also showed amply varying degrees of crystallinity and, possibly, presence of defects such as cationic vacancies and/or interstitial oxygen [17]. As already reported in the literature for other oxide systems [21], both crystallinity and defects content could strongly contribute to alter unit cell parameters, but not in a predictable way.

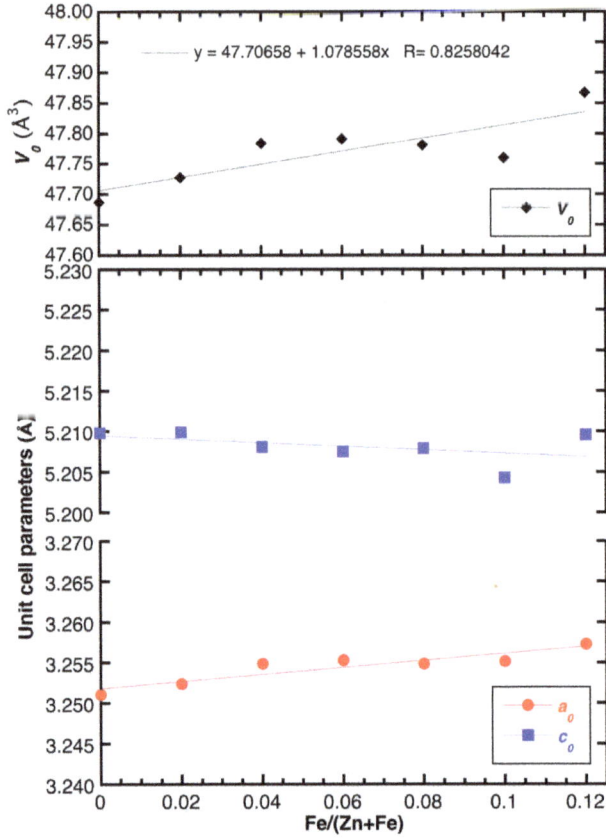

Figure 3. Refined unit cell parameters of the as synthesized samples (error bars shown within symbols) as a function of the Fe/(Fe + Zn) ratio. The linear fits for the refined data are also shown.

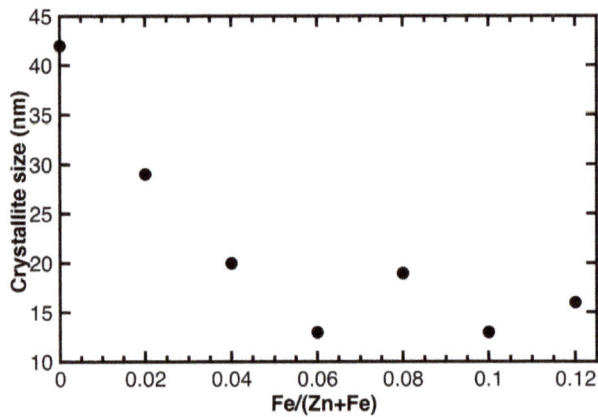

Figure 4. Crystallite size vs. Fe content of the as synthesized samples, revealing a marked decrease in crystallite size specifically for Fe concentrations as small as 0.02 and 0.04, whereas for higher Fe contents the decrease in crystallite size is less pronounced.

2.3. Fe K-Edge XANES Spectroscopy

X-ray absorption near edge structure (XANES) data collected at the Fe K-edges are presented in Figure 5. The XANES spectra were measured ex situ for various anode samples of carbon-coated $Zn_{0.9}Fe_{0.1}O$. The pristine sample, i.e., a non-cycled electrode, was comparable to that reported in Giuli et al. [17], whereas those ones discharged to 1.5 and 1.2 V vs. Li^+/Li displayed a slight shift of the edge energy toward lower energy. The background subtracted pre-edge peaks (shown in the inset of Figure 5) also had similar shapes and intensities. However, while the pre-edge peak of pristine $Zn_{0.9}Fe_{0.1}O$ displayed a major component at 7114.4 eV (typical of Fe^{3+}), the spectrum of $Zn_{0.9}Fe_{0.1}O/C$ discharged to 1.5 and 1.2 V had a clear component at ca. 7112.7 eV, revealing the presence of a significant fraction of Fe^{2+}.

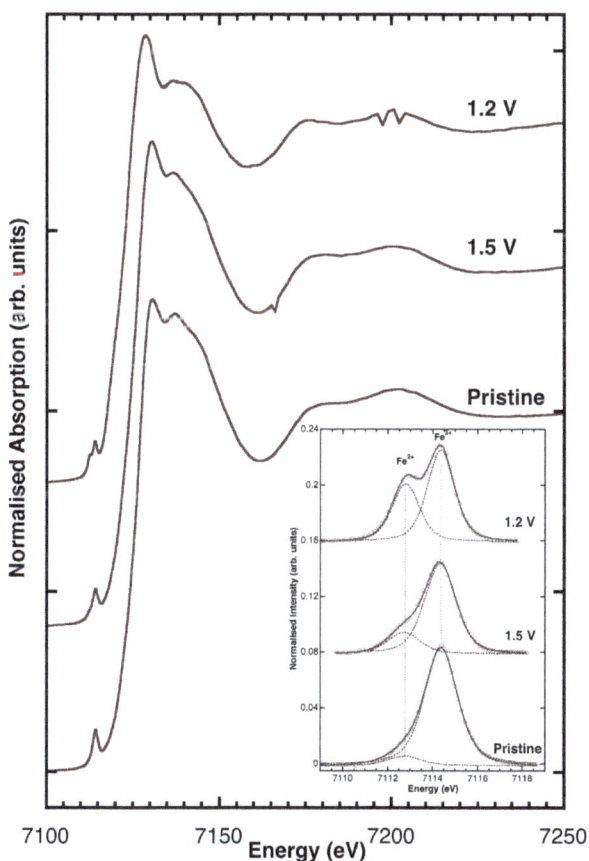

Figure 5. Fe K-edge XANES (X-ray absorption near edge structure) spectra measured ex situ for carbon-coated $Zn_{0.9}Fe_{0.1}O$ anodes before electrochemical testing (pristine) and discharged to 1.5 and 1.2 V vs. Li^+/Li.

Comparing the pre-edge peak integrated intensities and energy centroids with those of model compounds from literature [16,22–24] provides $Fe^{3+}/(Fe^{2+} + Fe^{3+})$ ratios = 0.95(5) for the pristine $Zn_{0.9}Fe_{0.1}O/C$., 0.75(7) and 0.5(7) for the samples discharged at 1.5 and 1.2 V, respectively.

2.4. Electrochemical Characterization

In order to complement these structural results with the electrochemical lithium storage behavior of $Zn_{1-x}Fe_xO$ as a function of the iron content, galvanostatic cycling was conducted. The results for the initial lithiation are depicted in Figure 6a. All samples essentially revealed the typical potential profile of Fe-doped ZnO, characterized by two distinct potential plateaus at ~0.8 V and ~0.5 V and a subsequent smooth potential decrease to 0.01 V, selected as the cathodic cut-off potential [12]. As determined earlier by means of in situ XRD, the lithiation mechanism includes the conversion of $Zn_{0.9}Fe_{0.1}O$ to Zn^0, Fe^0, and Li_2O as well as the alloying of Zn and Li [12].

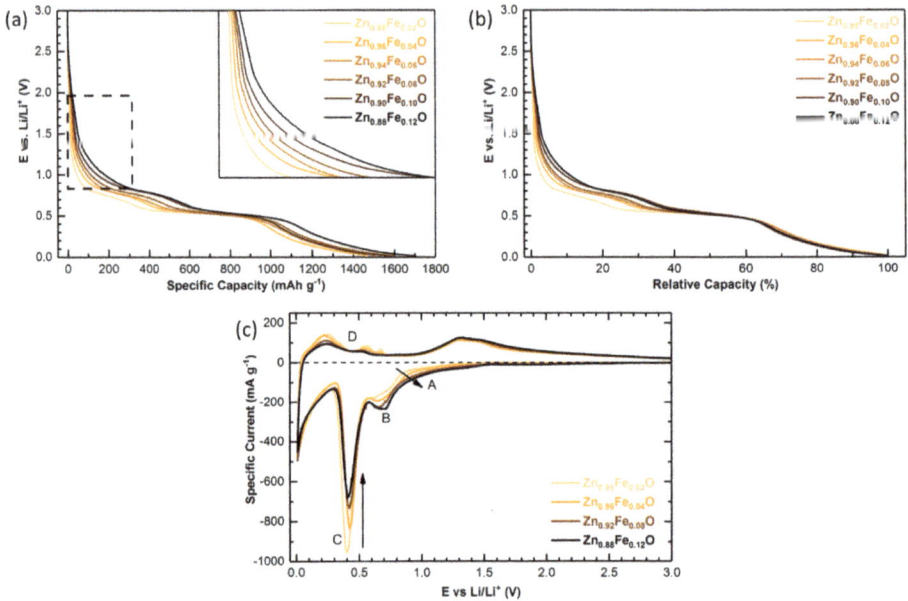

Figure 6. (**a**) Potential profiles for the first galvanostatic discharge of $Zn_{1-x}Fe_xO$-based electrodes (x = 0.02, 0.04, 0.06, 0.08, 0.10, and 0.12), applying a discharge rate of 0.05C (i.e., 48.3 mA g^{-1}); the inset shows a magnification of the area highlighted by the dashed box; (**b**) The same potential profiles as presented in (**a**), but plotted versus the normalized capacity; (**c**) Cyclic voltammetry conducted for selected samples (x = 0.02, 0.04, 0.08, and 0.12), applying a sweeping rate of 50 $\mu V\ s^{-1}$.

Interestingly, a significantly different behavior was observed for the high-potential region during the initial lithiation, i.e., the potential region from the open circuit voltage down to 0.8 V—or in terms of specific capacity for the first 100–350 mAh g^{-1} (indicated by the dashed box and magnified in the inset of Figure 6a). The Li-storage capacity in this region increased with increasing iron content. Combining the previously reported in situ XRD data [12], the herein presented XRD results (indicating an increasing amount of structural defects as a result of the substitution of Zn^{2+} by Fe^{3+}), and the XANES results (revealing the reduction of Fe^{3+} to Fe^{2+} at potentials of 1.5 and 1.2 V), the increasing capacity contribution observed at high potentials could be ascribed to Li^+ insertion into the wurtzite lattice favored by the presence of cationic vacancies and accompanied by the reduction of Fe^{3+} to Fe^{2+}, as schematically illustrated in Figure 7.

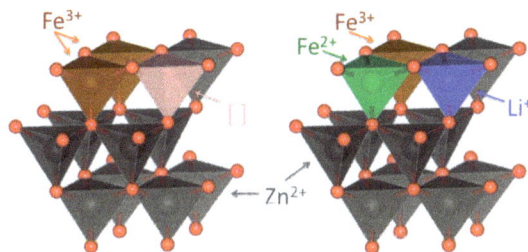

Figure 7. Schematic illustration of the cationic vacancies in the wurtzite lattice of the pristine material due to the Fe^{3+} for Zn^{2+} substitution (**left panel**) and the resulting initial lithium ion insertion in the vacant tetrahedral sites at higher potentials, leading to the partial reduction of Fe^{3+} to Fe^{2+} (**right panel**).

Interestingly, also the shape of the first potential plateau changed as a function of the Fe content (Figure 6a,b). While the $Zn_{1-x}Fe_xO$ samples with x from 0.04 to 0.12 showed a comparably well-defined plateau, though shifted to lower potentials for decreasing x, the sample with x = 0.02 revealed a rather sloped shape. Also, the overall capacity obtained at the end of this first potential plateau decreased with a decreasing iron content, providing about 425, 475, 550, 575, 630, and 660 mAh g^{-1} for x = 0.02, 0.04, 0.06, 0.08, 0.10, and 0.12, respectively. Considering that iron is completely reduced to the metallic state at this potential (as revealed by the in/ex situ XANES analysis, to be published soon) and that the overall contribution of the conversion reaction to the theoretical capacity is 666 mAh g^{-1}, these findings indicated that the conversion reaction was kinetically favored in case of high Fe concentrations, presumably as a result of the relatively larger amount of initially inserted lithium. For lower iron contents, the conversion reaction was accordingly completed along the second potential plateau, i.e., together with the occurring alloying reaction [12]. This is confirmed by the plot of the normalized capacity in Figure 6b, revealing the same capacity values for all samples at the end of this second plateau. As a matter of fact, such a mixed potential for the reduction of the transition metal dopant and zinc was observed also for Co-doped ZnO [13], which did not show any significant cationic vacancies, allowing for an initial Li^+ insertion, due to the divalent oxidation state of cobalt [17].

The general capacity excess upon discharge compared to the theoretical maximum for this reaction (ca. 966 mAh g^{-1}) was largely assigned to the relatively low first cycle coulombic efficiency of 61–64%, indicating a significant electrolyte decomposition, especially at lower potentials.

The general trend of a relatively increasing capacity at higher potentials (regions A and B in Figure 6c) and relatively decreasing capacity at lower potentials (region C in Figure 6c) for increasing x was very well observed when comparing the cyclic voltammograms (CVs) in Figure 6c. While we may refer to a previous publication [15] for the detailed discussion of the CVs, it appears noteworthy that $Zn_{0.98}Fe_{0.02}O$ reveals the typical de-alloying peaks (region D), commonly observed for pure ZnO [13,15], suggesting that the upon lithiation formed zinc nanograins are comparably larger for such a rather low iron content [13].

In conclusion, both galvanostatic cycling and cyclic voltammetry highly suggested that the defects in the wurtzite structure were cationic vacancies, allowing for the initial Li^+ insertion into these vacancies. This lithium insertion resulted in increasing capacities at higher potentials for increasing Fe content, accompanied by the reduction of Fe^{3+} to Fe^{2+}, and followed by the kinetically favored conversion reaction.

3. Materials and Methods

3.1. Material Synthesis

$Zn_{1-x}Fe_xO$ (with x = 0.02, 0.04, 0.06, 0.08, 0.10, 0.12) was synthesized according to a previously reported method [15]. Zinc (II) gluconate (Alfa Aesar, Lancashire, UK) and iron (II) gluconate

(Sigma Aldrich, St. Louis, MO, USA) were dissolved in ultra-pure water with respect to the desired dopant ratio (0.2 M total ion concentration). This solution was added dropwise to a second solution comprising 1.2 M sucrose, and the obtained solution was stirred for 15 min. Subsequently, the solvent was evaporated at 160 °C and the obtained material was further heated to 300 °C in order to decompose the sucrose and dry the precursor. Finally, the solid powder was grinded and calcined in a tubular furnace (Nabertherm, L9/12/P330, Lilienthal, Germany) at 450 °C for 3 h (3 K min^{-1} heating rate). ICP-OES was conducted in order to determine the metal ion concentrations by dissolving the samples in hot hydrochloric acid and via double determination in a Spectro Arcos from Spectro Analytical Instruments (Kleve, Germany) with axial plasma view. Scanning electron microscopy (SEM) was performed by means of a Zeiss LEO 1550 (Zeiss, Oberkochen, Germany).

3.2. Powder XRD Characterization and Structural Refinement

The crystal structure of the as-synthesized samples was characterized by powder XRD with an automated Philips Bragg-Brentano diffractometer equipped with a graphite monochromator. The long-fine focus Cu tube was operated at 40 kV and 25 mA. Spectra were recorded in the 2θ range 20–140° with a 0.03° step and 14 s counting time. The structures were refined with the program General Structure Analysis System (GSAS) [25]. The reflection shape was modeled with a Pseudo-Voigt function; the FWHM was refined as a function of 2θ taking into account both Gaussian and Lorentzian broadening. The refinement was carried out for the space group $P6_3mc$ and the starting atomic coordinates were those of Xu and Ching [26] with the initial values for isotropic temperature factors (*Uiso*), arbitrarily chosen as 0.025 Å^2. The O atom sites were designated as fully occupied, while constrains for fractional occupancies for Fe and Zn were used according to the stoichiometry of the synthesized samples. The background was modeled with a 9-terms polynomial function. Cell parameters, scale factor, and the background polynomial function were free variables during the refinement. Parameters were added stepwise to the refinement in the following order: 2θ zero-shift, peak shape, peak asymmetry, atomic coordinates, and isotropic thermal factor. The intensity cut-off for the calculation of the profile step intensity was initially set at 1.0% of the peaks maxima and were lowered to 0.1% in the final stages of the refinements. Final convergence was assumed to be reached when the parameter shifts were <1% of their respectively estimated standard deviation. Estimated errors, provided by the Rietveld refinement program, are ±0.0002 Å for the cell parameters and ±0.002 Å for the selected interatomic distances. However, the error calculation is probably over-optimistic, as it does not include the correlation among parameters and other error sources (like the overlapping of many diffraction reflections, for instance). In order to get an alternative estimate of the accuracy of the refined structural data, we have compared the set of structural parameters obtained using different refinement strategies for the same diffraction data. These comparisons show that realistic estimates of the error bars are ±0.0005 Å for the cell parameters and ±0.005 Å for the selected interatomic distances. Trials of refinements were also done assuming some of the iron could be located in interstitial sites. However, the resulting disagreement indexes were higher than in the case of Fe location in the Zn site. Thus, we assume that all the iron is substituting for Zn and that no significant amount of iron is located in interstitial sites [17].

3.3. XAS Data Collection and Analysis

Ex situ Fe K-edge XAS spectra were measured on cycled electrodes (section below) at beamline BM08 [27] of the European Synchrotron Radiation Facility (ESRF, Grenoble, France). A fixed-exit double-crystal Si(311) monochromator, operated in flat crystals mode, was used to select the energy of the beam delivered by the bending magnet source. Higher order harmonics were rejected using two Pd-coated mirrors working at an incidence angle of 3.6 mrad. The second mirror was left unbent. The beam size at the sample was about 2 mm × 2 mm FWHM. XAS spectra were measured in transmission mode using ionization chambers filled with N$_2$ and Ar gases at pressures tuned to achieve the optimal efficiency at the Fe K-edge absorption edge energy (20% and 80% of absorption of the incident beam respectively). The monochromator energy was calibrated by setting the first inflection point of the edge of metal Fe to 7112 eV. The spectrum of a metal Fe foil placed downstream of the main experimental chamber was collected simultaneously with any anode spectrum to monitor and correct the energy scale against possible monochromator instabilities. Fe K-edge pre-edge peak analysis was carried out following a standard procedure reported elsewhere [22,23]. The pre-edge peak was fitted by a sum of pseudo-Voigt functions and their intensities along with energy positions were compared with those of Fe model compounds from literature in order to extract the information on the absorber oxidation state in the cycled samples. More details on the pre-edge peak fitting method can be found in references [24,28].

3.4. Electrochemical Characterization

Electrodes were prepared with a dry composition of 75 wt % active material, 20 wt % Super C65 (Imerys, Paris, France) and 5 wt % sodium carboxymethyl cellulose (CMC; Dow-Wolff Cellulosics, Bollitz, Germany) dissolved in ultra-pure water. Slurries were homogenized in a planetary ball mill (Fritsch Vario-Planetary Mill Pulverisette 4, Fritsch GmbH, Idar-Oberstein, Germany) for 3 h using 12 mL zirconia jars and zirconia balls. Subsequently, the slurries were cast onto dendritic copper foil (Schlenk, Bitterfeld, Germany) at 50 mm s^{-1} with a doctor blade (BYK Additive & Instruments, Wesel, Germany) and a wet film thickness of 120 µm. The obtained electrode batches were dried at room temperature overnight before punching 12 mm disc electrodes and drying them in a vacuum glass oven (Büchi B585, Büchi, Rungis, France) at 120 °C for 24 h. All electrochemical experiments were conducted in three-electrode Swagelok-type cells. The prepared electrodes were used as working electrodes, while lithium foil (Honjo Metal Co., Higashi-Osaka, Japan) served as counter and reference electrode. Glass fiber sheets (Whatman GF/D, Whatman, Maidstone, UK), drenched in a 1 M solution of LiPF$_6$ in ethylene carbonate and diethyl carbonate (3:7 by wt), were used as separators. Galvanostatic cycling was conducted on a Maccor Series 4300 battery cycler, applying cut-off potentials of 3.0 V and 0.01 V vs. Li$^+$/Li. All applied currents refer to a theoretical specific capacity of 966 mAh g^{-1} (1C = 966 mA g^{-1}). For cyclic voltammetry experiments, a galvanostatic-potentiostatic VMP multichannel cycler (Biologic Science Instruments, Seyssinet-Pariset, France) was used, applying a sweeping rate of 50 µV s^{-1} and reversing potentials of 3.0 V and 0.01 V vs. Li$^+$/Li.

For the ex situ XAS measurements, electrodes based on carbon-coated Zn$_{0.9}$Fe$_{0.1}$O active material were prepared on carbon paper, serving as current collector, and, apart from the pristine sample, cycled in three-electrode Swagelok-type cells according to the same procedures as above. Subsequently, the electrodes were recovered at discrete stages (1.5 V and 1.2 V), washed with dimethyl carbonate to remove the electrolyte, and sealed within polyethylene (PE) foil in an Ar-filled glove box to avoid air contamination.

4. Conclusions

Fe-doped Zn$_{1-x}$Fe$_x$O samples (with x ranging from 0.00 to 0.12) have been successfully synthesized using a self-developed synthesis route. Powder XRD reveals all samples to be phase-pure, having the same wurtzite structure of pristine ZnO. Galvanostatic cycling and cyclic voltammetry data of Fe-doped ZnO electrodes highlight an increasing capacity contribution in the initial stages of lithiation, which is ascribed to the reduction of trivalent to divalent Fe, as also supported by the ex situ XANES data. In sum, the data reported herein indicate that increasing iron dopant contents lead to increasing cationic vacancies in the lattice, allowing for the electrochemical insertion of Li$^+$ into wurtzite-structured ZnO, thus, kinetically favoring the subsequent conversion reaction. We may, hence, anticipate that these results will further enlighten the role of non-divalent transition metal dopants in ZnO and, in general, aliovalent transition metal dopants for conversion/alloying materials (CAMs) in general, ideally paving the way for the development of new CAMs with further enhanced energy and power densities.

Acknowledgments: Financial support from the Vector Foundation within the NEW E^2 project (Tobias Eisenmann, Jakob Asenbauer, and Dominic Bresser) and the Helmholtz Association is kindly acknowledged. Moreover, the Italian CRG BM08 (LISA) @ESRF is acknowledged for granting beamtime within the frame of experiment No. 08-01-962.

Author Contributions: Gabriele Giuli, Angela Trapananti and Dominic Bresser conceived and designed the experiments. Franziska Mueller, Tobias Eisenmann and Jakob Asenbauer synthesized the samples and conducted the basic characterization. Gabriele Giuli, Angela Trapananti, Franziska Mueller, Tobias Eisenmann and Dominic Bresser performed the XRD, XAS, and electrochemical studies. Gabriele Giuli, Angela Trapananti, Tobias Eisenmann, Franziska Mueller and Dominic Bresser analyzed the data. Gabriele Giuli, Angela Trapananti, Dominic Bresser and Stefano Passerini supervised the project. All authors jointly wrote, discussed and revised the paper.

Conflicts of Interest: The authors declare no conflicts of interest.

Appendix A

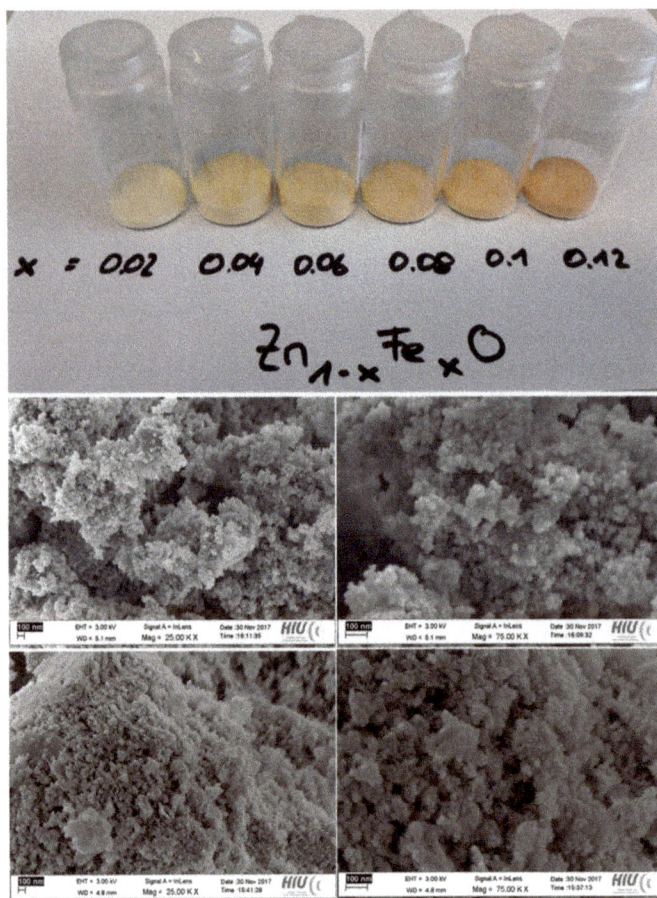

Figure A1. Top panel: Photograph of the $Zn_{1-x}Fe_xO$ samples with x being 0.02, 0.04, 0.06, 0.08, 0.10, and 0.12. The photograph nicely illustrates the increasing orange coloring with an increasing Fe content. **Lower panels**: Scanning electron microscopy (SEM) micrographs of selected $Zn_{1-x}Fe_xO$ samples with x being 0.12 (**middle left and right**) and 0.04 (**bottom left and right**).

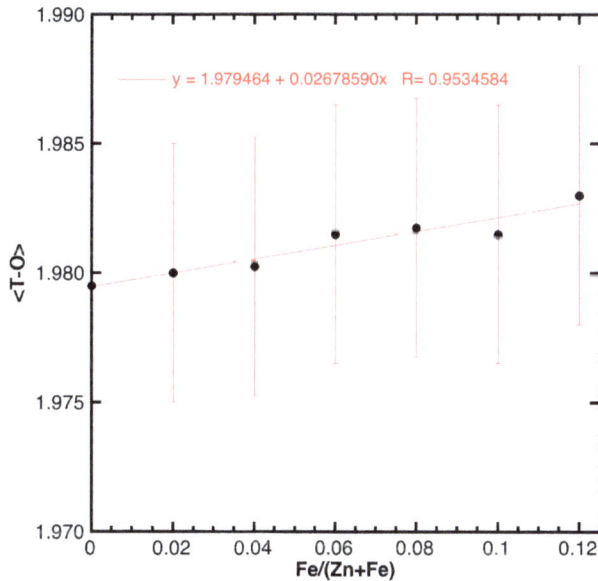

Figure A2. <T–O> distance as a function of the Fe concentration.

References

1. Tarascon, J.-M.; Armand, M. Issues and challenges facing rechargeable lithium batteries. *Nature* **2001**, *414*, 359–367. [CrossRef] [PubMed]
2. Armand, M.; Tarascon, J.-M. Building better batteries. *Nature* **2008**, *451*, 652–657. [CrossRef] [PubMed]
3. Scrosati, B.; Garche, J. Lithium batteries: Status, prospects and future. *J. Power Sources* **2010**, *195*, 2419–2430. [CrossRef]
4. Dunn, B.; Kamath, H.; Tarascon, J.-M. Electrical Energy Storage for the Grid: A Battery of Choices. *Science* **2011**, *334*, 928–935. [CrossRef] [PubMed]
5. Thackeray, M.M.; Wolverton, C.; Isaacs, E.D. Electrical energy storage for transportation—Approaching the limits of, and going beyond, lithium-ion batteries. *Energy Environ. Sci.* **2012**, *5*, 7854–7863. [CrossRef]
6. Aravindan, V.; Lee, Y.S.; Madhavi, S. Research Progress on Negative Electrodes for Practical Li-Ion Batteries: Beyond Carbonaceous Anodes. *Adv. Energy Mater.* **2015**, *5*. [CrossRef]
7. Yamada, Y.; Iriyama, Y.; Abe, T.; Ogumi, Z. Kinetics of Lithium Ion Transfer at the Interface between Graphite and Liquid Electrolytes: Effects of Solvent and Surface Film. *Langmuir* **2009**, *25*, 12766–12770. [CrossRef] [PubMed]
8. Zhang, W.J. A review of the electrochemical performance of alloy anodes for lithium-ion batteries. *J. Power Sources* **2011**, *196*, 13–24. [CrossRef]
9. Cabana, J.; Monconduit, L.; Larcher, D.; Palacín, M.R. Beyond Intercalation-Based Li-Ion Batteries: The State of the Art and Challenges of Electrode Materials Reacting Through Conversion Reactions. *Adv. Mater.* **2010**, *22*, 170–192. [CrossRef] [PubMed]
10. Obrovac, M.N.; Chevrier, V.L. Alloy Negative Electrodes for Li-Ion Batteries. *Chem. Rev.* **2014**, *114*, 11444–11502. [CrossRef] [PubMed]
11. Özgür, Ü.; Alivov, Y.I.; Liu, C.; Teke, A.; Reshchikov, M.A.; Doğan, S.; Avrutin, V.; Cho, S.J.; Morko, H. A comprehensive review of ZnO materials and devices. *J. Appl. Phys.* **2005**, *98*. [CrossRef]
12. Bresser, D.; Mueller, F.; Fiedler, M.; Krueger, S.; Kloepsch, R.; Baither, D.; Winter, M.; Paillard, E.; Passerini, S. Transition-Metal-Doped Zinc Oxide Nanoparticles as a new Lithium-Ion Anode Material. *Chem. Mater.* **2013**, *25*, 4977–4985. [CrossRef]

13. Mueller, F.; Geiger, D.; Kaiser, U.; Passerini, S.; Bresser, D. Elucidating the Impact of Cobalt Doping on the Lithium Storage Mechanism in Conversion/Alloying-Type Zinc Oxide Anodes. *ChemElectroChem* **2016**, *3*, 1311–1319. [CrossRef]

14. Bresser, D.; Passerini, S.; Scrosati, B. Leveraging valuable synergies by combining alloying and conversion for lithium-ion anodes. *Energy Environ. Sci.* **2016**, *9*, 3348–3367. [CrossRef]

15. Mueller, F.; Gutsche, A.; Nirschl, H.; Geiger, D.; Kaiser, U.; Bresser, D.; Passerini, S. Iron-Doped ZnO for Lithium-Ion Anodes: Impact of the Dopant Ratio and Carbon Coating Content. *J. Electrochem. Soc.* **2017**, *164*, A6123–A6130. [CrossRef]

16. Mote, V.; Purushotham, Y.; Dole, B. Williamson-Hall analysis in estimation of lattice strain in nanometer-sized ZnO particles. *J. Theor. Appl. Phys.* **2012**, *6*, 6–13. [CrossRef]

17. Giuli, G.; Trapananti, A.; Mueller, F.; Bresser, D.; Dácapito, F.; Passerini, S. Insights into the Effect of Iron and Cobalt Doping on the Structure of Nanosized ZnO. *Inorg. Chem.* **2015**, *54*, 9393–9400. [CrossRef] [PubMed]

18. Shannon, R.D. Revised Effective Ionic Eadii and Systematic Studies of Interatomic Distances in Halides and Chalcogenides. *Acta Crystallogr. Sect. A* **1976**, *32*, 751–767. [CrossRef]

19. Kumar, S.; Mukherjee, S.; Singh, R.K.; Chatterjee, S.; Ghosh, A.K. Structural and optical properties of sol-gel derived nanocrystalline Fe-doped ZnO. *J. Appl. Phys.* **2011**, *110*, 103508–103516. [CrossRef]

20. Reddy, A.J.; Kokila, M.K.; Nagabhushana, H.; Sharma, S.C.; Rao, J.L.; Shivakumara, C.; Nagabhushana, B.M.; Chakradhar, R.P.S. Structural, EPR, photo and thermoluminescence properties of ZnO:Fe nanoparticles. *Mater. Chem. Phys.* **2012**, *133*, 876–883. [CrossRef]

21. Hazen, R.M.; Jeanloz, R. Wüstite ($Fe_{1-x}O$): A review of its defect structure and physical properties. *Rev. Geophys.* **1984**, *22*, 37–46. [CrossRef]

22. Wilke, M.; Farges, F.; Petit, P.E.; Brown, G.E.; Martin, F. Oxidation state and coordination of Fe in minerals: An Fe K-XANES spectroscopic study. *Am. Mineral.* **2001**, *86*, 714–730. [CrossRef]

23. Giuli, G.; Pratesi, G.; Cipriani, C.; Paris, E. Iron local structure in tektites and impact glasses by extended X-ray absorption fine structure and high-resolution X-ray absorption near-edge structure spectroscopy. *Geochim. Cosmochim. Acta* **2002**, *66*, 4347–4353. [CrossRef]

24. Giuli, G.; Pratesi, G.; Eeckhout, S.G.; Koeberl, C.; Paris, E. Iron reduction in silicate glass produced during the 1945 nuclear test at the Trinity site (Alamogordo, New Mexico, USA). In *Large Meteorite Impacts and Planetary Evolution IV*; Gibson, R.L., Reimold, W.U., Eds.; Geological Society of America: Boulder, CO, USA, 2010; ISBN 9780813724652.

25. Larson, A.C.; Von Dreele, R.B. *General Structure Analysis System (GSAS)*; Los Alamos National Laboratory Report Laur 86–748; The Regents of the University of California: Oakland, CA, USA, 2004.

26. Xu, Y.-N.; Ching, W.Y. Electronic, optical, and structural properties of some wurtzite crystals. *Phys. Rev. B* **1993**, *48*, 4335–4351. [CrossRef]

27. D'Acapito, F.; Trapananti, A.; Torrengo, S.; Mobilio, S. X-ray Absorption Spectroscopy: The Italian Beamline GILDA of the ESRF. *Not. Neutroni Luce Sincrotrone* **2014**, *19*, 14–23.

28. Giuli, G.; Cicconi, M.R.; Paris, E. The $^{[4]}Fe^{3+}$–O distance in synthetic kimzeyite garnet, Ca_3Zr_2 $[Fe_2SiO_{12}]$. *Eur. J. Mineral.* **2012**, *24*, 783–790. [CrossRef]

![materials logo] *materials*

MDPI

Article

Surface Properties of Nanostructured, Porous ZnO Thin Films Prepared by Direct Current Reactive Magnetron Sputtering

Monika Kwoka [1,*], Barbara Lyson-Sypien [1], Anna Kulis [1], Monika Maslyk [2], Michal Adam Borysiewicz [2], Eliana Kaminska [2] and Jacek Szuber [1]

[1] Institute of Electronics, Silesian University of Technology, 44-100 Gliwice, Poland; Barbara.Lyson-Sypien@polsl.pl (B.L.-S.); Anna.Kulis@polsl.pl (A.K.); Jacek.Szuber@polsl.pl (J.S.)
[2] Institute of Electron Technology, 02-668 Warsaw, Poland; mmaslyk@ite.waw.pl (M.M.); mbory@ite.waw.pl (M.A.B.), eliana@ite.waw.pl (E.K.)
* Correspondence: Monika.Kwoka@polsl.pl; Tel.: +48-32-237-20-57

Received: 11 November 2017; Accepted: 8 January 2018; Published: 14 January 2018

Abstract: In this paper, the results of detailed X-ray photoelectron spectroscopy (XPS) studies combined with atomic force microscopy (AFM) investigation concerning the local surface chemistry and morphology of nanostructured ZnO thin films are presented. They have been deposited by direct current (DC) reactive magnetron sputtering under variable absolute Ar/O_2 flows (in sccm): 3:0.3; 8:0.8; 10:1; 15:1.5; 20:2, and 30:3, respectively. The XPS studies allowed us to obtain the information on: (1) the relative concentrations of main elements related to their surface nonstoichiometry; (2) the existence of undesired C surface contaminations; and (3) the various forms of surface bondings. It was found that only for the nanostructured ZnO thin films, deposited under extremely different conditions, i.e., for Ar/O_2 flow ratio equal to 3:0.3 and 30:3 (in sccm), respectively, an evident and the most pronounced difference had been observed. The same was for the case of AFM experiments. What is crucial, our experiments allowed us to find the correlation mainly between the lowest level of C contaminations and the local surface morphology of nanostructured ZnO thin films obtained at the highest Ar/O_2 ratio (30:3), for which the densely packaged (agglomerated) nanograins were observed, yielding a smaller surface area for undesired C adsorption. The obtained information can help in understanding the reason of still rather poor gas sensor characteristics of ZnO based nanostructures including the undesired ageing effect, being of a serious barrier for their potential application in the development of novel gas sensor devices.

Keywords: ZnO nanostructures; reactive magnetron sputtering; XPS; surface chemistry

1. Introduction

Zinc oxide (ZnO) is one of the most popular transparent conductive oxides (TCO) having unique optical and electronic properties (wide energy band gap of 3.37 eV and large exciton binding energy of -60 meV). Moreover, among the TCO materials, ZnO exhibits high charge carrier mobility (from several to hundreds cm^2/Vs). However, its precise value strongly depends on its forms and dimensionalities directly related to the preparation and deposition methods, what was nicely reviewed by Jagadish and Pearton [1].

The unique optical and electronic properties make zinc oxide a promising candidate mainly for selected optoelectronic devices [1] and solar cells [2]. In turn, having the electrical conductivity being sensitive to the composition of surrounding atmosphere, due to the adsorption/desorption processes on its surface, ZnO is a promising material for gas sensor application [3].

It is commonly known that the gas sensing performance of ZnO thin films strongly depends on their morphology, structure, and related surface nonstoichiometry. It is generally accepted that the high gas sensitivity can be achieved for the gas sensing material characterized by the internal structure exhibiting the largest surface-to-volume ratio. This condition can be fulfilled by using low dimensional nanostructures, especially 1D systems like nanowires, nanobelts, nanotubes, etc. [4,5]. In the literature, one can find numerous nanostructures reported for ZnO being fabricated using a wide array of techniques [6]. However, since the technology of 1D nanostructures elaboration is a time-consuming as well as relatively expensive task, their implementation is up to now rather unattractive for large-scale gas sensors device fabrication.

This is why special attention has recently been paid to other forms of ZnO with a large surface-to-volume ratio. From this point of view, ZnO porous thin films are well promising.

Apart from sol-gel [7–9], sol-gel combined with spin coating [10,11], sol-gel combined with dip coating [12,13], and low temperature electrochemical deposition [14,15], direct current (DC) magnetron sputtering can be successfully used to obtain nanoporous ZnO thin films [16–19] having, among others, improved gas sensor characteristics [20,21], as was reviewed by Kumar et al. [22].

Using magnetron sputtering technology, the various porous TiO_2 nanostructures, among other nanocolumnar and scaffold types, have also been recently obtained, however, mainly for their photovoltaics potential application [23,24], for which the expectations concerning the morphology and structure are rather different with respect to the gas sensors application, this being a main motivation of our studies.

Our recent Energy Dispersive X-ray (EDX) analysis of sputtered Zn/ZnO nanostructures fabricated at a constant total gas pressure, set argon-to-oxygen flow ratio, and changing gas flow values [19] confirmed that with increasing the gas flow, the respective oxygen content inside the obtained films also increased, probably as a result of the surface oxidation of the smaller grains. However, still, the relative O/Zn concentration was in the range of 0.42–0.80, meaning that even the final ZnO nanostructured porous thin films were rather very far from stoichiometry.

Independently, it was also confirmed by various groups [19,20,24] that with increasing gas flows during deposition, as well as when argon partial pressure in the plasma increased, the grain size of ZnO nanostructures decreased. This phenomenon was also observed in our recent studies [19], as, for the highest Ar/O_2 flow ratio (30:3), due to the increasing density of oxygen and related nucleation centers for new grains formation, dense nanoporous ZnO films were created.

This last fact is absolutely crucial from the point of the potential applications of porous ZnO nanostructured thin films as gas sensing materials, since the strongest gas sensing effect mainly appears just at the subsurface of the sensing layer at the depth related to the Debye length of a few nm [4,5].

However, this information is not enough when trying to understand the still poorly known characteristics of ZnO nanostructures [23,24]. This is related to the fact that the gas sensing mechanism involves surface chemisorption, i.e., charges exchange between adsorbed gaseous species and the surface of porous ZnO nanostructures. This is why surface analytical methods able to give the information about the surface chemistry (including undesired contaminations) of ZnO nanostructured forms are required.

Having all the above in mind, we have decided to perform comparative studies of the surface properties of the porous ZnO nanostructured thin films with various morphologies. In order to remove from the analysis the considerations related to different growth techniques, instead of collecting different ZnO morphologies grown by different methods, we chose to use different samples deposited by DC reactive magnetron sputtering, as developed by our group [19]. The application of specific deposition conditions, i.e., with a constant total gas pressure and a set argon-to-oxygen flow ratio of 10:1, while changing the Ar/O_2 gas flow in the range from 3:0.3 to 30:3 (in sccm), enabled us to achieve various nanostructured hierarchical morphologies, varying by the degree of surface development. The growth mechanism relies on a high plasma concentration to achieve low kinetic energy of the atoms ejected from the target surface, resulting in a low adatom mobility, with which the self-shadowing effects yield nanoporous morphologies, as was already discussed in detail in our recent paper [19].

In order to examine reliably the surface properties of the porous ZnO nanostructured thin films, in this study, we have decided to use the surface sensitive methods i.e., X-ray photoelectron spectroscopy (XPS) combined with atomic force microscopy (AFM), having the information depth related to the Debye length [4,5]. The proposed approach is absolutely crucial for deeper insight to the local surface properties of ZnO nanostructures with a special emphasis on surface chemistry (including undesired surface contaminations) directly related to their surface morphology.

Since the gas sensing effect involves surface chemisorption, a detailed characterization of the fundamental physico-chemical properties is required for the adequate design and construction of novel type gas sensor devices based on this material. The sensor effect appears just at the surface of the sensing material at the depth related to the Debye length [4,5], which is quite similar to the information depth for the XPS method

2. Results and Discussion

At the beginning, the AFM images were recorded for the nanostructured ZnO thin films deposited at the different conditions, i.e., for Ar/O$_2$ flow ratio ranging from 3:0.3 to 30:3 (in sccm). However, because the evident and most pronounced difference in AFM images has only been observed for the ZnO samples deposited at the extremely different conditions, i.e., for Ar/O$_2$ flow ratio equal to 3:0.3 and 30:3 (in sccm), respectively, the detailed analysis of surface morphology has been proposed for these two selected nanostructured ZnO thin films, on the base of respective AFM images shown in Figure 1.

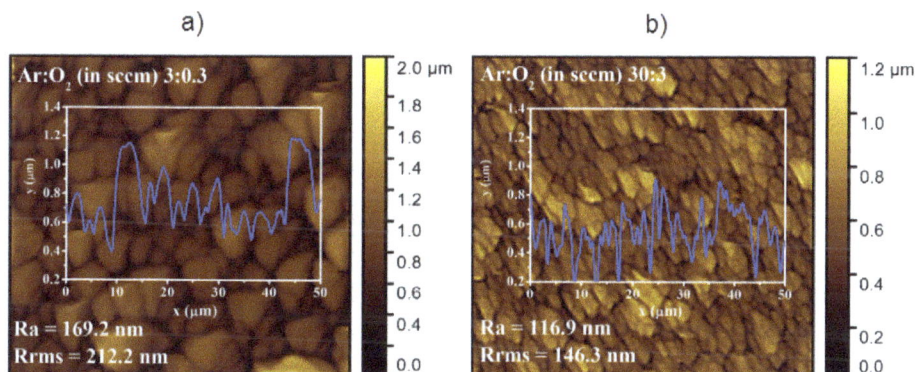

Figure 1. AFM images of nanostructured ZnO thin films deposited at the Ar/O$_2$ gas flow of 3:0.3 (**a**) and 30:3 (**b**), respectively (in sccm); R_a denotes arithmetical mean deviation of the assessed profile, whereas R_{rms} is a root mean square roughness parameter.

The visual information shown in Figure 1 has been deeply confirmed on the base of selected AFM analytical parameters, i.e., the commonly used root mean square roughness parameter, R_{rms}, as well as the arithmetical mean deviation of the assessed profile, R_a, defined as:

$$R_a = \frac{1}{L} \int_0^L |Z(x)| dx$$

where $Z(x)$ represents the function that describes the depth profile, whereas L denotes length taken into account.

From the respective AFM images shown in Figure 1, it is evident that for the highest Ar/O_2 flow ratio, the most dense ZnO nanostructures of highest porosity were obtained containing the well-recognized grains with a dimension of 100 nm, having a shape similar to nanoflowers, which in turn consist of densely packaged (agglomerated) nanograins having the average size in the range of 20–40 nm. It was related to the fact that only for the highest Ar/O_2 flow does the highest amount of oxygen inhibit the growth and coalescence of Zn crystallites by promoting new nucleation centers at the growth front in the presence of oxygen adatoms. This leads to a decrease in the crystallite size, directly causing the modification of samples morphology, as explained in details in [19] based on XRD, TEM, and SEM measurements. In turn, the XPS experiments, being the main point of this research, have been performed in order to verify the local surface chemistry of nanostructured ZnO thin films.

Within the XPS experiments, the survey spectra in the commonly used binding energy range (1200 eV) for the nanostructured ZnO thin films deposited at the different abovementioned Ar/O_2 flow ratios, have been recorded. In general, the respective XPS survey spectra were very similar, apart from the nanostructured ZnO thin films deposited under extremely different conditions i.e., for Ar/O_2 flow ratio equal to 3:0.3 and 30:3 (in sccm), respectively, for which an evident and the most pronounced difference has been observed, as for the case of AFM experiments. However, because the very large undesired background was observed—especially in the binding energy range 1200–600 eV, together with the contribution from undesired Auger electron emission line at 970 eV coming from the number of Auger transition O KLL—for the determination of relative concentration of main elements at the surface (in subsurface layers) of the nanostructured ZnO thin films, the survey spectra in the limited lower binding energy range (600 eV) were only taken into account. This is why Figure 2 only shows the XPS survey spectra in the limited lower binding energy range (600 eV) recalibrated with respect to XPS O1s spectral line obtained for our nanostructured ZnO thin films deposited under the abovementioned extremely different conditions.

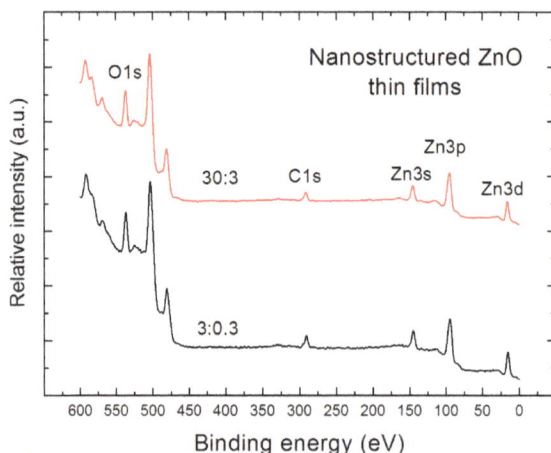

Figure 2. XPS survey spectra of nanostructured ZnO thin films deposited at the Ar/O_2 gas flow of 3:0.3 and 30:3 (in sccm), respectively.

The XPS survey spectra shown in Figure 2 mainly contain the contribution from XPS core level lines: O1s, Zn3s, Zn3p, and Zn3d, corresponding to the main elements of our nanostructured ZnO thin films. Moreover, what is crucial for our research, an evident undesired contribution of C1s XPS lines is observed, which confirmed the strong C contamination at the surface of nanostructured ZnO samples under investigation. However, apart from the above specific XPS lines in our spectra shown in Figure 2, one can also observe the additional peaks related to the Auger electron emission lines at

~570 eV, ~500 eV, and ~470 eV, corresponding to the Zn $L_3M_{23}M_{45}$, Zn $L_3M_{45}M_{45}$, and Zn $L_2M_{45}M_{45}$ Auger electron transitions, respectively.

As was mentioned above, using the XPS survey spectra shown in Figure 2, the relative concentration of main specific elements like O, Zn, and C, with respect to all the recognized surface atoms, in subsurface layers of our nanostructured ZnO thin films was determined using the following well-known analytical formulas [25,26]:

$$\frac{[O]}{[Zn]+[O]+[C]} = \frac{\frac{I_O}{ASF_O}}{\frac{I_{Zn}}{ASF_{Zn}}+\frac{I_O}{ASF_O}+\frac{I_C}{ASF_C}} \quad \text{and} \quad \frac{[C]}{[Zn]+[O]+[C]} = \frac{\frac{I_C}{ASF_C}}{\frac{I_{Zn}}{ASF_{Zn}}+\frac{I_O}{ASF_O}+\frac{I_C}{ASF_C}}$$

and using the abovementioned intensity (height) I_i of the O1s, C1s, and Zn3p core level lines (peaks), then corrected by the transmission function $T(E)$ of CHA PHOIBOS 100 of 1.00, 0.90, and 0.85, respectively, and finally after taking into account the atomic sensitivity factors *ASF* related to the height of peaks for O1s (0.66), C1s (0.25), and Zn3p (0.4), respectively. The obtained data are summarized in Table 1.

Table 1. The relative concentrations of all the main specific elements of nanostructured ZnO thin films in the subsurface layers.

Ar/O$_2$ Ratio at Deposition of ZnO Thin Films (in sccm)	Relative Concentration of the Main Specific Elements		
	[O]/([Zn] + [O] + [C])	[Zn]/([Zn] + [O] + [C])	[C]/([Zn] + [O] + [C])
3:0.3	0.25 ÷ 0.03	0.48 : 0.03	0.27 ÷ 0.03
30:3	0.29 ÷ 0.03	0.53 ÷ 0.03	0.18 ÷ 0.03

Of course, the above relative concentrations of the basic specific elements can be also be expressed as their total relative concentration (in %) in the subsurface region. On the basis of the information summarized in Table 1, one can notice that the relative concentration of O atoms with respect to all other surface atoms for our nanostructured ZnO thin films is rather similar because it only varies in the range of 0.25–0.29, being higher for the ZnO sample deposited at the highest gas flow ratio (30:3).

In turn, the relative concentration of Zn atoms with respect to all the surface atoms for our nanostructured ZnO thin films is rather similar because only varies in the range of 0.48–0.55, being higher for the nanostructured ZnO thin films deposited at highest gas flow ratio (30:3). Crucially, the respected difference in Zn concentration is more than two times larger with respect to the accuracy.

In contrary to the above, the relative concentration of C atoms with respect to all the surface atoms for our nanostructured ZnO thin films is evidently different varies in the range 0.27–0.18, being evidently lower for the nanostructured ZnO thin films deposited at highest gas flow ratio (30:3). It means that the respected difference in C concentration is three times larger with respect to the accuracy. This last information is crucial because it helps to recognize and then interpret the role of undesired C contaminations at the surface of our nanostructured ZnO thin films.

In general, the results described above prove that an evident nonstoichiometry appears in the surface/subsurface region of our nanostructured ZnO thin films, combined with the relatively high concentration of undesired C surface contaminations. This can be probably related to the existence of the specific additional forms of oxygen as well as carbon surface bondings.

In order to solve this issue, during the next step of our XPS research, we have focused on the more precise analysis of the local surface chemistry of our nanostructured ZnO thin films, with a special emphasis on the specific surface bonding. This analysis is based on the recorded XPS spectral lines: Zn2p, O1s, and C1s after their deconvolution. The deconvolution procedure was performed using the Casa XPS SPECS software. The obtained results are presented in Figures 3–5, respectively, and interpreted below.

Figure 3 shows the Zn2p$_{3/2}$ XPS lines for our both nanostructured ZnO thin films (having the highest intensity among all the specific XPS Zn lines).

a)   b)

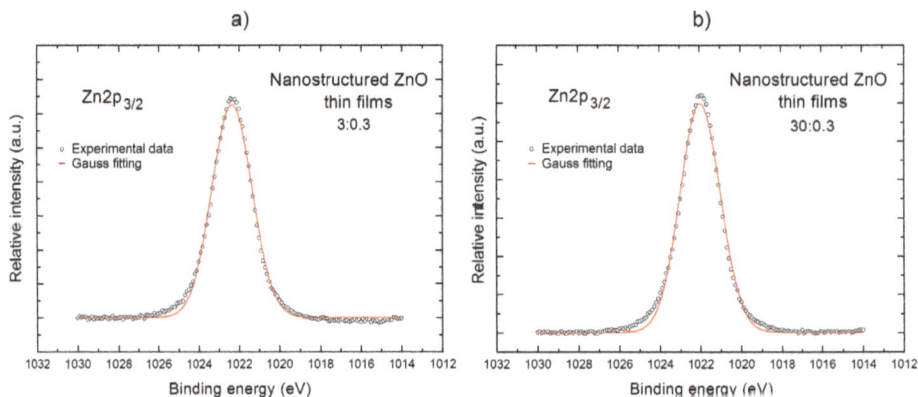

Figure 3. The XPS Zn2p$_{3/2}$ lines after deconvolution using Gauss fitting for the nanostructured ZnO thin films deposited at the Ar/O$_2$ gas flow of 3:0.3 (**a**) and 30:3 (**b**) (in sccm), respectively, having the most different Zn surface concentration.

It is evident that shape of XPS Zn2p$_{3/2}$ lines for our both samples look symmetrical. However, in order to verify the existence of any specific surface bonding, the decomposition procedure was performed for the XPS Zn2p$_{3/2}$ lines (after the respective linear background subtraction) using Gauss fitting, and the obtained results are shown as the red curves in Figure 3.

The deconvoluted XPS Zn2p$_{3/2}$ lines for our both samples are characterized by a very large line fitting parameter (RMS = 0.995) being close to 1, having similar line widths for both ZnO samples at the level of ~2.4 eV, which confirms that only one component is observed always at the binding energy ~1022 eV, which corresponds to the zinc atoms of ZnO lattice at the surface.

In turn, Figure 4 shows the O1s XPS lines for our both nanostructured ZnO thin films. It is evident that in contrary to XPS Zn2p$_{3/2}$ lines, the XPS O1s lines exhibit an evident asymmetry, which is probably related to the existence of different forms of oxygen bondings at their surface. The detailed verification of potential forms of O bondings at the surface of our nanostructured ZnO thin films was performed on the basis of deconvolution of XPS O1s (after the respective linear background subtraction) using the Gauss fitting procedure, and the obtained results are shown as the red curves in Figure 4.

a)   b)

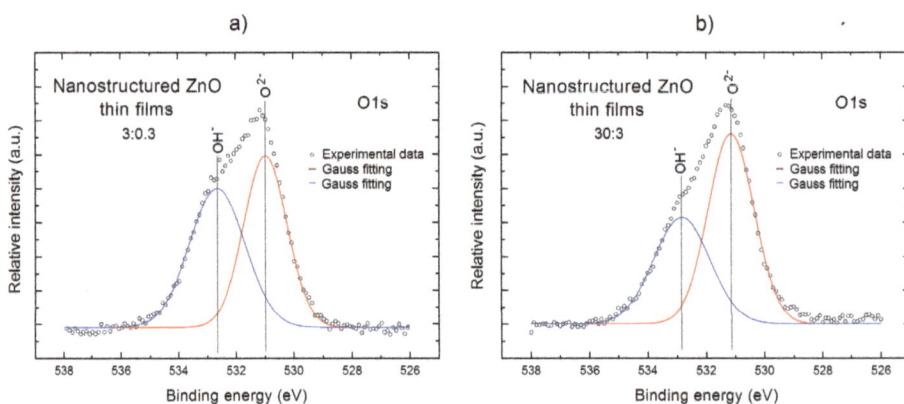

Figure 4. The XPS O1s lines after deconvolution using Gauss fitting for the two selected nanostructured ZnO thin films deposited at Ar/O$_2$ gas flow of 3:0.3 (**a**) and 30:3 (**b**), respectively, having the most different total relative O concentration.

For both nanostructured ZnO samples, the XPS O1s lines consist of two components, which are shown as the blue and red curves, respectively. The first one is located at the binding energy of ~531.0 eV and can be attributed to the O^{2-} ions in ZnO lattice, whereas the second one at binding energy ~532.5 eV corresponds to the oxygen atoms in OH^- groups at ZnO surface. The XPS line widths of recognized components for our both ZnO samples very similar and equal to 2.35 eV for the left component and 1.88 eV for the right component, respectively.

Similar results concerning the components of O1s XPS line were obtained by Gazia et al. [27] for the spongelike nanostructured ZnO films deposited from the sputtered nanostructured zinc films. Using the components of XPS O1s lines shown in Figure 4, the relative area under them corresponding to the O^{2-} ions and OH^- groups was determined. For the nanostructured ZnO thin films deposited at lower Ar/O_2 ratio (3:0.3), the contributions of OH^- groups and O^{2-} ions are almost comparable (~1.0). In turn, for the nanostructured ZnO thin film deposited at the highest Ar/O_2 ratio (30:3), the relative concentration of O^{2-} ions over OH^- groups increases reaching the value ~1.5.

In should be underlined at this moment that the abovementioned information related to the existence of OH^- groups at the surface of our both ZnO samples remains in agreement with the information obtained from the XPS C1s lines for our both nanostructured ZnO thin films, which are shown in Figure 5. These XPS C1s lines confirm the existence of undesired C contaminations appearing at the surface of our nanostructured ZnO thin films after their exposition to the air atmosphere during the transportation between the deposition chamber and XPS chamber.

It is evident that, in contrary to XPS $Zn2p_{3/2}$ lines, but similar to the XPS O1s lines, the XPS C1s lines look symmetrical. However, in order to verify the existence of any specific surface bonding, their decomposition was performed (after the respective linear background subtraction) using the Gauss fitting procedure, and the obtained results are shown as the red curves in Figure 5.

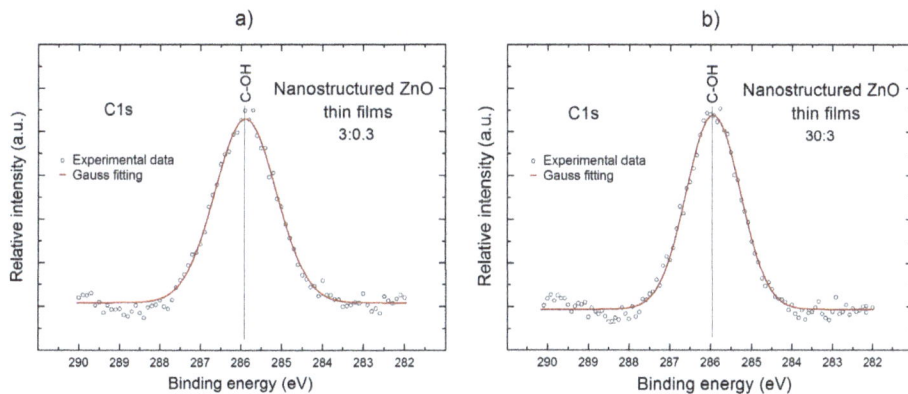

Figure 5. The XPS C1s lines after deconvolution using Gauss fitting for the selected nanostructured ZnO thin films deposited at Ar/O_2 gas flow of 3:0.3 (**a**) and 30:3 (**b**), respectively, having the most different C total relative concentration.

It is obvious that, for our both ZnO samples, after deconvolution of XPS C1s lines (with rather high line fitting parameter (RMS ~0.98)), only one component is observed at the binding energy of ~286 eV, having similar line widths of 1.84 eV for both ZnO samples, which can be attributed to the C–OH surface bondings [28].

This last information confirms that, in the case of our nanostructured ZnO thin films, one can observe the contribution from the two types of different hydroxyl groups (OH) at their surface, i.e., OH^- observed in XPS O1s peaks at binding energy ~532.5 eV, and C–OH groups observed in XPS

C1s peaks at the binding energy of ~286 eV. The presence of these hydroxyl groups can lead to the variation of local surface chemistry of our nanostructured ZnO thin films.

The different C concentrations at the surface of our nanostructured ZnO thin films are related to their nonstoichiometry, which can be directly correlated with their local surface morphology. As was mentioned above, in the case of the nanostructured ZnO thin films deposited at the highest Ar/O_2 ratio (30:3), having slightly higher nonstoichiometry (0.29/0.53), the highest porosity is observed, as the results of existence of densely packaged (agglomerated) nanograins having the average size in the range 20–40 nm, as shown in Figure 1. This is probably why in this case, the lowest (0.18) total relative C concentration at the surface was observed, what is related to the smallest surface area for carbon adsorption directly corresponding to the contribution of OH^- groups, because this ZnO sample adsorbs easier the OH^- groups at the surface.

This last observation was in opposite to the nanostructured ZnO thin films deposited at the lowest Ar/O_2 ratio (3:0.3), containing the well-recognized grains with a dimension of 100 nm, for which the highest relative C concentration (0.27) was observed, even having only slightly lower relative O concentration (0.25).

From the point of view of the possible gas sensing application of nanostructured ZnO thin films at this stage, one can conclude that ZnO nanostructures obtained at the highest Ar/O_2 ratio (30:3), having the lowest level of C contaminations can be promising candidate for the detection of mainly oxidizing gases, especially in the presence of water vapor (H_2O), because their nonstoichiometry corresponds to the higher concentration of oxygen vacancies, which always play a crucial role as the specific adsorption sites for various active oxidizing gases during the gas sensing process. It causes that these nanostructured ZnO thin films can be very sensitive mainly to the toxic gas species containing oxygen from the environment, like nitrogen dioxide (NO_2).

This is crucial because the high undesired concentration of C contamination including C–OH species always play a role of undesired barrier for instance toxic gas adsorption, especially at the lower working temperature, and can additionally strongly affect the uncontrolled sensor ageing effect.

3. Materials and Methods

ZnO nanostructures were obtained by DC reactive magnetron sputtering using zinc target at 80 W DC power under 4 mbar total pressure and Ar/O_2 gas flow at a set ratio (in sccm): 3:0.3; 8:0.8; 10:1; 15:1.5; 20:2; and 30:3. A 5′ long presputtering was performed before every deposition process to clean and stabilize the target surface. The thin films were prepared using a Surrey NanoSystems γ 1000 C reactor. The thicknesses of the films were in the range of 886 nm for the "3:0.3 sccm" sample (thickest film) to 570 nm for the "30:3 sccm" sample (thinnest film). With a constant deposition time of 1 h, this yielded deposition rates from 2.46 to 1.58 Å/s, respectively. The cathode voltages changed by 5%, from 400 V for 3:0.3 to 380 V for 30:3. The gas flows were controlled by dedicated Mass Flow Controllers and the system that utilized a capacitive Baratron manometer in a feed-back loop with a throttle valve (VAT, Haag, Switzerland) for pressure control. The deposition chamber was pumped by a cryogenic pump and the throttle valve was at the entrance to the pump. Such a setup enabled to obtain 10^{-5} Pa base vacuum and an independent control of the total gas pressure and the gas flow values. Si (100) wafers were used as the substrates. In order to remove any native oxides, prior to the ZnO nanostructured thin films deposition, the substrates were firstly degreased by boiling in selected organic solutions and bathed in the buffered HF solution to strip the oxide. The technological details concerning preparation of nanostructured ZnO thin films can be found elsewhere [19].

The surface chemistry together with the possible contaminations of the abovementioned nanostructured ZnO thin films have been examined by XPS method. In these studies, the commercial XPS spectrometer (SPECS, Berlin, Germany) equipped with the X-ray lamp (AlK$_\alpha$, 1486.6 eV, XR-50 model) and a concentric hemispherical analyzer (PHOIBOS-100 Model) was applied, pumped by oil-free pumping unit containing the Varian 110 model Scroll pump, Varian 551 model Turbo-pump, and Varian 300 model ion pump. The basic working pressure was below $\sim 10^{-8}$ hPa, controlled with the Granville Phillips 360 model gas pressure system

The binding energies (BE) of all the registered XPS spectra have been calibrated to Au4f peak at 84.5 eV. Other experimental details have been described elsewhere [29–31].

The nanostructured ZnO thin films' morphology was studied using AFM Bruker MultiMode 8 system (Bruker, Santa Barbara, CA, USA). It consists of MultiMode8 head completed with three scanners: AS-130VLR-2, AS-2VLR-2, and AS-05-2, having different scanning ranges (areas) and working with the NANOSCOPE V controller using the advanced original NanoScope V9.10 software. The MultiMode8 head is placed on a specific table (VT-102-2 model) equipped with pneumatic isolation system against vibrations, combined with the air compressor, which allows for the elimination of undesired mechanical vibrations of the surroundings.

4. Conclusions

In this paper, the information on the local surface chemistry of ZnO thin films, deposited by DC reactive magnetron sputtering under different Ar/O_2 gas flow ratio, was obtained using the XPS method. Basing on these experimental results, we were able to obtain the crucial information on: (1) the total relative concentrations of main elements combined with nonstoichiometry; (2) the existence of undesired C surface contaminations; and (3) the various forms of surface bondings. What is extremely important is that the lowest amount of undesired C contamination was observed at the surface of our nanostructured ZnO thin films deposited at the highest Ar/O_2 ratio, which can be directly correlated with their local surface morphology observed by SEM and related to the densely packaged (agglomerated) nanograins, yielding a smaller surface area for carbon absorption.

The information obtained in our studies can be very helpful in the interpretation of still rather poorly known gas sensor characteristics (mainly dynamic) of ZnO, which exhibits high electronic mobility (up to $2\,cm^2/V\cdot s$) and thus can be a very prospective gas sensor material, especially in the form of nanostructures. However, an exact gas sensor mechanism, including ageing effect in the case of various ZnO nanoforms, still remains unclear and requires further study.

Acknowledgments: This work was realized within the Statutory Funding of Institute of Electronics, Silesian University of Technology, Gliwice, and is partially supported by research grant of National Science Centre, Poland—OPUS 11, No. 2016/21/B/ST7/02244 and by National Centre of Research and Development within Program Lider V, No. LIDER/030/615/L-5/NCBR/2014.

Author Contributions: M.K. was involved in carrying out the XPS and AFM experiments, analyzing the experimental data, and drafting the manuscript. B.L.-S. conceived the XPS and AFM experiments, conducted data analysis, and verified the manuscript. A.K. carried out the AFM measurements. M.M. performed the samples. M.A.B. involved in the preparation of samples and analyzing the data. E.K. and J.S. conceived the study. All authors read and approved the final version of the manuscript.

Conflicts of Interest: The authors declare no conflict of interest.

References

1. Jagadish, C.; Pearton, S.J. *Zinc Oxide Bulk, Thin Films and Nanostructures—Processing, Properties and Application*, 1st ed.; Elsevier: Amsterdam, The Netherlands, 2006; ISBN 13 978-0080447223.
2. Zhang, Q.; Dandeneau, C.S.; Zhou, X.; Cao, G. ZnO Nanostructures for dyesensitized solar cells. *Adv. Mater.* **2009**, *21*, 4087–4108. [CrossRef]
3. Wan, Q.; Li, Q.H.; Chen, Y.J.; Wang, T.H.; He, X.L.; Li, J.P.; Lin, C.L. Fabrication and ethanol sensing characteristics of ZnO nanowire gas sensors. *Appl. Phys. Lett.* **2004**, *84*, 3654–3656. [CrossRef]
4. Eranna, G. *Metal Oxide Nanostructures as Gas Sensing Devices*; CRC Press: Boca Raton, FL, USA, 2012; ISBN 978-1-4398-6340-4.
5. Carpenter, M.A.; Mathur, S.; Kolmakov, A. *Metal Oxide Nanomaterials for Chemical Sensors*; Springer: New York, NY, USA, 2012; ISBN 978-1-4614-5395-6.
6. Wang, Z.L. Nanostructures of zinc oxide. *Mater. Today* **2004**, *7*, 26–33. [CrossRef]

7. Zhang, Y.; Lin, B.; Sun, X.; Fu, Z. Temperature-dependent photoluminescence of nanocrystalline ZnO thin films grown on Si (100) substrates by the sol–gel process. *Appl. Phys. Lett.* **2005**, *86*, 131910–131913. [CrossRef]
8. Srinivasan, G.; Kumar, J. Optical and structural characterization of zinc oxide thin films prepared by sol–gel process. *Cryst. Res. Technol.* **2006**, *41*, 893–896. [CrossRef]
9. Caglar, M.; Zlican, S.; Caglar, Y.; Yakuphanoglu, F. Electrical conductivity and optical properties of ZnO nanostructured thin film. *Appl. Surf. Sci.* **2009**, *255*, 4491–4496. [CrossRef]
10. Heredia, E.; Bojorge, C.; Casanova, J.; Cánepa, H.; Craievich, A.; Kellermann, G. Nanostructured ZnO thin films prepared by sol–gel spin-coating. *Appl. Surf. Sci.* **2014**, *317*, 19–25. [CrossRef]
11. Wagner, A.; Bakin, A.; Otto, T.; Zimmermann, M.; Jahn, B.; Waag, A. Fabrication and characterization of nanoporous ZnO layers for sensing applications. *Thin Solid Films* **2012**, *520*, 4662–4665. [CrossRef]
12. Kaneva, N.V.; Dushkin, C.D. Preparation of nanocrystalline thin films of ZnO by sol-gel dip coating. *Bulgarian Chem. Commun.* **2011**, *43*, 259–263.
13. Musat, V.; Rego, A.M.; Monteiro, R.; Fortunato, E. Microstructure and gas-sensing properties of sol–gel ZnO thin films. *Thin Solid Films* **2008**, *516*, 1512–1515. [CrossRef]
14. Gür, E.; Kiliç, B.; Coşkun, C.; Tüzemen, S.; Bayrakçeken, F. Nanoporous structures on ZnO thin films. *Superlattices Microstruct.* **2010**, *47*, 182–186. [CrossRef]
15. Bai, S.; Sun, C.; Guo, T.; Luo, R.; Lin, Y.; Chen, A.; Sun, L.; Zhang, J. Low temperature electrochemical deposition of nanoporous ZnO thin films as novel NO_2 sensors. *Electrochim. Acta* **2013**, *90*, 530–534. [CrossRef]
16. Hezam, M.; Tabet, N.; Mekki, A. Synthesis and characterization of DC magnetron sputtered ZnO thin films under high working pressures. *Thin Solid Films* **2010**, *518*, 161–164. [CrossRef]
17. Borysiewicz, M.A.; Dynowska, E.; Kolkovsky, V.; Dyczewski, J.; Wielgus, M.; Kaminska, E.; Piotrowska, A. From porous to dense thin ZnO films through reactive DC sputter deposition onto Si(100) substrates. *Phys. Status Solidi(a)* **2012**, *209*, 2463–2469. [CrossRef]
18. Shirazi, M.; Hosseinnejad, M.T.; Zendehnam, A.; Ghoranneviss, M.; Etaati, G.R. Synthesis and characterization of nanostructured ZnO multilayer grown by DC magnetron sputtering. *J. Alloys. Compd.* **2014**, *602*, 108–116. [CrossRef]
19. Masłyk, M.; Borysiewicz, M.A.; Wzorek, M.; Wojciechowski, T.; Kwoka, M.; Kaminska, E. Influence of absolute argon and oxygen flow values at a constant ratio on the growth of Zn/ZnO nanostructures obtained by DC reactive magnetron sputtering. *Appl. Surf. Sci.* **2016**, *389*, 287–293. [CrossRef]
20. Suchea, M.; Christoulakis, S.; Moschovis, K.; Katsarakis, N.; Kiriakidis, G. ZnO transparent thin films for gas sensor applications. *Thin Solid Films* **2006**, *515*, 551–554. [CrossRef]
21. Khalaf, M.K.; Chiad, B.T.; Ahmed, A.F.; Mutlak, F.A.-H. Thin film technique for preparing nano-ZnO gas sensing (O_2, NO_2) using plasma deposition. *Int. J. Appl. Innov. Eng. Manag.* **2013**, *2*, 178–184.
22. Kumar, R.; Al-Dossary, O.; Kumar, G.; Umar, A. Zinc oxide nanostructures for NO_2 gas-sensor applications: A Review. *Nano-Micro Lett.* **2015**, *7*, 97–120. [CrossRef]
23. González-García, L.; González-Valls, I.; Lira-Cantu, M.; Barranco, A.; González-Elipe, A.R. Aligned TiO_2 nanocolumnar layers prepared by PVD-GLAD for transparent dye sensitized solar cells. *Energy Environ. Sci.* **2011**, *4*, 3426–3435. [CrossRef]
24. Sanzaro, S.; Smecca, A.; Mannino, G.; Bongiorno, C.; Pellegrino, G.; Neri, F.; Malandrino, G.; Catalano, M.R.; Condorelli, G.G.; Iacobellis, R.; et al. Multi-Scale-Porosity TiO_2 scaffolds grown by innovative sputtering methods for high throughput hybrid photovoltaics. *Sci. Rep.* **2016**, *6*, 39509. [CrossRef] [PubMed]
25. Wagner, C.D.; Riggs, W.M.; Davis, L.E.; Moulder, J.F.; Mnilenberger, G.E. *Handbook of X-ray Photoelectron Spectroscopy*; Perkin-Elmer: Eden Prairie, MN, USA, 1979.
26. Watts, J.F.; Wolstenholme, J. *An Introduction to Surface Analysis by XPS and AES*; J. Wiley & Sons: Chichester, UK, 2003; ISBN 978-0-470-84713-8.
27. Gazia, R.; Chiodoni, A.; Bianco, S.; Lamberti, A.; Quaglio, M.; Sacco, A.; Tresso, E.; Mandracci, P.; Pirri, C.F. An easy method for the room-temperature growth of spongelike nanostructured Zn films as initial step for the fabrication of nanostructured ZnO. *Thin Solid Films* **2012**, *524*, 107–112. [CrossRef]
28. Moulder, J.F. *Handbook of X-ray Photoelectron Spectroscopy: A Reference Book of Standard Spectra for Identification and Interpretation of XPS Data*; Chastain, J., King, R.C., Eds.; Physical Electronics Division, Perkin-Elmer Corporation: Eden Prairie, MN, USA, 1995.

29. Kwoka, M.; Ottaviano, L.; Koscielniak, P.; Szuber, J. XPS, TDS and AFM studies of surface chemistry and morphology of Ag-covered L-CVD SnO$_2$ nanolayers. *Nanoscale Res. Lett.* **2014**, *9*, 260–269. [CrossRef] [PubMed]

30. Kwoka, M.; Krzywiecki, M. Rheotaxial growth and vacuum oxidation–Novel technique of tin oxide deposition—In situ monitoring of oxidation process. *Mater. Lett.* **2015**, *154*, 1–4. [CrossRef]

31. Kwoka, M.; Krzywiecki, M. Impact of air exposure and annealing on the chemical and electronic properties of the surface of SnO$_2$ nanolayers deposited by rheotaxial growth and vacuum oxidation. *Beilstein J. Nanotechnol.* **2017**, *8*, 514–521. [CrossRef] [PubMed]

materials

MDPI

Article

Scale-Up of the Electrodeposition of ZnO/Eosin Y Hybrid Thin Films for the Fabrication of Flexible Dye-Sensitized Solar Cell Modules

Florian Bittner [1,2], Torsten Oekermann [1,3] and Michael Wark [1,4,*]

1   Institute of Physical Chemistry and Electrochemistry, Leibniz University Hannover, Callinstr. 3a,
    30167 Hannover, Germany; florian.bittner@wki.fraunhofer.de (F.B.);
    torsten.oekermann@saftbatteries.com (T.O.)
2   Application Center for Wood Fiber Research HOFZET®, Fraunhofer Institute for Wood Research,
    Wilhelm-Klauditz-Institute WKI, Heisterbergallee 10A, 30453 Hannover, Germany
3   Friemann & Wolf Batterietechnik GmbH, Industriestr. 22, 63654 Büdingen, Germany
4   Institute of Chemistry, Chemical Technology 1, Carl von Ossietzky University Oldenburg,
    Carl-von-Ossietzky-Str. 9-11, 26129 Oldenburg, Germany
*   Correspondence: michael.wark@uni-oldenburg.de; Tel.: +49-441-798-3675

Received: 28 December 2017; Accepted: 29 January 2018; Published: 2 February 2018

Abstract: The low-temperature fabrication of flexible ZnO photo-anodes for dye-sensitized solar cells (DSSCs) by templated electrochemical deposition of films was performed in an enlarged and technical simplified deposition setup to demonstrate the feasibility of the scale-up of the deposition process. After extraction of eosin Y (EY) from the initially deposited ZnO/EY hybrid films, mesoporous ZnO films with an area of about 40 cm^2 were reproducibly obtained on fluorine doped tin oxide (FTO)-glass as well as flexible indium tin oxide (ITO)–polyethylenterephthalate (PET) substrates. With a film thickness of up to 9 μm and a high specific surface area of up to about 77 m$^2 \cdot$cm^{-3} the ZnO films on the flexible substrates show suitable properties for DSSCs. Operative flexible DSSC modules proved the suitability of the ZnO films for use as DSSC photo-anodes. Under a low light intensity of about 0.007 sun these modules achieved decent performance parameters with conversion efficiencies of up to 2.58%. With rising light intensity the performance parameters deteriorated, leading to conversion efficiencies below 1% at light intensities above 0.5 sun. The poor performance of the modules under high light intensities can be attributed to their high series resistances.

Keywords: electrochemical deposition; scale-up; zinc oxide; eosin Y; dye-sensitized solar cell; solar module

1. Introduction

Currently the generation of electricity by photovoltaics is almost completely limited to stationary solutions. Mobile applications, such as solar cells integrated in clothes for charging mobile electronic devices, have been largely unexploited, but promise huge potentials for environmental benefit and commercial success [1]. Flexible solar cells based on thin film technologies are especially suitable for mobile applications due to their low weight, their breaking resistance, and their adaptability. Furthermore, the manufacturing costs of thin film solar cells are potentially lower than those of conventional silicon-based solar cells. Roll-to-roll processes, which are applicable to flexible substrates, promise high production throughputs [1–3].

One of the thin film solar cell technologies suitable for the realization of flexible devices is that of dye-sensitized solar cells (DSSCs) [4,5]. The conversion efficiency record for DSSCs (reported for an active area of 0.36 cm^2 in 2011) of 12.3% on glass-based substrates [6] is higher than that recently reported for organic solar cells (11.2% [7]) and a-Si solar cells (10.2% [7]). The conversion efficiencies of inorganic thin film solar cells, such as 21.7% [8] for cells based on copper indium gallium diselenide, have not been achieved, but lower production costs are predicted and a much less quantity of low abundant or toxic metals is needed [3]. Another advantage of DSSCs in comparison to other solar cell technologies is the better utilization of diffuse or low-intensity illumination. As unique characteristics, DSSCs are semi-transparent, if both electrodes are based on transparent substrate materials, and their color can be varied [2,3]. The semi-transparency permits a versatile optical design of the solar cells.

Flexible DSSCs have been demonstrated in different configurations. Conversion efficiencies of up to 8.6% [9] have been reported for solar cells with photo-anodes based on metal substrates, which enable the sintering of the nanoparticular TiO$_2$ thin films [10–13]. Disadvantages of this configuration are the necessary back-side illumination that involves optical losses and the sacrifice of the semi-transparency of the solar cells. To preserve the semi-transparency, both electrodes have to be based on transparent plastic foil substrates. Most commonly used are foils of polyethylenterephthalate (PET) or polyethylennaphthalate (PEN) that are coated with indium tin oxide (ITO) as transparent conducting layer. Since plastic foils are not resistant against the high temperatures associated with the TiO$_2$ sintering process, low-temperature methods for the photo-anode fabrication have to be applied [4,5,14]. Several methods such as pressing [15], friction transfer [16], chemical sintering [17,18], or electrophoretic deposition [14,19] have been suggested to fabricate TiO$_2$ thin films on plastic foil substrates. A conversion efficiency of $\eta = 7.6\%$ for flexible DSSCs based on plastic foils has been reported using the pressing method for TiO$_2$ films on ITO-PEN substrates, where a pressure of 100 MPa was applied [15].

Besides TiO$_2$, ZnO has been evaluated as semiconductor material for DSSCs and organic perovskite based solar cells; with the organic perovskite CH$_3$NH$_3$PbI$_3$ recently conversion efficiencies up to 15.4% have been reported [20]. It features a higher electron mobility than TiO$_2$ and it can be synthesized easily at low temperatures in several morphologies and with a very high crystallinity [21,22]. Porous ZnO thin films obtained by electrochemical deposition with assistance of a structure directing agent (SDA) showed particularly favorable properties as DSSC photo-anode material such as fast electron transport [23]. The method utilizes the dye molecule eosin Y (EY) as SDA [24]. After removal of the SDA highly crystalline nanoporous ZnO remains. The temperature applied during film deposition is 70 °C, rendering the method feasible for application on plastic foil substrates. Using the indoline dye D149 as sensitizer, a conversion efficiency of 5.56% has been achieved with this kind of ZnO film on rigid glass substrates [24]. A comprehensive description of this deposition method is given by Yoshida et al. [24]. Briefly summarized, the deposition method is based on the cathodic reduction of dissolved oxygen in aqueous Zn^{2+} solutions, forming hydroxide ions. Zn^{2+} and OH$^-$ ions precipitate on the substrate as Zn(OH)$_2$, which rapidly dehydrates to ZnO at temperatures above 50 °C [24–27]:

$$O_2 + 2H_2O + 4e^- \rightarrow 4OH^- \tag{1}$$

$$Zn^{2+} + 2OH^- \rightarrow Zn(OH)_2 \rightarrow ZnO + H_2O \tag{2}$$

By carrying out the first deposition step without addition of an SDA, a dense layer of compact and highly crystalline ZnO is formed that can act as blocking layer. In the assembled DSSC it is supposed to prevent recombination between electrons in the conductive substrate layer and tri-iodide ions in the electrolyte [24,26–31]. When EY is added to the deposition solution as an SDA, the dye accelerates the reduction of O$_2$ and consequently the film growth. While both the non-reduced EY^{2-} and the electrochemically reduced EY^{4-} ions can be incorporated into the growing ZnO film, only the latter influences the morphology of the ZnO film significantly, resulting in a nanostructured hybrid film [24,30,32,33]. Therefore, the film deposition has to be performed at a potential more negative than the reduction potential of EY^{2-} (ca. −0.8 V vs. Ag/AgCl). The EY can be removed from the hybrid film by alkaline treatment, leaving a highly crystalline and porous ZnO structure [24,33–35].

For the electrodeposition of small-scaled ZnO films, rotating disc electrodes (RDEs) have been routinely used [24,36], since they provide fast and homogeneous mass transfer in the solution towards the electrode surface. This is a prerequisite for the preparation of uniform films. Since the technically complex RDE is limited to a deposition area of a few square centimeters, the suitability for scale-up of the deposition method has not yet been demonstrated. For this purpose we have developed a technically simplified electrodeposition setup that allows the deposition of ZnO/EY hybrid films on substrates up to about 60 cm^2. Being sized between the small laboratory scale and the pilot plant scale, the setup is referred to as a miniplant setup.

This paper presents a detailed description of the miniplant setup and its use for the fabrication of porous ZnO films on glass and plastic foil substrates. The structure of the deposited ZnO films has been compared to that of films prepared in an RDE setup and optimized by adjustment of deposition parameters such as deposition time, electrode potential, and substrate layout. To illustrate the suitability of the obtained ZnO electrodes for the use in flexible DSSCs, solar cell module demonstrators were prepared and characterized.

2. Experimental Methods

2.1. Electrodeposition

2.1.1. General Procedures

The electrolyte composition and conditions for the electrodeposition of ZnO and ZnO/EY hybrid films were based on the commonly used procedure [24,37]. Aqueous solutions of KCl ($c = 0.1$ mol·L^{-1}, $\geq$99.5%, Carl Roth, Karlsruhe, Germany), ZnCl$_2$ ($c = 5$ mmol·L^{-1}, $\geq$98%, ABCR, Karlsruhe, Germany) and, in the case of the deposition of hybrid films, additionally eosin Y ($c = 80$ µmol·L^{-1}, $\geq$80%, Acros Organics, Geel, Belgium) were used as electrolyte. Oxygen saturation of the solutions was achieved by gas bubbling. The solutions were kept at a temperature of 70 °C. Fluorine doped tin oxide (FTO)-coated glass (7 Ω·sq^{-1}, Pilkington TEC A8, NSG group, Minato-ku, Tokyo, Japan) or ITO-coated PET foil (50 Ω·sq^{-1}, CPFilms OC™50, or 15 Ω·sq^{-1}, CPFilms LR15, Canoga Park, CA, USA) served as substrate material. FTO-glass substrates and ITO-PET foil substrates LR15 were pre-treated in a mildly alkaline cleaning agent (1% solution of deconex® 12 BASIC, Borer Chemie AG, Zuchwil, Switzerland).

2.1.2. Fabrication of Small-Scaled ZnO Films

The electrodeposition of small-scaled ZnO and ZnO/EY hybrid films was performed on an RDE (Metrohm Autolab RDE-2, Metrohm Autolab, Utrecht, The Netherlands), rotating with 300 rpm. The deposition area was fixed to 1.54 cm^2. A zinc wire ($d = 1$ mm, $\geq$99.95%, Thermo Fisher (Kandel) GmbH, Karlsruhe, Germany) was used as the counter electrode. The deposition was performed at -0.91 V vs. Ag/AgCl. Compact ZnO films were deposited for 10 min, ZnO/EY hybrid films on top of the compact ZnO films were deposited for 45 min. The desorption of the EY was achieved by immersing the films for 24 h in an aqueous solution of KOH ($V = 500$ mL, pH = 10.5, $\geq$85%, Applichem, Darmstadt, Germany).

2.1.3. Up-Scaled Fabrication of ZnO Films

The miniplant setup that was developed for the up-scaled electrodeposition is schematically illustrated in Figure 1a. It consists of a deposition basin (1) including a substrate holder and a motor-driven paddle, a tempering basin (2), a diaphragm pump (3), and a potentiostat (4). Polypropylene was used as construction material for the basins. The electrodes are connected to the potentiostat (Amel Instruments Mod. 7050, Milan, Italy) in a 3-electrode-setup. The pump (Liquiport NF300KT.18S, KNF Neuberger, Balterswil, Switzerland) forwards the deposition solution from the tempering basin to the deposition basin, from where it flows back into the tempering basin. A flow rate of 2.21 L·min^{-1} was determined as optimum setting of the diaphragm pump, giving a satisfactory electrolyte flow while keeping vibrations low. The volume of the solution necessary for the electrodeposition is approximately 6.5 L.

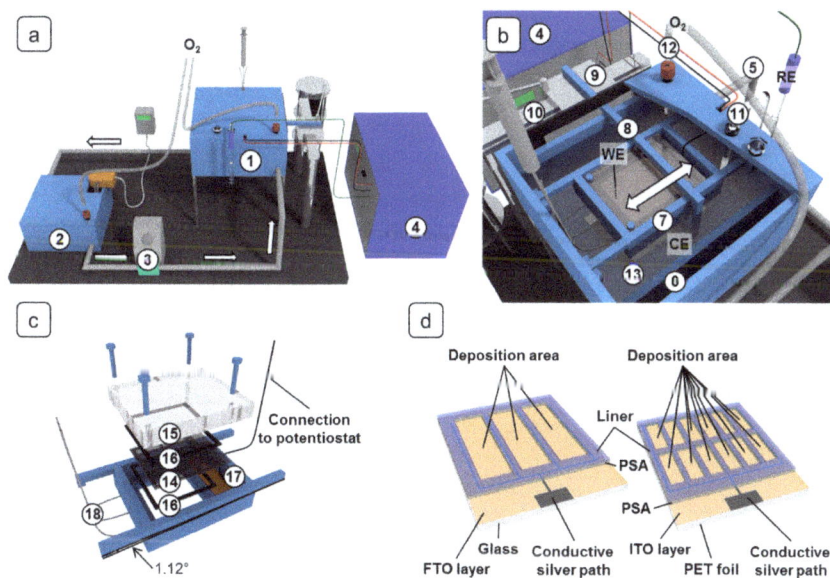

Figure 1. Schematic illustration of the miniplant setup. (**a**) Total view: (1) Deposition basin, (2) Tempering basin, (3) Diaphragm pump, (4) Potentiostat. (**b**) Deposition basin: (5) Inlet, (6) Overflow, (7) Substrate holder, (8) Paddle, (9) Slide, (10) Linear guide, (11) Salt bridge, (12) Glass frit, (13) Zinc foil. (**c**) Substrate holder: (14) Substrate, (15) pressure plate, (16) sealing mats, (17) connection area, (18) gas suction tubes. (**d**) Layout of the FTO-glass (left) and ITO-PET (right) substrates.

To heat the solution, the basin contains a bended heating element encapsulated in polytetrafluoroethylene (PTFE) (custom build Nuega PTFE heating element, 800 W, 13 mm diameter and 1300 mm length) and a combined temperature/liquid level probe (custom build Nuega PTFE/graphite probes, Nuega, Georgensgmuend, Germany).

Figure 1b depicts the deposition basin. The deposition solution is pumped into the basin via the inlet (5) and flows across an overflow (6) back into the tempering basin (2). The overflow is necessary to maintain a constant level of the solution.

During electrodeposition the substrate in the substrate holder (7) serves as the working electrode (WE) and is immersed into the solution with the conductive side pointing downwards. The substrate holder is described below in more detail.

To ensure a homogeneous convection of the solution, a paddle (8) performs a forward and backward movement underneath the substrate. For this purpose the paddle is attached to a slide (9) on a linear guide (10). The surface of the paddle in the solution has dimensions of about 15.5×2.5 mm^2. The movement of the slide is promoted by an electric motor with planetary gearhead via an eccentric disc. The operation of the electric motor at its power limit results in 42.5 cycles per minute for the paddle movement.

A salt bridge (11) (KCl, $c = 0.1$ mol·L^{-1}) connected to an Ag/AgCl reference electrode (RE, XR300, Radiometer analytical, Lyon, France) and a glass frit (12) for oxygen influx are placed in the solution. A zinc foil (13) (ABCR, $\geq$99.9%, thickness 0.62 mm) with dimensions of 150×100 mm^2 acts as counter electrode (CE). The WE and CE wires, which are in contact with the deposition solution, are sheathed with PTFE.

The design of the substrate holder is depicted in Figure 1c. A cut-out of 75×75 mm^2 allows the contact of the substrate with the deposition solution. The substrate (14) is pressed onto the base plate through a pressure plate (15). Tailored sealing mats (16) (EPDM 65, thickness 1.5 mm, Eriks NordOst,

Hannover, Germany) on both sides of the substrate ensure a tight contact between substrate and base plate. The substrate is connected to the potentiostat on the end (17) which is not covered by the sealing mats and which features a dedicated area coated with conductive silver. The WE cable is soldered on a copper plate, against which the substrate is pressed. Additionally, conductive copper tape promotes electrical contact between the substrate and the copper plate. To remove gas bubbles underneath the substrate, the substrate holder has a tilt of 1.12° to collect the gas bubbles on one side of the cut-out. Via three PTFE tubes (18), which are incorporated into the substrate holder, the gas is sucked off before starting the electrodepositions and, if the deposition duration exceeds 20 min, during the deposition as well.

As illustrated in Figure 1d, the substrates are equipped with conducting silver paths to lower the ohmic voltage drop. On FTO-glass substrates the silver paths were prepared manually with silver paste (Acheson Silver DAG 1415, Agar Scientific, Essex, UK), on ITO-PET substrates they were applied by screen printing (mesh: PET 120-34 Y; silver paste: InkTec TEC-PA-040, Ansan-city, Kyungki-do, South Korea). To protect the silver from contact with the deposition solution (and eventually with the redox electrolyte in the DSSC), the paths are masked with tailored pressure-sensitive adhesives (PSA) (tesa® 61562, Hamburg, Germany). At the same time, the PSAs served as spacers between the two electrodes in the assembled DSSCs. During the deposition process the PSAs were protected with liners.

Details about the implications of the electrodeposition scale-up—influencing the substrate layout and the deposition procedure—are given in the results Section 3.1.1.

To ensure a sufficient oxygen content of the solution the depositions were started not earlier than 15 min after the substrate holder was inserted into the deposition basin. The oxygen content prior to the deposition was determined with a colorimetric test kit (Macherey-Nagel VISOCOLOR® ECO Oxygen, Dueren, Germany), proving saturation with oxygen. Directly after the deposition of ZnO/EY hybrid films the substrates were immersed in an aqueous KOH solution (V = 3 L, pH = 11, $\geq$85%, Applichem, Darmstadt, Germany) to desorb the EY. They were kept in the solution for 24 h while the solution was stirred at a rotation speed of 150 rpm.

2.2. Characterization of ZnO Thin Films

The morphology of the electrodeposited ZnO films was investigated with a scanning electron microscope (SEM) JSM 6700F (JEOL, Akishima, Tokyo, Japan). Profilometry was used to determine the thickness of ZnO films, using a Veeco Instruments Dektak 6M Stylus Profiler (Veeco Instruments Inc., Dornach Munich, Germany). The thickness of ZnO films on ITO-PET substrates that were fabricated in the miniplant setup could not be determined by profilometry because of the adjacency of the PSAs to the film edges and the impossibility to scratch films on plastic foils. The thickness of these films was derived from SEM images instead.

The adsorption isotherm of Kr at about 77 K was determined by an ASAP 2010 volumetric adsorption unit (Micromeritics), liquid nitrogen being used as a coolant. The sample to be measured was placed in a tailor-made adsorption cell. Prior to the adsorption experiment, the sample was outgassed for 48 h at 150 °C. The specific surface area was determined by the BET method using the molecular cross-sectional area of Kr of 0.21 nm^2 and the saturation pressure of solid Kr of about 1.6 Torr. Details are given in [38].

2.3. Device Fabrication

2.3.1. Fabrication of Small-Scaled Flexible DSSCs

The photo-anodes of small-scaled DSSCs featured ZnO thin films consisting of a dense ZnO film and a porous ZnO film. Both, the photo-anodes and the counter electrodes were based on ITO-PET foil substrates LR15. The counter electrodes were prepared by platinum sputtering (Cressington Scientific Instruments Inc. sputter coater 108auto, Watford, UK) for 120 s at a current of 30 mA and an argon pressure of 0.1 mbar.

After drying for 2 h at 100 °C, the ZnO films were sensitized for 1 h in a solution of the dye D149 (c = 0.5 mmol·L^{-1}, Mitsubishi Paper Mills Ltd., Tokyo, Japan) and the additive cholic acid (c = 1 mmol·L^{-1}, Carl Roth, ≥99.0%) in a 1:1 mixture of acetonitrile (Carl Roth, ≥99.9%, ≤10 ppm H$_2$O) and tert-butanol (ABCR, ≥99.9%). The sensitized ZnO films were subsequently rinsed with acetonitrile and dried for 1 h at 80 °C. PSAs (tesa® 61562) with a circular cut-out, defining the active area to 1.33 cm^2, were used as spacers and, at the same time, as sealants between the two electrodes. The assembled cells were filled with electrolyte using the backfilling method through a hole in the counter electrode. A special syringe (Solaronix Vac'n'Fill Syringe, Aubonne, Switzerland) was used for this procedure. Afterwards the hole was sealed with PSA. The electrolyte was a solution of tetrapropylammonium iodide (c = 1 mol·L^{-1}, ≥98% Sigma-Aldrich, St. Louis, MO, USA) and iodine (c = 0.1 mol·L^{-1}, Sigma-Aldrich, ≥99.999%) in a 1:4 mixture of acetonitrile and ethylene carbonate (≥99.0%, Acros Organics, Geel, Belgium).

2.3.2. Fabrication of Up-Scaled Flexible DSSC Modules

The procedure for the assembly of up-scaled DSSC modules was analogous to that of small-scaled DSSCs, with the following differences: The platinum coating of the counter electrodes was performed twice with different orientations of the substrates to achieve a coating as homogeneous as possible on this comparatively large area. The PSAs used for sealing the cells were those already present on the photo-anode substrates during ZnO deposition (see Section 2.1.3). To fill the individual cells with electrolyte, the counter electrodes contained two holes per segment. With a syringe the electrolyte was pushed through one hole into the segment, until it emerged from the second hole.

2.4. Photovoltaic Characterization

2.4.1. I-V Characteristics

I-V characteristics of the DSSC modules were recorded under real sunlight. The sweep rate was 50 mV·s^{-1}. A pyranometer (Kipp & Zonen CMP 21, Delft, The Netherlands) was used to simultaneously measure the incident light intensity. Directly after the measurements the module temperatures were determined using a contact thermometer (Testo 905-T2, Sparta, USA). Besides the characterization under real sunlight, the small-scaled DSSCs were also characterized under simulated sunlight with an irradiance of 1000 W·m^{-2}. A Xenon arc lamp (Oriel Instruments 66901/69911, Newport Spectra Physics, Darmstadt, Germany) equipped with filters to generate AM 1.5D conditions served as the light source. The light intensity was adjusted with a thermopile (Kipp & Zonen CA2, Delft, The Netherlands). The reported current densities and conversion efficiencies refer to the photoactive area of the solar cells.

2.4.2. Electrochemical Impedance Spectroscopy (EIS)

Electrochemical impedance spectra were recorded under illumination with simulated sunlight (AM 1.5D conditions) of about 550 W·m^{-2} being the highest light intensity at which an almost homogeneous illumination of the DSSC modules could be maintained using the given light source. The measurements were performed with an electrochemical workstation IM6e (ZAHNER-Elektrik, Kronach, Germany). For the analysis of the data the software ZVIEW (Version 3.3b) from Scribner Associates Inc. (Southern Pines, NC, USA) was used.

3. Results and Discussion

3.1. Electrodeposition of ZnO Films

3.1.1. Adaption of the ZnO Electrodeposition Process to the Miniplant Setup

Effects associated to the up-scaling of the ZnO electrodeposition process required the following adjustments of the deposition method:

(1) The increased dimensions of the setup in comparison to the RDE setup cause increased ohmic voltage drops when a deposition voltage is applied. Predominantly this applies to the substrates with their relatively high ohmic resistances. Without modifications of the substrates these voltage drops can result in considerably more positive potentials at the centers of the deposition areas compared to their edges. Consequently the ZnO growth in the centers is slower than at the edges, giving non-uniform films. If a more negative voltage is applied to obtain an adequate film growth in the centers, this can cause the deposition of elemental zinc at the edges.

To reduce the described voltage drops on the substrates and thereby permit the deposition of uniform ZnO films, the substrates were divided into several smaller segments surrounded by conductive silver grids. On the FTO-glass substrates this resulted in three deposition segments with a cumulative area of 38.88 cm^2.

Since the sheet resistances of the ITO-PET substrates are higher than those of the FTO-glass substrates, a pattern with 12 smaller deposition segments, giving a cumulative deposition area of 34.56 cm^2 was necessary for these substrates. The latter layout follows simulation results by Zhang et al. [39] for optimized DSSC module performances. The limitation of the series resistance of a module is crucial to reduce the ohmic power dissipation, as is discussed in Section 3.2.

(2) When the electrodeposition was always performed at the same potential vs. the reference electrode as is usually the case in the RDE setup, this actually resulted in varying film properties. Equal deposition conditions delivered either films of ZnO or films containing elemental Zn. It was observed that prior to the deposition the rest potentials of the substrates drift towards constant values, which, however, differ from substrate to substrate. For example in the miniplant setup with an FTO-glass substrate the rest potential drifted in one case from about 30 mV vs. Ag/AgCl to about 60 mV vs. Ag/AgCl during a time span of about 20 min. Although the drift velocity decreased, a constant value even was not reached after 20 min. In contrast, in the RDE setup the rest potential drifts by about 60–70 mV in the first 10 min, reaching a constant value of about 200 mV vs. Ag/AgCl.

The observed drifts indicate changing properties of the substrate surface, caused for instance by adsorption processes. In the miniplant setup this process appears to occur on a longer time scale than in the RDE setup. Consequently the conditions present at the beginning of the deposition have a significant influence on the deposition process. To take into account the observed on-going drift of the rest potential, electrodeposition in the miniplant setup was performed potentiostatically vs. the rest potential of the substrate after an equilibration time of about 20 min. The rest potential was measured for 5 s immediately before the start of the deposition. Only by this procedure was a high reproducibility of the deposition process possible, especially on ITO-PET substrates.

The applied deposition voltage depended on the substrate type and was adapted to allow the highest deposition current possible without formation of elemental zinc. For the deposition on FTO-glass substrates a voltage of −1.0 V vs. the rest potential vs. Ag/AgCl was applied. With ITO-PET substrates OC^{TM}50 and LR15, the deposition voltage was set to −0.93 V and −0.95 V vs. the rest potential vs. Ag/AgCl, respectively. The rest potentials in the miniplant setup, on which the deposition voltages relied, were in the range between −80 mV to 10 mV vs. Ag/AgCl for FTO-glass substrates and in the range between −60 mV to 35 mV vs. Ag/AgCl for ITO-PET substrates.

(3) In the miniplant setup lower current densities than in the RDE setup were obtained. When for example ZnO/EY hybrid films were deposited, current densities of up to 1.5 mA·cm^{-2} were observed in the RDE setup, while the current densities in the miniplant setup were limited to 0.5 to 0.8 mA·cm^{-2}, depending on the substrate layout and the sheet resistance (compare Figure 2). As lower current densities cause a slower film growth, longer deposition times are needed in the miniplant setup to obtain ZnO thicknesses comparable to those obtained in the RDE setup.

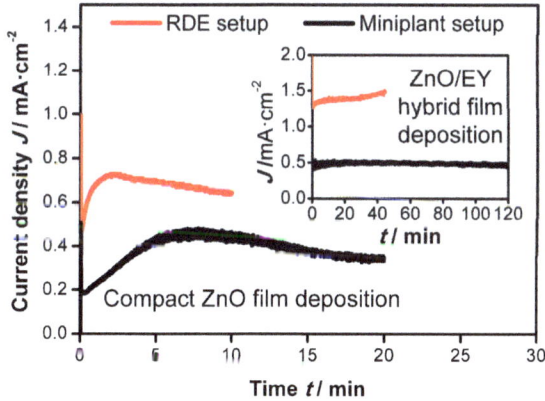

Figure 2. Comparison of current-time behavior between rotating disc electrode (RDE) setup (red lines, at −0.91 V vs. Ag/AgCl) and miniplant setup (black lines, at −1.0 V vs. the rest potential vs. Ag/AgCl) during the electrodeposition on FTO-glass substrates: Deposition of compact ZnO films (full scale) and subsequent deposition of ZnO/EY hybrid films (inset).

Probably the current density in the miniplant setup is decreased by the higher ohmic resistances due to the enlarged dimensions in comparison to the RDE setup. This can include resistances of cables, substrates, and of the electrolyte solution. As mentioned before, the substrate resistance constitutes a major obstacle for the ZnO deposition that was improved by conductive silver paths. The resistance of the electrolyte solution is expected to be higher than in the RDE setup because of the higher distance between working electrode and counter electrode. The contribution of the cable resistance to the reduced current density in the miniplant setup is expected to be small in comparison to the two other quoted factors.

The convection strength in the solution certainly differs between the RDE and the miniplant setup because of the different dimensions and mixing concepts. This probably has an influence on the reagent transport from and to the substrate and consequently the rate of film growth, too. An optimization of the paddle movement frequency might lead to a faster film growth and shorter deposition times in the miniplant setup.

The reproducibility of the deposition of ZnO/EY hybrid films in the miniplant setup was generally sufficient. On all substrate types the film thickness correlated to the transferred charge per area, indicating constant deposition efficiency. The deposition rate of the hybrid films amounts to about $1.4\ \mu m \cdot C^{-1} \cdot cm^{-2}$, which is similar to deposition rates reported for small-area depositions in RDE setups [40].

3.1.2. Electrodeposition of ZnO Films on FTO-Glass Substrates

Depositions on FTO-glass substrates were first used to demonstrate that the miniplant setup is suitable to produce ZnO films comparable to those obtained on small substrates in an RDE setup. In Figure 2 the current-time curves are shown for both deposition setups. During the deposition of compact ZnO films in both setups an initial increase of the deposition current is observed, which can be attributed to the three-dimensional growth of the individual ZnO crystals. The maximum current density is obtained when the crystals start merging, leading to a transition from three-dimensional to one-dimensional growth and a decrease in the current density [27,33].

When ZnO/EY hybrid films are deposited (inset in Figure 2) on top of the compact ZnO layer, the current density remains nearly constant over time, meaning that the one-dimensional film growth seen towards the end of the deposition of the compact ZnO layer is continued. Due to the catalytic effect of EY on the oxygen reduction the current densities are increased in comparison to the deposition of compact ZnO films in both setups [24,30].

The photograph in Figure 3a presents the homogeneous deposition characteristic of the ZnO film which is typical for an RDE deposition process. As seen in the photograph in Figure 3b, the ZnO films deposited on the enlarged substrate in the miniplant setup are also rather homogeneous. The cross section image of the film deposited in the RDE setup (Figure 3c) shows a rather inhomogeneous surface of the film, which also explains the slight increase of the current density over time for the hybrid film in this setup, as some crystals grow higher than others and revert to three-dimensional growth. Irregularities of the substrate surface and the compact ZnO film surface are supposed to influence the hybrid film growth in a way that becomes more distinct with increasing film thickness. The homogeneity of the film thickness from the miniplant setup is proven by the SEM image in Figure 3d. Both, compact ZnO films as well as porous ZnO films show a high degree of substrate coverage. The thickness of the film obtained from the miniplant setup amounts to about 5 µm after deposition of the hybrid film for 120 min. This is less than the thickness of about 7.6 µm of the film from the RDE setup after deposition for 45 min. The slower film growth in the miniplant setup has already been mentioned in Section 3.1.1. A layout with more, but smaller cells, as it was applied on the ITO-PET foil substrates, would probably lead to a lower substrate resistance and therefore a faster film growth. Anyhow, a film thickness of about 5 µm approaches the thickness of films from the RDE setup sufficiently to demonstrate the successful operation of the miniplant setup. The Figure 3c,d document the comparable microscopic growth of the films in both setups. The nanostructure of the porous films can be observed clearly in Figure 3e,f. Films from both setups contain stress cracks in the porous films.

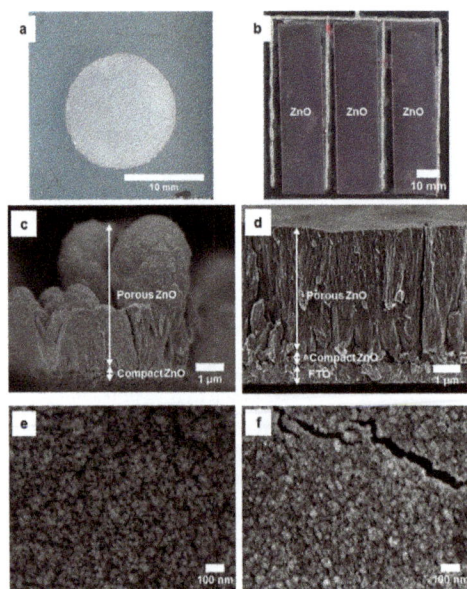

Figure 3. Morphology of ZnO films electrodeposited on FTO-glass substrates in the RDE setup (deposition of compact layer for 10 min and of porous layer for 45 min) and in the miniplant setup (deposition of compact layer for 20 min and of porous layer for 120 min): Photographs of ZnO films after deposition in the RDE setup (**a**) and miniplant setup (**b**); scanning electron microscopy (SEM) cross section images of compact and porous ZnO film deposited in the RDE setup (**c**) and miniplant setup (**d**); SEM top view images of compact and porous ZnO film deposited in the RDE setup (**e**) and miniplant setup (**f**).

3.1.3. Electrodeposition of ZnO Films on ITO-PET Foils

As mentioned in Section 3.1.1 a modification of the substrate layout was necessary to apply the electrodeposition process on ITO-PET foil substrates. However, despite the change in the layout the current densities during depositions of compact ZnO on these substrates were lower than those on FTO-glass substrates as depicted in Figure 4. No maximum is observed in the current-time curves, indicating that no coalescence of the ZnO crystals is obtained. Considerably longer deposition times or different pre-treatment methods (for example treatment with various reagents, corona discharge, pre-electrolysis etc.) of the substrates might lead to the deposition of dense compact ZnO films on the given ITO-PET substrates.

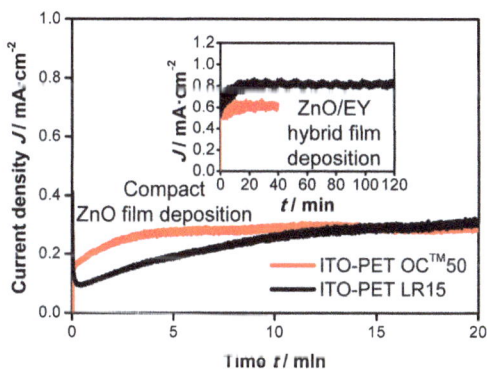

Figure 4. Comparison of current-time behavior during the electrodeposition of compact ZnO films (full scale) and subsequent ZnO/EY hybrid films (inset) on ITO-PET substrates OC™50 (at −0.93 V vs. the rest potential vs. Ag/AgCl) and LR15 (at −0.95 V vs. the rest potential vs. Ag/AgCl) in the miniplant setup.

Again, the current densities during the deposition of ZnO/EY hybrid films on top of the initially deposited ZnO crystals are markedly higher due to the catalytic effect of the EY, as illustrated in the inset of Figure 4. In addition, a comparison with Figure 2 shows that the current density is even higher than that observed during ZnO/EY hybrid film deposition on FTO-glass substrates. This observation leads to two important conclusions: First, it shows the positive effect of the changed layout on the film deposition. Second, it proves that the low current density during the initial ZnO deposition does not occur due to limitation by the electrical resistance of the substrate but probably due to the surface properties of the ITO.

On the substrate type OC™50 the deposition time for the ZnO/EY hybrid films had to be limited to 40 min. In the case of longer deposition times the current density steadily decreased and compact ZnO was deposited on top of the hybrid film. This upper compact film sealed the hybrid film and therefore prevented desorption of the EY from the film and the dye loading. The reason for the transition from ZnO/EY to pure ZnO being deposited is the increasing potential drop caused by the cumulative resistance of the ITO layer (50 $\Omega\cdot sq^{-1}$) and the growing ZnO film, which eventually leads to a potential >−0.8 V vs. Ag/AgCl at the electrode surface. In this potential region, EY is not electrochemically reduced, preventing the formation of the nanostructured hybrid film. When a more negative potential than −0.93 V vs. the rest potential vs. Ag/AgCl was used to force the reduction of EY, this resulted in the formation of elemental zinc on the substrate. When the deposition time of the hybrid film at a deposition potential of −0.93 V vs. the rest potential was limited to 40 min, the deposition of the compact ZnO top layer could be avoided.

When ZnO/EY hybrid films are deposited on the substrate type LR15 (again on top of a layer of compact ZnO crystals), the lower sheet resistance (15 $\Omega\cdot sq^{-1}$) compared to OC™50 leads to a higher current density (Figure 4). Note that the deposition current on LR15 is actually lower than on OC™50 during the first 15 min of the ZnO bottom layer deposition, which again proves that this process is not limited by the resistance of the substrate, but by the surface properties of the ITO. In this respect, the LR15 material seems to have a disadvantage, which is, however, overcome after 15 min of ZnO deposition. For the deposition of the ZnO/EY layer, the higher conductivity of the LR15 substrate is clearly advantageous, allowing a deposition time of 120 min.

Both, the film deposited on OC™50 and the film deposited on LR15 display a macroscopic homogeneity (Figure 5a,b). As the photography in Figure 5b shows, the film deposited on LR15 exhibits a light pink color, which evidences non-desorbed residues of EY, although no evidence for the deposition of compact ZnO above the ZnO/EY layer could be found. The incomplete dye desorption therefore seems to arise from the high film thickness, making technical optimization of the desorption process seem necessary to remove the EY completely. Due to the entrapment of the eosin Y inside the ZnO pores the dye residues are assumed to have no direct accessibility by the electrolyte, hence they do not contribute significantly to the photocurrent in the assembled solar cells. However, it is very likely that the entrapped eosin Y decreases the ZnO surface area available for adsorption of the photosensitizer dye to some extent.

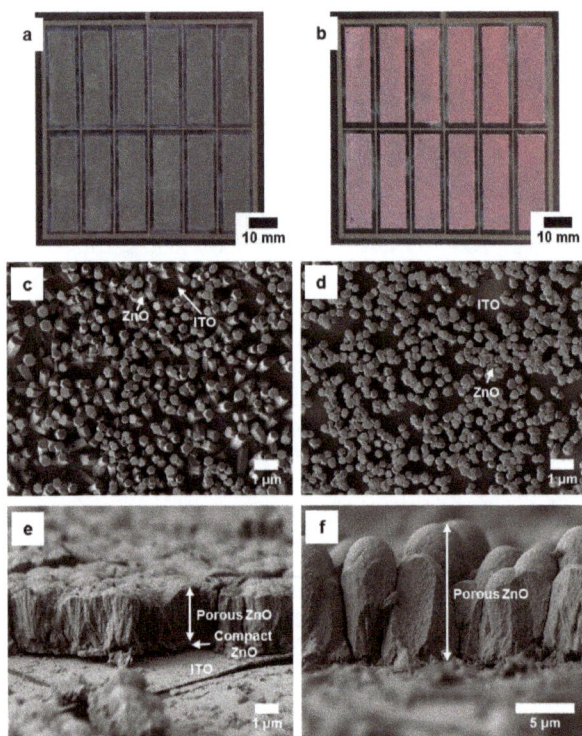

Figure 5. Morphology of ZnO films electrodeposited in the miniplant setup on ITO-PET OC™50 substrates (deposition of compact layer for 20 min and of porous layer for 40 min) and on ITO-PET LR15 substrates (deposition of compact layer for 20 min and of porous layer for 120 min): Photographs of ZnO films after deposition on OC™50 (**a**) and LR15 (**b**); SEM top view images of compact ZnO films deposited on OC™50 (**c**), and LR15 (**d**); SEM cross section images of compact and porous ZnO films deposited on OC™50 (**e**), and LR15 (**f**).

The SEM images in Figure 5c,d confirm that neither on the OC™50 nor on the LR15 substrate were dense compact ZnO films obtained, as already predicted from the current-time behavior of the electrodeposition.

On the OC™50 substrate ZnO/EY films with a similar morphology to those on FTO-glass substrates were obtained, having a thickness of about 2.5 μm (Figure 5e). Resulting from the longer deposition time the ZnO/EY films deposited on the LR15 substrate have a film thickness of about 9 μm (Figure 5f). As mentioned above no evidence for the deposition of compact ZnO on top of the ZnO/EY film can be found.

The porosity of the porous ZnO films obtained from the miniplant setup after desorption of EY was verified exemplary by Kr adsorption measurements with a film deposited on an ITO-PET LR15 substrate. This flexible substrate type was chosen for the adsorption measurement because the ZnO film morphology on ITO or FTO glasses has already proven [35,40] and appeared to be promising for flexible DSSC photo-anodes as well. The resulting isotherms are shown in Figure 6. The well-developed hysteresis proves the presence of pores smaller than about 10 nm in the sample. The specific surface area related to the film area, also called roughness factor (RF), was determined as 373 $cm^2 \cdot cm^{-2}$. The specific surface area related to the film volume amounts to 77 $m^2 \cdot cm^{-3}$. Elsewhere a RF of 220 for a film thickness of 3 μm has been reported [41]. This corresponds to a specific surface area related to the film volume of 73 $m^2 \cdot cm^{-3}$, and is therefore in good agreement with our value.

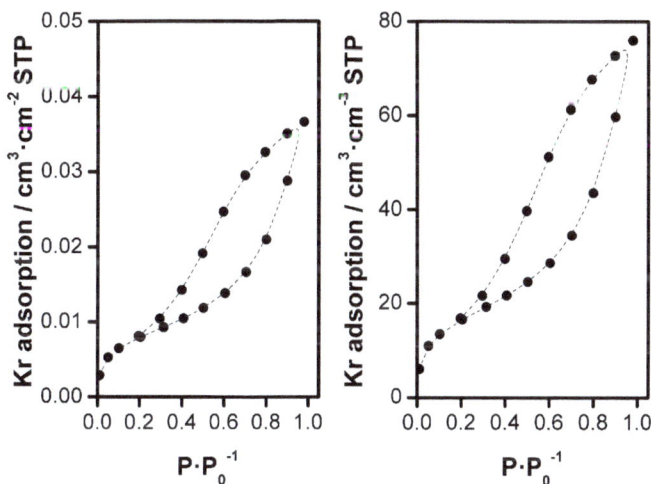

Figure 6. Adsorption isotherms of Kr at about 77 K of an up-scaled porous ZnO film on an ITO-PET substrate LR15 with a thickness of about 4.8 μm. The adsorption is related to 1 cm^2 of the support and 1 cm^3 of the ZnO film, respectively.

Among the porous ZnO films, those prepared on the ITO-PET LR15 substrates were regarded as being most suitable for the use as photo-anodes in modules due to their higher thickness and acceptable porosity and stability. These substrates were therefore used to prepare and test flexible DSSC modules as described in the following section. The presence of a dense compact ZnO bottom layer would be preferable to suppress recombination reactions between substrate and electrolyte but it is not a prerequisite for the operation of a DSSC.

3.2. Flexible DSSC Modules

A typical module as prepared and tested in this study is depicted in Figure 7. Due to the electrical contact between the active areas on each electrode a parallel connection of the individual cells results. The photovoltaic parameters of two equally prepared modules (flexible DSSC modules "1" and "2") are depicted in Figure 8 in comparison to a small-scaled DSSC. At low light intensities ϕ the *I-V* characteristics of the modules exhibit distinctive diode behavior. With increasing ϕ values, however, the fill factor (FF) strongly decreases towards 25%, converting the shape of the *I-V* curve to a straight line.

Figure 7. Functional flexible DSSC module with a photoactive area of 34.56 cm^2. ITO-PET foils are used as substrates for the photo-anode and the counter electrode. The photo-anode contains electrodeposited ZnO films with a thickness of about 9 μm.

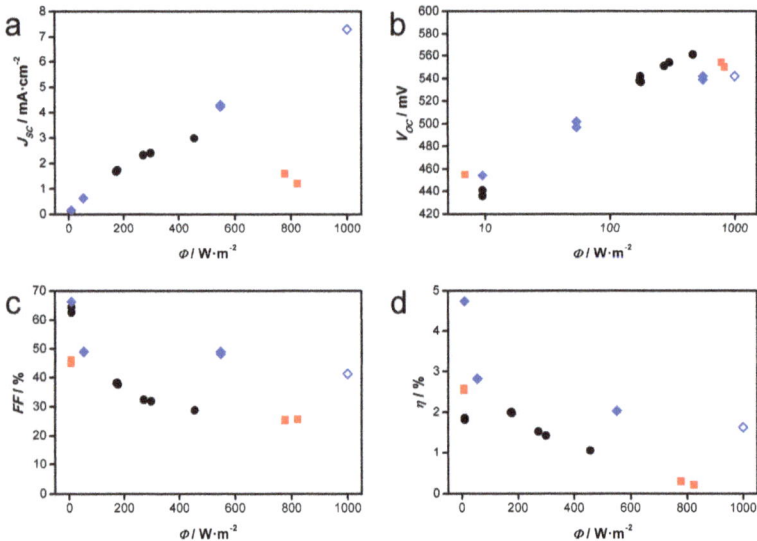

Figure 8. Dependence of photovoltaic parameters on light intensity ϕ. (■) Flexible DSSC module "1". (●) Flexible DSSC module "2". (♦) Flexible small-scaled DSSC. Open symbol: Measurement performed in solar simulator. All other measurements were performed under real sunlight. (**a**) short circuit current J_{SC}; (**b**) open circuit voltage V_{OC}; (**c**) fill factor (FF); (**d**) power conversion efficiency η.

The low fill factor, i.e., the small slope of the *I-V* curve starting from V_{oc} towards lower voltages, also seems to restrict the short-circuit current density J_{SC} of the modules at higher ϕ values. While a linear increase of J_{SC} to 7.29 mA cm^{-2} at 1000 W·m^{-2} is seen for a small-scaled flexible DSSC based on a comparable ZnO film (blue symbols in Figure 8), the J_{SC} seen for the DSSC module "2" (black symbols) starts to deviate from the linear behavior at ϕ >200 W·m^{-2}. Compared to module "2" with 3.00 mA cm^{-2} at 456 W·m^{-2}, module "1" even gives lower J_{SC} values of 1.60 and 1.20 mA cm^{-2} (red symbols), although ϕ was further increased to 777 and 823 W·m^{-2}, respectively, which is further discussed below in conjunction with the EIS results. The decrease in both values, J_{SC} and FF, is the reason for the considerably lower power conversion efficiencies η of the modules at higher ϕ values, reaching 1.06% at 456 W·m^{-2} for module "2" and only 0.21% at 823 W·m^{-2} for module "1". V_{oc} however shows the typical increasing trend with the logarithm of the light intensity for the small-scaled cell as well as the modules. The highest η value of 2.58% for module "2" is obtained at ϕ = 6.86 W·m^{-2}. In contrast, the small-scaled flexible DSSC shows a far less pronounced dependence of FF, J_{SC} and η on the light intensity. While at a low light intensity ϕ of 9.48 W·m^{-2} the FF and the η of this cell amount to 66.2% and 4.74%, respectively, relatively high values of FF = 41.1% and η = 1.63% are retained at ϕ = 1000 W·m^{-2}.

Further investigations of the modules and the small-scaled cell were conducted by EIS. The model used for fitting of the EIS spectra is adapted from a transmission line model suggested by Fabregat-Santiago et al. [42]. A transmission line model is suitable to describe systems that contain porous electrodes like the ZnO films in this study. The elements of the model describe the properties of individual cell components and processes. The applied model and exemplary EIS spectra are shown in Figure 9. At high frequencies—at the left-hand side of the spectra—the impedance is controlled by the series resistance R_S of the cells and by the charge transfer resistance R_{CT} and the double-layer capacity C_{DL} of the counter electrode. The latter two form the first semicircle. The second semicircle at lower frequencies is constituted by the transport resistance r_{tr} and the chemical capacitance c_μ of the ZnO film as well as the recombination resistance r_{rec} of the ZnO/electrolyte interface. The model uses constant phase elements (CPEs) to describe the capacities. The formula that was used to convert the CPE parameters to equivalent capacities is described in Reference [42] Equation S1.

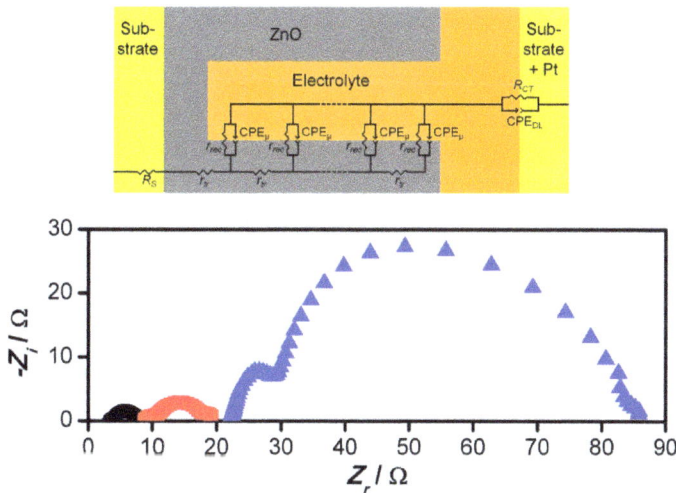

Figure 9. Applied electrochemical impedance spectroscopy (EIS) model for fitting (**top**) and exemplary EIS spectra recorded at V = 300 mV (**bottom**). (■) Flexible DSSC module "1". (●) Flexible DSSC module "2". (♦) Flexible small-scaled DSSC. R_S: Series resistance. r_{tr}: Transport resistance. r_{rec}: Recombination resistance. CPE$_\mu$: Constant phase element describing the chemical capacitance. R_{CT}: Charge-transfer resistance. CPE$_{DL}$: Constant phase element describing the double-layer capacitance.

The fitted parameters are presented in Figure 10. The values for the chemical capacitance c_μ of the ZnO film show no major differences between the modules on the one hand and the small-scaled cell on the other hand. The distinct increase of the chemical capacitances between 100 and 500 mV can be explained by a rise of the quasi-fermi level towards more negative applied potentials, increasing the number of energetic states that can be occupied by electrons. Also the recombination resistances r_{rec} at the ZnO/electrolyte interface are comparable for the examined modules and the small-scaled cell. The somewhat higher values seen for the modules in a part of the voltage range most probably arise from the fact that some of the pores in the films are not electrochemically accessible as seen in the incomplete eosin Y desorption (see Figure 5 and related discussion). On the other hand, they confirm that the lack of a completely dense compact ZnO layer in these films does not lead to significantly more recombination, since the latter would be expected to decrease the r_{rec} value. The differences observable in the transport resistances r_{tr} of the ZnO films are not significant, because the corresponding part of the EIS spectra is difficult to fit. In conclusion, these three parameters related to the porous ZnO film support the assumption that the ZnO films fabricated in the miniplant setup behave similarly to ZnO films from the RDE setup.

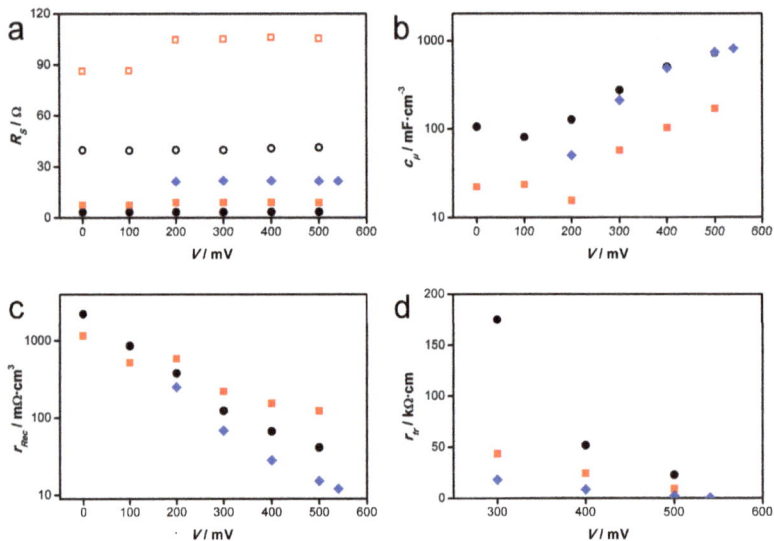

Figure 10. EIS fitting results. (■) Flexible DSSC module "1". (●) Flexible DSSC module "2". (♦) Flexible small-scaled DSSC. Open symbols: Estimated series resistances R_S per module cell. (**a**) series resistance R_S; (**b**) chemical capacitance c_μ; (**c**) recombination resistance r_{rec}; (**d**) transport resistance r_{tr}.

Low fill factors in solar cells and modules often result from high series resistances R_S, because the slope of an I-V curve at its onset is limited by the overall cell resistance. Too high series resistances diminish the FF and subsequently the obtainable J_{SC}, as observed for the modules in this study and also previously in other studies concerning the scale-up of DSSCs [43–46]. In DSSCs the resistance of the transparent conducting substrate primarily contributes to R_S [42]. With about 8 Ω and 3.3 Ω the series resistances of the modules are lower than that of the small-scaled cell with about 21.5 Ω. However, 12 cells in parallel as present in the modules should only have 1/12 of the resistance of a single cell, meaning that the series resistances per cell in the modules can be estimated at 40 Ω and 96 Ω. At the same current per cell, this leads to a lower FF of the module. Consequently much lower R_S values (<1 Ω) are required to obtain higher fill factors and hence higher conversion efficiencies of the modules under high light intensities.

One reason for the high R_S values of the modules can be seen in the sheet resistance of the used ITO-PET foil substrates (15 $\Omega \cdot sq^{-1}$), which is twice as high as that of FTO-glass substrates (7 $\Omega \cdot sq^{-1}$). Furthermore, the silver paths may exhibit high contact resistances to the ITO layers, which, however, could not be measured. Also the electrolyte resistance related to the relatively high distance of 50 μm between the two electrodes—necessary to prevent contact between the silver paths of the two electrodes—might have an influence. Besides reduction of the substrate resistances, the connection of the individual cells in series might be helpful to reduce the ohmic power dissipation [5,21]. Furthermore it should be noted that in the case of the modules the charge-transfer resistances R_{CT} at the counter electrode were relatively high (29.3 and 7.3 $\Omega \cdot cm^2$, respectively) because for technical reasons the substrates could not be coated homogeneously enough with platinum.

It should be noted at this point that the R_S value of module "1" is even considerably higher than that of module "2". This explains the lower J_{SC} value of module "1", since it leads to a smaller slope of the of *I-V* curve form the photocurrent onset at V_{OC} towards the *I* axis. Since both modules had the same design and were made from the same materials, the higher R_S value of module "1" is most probably due to corrosion of silver paths by iodine from the electrolyte in this module, indicating that the pastes used for sealing have to be further optimized.

4. Conclusions

The electrochemical fabrication of porous ZnO films is an attractive low-temperature method for the preparation of photo-anodes for flexible DSSCs and perovskite solar cells. We demonstrated the scalability of this process from RDE setups for small-scaled films to a technically simplified miniplant setup that was developed for this purpose. The application of conductive silver paths was necessary to limit the voltage drop across the enlarged substrate areas. ZnO electrodeposition was performed successfully on FTO-glass substrates on deposition areas of 38.88 cm^2. The up-scaled electrochemical deposition process can be applied also on flexible ITO-PET foil substrates, yielding ZnO films with an overall area of 34.56 cm^2. The properties of the substrates have a strong influence on the success of the deposition. Sheet resistances of the substrates lower than 50 $\Omega \cdot sq^{-1}$ are necessary to enable the deposition of hybrid films with thicknesses above 2.5 μm. Furthermore, the deposition of dense compact ZnO films proved to be difficult on the ITO-PET substrates, most probably due to unfavorable surface properties. Under the prerequisite of sufficiently low sheet resistances of the ITO-PET foil substrates (<50 $\Omega \cdot sq^{-1}$), ZnO/EY hybrid films with thicknesses of up to 9 μm have been realized. The high porosity of the films after desorption of the EY was proven by means of SEM and Kr sorption. Due to the successful scale-up of the deposition process, also a roll-to-roll process, which would enable high production throughputs, appears realizable.

Operative flexible DSSC modules (η = 2.58% at ϕ = 6.86 $W \cdot m^{-2}$) proved the suitability of the up-scaled porous ZnO films as DSSC photo-anodes. The analysis of the EIS spectra revealed similar photo-anode properties for up-scaled modules and the small-scaled cells, but also too high series resistances of the modules. These high series resistances cause high ohmic power dissipations and consequently a significant decrease of the fill factor and conversion efficiency of the modules under high light intensities. A technical solution of this problem, for example by the application of different silver pastes and an optimized sealing to avoid corrosion of the silver by the electrolyte, seems possible.

Acknowledgments: The financial support for this collaborative research by the BMBF (German Federal Ministry of Education and Research, Funding code: 03X3519F, project name "FSZ Industrie"), implemented by the PTJ (Project Management Juelich, Germany), is gratefully acknowledged. We thank our project partners Jan Ellinger, Judith Gruenauer, and Klaus Keite-Telgenbuescher (tesa SE, Hamburg, Germany) for providing the flexible substrates. Furthermore, we thank LNQE (Laboratory of Nano and Quantum Engineering, Hannover, Germany), Juan Du, Juergen Caro, and Carla Vogt (all Leibniz University Hannover, Hannover, Germany) for providing technical equipment, deposition of ZnO on some flexible substrates, fruitful discussions, resources and access to a sputter coater, respectively. Finally, we grateful acknowledge recording of adsorption isotherms by Jiri Rathousky (Heyrovsky Institute of Physical Chemistry, Prague, Czech Republic).

Author Contributions: Florian Bittner constructed the miniplant set-up for electrochemical deposition, fabricated the films, characterized the DSSC photo-anodes, analyzed all the data, and wrote the manuscript. Torsten Oekermann and Michael Wark planned and supervised the project. Both authors contributed significantly to the discussion of the results and to the writing of the manuscript.

Conflicts of Interest: The authors declare no conflict of interest.

References

1. Schubert, M.B.; Merz, R. Flexible solar cells and modules. *Philos. Mag.* **2009**, *89*, 2623–2644. [CrossRef]
2. Lu, H.; Zhai, X.; Liu, W.; Zhang, M.; Guo, M. Electrodeposition of hierarchical ZnO nanorod arrays on flexible stainless steel mesh for dye-sensitized solar cell. *Thin Solid Films* **2015**, *586*, 46–53. [CrossRef]
3. Baker, J.A.; Worsley, C.; Lee, H.K.H.; Clark, R.N.; Tsoi, W.C.; Williams, G.; Worsley, D.A.; Gethin, D.T.; Watson, T.M. Development of Graphene Nano-Platelet Ink for High Voltage Flexible Dye Sensitized Solar Cells with Cobalt Complex Electrolytes. *Adv. Eng. Mater.* **2017**, *19*, 1600652. [CrossRef]
4. Hashmi, G.; Miettunen, K.; Peltola, T.; Halme, J.; Asghar, I.; Aitola, K.; Toivola, M.; Lund, P.D. Review of materials and manufacturing options for large area flexible dye solar cells. *Renew. Sustain. Energy Rev.* **2011**, *15*, 3717–3732. [CrossRef]
5. Casaluci, S.; Gemmi, M.; Pellegrini, V.; Di Carlo, A.; Bonaccorso, F. Graphene-based large area dye-sensitized solar cell modules. *Nanoscale* **2016**, *8*, 5368–5378. [CrossRef] [PubMed]
6. Yella, A.; Lee, H.-W.; Tsao, H.N.; Yi, C.; Chandiran, A.K.; Nazeeruddin, M.K.; Diau, E.W.G.; Yeh, C.-Y.; Zakeeruddin, S.M.; Graetzel, M. Porphyrin-Sensitized Solar Cells with Cobalt (II/III)–Based Redox Electrolyte Exceed 12 Percent Efficiency. *Science* **2011**, *334*, 629–634. [CrossRef] [PubMed]
7. Green, M.A.; Hishikawa, Y.; Dunlop, E.D.; Levi, D.H.; Hohl-Ebinger, J.; Ho-Baillie, A.W.Y. Solar cell efficiency tables (version 51). *Prog. Photovolt. Res. Appl.* **2018**, *26*, 3–12. [CrossRef]
8. Jackson, P.; Hariskos, D.; Wuerz, R.; Kiowski, O.; Bauer, A.; Friedlmeier, T.M.; Powalla, M. Properties of Cu(In,Ga)Se$_2$ solar cells with new record efficiencies up to 21.7%. *Phys. Status Solidi (RRL)* **2015**, *9*, 28–31. [CrossRef]
9. Park, J.H.; Jun, Y.; Yun, H.-G.; Lee, S.-Y.; Kang, M.G. Fabrication of an Efficient Dye-Sensitized Solar Cell with Stainless Steel Substrate. *J. Electrochem. Soc.* **2008**, *155*, F145–F149. [CrossRef]
10. Fang, X.; Ma, T.; Guan, G.; Akiyama, M.; Kida, T.; Abe, E. Effect of the thickness of the Pt film coated on a counter electrode on the performance of a dye-sensitized solar cell. *J. Electroanal. Chem.* **2004**, *570*, 257–263. [CrossRef]
11. Miettunen, K.; Ruan, X.; Saukkonen, T.; Halme, J.; Toivola, M.; Guangsheng, H.; Lund, P.D. Stability of Dye Solar Cells with Photoelectrode on Metal Substrates. *J. Electrochem. Soc.* **2010**, *157*, B814–B819. [CrossRef]
12. Chang, H.; Chen, T.L.; Huang, K.D.; Chien, S.H.; Hung, K.C. Fabrication of highly efficient flexible dye-sensitized solar cells. *J. Alloys Compd.* **2010**, *504*, S435–S438. [CrossRef]
13. Watson, T.; Mabbett, I.; Wang, H.; Peter, L.M.; Worsley, D. Ultrafast near infrared sintering of TiO$_2$ layers on metal substrates for dye-sensitized solar cells. *Prog. Photovolt. Res. Appl.* **2011**, *19*, 482–486. [CrossRef]
14. Miyasaka, T.; Kijitori, Y. Low-Temperature Fabrication of Dye-Sensitized Plastic Electrodes by Electrophoretic Preparation of Mesoporous TiO$_2$ Layers. *J. Electrochem. Soc.* **2004**, *151*, A1767–A1773. [CrossRef]
15. Yamaguchi, T.; Tobe, N.; Matsumoto, D.; Nagai, T.; Arakawa, H. Highly efficient plastic-substrate dye-sensitized solar cells with validated conversion efficiency of 7.6%. *Sol. Energy Mater. Sol. Cells* **2010**, *94*, 812–816. [CrossRef]
16. Yang, L.; Wu, L.; Wu, M.; Xin, G.; Lin, H.; Ma, T. High-efficiency flexible dye-sensitized solar cells fabricated by a novel friction-transfer technique. *Electrochem. Commun.* **2010**, *12*, 1000–1003. [CrossRef]
17. Park, N.G.; Kim, K.M.; Kang, M.G.; Ryu, K.S.; Chang, S.H.; Shin, Y.-J. Chemical Sintering of Nanoparticles: A Methodology for Low-Temperature Fabrication of Dye-Sensitized TiO$_2$ Films. *Adv. Mater.* **2005**, *17*, 2349–2353. [CrossRef]
18. Li, X.; Lin, H.; Li, J.; Wang, N.; Lin, C.; Zhang, L. Chemical sintering of graded TiO$_2$ film at low-temperature for flexible dye-sensitized solar cells. *J. Photochem. Photobiol. A* **2008**, *195*, 247–253. [CrossRef]
19. Chiu, W.-H.; Lee, K.-M.; Hsieh, W.-F. High efficiency flexible dye-sensitized solar cells by multiple electrophoretic depositions. *J. Power Sources* **2011**, *196*, 3683–3687. [CrossRef]

20. Qin, F.; Meng, W.; Fan, J.C.; Ge, C.; Luo, B.W.; Ge, R.; Hu, L.; Jiang, F.Y.; Liu, T.F.; Jiang, Y.Y.; et al. Enhanced Thermochemical Stability of $CH_3NH_3PbI_3$ Perovskite Films on Zinc Oxides via New Precursors and Surface Engineering. *ACS Appl. Mater. Interfaces* **2017**, *9*, 26045–26051. [CrossRef] [PubMed]

21. Hagfeldt, A.; Boschloo, G.; Sun, L.; Kloo, L.; Pettersson, H. Dye-Sensitized Solar Cells. *Chem. Rev.* **2010**, *110*, 6595–6663. [CrossRef] [PubMed]

22. Abd-Ellah, M.; Moghimi, N.; Zhang, L.; Thomas, J.P.; McGillivray, D.; Srivastava, S.; Leung, K.T. Plasmonic gold nanoparticles for ZnO-nanotube photoanodes in dye-sensitized solar cell application. *Nanoscale* **2016**, *8*, 1658–1664. [CrossRef] [PubMed]

23. Oekermann, T.; Yoshida, T.; Minoura, H.; Wijayantha, K.G.U.; Peter, L.M. Electron Transport and Back Reaction in Electrochemically Self-Assembled Nanoporous ZnO/Dye Hybrid Films. *J. Phys. Chem. B* **2004**, *108*, 8364–8370. [CrossRef]

24. Yoshida, T.; Zhang, J.; Komatsu, D.; Sawatani, S.; Minoura, H.; Pauporté, T.; Lincot, D.; Oekermann, T.; Schlettwein, D.; Tada, H.; et al. Electrodeposition of Inorganic/Organic Hybrid Thin Films. *Adv. Funct. Mater.* **2009**, *19*, 17–43. [CrossRef]

25. Peulon, S.; Lincot, D. Cathodic electrodeposition from aqueous solution of dense or open-structured zinc oxide films. *Adv. Mater.* **1996**, *8*, 166–170. [CrossRef]

26. Peulon, S.; Lincot, D. Mechanistic Study of Cathodic Electrodeposition of Zinc Oxide and Zinc Hydroxychloride Films from Oxygenated Aqueous Zinc Chloride Solutions. *J. Electrochem. Soc.* **1998**, *145*, 864–874. [CrossRef]

27. Pauporté, T.; Lincot, D. Electrodeposition of semiconductors for optoelectronic devices: Results on zinc oxide. *Electrochim. Acta* **2000**, *45*, 3345–3353. [CrossRef]

28. Pérez-Hernández, G.; Vega-Poot, A.; Pérez-Juárez, I.; Camacho, J.M.; Arés, O.; Rejón, V.; Peña, J.L.; Oskam, G. Effect of a compact ZnO interlayer on the performance of ZnO-based dye-sensitized solar cells. *Sol. Energy Mater. Sol. Cells* **2012**, *100*, 21–26. [CrossRef]

29. Pauporté, T.; Lincot, D. Heteroepitaxial electrodeposition of zinc oxide films on gallium nitride. *Appl. Phys. Lett.* **1999**, *75*, 3817–3819. [CrossRef]

30. Yoshida, T.; Pauporté, T.; Lincot, D.; Oekermann, T.; Minoura, H. Cathodic Electrodeposition of ZnO/Eosin Y Hybrid Thin Films from Oxygen-Saturated Aqueous Solution of $ZnCl_2$ and Eosin Y. *J. Electrochem. Soc.* **2003**, *150*, C608–C615. [CrossRef]

31. Canava, B.; Lincot, D. Nucleation effects on structural and optical properties of electrodeposited zinc oxide on tin oxide. *J. Appl. Electrochem.* **2000**, *30*, 711–716. [CrossRef]

32. Zhang, J.; Sun, L.; Yoshida, T. Spectroelectrochemical studies on redox reactions of eosin Y and its polymerization with Zn^{2+} ions. *J. Electroanal. Chem.* **2011**, *662*, 384–395. [CrossRef]

33. Goux, A.; Pauporté, T.; Yoshida, T.; Lincot, D. Mechanistic Study of the Electrodeposition of Nanoporous Self-Assembled ZnO/Eosin Y Hybrid Thin Films: Effect of Eosin Concentration. *Langmuir* **2006**, *22*, 10545–10553. [CrossRef] [PubMed]

34. Yoshida, T.; Iwaya, M.; Ando, H.; Oekermann, T.; Nonomura, K.; Schlettwein, D.; Woehrle, D.; Minoura, H. Improved photoelectrochemical performance of electrodeposited ZnO/EosinY hybrid thin films by dye re-adsorption. *Chem. Commun.* **2004**, 400–401. [CrossRef] [PubMed]

35. Rathouský, J.; Loewenstein, T.; Nonomura, K.; Yoshida, T.; Wark, M.; Schlettwein, D. Electrochemically self-assembled mesoporous dye-modified zinc oxide thin films. *Stud. Surf. Sci. Catal.* **2005**, *156*, 315–320.

36. Pauporté, T.; Yoshida, T.; Cortès, R.; Froment, M.; Lincot, D. Electrochemical Growth of Epitaxial Eosin/ZnO Hybrid Films. *J. Phys. Chem. B* **2003**, *107*, 10077–10082. [CrossRef]

37. Guerin, V.-M.; Magne, C.; Pauporté, T.; Le Bahers, T.; Rathouský, J. Electrodeposited Nanoporous versus Nanoparticulate ZnO Films of Similar Roughness for Dye-Sensitized Solar Cell Applications. *ACS Appl. Mater. Interfaces* **2010**, *2*, 3677–3685. [CrossRef] [PubMed]

38. Rathouský, J.; Kalousek, V.; Yarovyi, V.; Wark, M.; Jirkovský, J. A low-cost procedure for the preparation of mesoporous layers of TiO_2 efficient in the environmental clean-up. *J. Photochem. Photobiol. A Chem.* **2010**, *216*, 126–132. [CrossRef]

39. Zhang, Y.D.; Huang, X.-M.; Gao, K.-Y.; Yang, Y.-Y.; Luo, Y.-H.; Li, D.-M.; Meng, Q.-B. How to design dye-sensitized solar cell modules. *Sol. Energy Mater. Sol. Cells* **2011**, *95*, 2564–2569. [CrossRef]

40. Michaelis, E.; Woehrle, D.; Rathousky, J.; Wark, M. Electrodeposition of porous zinc oxide electrodes in the presence of sodium laurylsulfate. *Thin Solid Films* **2006**, *497*, 163–169. [CrossRef]

41. Loewenstein, T.; Nonomura, K.; Yoshida, T.; Michaelis, E.; Woehrle, D.; Rathouský, J.; Wark, M.; Schlettwein, D. Efficient Sensitization of Mesoporous Electrodeposited Zinc Oxide by cis-Bis(isothiocyanato)bis(2,2′-bipyridyl-4,4′-dicarboxylato)-Ruthenium(II). *J. Electrochem. Soc.* **2006**, *153*, A699–A704. [CrossRef]

42. Fabregat-Santiago, F.; Garcia-Belmonte, G.; Mora-Sero, I.; Bisquert, J. Characterization of nanostructured hybrid and organic solar cells by impedance spectroscopy. *Phys. Chem. Chem. Phys.* **2011**, *13*, 9083–9118. [CrossRef] [PubMed]

43. Xia, J.; Yanagida, S. Strategy to improve the performance of dye-sensitized solar cells: Interface engineering principle. *Sol. Energy* **2011**, *85*, 3143–3159. [CrossRef]

44. Ramasamy, E.; Lee, W.J.; Lee, D.Y.; Song, J.S. Portable, parallel grid dye-sensitized solar cell module prepared by screen printing. *J. Power Sources* **2007**, *165*, 446–449. [CrossRef]

45. Ito, S.; Nazeeruddin, M.K.; Liska, P.; Comte, P.; Charvet, R.; Péchy, P.; Jirousek, M.; Kay, A.; Zakeeruddin, S.M.; Graetzel, M. Photovoltaic characterization of dye-sensitized solar cells: Effect of device masking on conversion efficiency. *Prog. Photovolt. Res. Appl.* **2006**, *14*, 589–601. [CrossRef]

46. Miyasaka, T.; Kijitori, Y.; Ikegami, M. Plastic Dye-sensitized Photovoltaic Cells and Modules Based on Low-temperature Preparation of Mesoscopic Titania Electrodes. *Electrochemistry* **2007**, *75*, 2–12. [CrossRef]

materials

MDPI

Review

Chemical Sensing Applications of ZnO Nanomaterials

Savita Chaudhary [1,*], Ahmad Umar [2,3,*], K. K. Bhasin [1] and Sotirios Baskoutas [4]

[1] Department of Chemistry and Centre of Advanced Studies in Chemistry, Panjab University, Chandigarh 160014, India; sav66hooda@gmail.com
[2] Department of Chemistry, College of Science and Arts, Najran University, Najran 11001, Saudi Arabia
[3] Promising Centre for Sensors and Electronic Devices (PCSED), Najran University, Najran 11001, Saudi Arabia
[4] Department of Materials Science, University of Patras, Patras GR 26504, Greece; bask@upatras.gr
[*] Correspondence: chemsavita@gmail.com (S.C.); ahmadumar.aspu@gmail.com (A.U.)

Received: 18 November 2017; Accepted: 6 February 2018; Published: 12 February 2018

Abstract: Recent advancement in nanoscience and nanotechnology has witnessed numerous triumphs of zinc oxide (ZnO) nanomaterials due to their various exotic and multifunctional properties and wide applications. As a remarkable and functional material, ZnO has attracted extensive scientific and technological attention, as it combines different properties such as high specific surface area, biocompatibility, electrochemical activities, chemical and photochemical stability, high-electron communicating features, non-toxicity, ease of syntheses, and so on. Because of its various interesting properties, ZnO nanomaterials have been used for various applications ranging from electronics to optoelectronics, sensing to biomedical and environmental applications. Further, due to the high electrochemical activities and electron communication features, ZnO nanomaterials are considered as excellent candidates for electrochemical sensors. The present review meticulously introduces the current advancements of ZnO nanomaterial-based chemical sensors. Various operational factors such as the effect of size, morphologies, compositions and their respective working mechanisms along with the selectivity, sensitivity, detection limit, stability, etc., are discussed in this article.

Keywords: zinc oxide; synthesis; chemical sensing; sensitivity; selectivity; morphology

1. Introduction

The rapid growth of industries and frequent use of chemicals in textile, pharmaceutical, food and automobile industries have contributed to a major threat to the survival of living beings on the Earth [1–3]. The emission of harmful toxins from automobile exhaust and factory outlets has become a major source of environmental pollution. Therefore, an authentic means for the effectual recognition of harmful chemicals by using chemical and biological sensors is in urgent need of the present [4,5]. Of all available types of semiconductor sensors for different types of chemicals and biological toxins, zinc oxide (ZnO)-based sensors have gained extensive attention around the world. The presence of a good response rate towards the chemical toxins with outstanding selectivity and sensitivity makes it one of the most significant materials for preparing low cost sensors [6]. The existence of a diverse range of morphologies such as nanorods, wires, needles, ellipsoids, urchins, helices, combs, flowers and disk shapes of ZnO materials has provided good control over the surface to volume ratio for the prepared nanomaterials and enhances their utility in sensing devices [7].

ZnO nanomaterials have been widely considered for their significant applications in different classes of nanoscale serviceable tools used in chemical, medical, diagnostics, food and nationwide defense-based equipment [8]. Being an n-type semiconductor with a wide band gap of 3.37 eV and a large exciton binding energy of 60 meV, the electron mobility in ZnO nanomaterials is enhanced.

The existence of a high, photoelectric reaction with an admirable chemical and thermal stability makes ZnO nanomaterials among the potential contenders for the preparation of effective chemical and biological sensors [9]. In addition, ZnO nanomaterials have the advantages of a low cost of production, a harmless nature and a simple mode of large-scale production [10]. Furthermore, ZnO nanomaterials are chemo-resistive in nature, and their sensing aptitude is principally restricted by the change in the chemical signal when the respective analyte molecules encounter its exterior surface [11].

In recent years, much attention has been given to ZnO-based nanostructures for sensing. The presence of the numerous properties of ZnO has been utilized for the development of effective and highly selective sensors. For instance, gas sensors are prepared due to the variations in the conductance with the reversible chemisorption process of reactive gases on the surface of ZnO [12]. The non-lethal nature of ZnO nanoparticles has been used for the generation of effective biosensors [13]. Although to date, a large number of literature works has been produced for the fabrication of different types of sensing devices based on ZnO nanostructures, the challenge of an effective and selective sensing was still not discussed in detail [14,15].

The present review meticulously introduces the current advancements of ZnO nanostructure-supported sensors with the main emphasis on chemical and biosensors for different analytes. The different types of operational factors such as the effect of size, morphology and the respective working mechanisms of nano-ZnO-based sensors, along with their selectivity and sensitivity behavior will be considered in detail.

2. ZnO Nanomaterials for Sensing Applications

The presence of high surface area, the biocompatible nature, thermal stability, wide band gap and superior response towards the photoelectric reaction makes ZnO nanomaterials among the candidates for the manufacture of useful chemical and biological sensors for a diverse range of moieties [16]. The existence of a minute size range with great variations in the surface to volume ratio makes them very effective for the adsorption of harmful analytes on the exterior surface of particles. The higher surface area of ZnO has also provided an additional amount of surface active sites for the analytes [17]. The presence of more surface atoms has generated the active sensing layer for the materials to be sensed from the surrounding environment. Moreover, the small grain size, i.e., as small as the depth of the space-charge layer in the ZnO-based nanostructure, has greatly controlled the sensing response to different types of toxins [18]. Due to such a behavioral aspect, the response rate of the ZnO-based sensor has been exponentially amplified with the reduction of the size of the particles formed. The size variations in the formed materials also influenced the van der Waals force of the particles. These forces were decreased with the reduction in the size of ZnO-based nanostructures and influenced the sensing aptitude. In addition to the size, the surface morphology of the particles has also influenced the sensing behavior of ZnO-based sensors [19]. These surface morphologies of the particles have a direct influence on the number of surface defects and the porosity ZnO nanostructures. These factors have a direct influence on the electrochemical sensing aptitude for various types of biological, as well as chemical species. The existence of different types of shapes of ZnO nanostructures has provided a diverse range of spatial structures and specific areas for the particles formed. These particles have also provided a diverse range of capabilities for the circulation of analytes during adsorption-desorption of different types of moieties [20]. Furthermore, these different types of morphologies have a direct influence on the amount of surface defects, involving the concentration of oxygen vacancies to modulate the conductivity of ZnO nanoparticles, which is quite essential for the detection of chemical analytes by using electrochemical sensing [21]. Theses factor have further affected the surface sensing aptitude of the particles formed and decreased the required temperature for the detection of the analyte. For instance, one-dimensional nanostructures of ZnO possessed an excessive amount of electrical conductivity as compared to other structures. Their superior electron mobility encouraged the effectual division among the electrons and holes in the nanostructures formed and helped to decrease the amount of generated resistance in the particles formed. Under normal atmospheric conditions, oxygen molecules from the atmosphere have the ability to be adsorbed on the surface of ZnO and then converted into reactive oxygen species by taking the electrons from the conduction band of ZnO nanomaterials, escorting them to the generation of the surface depletion layer on the surface of the particles and, thus, enhancing the sensor resistance with respect to different analytes [22]. When reactive analytes come near the surface of ZnO, these reactive oxygen molecules will interact with them and release the ensnared electrons towards the conduction band and producing a change in the signal [23]. In addition, the sensing aptitude of ZnO-based chemical and biological sensors is chiefly reliant on the working temperature ranges, which further modulate the reaction kinetics, the conductive nature and the electron mobility of these nanomaterials [24]. The ZnO-based nano-/micro-structures also possess high crystal quality with periodical structures in their geometry and smooth exterior surfaces with roughly the same wavelength-level size, which can be utilized for the generation of optical microcavities on the surface of nanomaterials. These microcavities provide an intermediate path for the sensing of the external analyte. Further, the interactions of the external materials can be accurately regarded in the recommended manner, which is very supportive of the analysis of the external moieties [25]. Using the external templating agents and prescribed assembly of ZnO materials, these nanomaterials can be simply accumulated on the surface of the electrode for the manufacturing of sensing electrodes for the effective and highly receptive detection of harmful heavy metal ions and gases [26,27]. The selectivity of the sensor is also modulated by using the incorporation of functional groups on the surface of ZnO nanostructures. The surface functionalization has the ability to improve the surface to volume ration and has provided extra sites for the adsorption of analyte during the sensing process. The presence of piezoelectric properties in the ZnO nanostructures has provided new dimensions for the generation of pressure sensors. Therefore, it is not wrong to say that these ZnO nanostructures have resulted in different devices with very high sensing capabilities. However, the question of a highly selective reaction still remains a great challenge. Therefore, the current review aims to present the current achievements of ZnO nanostructure-based chemical and biological sensors.

3. Chemical Sensing Applications of ZnO Nanomaterials

The estimation of the harmful chemicals present in the ecosystem is one of the prime issues to keep our environment hygienic and secure. Therefore, engineered materials for the preparation of chemical sensors capable of recognizing harmful toxins has received substantial attention from the scientific community. For a particular kind of chemical sensor, one can recognize a transformation in electrical or optical signals as an effect of chemical and physical associations with external toxins. In general, these chemical sensors [28] are very constructive in the fabrication of different types of security-based devices, where one can estimate any kind of leakage of toxins. It is well-known that ZnO nanoparticles are employed as in chemical sensors due to their extensive range of stability under thermal and chemical variations. The alteration in resistance due to the presence of external adsorbed surface species is mainly attributable to the oxygen vacancy in these ZnO materials [28].

In early investigations, chemical sensors were based on the inherent properties of the electrode materials used. As a result, studies were mostly centered on the choice of an appropriate material for the development of effective sensors, and the production of novel materials for sensing became a main focus of research. With the latest progress in the area of nanoscience, the above-mentioned paradigm has totally changed. Currently, the utilization of these engineered materials has provided a better substitute for conventional electrode materials, giving more control over the sensitivity, selectivity, and stability of the developed sensors [29]. The regular types of ZnO-based nanostructure with uniform size distribution can markedly advance the detection ability, as evaluated against disordered structures of different materials. These behavioral variations were explained due to the high crystalline nature of ZnO particles with large contact area. ZnO nanostructures have more catalytic sites and better control over the electron transfer resistance in the presence of external analytes [30]. Therefore, the internal properties of ZnO-based nanostructures are considered as one of the potential factors for the effective performance of sensors.

3.1. Hydrazine and Phenyl Hydrazine Chemical Sensor

Hydrazine and phenyl hydrazine are chemicals which are mainly employed in textiles, pesticides, aerospace fuel, and pharmaceutical industries. The excessive utilization of these chemicals has created the problem of their discharge in the surrounding environment, producing toxic effects in living beings [31]. Their adverse effects at minute concentrations have produced undesirable consequences for flora and fauna. Scientists have prepared a large number of metal oxide-based chemical sensors due to their potential properties of easy preparation with low processing costs and eco-friendly nature [32]. Among the different types of metal oxide materials, ZnO is extensively employed for the fabrication of sensors for perilous chemicals. This is due to the conductive nature of the ZnO nanostructure and its high chemical and thermal strength under the operation conditions of the developed sensor. For instance, Ameen et al. used vertically aligned nanorods of ZnO as a modification of the electrode surface for the estimation of hydrazine [33]. Additionally, Ibrahim et al. prepared a highly sensitive and selective electrochemical chemical sensor for phenyl hydrazine by using Ag-doped ZnO nanoflowers [34]. Umar et al. have reported the application of nano-urchins of ZnO particles for preparing sensory electrodes for the recognition of phenyl hydrazine by using the current-voltage (I–V) technique. [35]. The preparation of the nano-urchins was achieved by using the hydrothermal method at low-temperature conditions of ~165 °C. The urchin structures of ZnO are mainly generated by the gathering of numerous nanoneedles which are begun from a distinct center. These nanoneedles display sharp tips with broad bottoms. The characteristic diameter of each nanoneedle at their tips and bases were 45 ± 10 nm and 180 ± 20 nm, respectively. It was also found that the ZnO-based nano-urchins were independently developed with high density under normal synthetic conditions with a size ranging to 2 μm. The electrode was mainly prepared by using a slurry of nano-urchin-shaped ZnO particles on the surface of a glassy carbon electrode with a surface area of 0.0316 cm^2 (Figure 1). The respective measurements of the current response were done from 0.0 to +1.5 V with the time delay and response times of 1.0 and 10.0 s, respectively, for the developed sensor in phosphate buffer of 0.1 M. The sensing performance of the ZnO was tested against the wide range of concentration of phenyl hydrazine concentration ranging from 98 μM to 25 mM. The estimation of phenyl hydrazine with ZnO-based nano-urchins can be achieved due to the oxidation and reduction characters of ZnO particles. The presence of ZnO nanostructures has shown the enhancement of current which is mainly associated with the excellent electro-catalytic performance and superior sensitivity of as-prepared ZnO nano-urchins for the estimation of phenyl hydrazine.

The adsorbed oxygen molecules play an imperative function in the sensing process. At the beginning, the molecules of O_2 are mainly adsorbed in their ionic sate on the surface of ZnO nano-urchins (Figure 1a). These adsorbed oxygen ions have the tendency to eradicate the available electrons in the conduction band. The resultant depletion region possesses a low conductivity range close to the surface of ZnO, which significantly reduces the overall conductivity of the formed ZnO. The as-formed ionic oxygenated species has the ability to react with phenyl hydrazine to generate molecules of diazenyl benzene (Figure 1d). Then, the electrons are sent back to the respective conduction band of ZnO and there is a significant enhancement in current (Figure 1b,c). Similarly, composites of ZnO particles with SiO_2 have also been utilized for the detection of phenyl hydrazine by using current-potential (I–V) measurements. The detection limit of the formed sensor was found to be 1.42 μM with sensitivity values reaching 10.80 μA·cm^{-2}·mM^{-1} as a result of good adsorption ability and large surface area of the prepared sensor [36]. The main purpose of incorporating of SiO_2 with ZnO nanoparticles is to enhance the stability of the formed particles during the sensing runs. The obtained composite was reusable up to five times without any decrease in the sensitivity value. In addition, there was no decline in the I–V characteristics after three months of storage, showing the long-term stability of the formed sensor. Kumar and coworkers [37] have used Ag doping on the surface of nanoellipsoids of ZnO particles for the development of effective chemical sensors for hydrazine. The presence of Ag ion doping has the ability to enhance the interfacial charge transfer ability of the formed ZnO particles. The metal ion doping has a direct influence over the absorption characteristics of ZnO particles, and hence modifies their catalytic behavior in aqueous media. The presence of Ag in the lattice of ZnO particles has a direct influence on the light absorption ability of ZnO nanomaterial, improving its sensing aptitude towards hydrazine molecules. Figure 2 displays the respective representation of cyclic voltammetry (CV) data showing the effect of hydrazine in the presence and absence of Ag-doped ZnO nanoellipsoids at a scan rate of 100 mV/s.

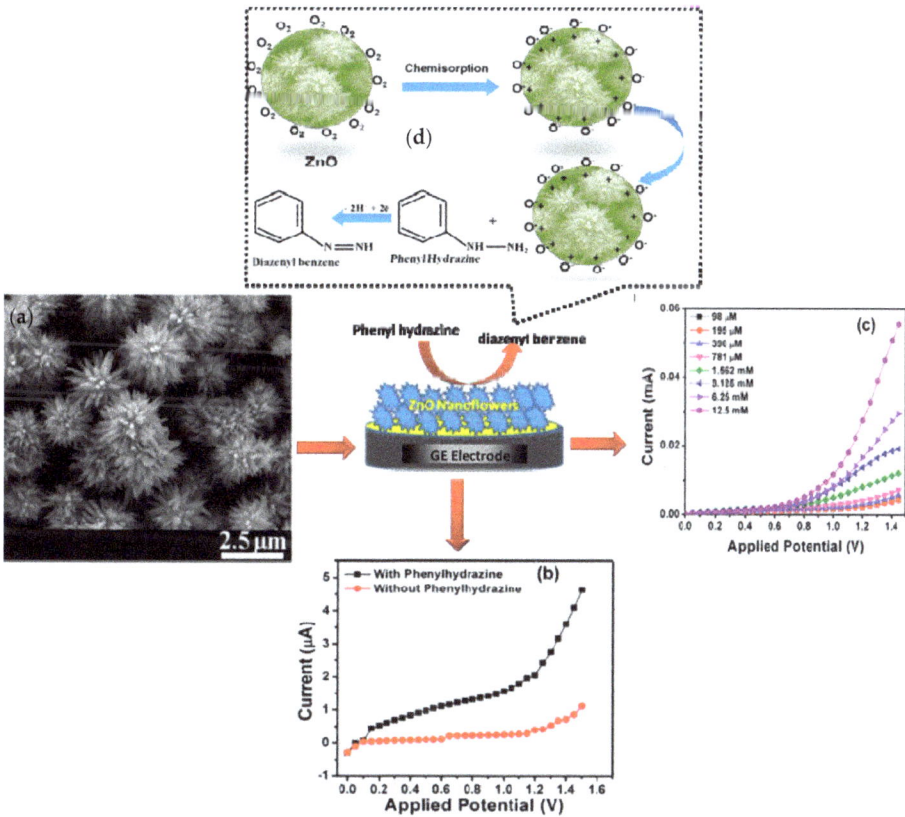

Figure 1. (**a**) Electrochemical measurement of fabricated phenyl hydrazine chemical sensor by using ZnO nano-urchins. (**b,c**) current-voltage (I–V) response in the presence and absence of phenyl hydrazine by employing the modified GCE in 10 mL, 0.1 M phosphate-buffered saline (PBS) solution. (**d**) Schematic mechanism of sensing. Adapted figure from [35] with permission from copyright, (2015), Elsevier.

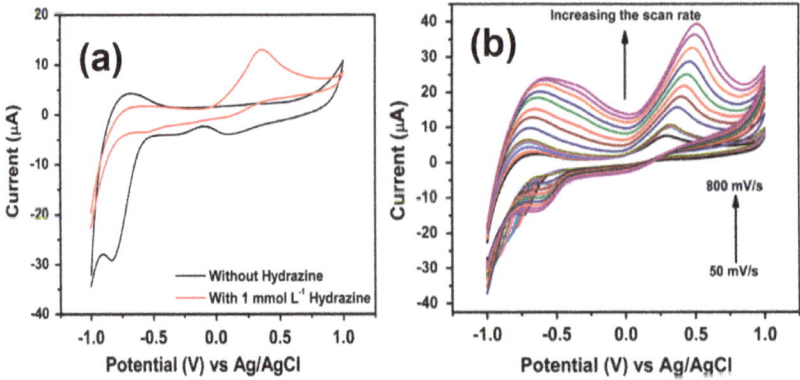

Figure 2. (a) Typical cyclic voltammetry (CV) sweep curve for Ag-ZnO nanoellipsoids/Au modified electrode with and without 11.0 mmol·L^{-1} hydrazine in 0.1 mol·L^{-1} phosphate buffer solution (PBS; pH ~7) at scan rate of 100 mV/s; (b) CV sweep curves at different scan rates (50, 60, 70, 80, 90, 100, 200, 300, 400, 500, 600, 700, and 800 mV/s) of Ag-ZnO nanoellipsoids/Au modified electrode. Adapted figure from [37] with permission from copyright (2015), Elsevier.

The existence of a clear oxidation peak at 0.357 V was observed for 1.0 mmol·L^{-1} of hydrazine molecules in buffer solution. The peak also displayed the significant enhancement in height in presence of ZnO molecules. This aspect was mainly due to the fast exchange of electrons between the nanomaterials and analyte molecules. Moreover, the anodic peak current showed significant enhancement with the variation of scan rate from 50 to 800 mV/s. There was a peculiar shift in the peak potential toward the positive side with varying scan rates. The effect of annealing temperature has also been used to investigate the performance of prepared chemical sensors for phenyl hydrazine molecules. By varying the temperature, it was found that the obtained size and crystalline nature of ZnO particles was varied (Figure 3). There was a significant effect on the surface of the formed particles, and their corresponding electron transport ability during the sensing was affected by the change in the particles' surface area [38].

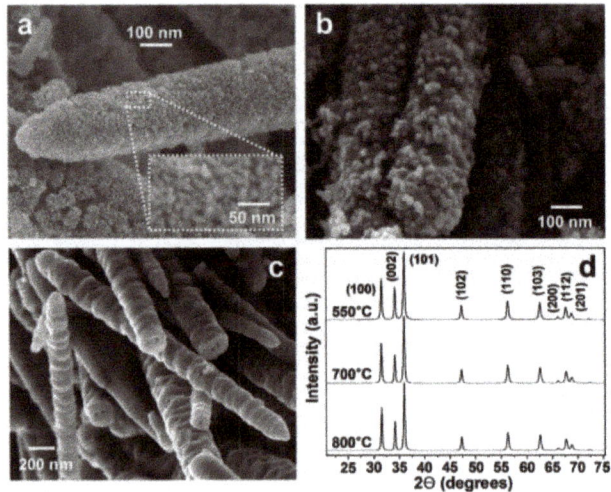

Figure 3. *Cont.*

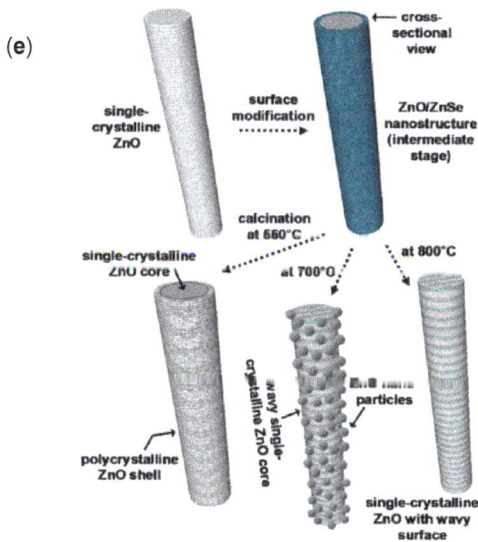

Figure 3. SEM images showing the effect of calcinations on ZnO particles in air for 3 h at (**a**) 550 °C; (**b**) 700 °C; or (**c**) 800 °C; (**d**) XRD patterns of the powders after calcination of the powders in air for 3 h; (**e**) Schematic illustration of the procedure for achieving surface texturing of a single-crystalline ZnO nanorod. Figure adapted from [38] with permission from copyright (2011), American Chemical Society (Washington, DC, USA).

The electron exchange rate and electrocatalytic oxidation characteristics of ZnO particles were affected with the variations of temperature [39]. Zhao et al. have used nanowires of ZnO for the preparation of a wide-linear range sensory device for the detection of hydrazine [40]. Ni et al. and Fang et al. have employed hierarchical micro/nanoarchitectures of ZnO and carbon nanotube-modified nanoflowers of ZnO for the detection of hydrazine [41,42]. The microstructures were mainly prepared by using the structure-directing group (i.e., hexamethylenetetramine) during the synthesis of ZnO nanomaterials. In addition, the presence of $NH_3 \cdot H_2O$ has played a critical role during the generation of hierarchical structures of ZnO-based micro/nanoarchitectures. It was found that when there was no addition of $NH_3 \cdot H_2O$ molecules, the formation of irregular nanosheets was observed. Regarding the addition of $NH_3 \cdot H_2O$ in the reaction media, the respective columnar-shaped hexagonal microcrystals of ZnO were formed with high density. Moreover, electrochemical testing results have shown that flower-shaped particles of ZnO possessed more potential to support the oxidation process of hydrazine as compared to other structures. The only disadvantage of the formed sensor was the poor selectivity towards hydrazine. Nanocones of ZnO particles were also used for the selective detection of hydrazine with a sensitivity of 50×10^4 $\mu A \cdot \mu M^{-1}$ cm^{-2} and detection limit 0.01 mM. The oxidation peak for hydrazine was observed at 424 mV with the peak current value of 380 mA. The C-V data clearly demonstrated highly responsive behavior of ZnO particles for hydrazine, and hence substantiate that the nanocones of ZnO are quite efficient in the electron transportation between the particles and external analyte [43]. Liu et al. used pristine nanorods of ZnO for modifying the surface of an alloy-based electrode for the sensing of hydrazine molecules [44]. The fabrication of the sensor was mainly achieved by the simple immersion-calcination process on the surface of inert alloy, and was employed for the preparation of a working electrode to build an effective sensor for hydrazine. The enhancement of electrocatalytic activity was mainly explained on the basis of synergistic behavior among the inert layer of carbon and ZnO nanorods. The electron transport rate was found to be significantly affected along the growth direction of prepared nanoparticles. For instance, these nanorods have the ability to grow in a one-dimensional (1D) pathway and possessed a greater electrochemical sensing aptitude along the length direction. It was observed that these rods had the tendency to align directly on the surface of electrode substrates; the rate of electron transport was affected, and hydrazine sensing was found to be improved. These aligned nanorods/nanowires can provide abundant "nanoelectrodes" on the surface to improve the electrochemical action. These rods provide more surface area for the transportation of electrons, and hence affect the sensing ability of the electrodes. Moreover, these ZnO nanorods have the ability to construct binder-free, stable, and highly-active chemical sensors for hydrazine. The incorporation of metallic silver with vertically-aligned zinc oxide nanorods has the tendency to modulate the electron transport and hence affect the sensing performance of the developed sensor [45]. The obtained sensitivity for the Ag@ZnO nanorods was found to be 105.5 $\mu A \cdot \mu M^{-1} \cdot cm^{-2}$, with a working concentration of 98.6 μM. The sensor had the ability to detect concentrations as low as 0.005 μM. The higher stability, its reproducible nature, and its selectivity for hydrazine make it more effective for the detection of hydrazine. Spiked samples from different water sources have also shown effective results for the estimation of hydrazine with the help of Ag@ZnO nanorods. The performance of the formed sensor was mainly explained due to the presence of directly growing nanostructures of ZnO particles, which further affected the kinetics for electron transference. Thus, the formed sensor has future prospective roles in binder-free, economical, and highly stable and reproducible sensors for harmful analytes.

On the other hand, electrodeposition methods have also been used for the fabrication of ZnO nanostructures on the surface of gold or glass electrodes for the selective sensing of hydrazine molecules. These methods have direct control over the morphology and size of the formed nanoparticles. The respective surface area of the prepared nanoparticles was also modulated by using the electrodeposition process. The variations in the deposition parameters included changes in the concentration of external agent, the potential value used, as well as the applied current, and directly influenced the deposition layer of ZnO particles. The electrocatalytic activity of the formed nanoparticles was also varied with the concentration of hydrazine moieties, and displayed a different sensitivity and detection limit than other morphologies (Table 1).

Table 1. A response comparison of as-fabricated sensors in this work with various ZnO nanostructure-based hydrazine sensors.

Electrode Material	Sensitivity ($\mu A \cdot \mu M^{-1} \cdot cm^{-2}$)	Limit of Detection (μM)	Linear Range (μM)	Response Time (s)	Ref.
Nano-nails of ZnO	-	0.2	0.1–1.2	<5	[46]
Nanowires of ZnO with high aspect ratio	-	0.0847	0.5–1.2	<5	[47]
Nanorods of ZnO	4.76	2.2	0.2–2.0	<10	[48]
Hierarchical micro/ nanostructures of ZnO	0.51	0.25	0.8–200	<3	[49]
Nanorods of ZnO/FTO	0.44	515.7	-	<10	[50]
ZnO nanorods/alloy	4.48	0.2	-	<8	[51]
ZnO nanorods/Single walled carbon nanotube	0.1	0.17	0.5–50	-	[52]
Reduced graphene oxide/ZnO-Au	5.54	0.018	0.05–5	<3	[53]
Cetyl pyridine chloride/ZnO	0.172	24	1000–60,000	-	[54]
ZnO-1/Au/gold electrode	1.70	0.25	0.8–251	<3	
ZnO-2/Au/gold electrode	2.97	0.17	201–851	<3	[55]
ZnO-3/Au/gold electrode	1.25	0.08	101–851	<3	

Recently, Zhang et al. used a combination of zeolitic imidazolate framework-8 (ZIF-8)-derived N-doped carbon film-immobilized gold nanoparticles (AuNPs) on a ZnO jungle for the estimation of hydrazine [56]. The in-situ oxidation was mainly responsible for the synthesis of N-doped carbon film-coated ZnO particles used for the sensor fabrication. The jungle of ZnO structures was mainly formed by nanorods of ZnO particles generated by the electrodeposition process. The presence of Au in combination with ZnO provided better control over the catalytic activity of hydrazine and provided a distinguished current peak at 0.66 V with a current of 6.19 µA. The detection of hydrazine mainly involves the transference of four electrons in the system [57]. The low-temperature-based synthesis of ZnO particles with high aspect ratio has also been used for the fabrication of chemical sensors for hydrazine molecules with the ultra-high sensitivity of ~97.133 A·cm^{-2}·M^{-1} and detection limit as low as of 147.54 nM. The prepared sensor was quite selective and worked selectively in the presence of interfering ions. Multiphase solid materials with nano-dimensional ZnO particles have also been used for the development of effective sensors for hydrazine molecules. These multiphase solid materials with ZnO have a direct influence on their size, interfacial properties, and electrocatalytic activities against harmful analytes [58]. It was found that most of the electrochemical-based sensing of hydrazine molecules using ZnO nanoparticles have mainly used nafion for the tight attachment of ZnO nanomaterials on the surface of gold electrodes. These nafion molecules have the ability to form net-like layered structures on the surface of electrodes which form a partially blocked array system on the electrode surface [59]. These nafion molecules further coat the ZnO particles on the surface of bare gold electrode, and are helpful in the passivation of the electrode surface and further decrease the electroanalytical performance against external agents. In order to overcome this problem, researchers have recently used carbon nanotubes (CNTs) along with ZnO nanomaterials for the modification of the electrode surface in chemical sensing applications [60]. The presence of CNTs further affects the availability of accessible surface area for electrocatalytic activities. Moreover, the presence of CNTs further lowers the electrical resistance and enhances the mechanical strength and stiffness of the modified electrode during the analysis. The charge transport ability and chemical stability of the ZnO particles were further enhanced by using the CNTs.

3.2. Nitrophenol Chemical Sensor

4-nitrophenol (4-NP) is a distinguished type of aromatic organic pollutant, considered as a noxious and bio-obstinate complex which is harmful to various vital organs, including the central nervous system, liver, kidney, and blood of living organisms, and produces several diseases [61]. The higher stability and lower solubility in aqueous media makes it difficult to degrade to non-hazardous products. In this regard, the catalysis of this organic pollutant by employing the ZnO nanomaterials have been proven as an effective environmentally-friendly and economical means of detecting and removing these harmful analytes. For instance, Thirumalraj et al. [62] produced a very easy and responsive electrochemical means for the estimation of 4-nitrophenol (4-NP) in different types of water samples by the application of ZnO nanoparticles. The used particles were functionalized using chitosan (CHT) molecules, and the obtained nanoneedles of ZnO were coated on the surface of screen-printed carbon electrode for the estimation of 4-nitrophenol. The formed sensor has several advantages, such as high sensitivity, low working cost, and good reproducibility as compared to conventional methods. The working procedure for the developed sensor was quite easy and economical with very rapid response to the analyte. The presence of chitosan provided extraordinary properties to the ZnO nanostructures, including the presence of high surface area with complimentary electronic properties with higher biocompatibility with the surface of nanoparticles. The detection measurements for 4-nitrophenol were carried out in 0.05 M acetate buffer media at pH ~5 with scan rate of 50 mV/s. The comparative studies were also carried out with bare, ZnO nanoparticles-, and chitosan-functionalized nanoneedles at a scan rate of 50 mV/s. The modified electrode displayed a weak cathodic peak at −0.793 V which was associated with the reduction of 4-NP to hydroxylaminophenol. The conversion mainly involves a four electron and proton transfer electrochemical process. The corresponding reversible peaks were observed at 0.373 and 0.150 V. The surface modification with the chitosan provided the appearance of small sub-units on the surface of the electrodes which acted as electro-active centers for the transference of electrons and provided a better material for the electrocatalytic measurements of 4-nitrophenol. The nature of the reaction media (e.g., the chosen buffer and the pH of the reaction media) played a significant role in the detection ability of ZnO-modified electrode against nitrophenol. To optimize these parameters, three different types of electrolytic solutions (0.05 M PBS, 0.05 M citrate buffer, and 0.05 M acetate buffer) were chosen and optimized for the catalytic detection of nitrophenol molecules. Among the chosen electrolytic solutions, it was found that the CV response of nitrophenol was maximum for the 0.05 M acetate buffer solution as compared to the other buffers. Therefore, studies varying scan rate and concentration were performed in the presence of 0.05 M acetate buffer. The other crucial parameter (i.e., pH) was also optimized by varying it from 2 to 14. The outcomes have pointed to the presence of a strong peak at pH ~5. The peak strength was found to decrease by decreasing or increasing the pH of the reaction media. Therefore, the pH 5, 0.05 M acetate buffer is an optimized buffer solution for the electrochemical measurements of nitrophenol.

These ZnO-based nanostructures have the ability to act as an effective sensitizer for light-mediated redox reactions. The electronic structure of ZnO molecules was mainly associated with these light-induced redox processes. These ZnO nanostructures possess well-filled valence bands and an empty conduction band. When an external photon with threshold energy equivalent to or greater than the band energy of ZnO was applied to the particles, there was the probability of electron transference from the valence band to the conduction band in these nanostructures. This kind of transference has the ability to produce a hole in the valence band and an extra electron in the conduction band. Being unstable in nature, these excited conduction and valence bands have the ability to recombine these electrons and holes and liberate a respective amount of energy in the form of external heat and get ensnared in the respective metastable surface states in nanostructures. These surface states were further reacted with the electron donor and electron acceptor species and further adsorbed on the external surface of ZnO. The corresponding electrical double layer of the adsorbed species was formed on the surface of ZnO materials. These charged species have the ability to act as scavengers or surface defect mediators to catch the available electrons or holes in these nanostructures. Thus, the corresponding recombination of holes and electrons is averted and following redox reactions have the chance to occur on the ZnO surface. Undoped ZnO has n-type conductivity, which can be assigned to the asymmetric doping restraints and tendency towards defects or impurities. *p*-type doping with Ag, P, N, etc. has the tendency to modulate the conductivity and electrical and catalytic activities of ZnO nanoparticles. It has been well-established in the literature that the photocatalytic performance of ZnO was greatly improved by doping the surface of ZnO nanostructures with Ag. The electron transference rate was found be highly influenced in the presence of Ag [63]. For instance, Divband et al. [63] used the efficacy of Ag-doped ZnO nanostructures for the photocatalytic degradation of 4-nitrophenol (4-NP). The measurements were mainly done with the 50 mL solution of 10 ppm concentration of 4-nitrophenol. The prepared solution was thoroughly stirred with the help of magnetic stirring in the dark for at least 15 min. This stirring is helpful for the establishment of an effective adsorption/desorption equilibrium between the 4-nitrophenol on the external surface of ZnO. After equilibration in darkness, the respective solution was placed in ultraviolent light under stirring conditions. The samples were removed after a particular time frame and centrifuged at 1500 rpm to remove the catalyst particles. The respective measurements for the concentration of 4-nitrophenol were made by using UV-visible spectroscopic techniques. The absorption intensity appeared to be decrease at 315 nm with time. This behavioral aspect was related to the oxidative degradation of 4-nitrophenol in the presence of ZnO nanoparticles.

With the introduction of microreactor technology in photocatalysis, 1D nanomaterials of ZnO have been comprehensively investigated as a new method to amend the internal wall of capillary microchannels with nanostructures. Being a wide band gap material, ZnO has shown tremendous potential as an effective catalyst to degrade unrelenting organic pollutants from the environment. 1D ZnO micro-nanostructures have the ability to form an effective microchannel inside the internal wall of a capillary. The formed nanorod of ZnO has the tendency to uniformly arrange on the internal wall of capillary microchannels and provides a better composite catalyst or device for the degradation studies. Conversely, the one main disadvantage of these 1D nanorods and nanotubes of ZnO-based nanostructures with vertical heights of around 2–5 μm is the lesser tendency to make complete utilization of the full internal space inside capillaries of microchannels [64]. Therefore, by amending the properties of as-prepared ZnO-based nanostructures, there is a great deal of room for the improvement of the sensitivity of the as-prepared sensor. In this regard, Zhang et al. [64] prepared grass-like double-layer micro-nanostructures of ZnO in restricted microchannels by using a fluid construction process. The developed height of the grass-like structures was found to be 50 μm. The potential use of these grass-like catalysts was checked against the degradation of o-nitrophenol molecules. The outcome of the work revealed that capillary microreactors of ZnO have the tendency to solve the problem of the stability of the developed sensor, and the formed sensor was quite selective and sensitive against the chemical reduction of o-nitrophenol. The association of surface-connected processes with the quantum confinement effects in ZnO-based nanostructures resulted in admirable optical and electrical properties of nanostructures that can be adjusted to tackle particular requirements. In addition, the porous morphologies of these ZnO nanowires further assists the quick distribution of analyte species to binding locations, leading to a quicker reaction rate in the developed sensor. Additionally, the zinc oxide (ZnO) nanowires have the advantages of an easy and cost-effective growth process on the surface of insulating materials. These properties have further facilitated sensor fabrication, as the growth of ZnO nanowires can be simply managed to produce a conductive mesh-like arrangement on the external templates, thereby decreasing the production cost of sensors. Moreover, simplicity in fine-tuning the optical properties of ZnO nanostructures by using the defined nanowire diameters might afford added benefits in terms of a composite opto-electronic podium for the sensor development.

Gupta et al. [65] used heterostructures of a ZnO nanowire with surface modifications with pyrenebutyric acid (PBA) for the effective sensing of trace amounts of p-nitrophenol in biological systems. It was found that the fluorescence intensity of the prepared ZnO nanowires was significantly reduced in the presence of p-nitrophenol. Therefore, the detection was based on the degree of the fluorescence quenching in the presence of external analyte. The collision quenching of ZnO with the external moiety is commonly acknowledged as the central de-excitation path, where energy transmission from the π^* orbital of pyrenebutyric acid to the π orbital of p-nitrophenol occurs. This process permits exceptionally high receptive recognition of analyte that could facilitate the detection of very minute amounts of p-nitrophenol in biological systems. These ZnO-based nanowires have offered a strong sustainable configuration for implanting the receptor molecules, while in parallel it is advantageous for the fabrication of sensory devices due to the economical processing method. Similarly, Singh et al. have employed composite structures of ZnO with CeO_2 nanoparticles for the detection of p-nitrophenol in aqueous media [66]. The formed composite nanostructures had good control over the crystallinity and optical properties of ZnO nanoparticles, and acted as a good electron mediator for the detection of external analytes. The as-prepared chemical sensors displayed sensible, selective, and reproducible sensitivity of around 0.120 $\mu A/(nM \cdot cm^2)$ with a detection limit of 1.163 μM. This kind of analysis unfolds the ways in which the simply synthesized CeO_2–ZnO can competently be used for sensor fabrication. The nanoplates of ZnO nanostructures on the surface of ITO substrate also provided a better catalytic ability for the detection of 4-nitrophenol [67]. The analysis was mainly achieved in the absence of reducing agent. The catalysis was found to be influenced by the ultrasonication waves under room-temperature conditions. The respective conversion rate for 4-nitrophenol to 4-aminophenol was found to be as high as 4.483×10^{-2} mol·min^{-1}. The presence of a high content of oxygen vacancies on the surface of (001) faceted nanoplate of ZnO was assumed to be the driving force for catalytic conversion of 4-nitrophenol to 4-aminophenol.

3.3. Nitroaniline Chemical Sensor

The derivatives of anilines such as nitroaniline are extensively employed in pharmaceutical, polymer, rubber, dyes, paints, and explosive industries. The surplus discharges of these pollutants in surroundings have destructive consequences on living beings and the environment. These chemical substances are highly lethal towards aquatic life, lead to adverse effects on the liver, and cause methemoglobinemia in humans. Additionally, ingestion and absorption of these toxins through the skin produces an allergic reaction. As a result, there is an increasing awareness among scientists of the necessity to develop an uncomplicated, time effective, highly stable, selective, and sensitive method for the estimation of this harmful chemical. In this area, ZnO has received more attention due to its unique and distinct optical and electrical properties. The availability of high crystallinity at low working temperature with modifiable electrical properties with good control over the biocompatibility of ZnO makes it a promising material for the detection of nitroaniline molecules with great accuracy and precision.

Recently, Ahmad et al. developed a binder-free, highly reactive, and sensitive chemical for the detection of p-nitroaniline by using ZnO nanoparticles [68]. Nanorods of ZnO were used for the modification of fluorine doped tin oxide (FTO) electrode. The main advantage of the method is that the analysis was performed at low-temperature conditions in aqueous medium. The developed sensor displayed incredibly high stability and reproducibility. Cyclic voltammetry (CV) peaks with nanorods of ZnO on the surface of FTO electrode in PBS with pH ~7.0 were observed in the potential range from 0.1 to +1.0 V. The scan rate for the measurements was kept at 100 mV/s. The reference electrode during the analysis was Ag/AgCl (saturated KCl) solution. The modified electrode in the absence of p-nitroaniline showed no current peak. The well-defined peak was observed in the presence of p-nitroaniline at +0.55 V. The obtained current peak was associated with the electrocatalytic oxidation of p-nitroaniline on the surface of modified FTO electrode with ZnO nanorods. The influence of the presence of external binder was also checked in the CV response for modified electrode in presence of p-nitroaniline. Figure 4b shows that in the absence of any kind of external binder, the electro-catalytic response against p-nitroaniline was found to be better as compared to the results in the presence of binders. The reactive ionic species which were converted from dissolved oxygen molecules in the reaction media in the presence of ZnO nanoparticles have the ability to absorb electrons from the conduction band of the particles. These reactive species were further reacted with the p-nitroaniline and oxidized it to respective CO_2 and H_2O molecules after transitory movements of numerous intermediary responses.

The potential catalytic properties of ZnO nanostructures were further enhanced by doping the nanostructures with other lanthanide oxide materials. The doping of the nanostructures lends the ability to increase the reactive surface area of the particles, the number of defects, and the respective amount of oxygen vacancies on the particles. The effective diffusion rate for the electron transportation is also influenced by the presence of dopants. For instance, the combination of CeO_2 with ZnO has the ability to form n–n type hetero-junctions for the electron transference reactions (Figure 4). The transference of electrons results in the formation of depletion around the interface of the nanocomposites, and improves the catalytic properties of the formed sensor [69]. A similar analysis was also performed with the help of the bi-composites of CdO-ZnO hexagonal nanocones for the detection of nitroaniline. The prepared sensor displayed a sensitivity of ~129.82 $\mu A \cdot mM^{-1}\, cm^{-2}$. The distinctive hexagonal nanocones of ZnO provided a high surface-to-volume ratio, which was found to be accountable for producing a huge amount of reactive oxygen groups on the surface of the CdO-ZnO nanocones and eased the oxidation process of nitroaniline (Figure 5) [70].

The role of Sm_2O_3-doping on the hierarchical structures of ZnO was also used for the generation of an electrochemical sensor for nitroaniline. The synthesis of the biocomposites was done by employing a simple, very economical, and fast hydrothermal method. The starting materials for the preparation (i.e., $Zn(CH_3COO)_2 \cdot 2H_2O$ and $Sm(NO_3)_3 0 \cdot 6H_2O$) are easily available. The synthesis was mainly carried out at 155 °C for 7 h in basic media with pH of 9.5, respectively. The formed composite mainly displayed needle-shaped and leaf-shaped morphology, with particular outlines of ovate or triangular-ovate structures resembling ferns. The high surface area with good control over the electron transportation provided a better pathway for the electrochemical detection of nitroaniline. The sensitivity of the formed sensor was 1.71 $\mu A \cdot \mu M^{-1} \cdot cm^{-2}$ with detection limit of 15.6 μM [71]. Figure 6 exhibits the schematic representation of the proposed sensing mechanism for nitroaniline sensing using Sm_2O_3-doped ZnO beech fern hierarchical structures.

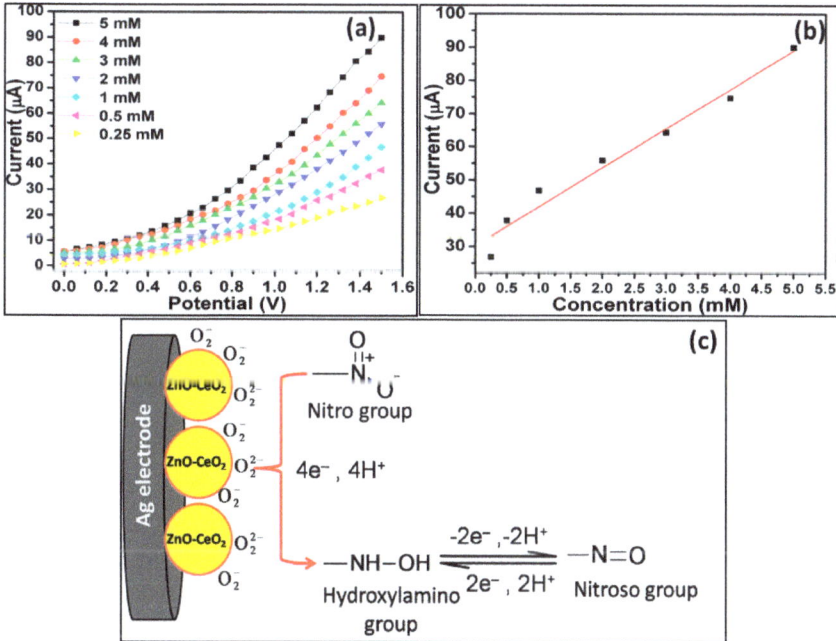

Figure 4. (**a**) Current-voltage responses for various concentrations of nitroaniline; and (**b**) Calibration curve for nitroaniline using ZnO-doped CeO$_2$ nanoparticles-modified silver electrode (AgE); (**c**) A proposed sensing mechanism for the ZnO-doped CeO$_2$ nanoparticles-modified AgE toward nitroaniline sensing. Adapted figure from [69] with permission from copyright (2016), Elsevier.

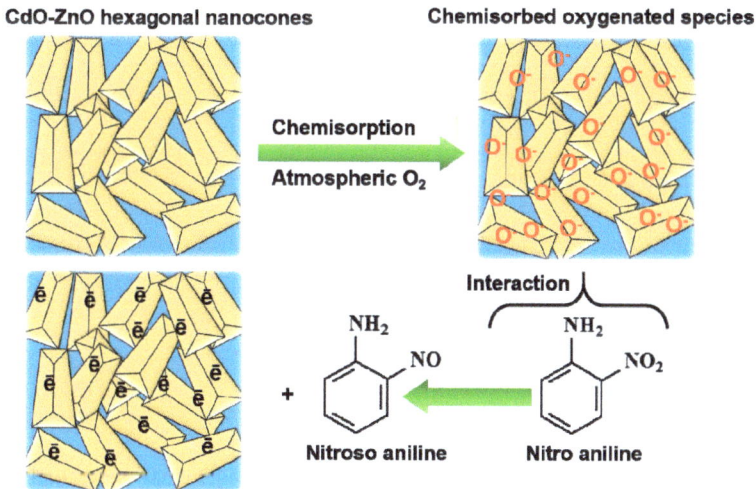

Figure 5. A sensing mechanism for nitroaniline sensing using modified GCE with CdO-ZnO hexagonal nanocones. Adapted figure from [70] with permission from copyright (2017), Elsevier.

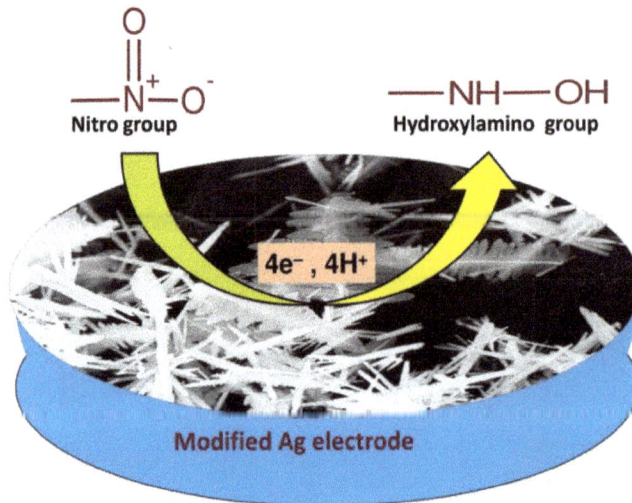

Figure 6. A schematic representation of the electrochemical sensing mechanism for the Sm_2O_3-doped ZnO beech fern hierarchical structures-modified AgE toward nitroaniline sensing. Adapted figure from [71] with permission from copyright (2017), Elsevier.

3.4. Ethanol, Methanol, and Propanol Chemical Sensor

The application of ZnO-based nanostructures for the recognition of venomous pollutants including inflammable complexes and organic compounds are an area of growing interest in both household and workplace atmospheres. For instance, Sahay and coworkers [72] investigated the application of thin films of ZnO nanostructures obtained by the spray method for the analysis of ethanol and acetone molecules. The corresponding effect of external dopants such as Fe, Al, In, Sn, and Cu on the response rate of a thin film-based ZnO sensor for ethanol was analyzed [73]. The sensing abilities of thin films of ZnO-based nanoparticulate were tested by Cheng et al. for the-sensing of ethanol and propyl alcohol molecules [74]. The developed sensors displayed higher sensitivity and selectivity for the tested molecules. The ability of thick films of ZnO-based nanostructures to simultaneously detect ethanol and propanol was achieved by Arshak and Gaiden. As for the sensing utilities of ZnO nanostructures, the presence of high Brunauer Emmett Teller (BET) surface area of the nanomaterial is a prime requirement for the analysis. In order to detect the optimum BET surface area for the preparation of sensor, a nanomaterial with highly porous morphology is necessary. In the current era of research, various ZnO morphologies, such as rods, belts, tubes, fibers, sheets, films, microflowers, and cones have been prepared with optimized porous character. For instance, the sensory reaction for nanoporous microbelts of ZnO was found to be very high for a diverse range of ethanol concentrations at a temperature of 300 °C. For a low concentration of around 1 ppm, the response rate was found to be 4.7. Upon increasing the amount of ethanol to 100 ppm in the aqueous media, there was an amazingly higher response value of 38.4. Moreover, the formed sensor showed a higher reversibility for ethanol [75].

These developed sensors work on the principle of variations in the values of conductance brought by the adsorption of an analyte on the exterior surface of the ZnO-based sensor. The atmospheric oxygen group has the ability to be adsorbed on the microporous structures of ZnO nanobelts. By the adsorption of electrons from the conduction band of nanomaterials [76], these oxygen molecules are converted into their oxidized form on the exterior surface of the sensor and form a substantially thick space-charge layer on the nanoparticles and amplify the potential barrier with an enhanced resistance during the analysis. These ethanol molecules have the tendency to respond against the adsorbed oxygen molecules and form CO_x and H_2O molecules with the discharge of electrons at a reasonably high temperature of 300 °C. During the analysis, the decrease in the oxygen coverage results in the thinning of the depletion layer and the enhancement of conductivity in the presence of ethanol. On the basis of the resistivity of nanofilms of ZnO-based materials, the sensitivity of the formed sensor was varied drastically. The reaction temperature reversibly modulated the resistances of the nano films. The concentrations of external dopant on the surface of ZnO nanostructures further modified the resistivity of the sensor. It was verified that the higher concentration of dopant had a higher scattering efficiency, which in turn produced the enhancement of resistivity during the measurements. The use of three-dimensional (3D) nanoscale ZnO materials has received substantial consideration due to their astonishing features as applied in sensory devices. It is a well-established fact that the sensory aptitude of particles mainly depends on the microstructural properties of ZnO materials, which are affected by the method of synthesis, which latter affects the detection rate and sensitivity of the formed sensor. The growth behaviors of these three-dimensional (3D) nanoscale structures of ZnO are also affected by the variations in the external conditions of temperature, pH, and concentrations of the starting materials. The comparative studies have clearly shown that the sensing aptitude of ZnO nanoflowers was found to be much higher as compared to the nanowires, rods, or plates of ZnO materials. In addition, the sensor prepared by using these flowerlike structures of ZnO materials has shown considerably higher response of 24.1, 14.6, 14.2, and 13.8 for methanol, propanol, n-butanol, and acetone, respectively. The higher rate of diffusion of analytes in these 3D flowerlike ZnO nanostructure films is mainly responsible for the higher sensitivity. The resultant rate of adsorption and desorption was also found to be higher in these flowerlike nanostructures.

The use of the easy and consistent means of I–V techniques has shown the tremendous potential for the detection of methanol molecules with short response time. When methanol molecules were present in aqueous media, the obtained current response of ZnO nanoparticles was found to be changed significantly. This is mainly due to the adsorption of methanol molecules on the surface of ZnO nanoparticles [77]. The liquid phase analysis of the methanol molecules was mainly performed by using calcined samples of ZnO nanoparticles. The sensor was prepared by coating the surface of a glass electrode with a combination of a calcined ZnO-based structure and conducting agents. The formation of an even layer with high stability was achieved by drying the modified electrode in an electric oven at 60 °C. The respective I–V responses of for the liquid methanol were tested by ZnO thin-film as a function of current against potential. The corresponding time delay for the measuring device was 1.0 s. The obtained results clearly show that the current response of ZnO was increased with increase in the concentration of methanol at room temperature. These results have pointed out that the sensing mechanism for methanol molecules is a surface-based procedure. The obtained detection limit was tested by using a broad range of methanol concentrations. The obtained sensor displayed a methanol detection range varying between 0.25 mM to 1.8 M with a detection limit of 0.11 mM. The results clearly demonstrate that the exceptionally high surface area contributes greatly to the adsorption of external analyte and had a direct influence on the catalytic activity of the ZnO NPs for the estimation of methanol molecules. cn. ZnO molecules have the tendency to generate a constructive microenvironment on the exterior surface for the estimation of methanol molecules by the adsorption process. The enhanced sensitivity of ZnO-based nanostructures is attributable to the elevated rate of electron communication between the analyte and the used nanostructures. Higher stability with better reproducibility and shelf life were the additional benefits of the produced sensor. Methanol molecules have the ability to react with the adsorbed oxygen species on the exterior surface of nanostructures and are oxidized to formaldehyde and subsequently converted to formic acid by the release of electrons into the conduction band of the ZnO particles. As a result, there was a reduction in the resistance of the nonmaterial upon contact with the methanol molecules.

The thermodynamic studies for detecting the sensory mechanism for methanol and ethanol molecules have clearly demonstrated that the alcohols under investigation with identical molar masses have shown a reliance on packing arrangements. The more strongly-packed atoms on the exterior molecular structure of the nanoparticles displayed lower response time for the analyte being studied. This behavioral aspect can be elucidated by a rapid rate of adsorption on the exterior surface of semiconducting nanomaterials [78]. In addition, the boiling point of the adsorbed analytes also influenced the adsorption rate of the sensor. Molecules with higher boiling point had a greater tendency to adsorb onto the surface of the developed sensor. These phenomena were quite similar to the condensation process. Mainly, the adsorption was achieved by the breaking of respective O-H and C-H bonds in the alcohols, and the formed charged species were adsorbed onto the surface of the sensor. The adsorption process was well in accordance with the Freundlich isotherm equation. These isotherms facilitate the construction of the curves for the detection of sensitivity and the respective kinetics for the process. Being a well-liked drink, liquor has been consumed worldwide for thousands of years. With advancements in science and technology, there is a progressively greater diversity of liquors in the marketplace. In the interest of profit, the adulteration of liquor products is a known phenomenon, often with unsafe elements that harm the vital body organs. Therefore, means of differentiating the damaging components in liquors are catching the interest of scientists. It was found that these liquors are a complex combination of water and alcohol molecules as the major component, with hundreds of other small additional compounds. By detecting the harmful molecules in the liquor, we can easily judge the quality of the liquor. In this regard, a novel coplanar ZnO sensor was used for checking the purity of liquor. For the analysis, liquor was injected on the surface of a pure ZnO nanofilm which was previously screen-printed. The presence of external dopants such as transition metals was checked for the investigation of analytes [79]. For the sensor preparation, the bare gold electrode was first stamped on the dirt-free clean surface of an alumina ceramic chip by screen printing methodology. Then, the formed electrode was sintered at 850 °C for 15 min. The resistor paste RuO$_2$ was printed on the modified electrode by screen printing. After that, the corresponding paste of ZnO was printed on the formed chip. The obtained area for the detection of analyte was about 0.5 mm × 1 mm, with a thickness of about 10 μm. The obtained sensor displayed a highly porous sponge-like thick film of ZnO for the detection of harmful pollutants. This porous nature of ZnO nanostructures provides different paths and enhanced surface area for the mass movement of target species. In addition, the hierarchically porous nanorod-like ZnO structures have displayed good response towards ethanol and acetone molecules. At low ethanol and acetone concentrations of around 1 ppm, the obtained responses were about 3.2 and 3.3, respectively, as compared to 24.3 and 31.6 at concentration of 100 ppm of ethanol and acetone in the reaction media. In order to study the repeatability of the sensor, the response analysis was performed by using the exposure to 100 ppm ethanol at 320 °C up to ten times. It was observed that the average value of the obtained sensitivities was 24.5, with the changeability of the sensitivity being less than 2.1%. All of these outcomes clearly show the good repeatability of the ZnO-based sensors. The operating temperature also influenced the working potential of the ZnO-based sensor for ethanol and methanol molecules. At an operating temperature of 200 °C for diverse range of concentrations of ethanol and methanol, the developed sensor did not show any considerable response against methanol. This was mainly explained by the presence of insufficient thermal energy in the system to respond with the adsorbed oxygen species on the exterior surface of the nanoparticles. The presence of a space-charge layer on the surface of nanoparticles has the tendency to hinder the charge transportation at low-temperature conditions. On the other hand, as the working temperature was enhanced during the measurements, the response rate of the ZnO-based film to methanol was augmented for all studied concentrations. The electron transport and the respective surface reactions progressed quickly with the enhancement of thermal energy. Among the various temperature ranges evaluated, the response rate was maximum in the region of 275–300 °C. This is mainly explained by the presence of an adequate amount of adsorbed oxygen species on the surface of the ZnO film, which respond most efficiently to the methanol molecules. On the other hand, for the ethanol molecules, the enhancement of the response rate with the operating temperature was comparatively less for concentrations up to 200 ppm. Upon further increasing the concentration to 250 ppm, the response rate was found to be enhanced up to 250 °C and then started to decline.

This is mainly explained by the enhancement of surface coverage by ethanol molecules on the surface of ZnO films, which stopped the successive adsorption of atmospheric oxygen on the surface of the films, causing the chemical reaction to progress with a slow rate, and thus a decline in response occurred.

In addition, the pH of reaction media also played a significant role in the sensing ability of nanostructures. The pH of the reaction media is considered as one of the primary components and is an easily computable and amendable factor during the fabrication of nanostructures that influences the structural and morphological behavior of particles. Studies have shown that the variation of pH values during the synthesis of particles affected the shape and morphology of the generated structures. By simply varying the pH, one is able to achieve particle morphologies ranging from plate-like to flower-like. The surface area of the particles is also affected by varying the pH of the reaction media, and hence the sensing ability was influenced by varying the reaction conditions. In the case of nanorods of ZnO-based particles, it was observed that at pH ~8, the formation of nanorods was well-defined. This was mainly explained by the minimization of the total energy of the system by the unstructured and spontaneous polarization of the reaction system. This polarization further affected the non-centrosymmetric crystal structures of the obtained nanorods of ZnO in the reaction media. On further enhancing the pH of the reaction media to 10, there was a tendency of these formed nanorods to agglomerate. This may be explained by taking the effect of reaction solution super-saturation with enhancement of the system pH. The nanorods then lost their shape to form joined agglomerates, minimizing the overall energy of the reaction system. The obtained particles displayed a higher sensitivity for methanol, ethanol, and propanol at different reaction temperatures [80]. On interpreting the results, it was found that the response rate of the prepared sensor varied as the pH of the reaction media varied. The particles prepared at pH 11 had a greater response than particles prepared at pH ~8. The explanation for this augmentation in the sensitivity at higher pH values may be attributed to the morphologies of the formed particles.

The bi-composites of p-Co_3O_4/n-ZnO particles were also used for the fabrication of an effective sensor for ethanol molecules. The advantage of this method is that it has also been applied for the detection of acetone and nitrogen dioxide molecules with great accuracy and efficiency. For the synthesis of biocomposites, the $\langle 001 \rangle$ oriented nanorods of ZnO nanoparticles were first raised on the surface of alumina substrates by using the plasma-enhanced chemical vapor deposition (PECVD) technique. The as-produced templates were further employed for the growth of nanograins of Co_3O_4 molecules. To use the prepared particles for practical sensing applications, the selectivity of the formed sensor was optimized as a function of time, concentration, and temperature. Out of all the studied parameters, the working temperature was established as one of the crucial factors for detecting the harmful analyte. The sensing ability of the developed sensor was found to be superior to those of earlier representative examples of Au-doped or F-doped Co_3O_4 nanoparticles. Mainly, the p-Co_3O_4/n-ZnO junctions have the ability to generate a better charge partition on the interface of the two used oxides (Figure 7). The formed junctions further affected the conductance of the system and modulated the interaction of the target species with the nanoparticles [81].

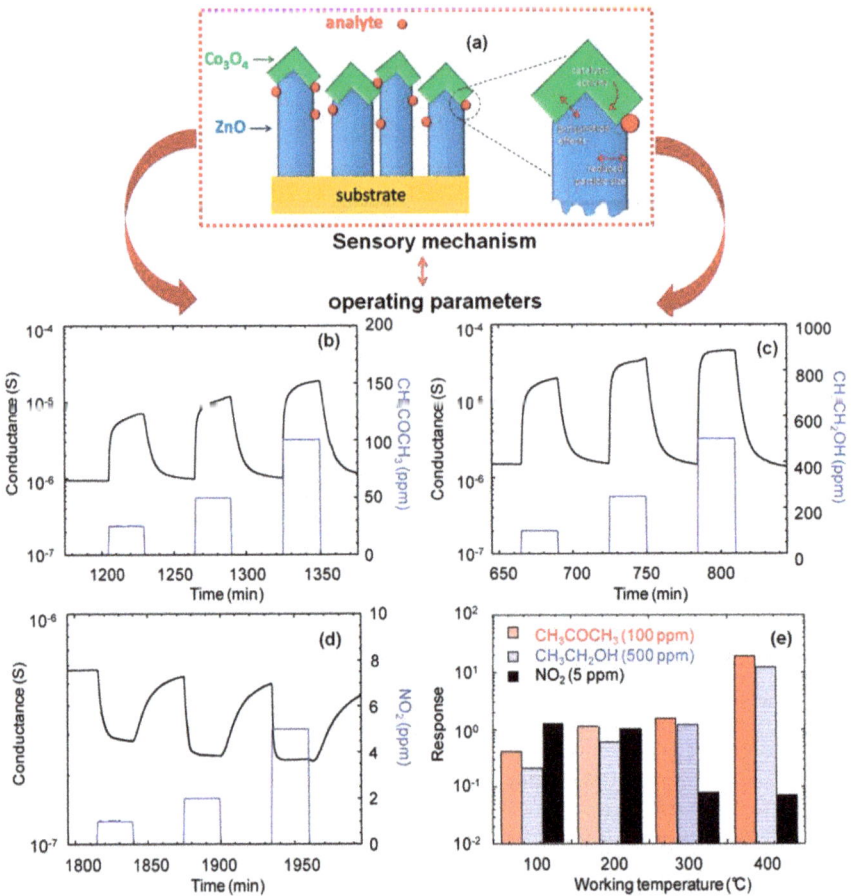

Figure 7. (**a**) Schematic representation of the main phenomena beneficially affecting the sensing behavior of the present Co_3O_4/ZnO nanocomposites. Sensing responses (black) of a Co_3O_4/ZnO sensor (specimen ZnCo10) toward square concentration pulses (blue) of (**b**) CH_3COCH_3; (**c**) CH_3CH_2OH; and (**d**) NO_2. Working temperatures were (a) 400 °C and (b) 200 °C; (**e**) Dependence of the response on the operating temperature for selected analyte concentrations (specimen ZnCo10). Adapted figure from [81] with permission from copyright (2012), American Chemical Society (Washington, DC, USA).

It was also observed that hierarchical microspheres of ZnO particles demonstrated higher potential for carrying out electron transportation and response against external analytes. The higher rate of kinetics was mainly due to the presence of high surface areas in these hierarchical structures. The analyte molecules had a higher rate of diffusion on the surface of the particles and enhanced the sensitivity of the developed sensor. The applications of these hierarchical microspheres of ZnO particles with the cataluminescence responses against ethanol molecules have shown tremendous importance in the measurements. The prepared sensor showed a lesser consumption of energy and superior response, with high sensitivity and selectivity for ethanol molecules. The developed sensor could be one of the potential ways to solve the problem of detection in lower concentration ranges. The flow rate of analyte also influenced the sensitivity of the formed sensor. It was found that in the presence of a high concentration of ethanol molecules, the flow rate was high and the subsequent catalytic reaction was greater, enhancing the signal intensity during the measurement. The sensing mechanism was explained by two connected steps: one of the processes involves the formation of excited species during the analysis, followed by their relaxation after losing radiation. The other process comprised the recombination of charge carriers initiated from the external surface states of the ZnO nanostructures. The chemically adsorbed species on the external surface of the particles had the tendency to be converted into reactive centers, and enhanced the catalytic rate of the reaction. On the other hand, nanowires of ZnO particles had a higher response rate against ethanol molecules. The fabrication of the nanowires was easily achieved by the chemical vapor deposition method. The microelectromechanical system (MEMS) technology also affected the sensory aptitude of the ZnO particles. The sensor displayed high sensitivity with rapid response towards ethanol molecules at a working temperature of 300 °C. It was also seen that at very low concentration of 1 ppm of ethanol and n-butanol, the sensitivity was around 2.2 and 3.4, respectively. At a very high concentration of 100 ppm of these analytes, the virtual responses of the formed sensor were found to be 22.6 and 18.2, respectively. The formed sensor was found to be quite stable, with higher reproducibility. At the same working temperature, nanoflowers of ZnO displayed higher sensitivity than the ZnO nanowires. The primary reason for this variation was the difference in the surface area of the formed particles and differences in the reactive centers on the exterior surface of particles. The recovery time for the prepared sensors was 5 s and 18 s, respectively, for nanowires and nanoflowers of ZnO [82]. On the other hand, if porosity was introduced in ZnO nanowires, the respective response rate was significantly enhanced in the presence of ethanol or methanol molecules. The introduction of porous nature in these nanowires was mainly done by calcining the particles at a known temperature. This kind of heat treatment has the ability to decompose the used starting materials into wire-like zinc hydroxide carbonate molecules. The BET analysis showed the development of 39.1 $m^2 \cdot g^{-1}$ of surface area of ZnO nanowires. These porous structures showed a higher rate of diffusion of molecules with a higher rate of mass transportation in the sensing material. The sensitivity and the response time were further improved by varying the reaction conditions such as pH, temperature, and analyte concentration. The development of an ethanol sensor by using flowerlike bundles of ZnO nanorods was done by Zeng et al. [83]. The formed sensor showed a higher sensitivity of around 154.3 and superior reversible response rate towards ethanol for concentration ranges varying from 1 ppm to 100 ppm. The sensing mechanism was checked at various operational temperature ranges as a function of ethanol concentration. Figure 8 shows the response curves for ZnO particles in the presence of 10 ppm of ethanol molecules for different temperatures (220, 250, and 300 °C). From the data, it is seen that the sensitivity of the sensor had a maximum of about 15.6 at the working temperature of 250 °C. Upon increasing the temperature of the system, the response and recovery rate of the developed sensor was found to decrease. This can be explained by considering the kinetics and mechanism of ethanol adsorption on the surface of ZnO particles. Thus, it is clearly verified that the nanorods of ZnO particles have the potential to act as an effective chemical sensor against ethanol molecules with the use of low power. The higher response rate of the developed sensor at low operating temperature has further enhanced the scope of ZnO particles in rapid sensing applications.

Fan and co-workers synthesized dandelion-like hollow ZnO particle hierarchitectures, and used them for the sensing of ethanol molecules. In total, three different structure types were prepared by heating the precursors at 350, 450, and 550 °C in the presence of air for approximately 2 h. The heating was controlled with a heating rate of 5 °C·min^{-1} during the synthesis of particles termed as ZnO-350, ZnO-450, and ZnO-550, respectively (Figure 9A–F). In order to compare the results, the flower-shaped ZnO particles were also prepared by hydrothermal methods, and their sensing ability was tested against dandelion-like hollow hierarchitectures of ZnO. The development of the sensor was done by dispersing the formed particles in the presence of ethanol molecules to form a homogenous colloidal solution. The as-formed dispersing solution was then encrusted on a ceramic tube having a pair of previously casted electrodes. The heating was performed by using a Ni–Cr alloy. The developed sensor was further aged at 200 °C for around 7 days before the analysis. Figure 9G illustrates the response of the ZnO particles used for the sensing of 50 ppm of ethanol molecules as a function of temperature.

Form the data, it was observed that the response showed irregular behavior. First, the response rate increased, followed by a decrease in the response activity from 150 to 400 °C. It is clear that the detection of ethanol molecules was achieved by the adsorption and desorption of molecules on the available surface of ZnO particles.

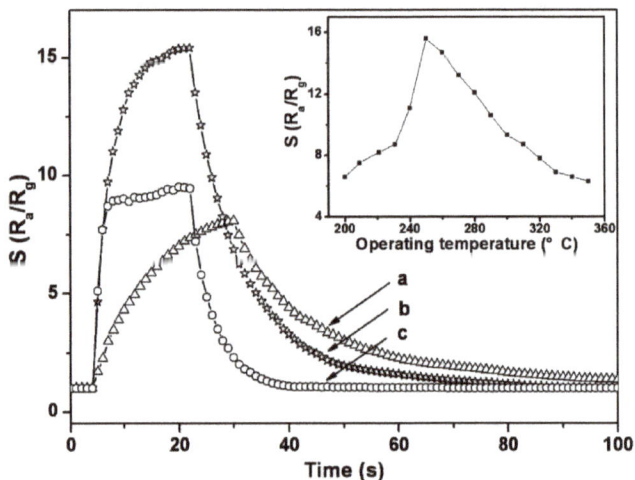

Figure 8. Response of the present ZnO to 10 ppm ethanol at various operating temperatures of (a) 220; (b) 250; and (c) 300 °C. The inset shows the sensitivity to 10 ppm ethanol at operating temperatures in the range of 200–350 °C. Adapted figure from [83] with permission from copyright (2009), American Chemical Society (Washington, DC, USA).

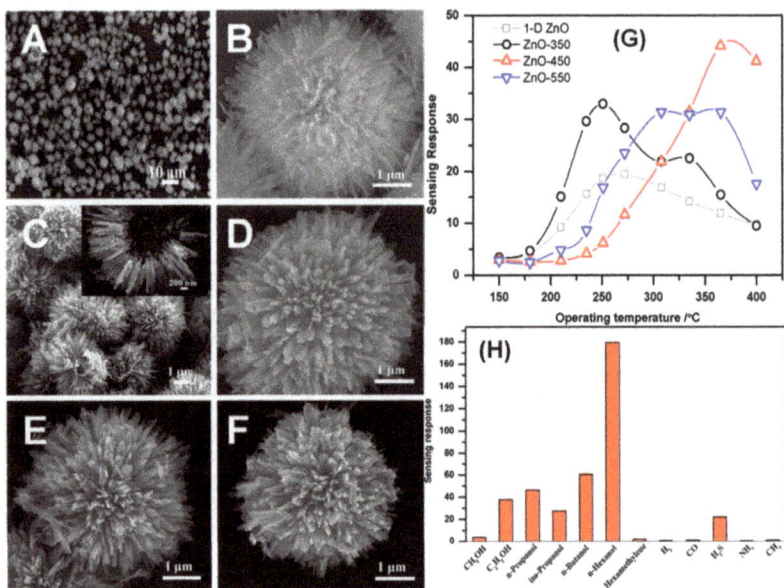

Figure 9. *Cont.*

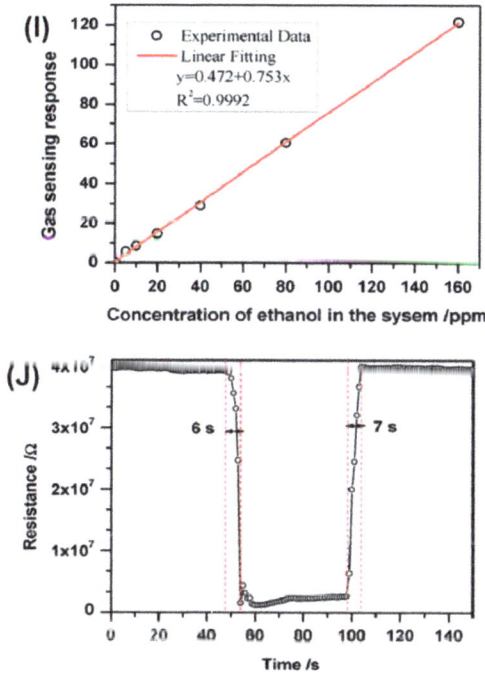

Figure 9. SEM images of the samples: precursor (**A,B**); ZnO-350 (**C,D**); ZnO-450 (**E**); and ZnO-550 (**F**), the inset in C presents a hollow structure of the ZnO-350; (**G**) ZnO sensors response to 50 ppm ethanol under different operating temperature; (**H**) Response of ZnO-350 dandelion-like hierarchitectures toward 50 ppm of different interfering molecules at the optimum operating temperature of 250 °C; (**I**) Response curve and linear fitting curve of the sensing response of ZnO-350 to different concentrations of ethanol at the operating temperature of 250 °C; (**J**) response and recovery time of ZnO-350 to 50 ppm ethanol at the operating temperature of 250 °C. Adapted figure from [84] with permission from copyright (2014), American Chemical Society, (Washington, DC, USA).

An adequate amount of thermal energy provided by external heating is crucial to conquering the activation energy barrier for the chemically adsorbed molecules of ethanol on the surface of ZnO. The quantity of chemically adsorbed molecules of ethanol was enhanced with increases in the working temperature of the reaction media. On further enhancing the temperature, the process of desorption was the main factor decreasing the response behavior of the developed sensor. The optimum working temperature for the sensor was 250, 370, and 330 °C for ZnO-350, whereas the 1-D ZnO-based nanorods had the working temperature value of 450, respectively [85]. Moreover, the sensing response of the used ZnO particles was found to be highest for ethanol molecules as compared to other interfering molecules (Figure 9H). Figure 9I,J shows the variations of response as a function of ethanol concentration varying from 5 to 160 ppm for ZnO-350 at the temperature of 250 °C. In addition, the three-dimensionally highly ordered macroporous nanostructure of ZnO particles also showed advantages for the detection of ethanol molecules. These macroporous structures had the ability to provide good control over the surface area of particles with high electron transference rate for these molecules, which enhanced the scope of particles for developing effective sensors. The external doping of ZnO particles with indium (In) is one way to improve the sensing properties of ZnO particles. The composites were mainly prepared by using the colloidal crystal templating method (Figure 10). While varying the percentage doping rate, it was found that 5% doping of in provided the highest sensitivity of ~88% for 100 ppm ethanol molecules at 250 °C.

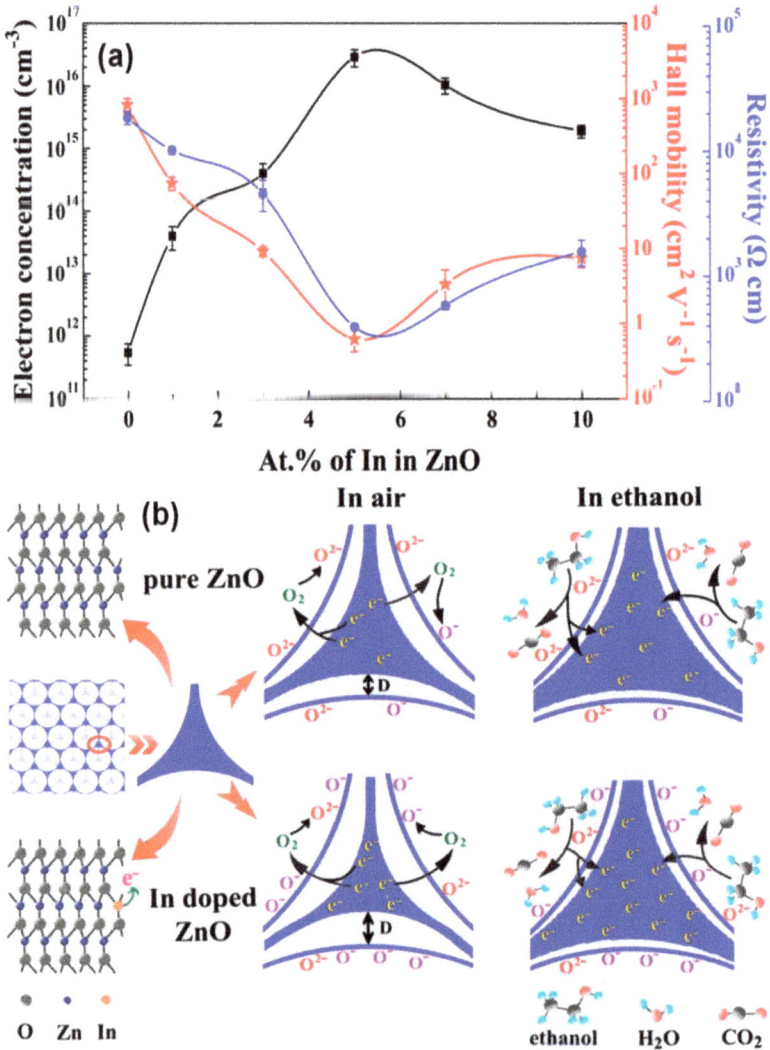

Figure 10. (**a**) Electron carrier concentration, resistivity, and Hall mobility of the 3 dimentional macroporous structures; the error bars represent the SD of the determinations for three independent samples; (**b**) Schematic diagram of ethanol sensing on the surface of pure and In-doped 3DOM ZnO. Adapted figure from [85] with permission from copyright (2016), American Chemical Society, (Washington, DC, USA).

In comparison with the bare macroporous nanostructure of ZnO, the sensitivity was found to be three times higher. The colossal enhancement of the sensitivity to ethanol was associated with the amplification in the surface area and the electron carrier concentration in the prepared particles. The introduction of in provided additional electrons in the medium, which is supportive for escalating the quantity of adsorbed oxygen, guiding the high sensitivity of the sensor. The tetrapodal-like morphology with the presence of leg-to-leg linking of ZnO nanostructures was also revealed to be important in the investigation of very low concentrations of ethanol [86]. The electrical properties of ZnO nanoparticles were mainly used for the sensing of ethanol molecules. It was found that the use of UV light was quite effective for the excitation of electrons from the conduction band. Therefore, the use of external heating was avoided during the measurements. The respective plots of resistance in the presence of air and ethanol molecules were compared under UV illumination at room temperature. The excitation of electrons from the conduction band was generated due to the increase of photoelectrons by using the external photon absorption from UV light illumination. After switching off the UV illumination, the resistance value for the developed sensor returned to its usual value at the starting point. Conversely, in the presence of ethanol, the resistance values for the ITN-ZnO sensor were found to be astonishingly amplified, which is different from the case at elevated functioning temperature values. This enhancement of resistance was mainly associated with the *p* type behavior of ZnO particles. The enhancement of resistance was also noted for different concentration ranges from 10 to 1000 ppm. The outcomes of the work have clearly illustrated the importance of the ability of a sensor to act at room temperature conditions for a wide range of concentrations.

3.5. Hydroquinone Chemical Sensor

Out of various types of phenolic molecules, p-hydroquinone (HQ) is considered as an essential compound in various types of biological and industrial practices involving the manufacture of coal-tar molecules, paper textile dyes, makeup products, and graphic developers. The excess breathing of HQ chemical has adverse effects on various vital organs, such as lungs, liver, kidney, and genito-urinary tract of human beings. In addition, the degradation of HQ is not simple in the current environmental circumstances. In this regard, ZnO nanostructures have received tremendous interest for the fabrication of effective sensors for HQ. For instance, Ameen and coworkers [87] used ZnO nanowhiskers for the detection of HQ by using the I–V technique. For the development of the sensor, a bare glassy carbon electrode was modified by using the ZnO nanowhiskers. It was observed that the presence of 10 mM of HQ in the reaction media considerably augmented the current value. The saturation current value was obtained after the concentration of 200 µM of HQ. The saturation is mainly explained to be due to the absence of free active sites on the surface of ZnO particles for the effective adsorption of HQ molecules. The developed sensor displayed a sensitivity of 99.2 $\mu A \cdot \mu M^{-1} \cdot cm^{-2}$ and a limit of detection of 4.5 µM, with a correlation coefficient of 0.98144 and response time of less than 10 s. The available active oxygen species on the surface of nanoparticles have the ability to react with HQ molecules, and oxidized HQ to 1,4-benzoquinone in the electrochemical analysis. The presence of the distinctive whiskers morphology of ZnO is useful for the production of an effective electrode material for the quick recognition of HQ.

In addition, the high electrocatalytic activity of ZnO-based nanoparticles for the electrochemical oxidation of hydroquinone was expressed by Freire et al. [88]. The sensing aptitude was tested by attaching the ZnO nanowires to carboxylic acid-functionalized multi-walled carbon nanotubes (MWCNTs). Comparison of the catalytic activity based on the preparation of ZnO particles was done by using the three different conditions involving the temperature variation in a hydrothermal microwave-based method. These variations in the synthetic conditions had the tendency to generate spherical, flower-like, and non-structured ZnO particles. From the CV analysis, it was found that due to the presence of high surface area, the flower-like morphologies of ZnO had better sensitivity towards HQ molecules. The lowest value of response rate was observed in irregularly-shaped ZnO particles. This was explained to be due to the inferior electron transport kinetics in contrast with the other composites of ZnO. In addition, these ZnO nanoflowers possessed a porous nature, which further augmented the surface area and supplied a minor mass transport obstruction and produced a rapid diffusion of HQ compounds from electrolytic solution to the modified electrode surface and enhanced the current intensity at this modified electrode. Recently, Ahmed et al. investigated the application of ternary metal oxides of ZnO with SrO and NiO (TMO) for the preparation of an HQ sensor [89]. The detailed spectroscopic studies have shown that the as-formed TMO displayed very unusual electronic and catalytic features compared to the individual oxides used for the preparation of mixed oxides. The formed TMO particles on the surface of glassy carbon electrode displayed the finest reactivity at a neutral pH range of 7.0. The mechanism mainly involved the transference of a coupled two-electron two-proton reversible reaction for the estimation of HQ molecules. Due to the presence of an adequate amount of active sites on the surface of TMO, molecules could provide a supporting nano-environment during the measurement of HQ molecules. The sensitivity, limit of detection, response rate, reproducibility, and stability of the developed sensor were found to be quite high for HQ molecules. Hollow and highly porous nanospheres of ZnO prepared by using the hard template process were used to enclose inside the nanosheets of graphene oxide for the development of effective engineered material for the detection of HQ molecules [90]. The effective electrostatic interactions between the ZnO particles and graphene oxide (GO) nanosheets were achieved due to the amino-functionalization of the particles. The presence of the highly porous structure of functionalized ZnO with tetra phenyl hydroxyl sulphate (TPHS) and GO sheets provided a large surface area for the adsorption of ethanol molecules. The electron transfer activity of the developed material was also found to be quite high for the estimation of analyte. For the HQ molecules, the respective peak was observed at 0.43 V, associated with the oxidation peak for HQ molecules, and the respective reduction peak was observed at 0.033 V on the surface of bare GCE. The engineered ZnO particles with GO particles on the surface of a GCE electrode displayed redox peak currents with 2.47-fold and 3.99-fold higher peak current values for the oxidation of HQ molecules. In addition, the peak-to-peak separation value was found to be 0.37 V, which was less than that of bare GCE.

On the other hand, the combination of ZnO particles as working electrode and Ag/AgCl with saturated KCl solution as reference electrode was employed for the estimation of hydroquinone [91]. The formed sensor displayed anodic (*Epa*) and cathodic (*Epc*) peak potential values at 0.431 V and 0.245 V for the 0.78 mM amount of hydroquinone. The obtained ratio of anodic (*Ia*) and cathodic (*Ic*) peaks was found to be 2.2. The ratio clearly points out the signals for the quasi-reversible process during the estimation of hydroquinone molecules. The prepared sensor also displayed higher conductivity, as the peak potential difference for the anodic and cathodic peak was found to be quite low. The effect of different scan rates has been visualized in Figure 11 for the developed sensor in the presence of 5 mM hydroquinone, showing its high sensitivity.

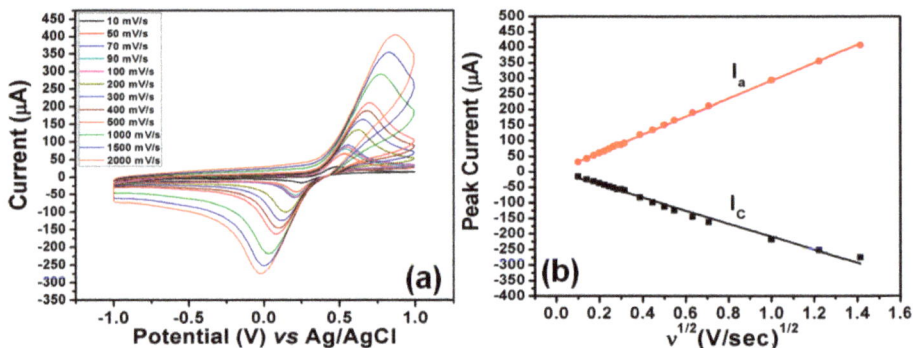

Figure 11. *Cont.*

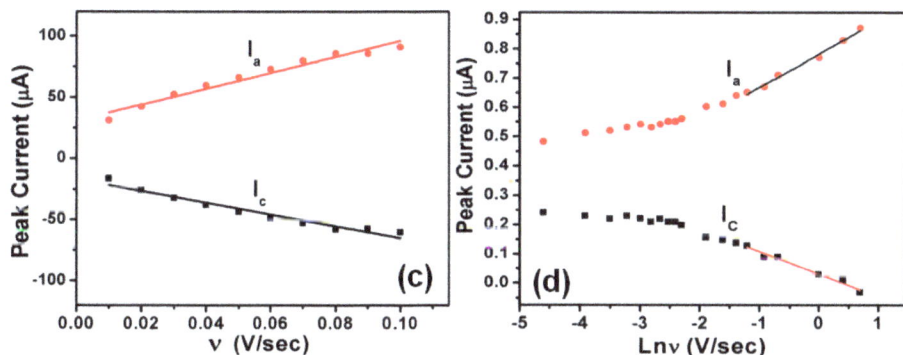

Figure 11. (a) Cyclic voltammograms obtained for ZnO nanoparticles/GC electrode in 0.1 M PBS (pH = 7.4), containing 5 mM hydroquinone at various scan rates of 10, 50, 70, 90, 100, 200, 300, 400, 500, 1000, 1500, and 2000 mV/s; (b) Plot for the anodic and cathodic peak current versus the square root of the scan rates in the same solution; (c) Plot for the anodic and cathodic peak current versus the scan rates in same solution; and (d) Plot for the anodic and cathodic peak current versus the natural Log of scan rates in the same solution. Adapted figure from [91] with permission from copyright (2014), American Scientific Publishers (Los Angeles, CA, USA).

3.6. Acetone Chemical Sensor

Among the various types of organic solvents, acetone is considered to be an effectual biomarker for the non-persistent identification of type-I diabetes in human beings, since this type of ailment has a direct effect on the concentration of acetone in the breath of human beings. It was found that the concentration of acetone was less than 900 ppb in healthy human beings, whereas the concentration rose to 1.8 ppm for type-I diabetes-affected persons. Therefore, preparing an effective method for the detection of acetone at very low concentrations is in high demand. In this regard, the effective electrical and conductive nature of ZnO makes it an efficient material for the preparation of chemical sensors for acetone. In addition, its diverse range of physical properties including electro-optical, acoustic, piezoelectric, wide band gap, and optimized luminescence properties make ZnO a promising material for the preparation of chemical sensors for acetone molecules. For instance, Xiao et al. [92] have used highly-porous and single-crystalline nanosheets of ZnO particles for the preparation of sensors. The synthesis of nanosheets was achieved by employing the annealing of hydrozincite (i.e., $Zn_5(CO_3)_2(OH)_6$) nanoplates. The solvothermal process was employed in a mixture of water/ethylene glycol as the solvent media. The effect of Pd coating on the surface of ZnO nanosheets was also evaluated to compare the efficacy of the prepared sensor. The doping of the Pd particles were mainly achieved by using the self-assembly method. In order to test the chemical sensing, the effects of the ZnO nanosheets were explored cautiously before and after the surface alterations with Pd nanoparticles. From the data, it was found that the chemical activity of the porous ZnO nanosheets was mainly due to the superior selectivity of the formed sensor with high response kinetics. This was mainly due to the higher contact of 2D nano-crystals of ZnO via the presence of (100) facets. The formation of the sensor was mainly done by dispersing the ZnO materials on the surface of adhesive Terpineol. The dispersion was prepared by the gentle grinding of particles in an agate mortar to form a slurry of the used materials. Then, this prepared suspension was cast on the exterior surface of a tube made up of ceramic material. The used ceramic tube also possessed two electrodes made up of Pt. As-coated material was dried at 80 °C. Subsequently, the dried tube was subjected to calcinations at 500 °C for around 1 h to generate thick films on the surfaces of the ceramic tube. The importance of the method was that no external conductive binder was mixed with the used material for the preparation of the sensor. The heating was done by using a small coil made up of

Ni–Cr alloy. In order to attain stability, the prepared sensor was aged at 300 °C for one week. Figure 12 shows the responses to acetone as a function of temperature for the formed sensors with ZnO and Pd@ZnO, respectively. Upon looking at the plots, it can be seen that the response rate of the sensor to acetone molecules varied noticeably in the temperature range between 200 to 500 °C. The sensor made up of nanosheets of ZnO particles displayed the highest response of 37.5 to 100 ppm of acetone at 420 °C, and the responses declined further on enhancing the temperature. Modifying the surface of the ZnO particles with 0.5 wt % Pd nanoparticles led to a significant enhancement of 70 times as compared to ZnO nanosheets at the working temperature of 340 °C. Figure 12 displays the dynamic response and recovery of chemical sensors for acetone molecules prepared using nanosheets of ZnO and Pd-doped nanosheets of ZnO, respectively. From the data, one can easily visualize that Pd@ZnO-based sensors showed superior responses to the different concentrations The sensors prepared with Pd doping on the ZnO nanosheets had the tendency to achieve a reaction of ~222 to acetone molecules with concentrations as low as 500 ppm. The response for doped particles was found to be around three times higher than that of nanosheets of ZnO without any doping. The response rate and recuperation rate for Pd@ZnO sensors for 100 ppm acetone was around 9 and 6 s, respectively, compared to 10 and 7 s for nanosheets of ZnO-based sensors. The highly porous nature of Pd@ZnO sheets have a better ability to assist in the diffusion of acetone molecules, and enhanced the mass transport during the analysis.

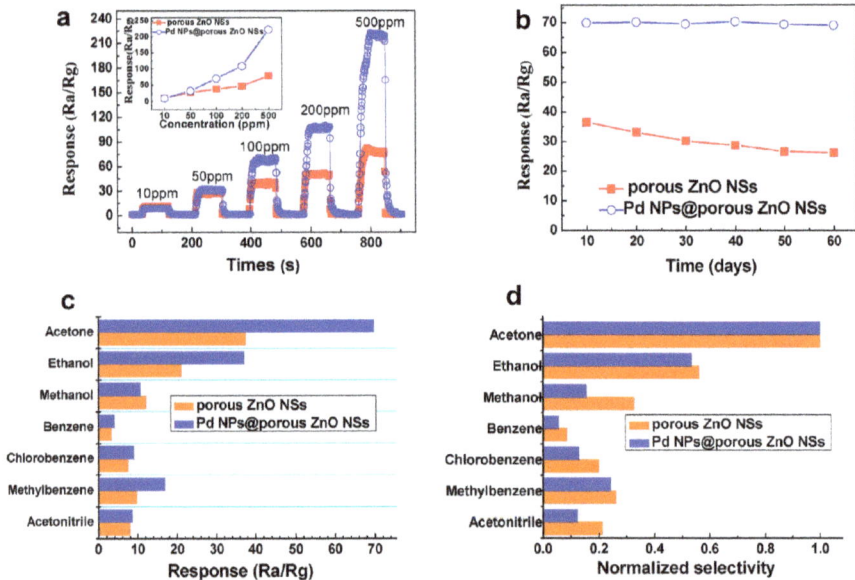

Figure 12. (**a**) Dynamic responses to acetone at concentrations ranging from 10 to 500 ppm of acetone sensors made with Pd−ZnO−Nnanosphere and ZnO−nanosphere at optimized operating temperatures. The inset shows response vs concentration curves of the corresponding sensors; (**b**) Stability studies of sensors exposed to 100 ppm acetone; (**c**) The selectivity of the acetone sensors to reducing gases; and (**d**) The corresponding normalized selectivity of sensors from (**c**). Adapted figure from [92] with permission from copyright (2012), American Chemical Society (Washington, DC, USA).

The sensing mechanism for the acetone molecules on the surface of ZnO particles was explained as follows. It was observed that the atmospheric oxygen molecules were primarily adsorbed on the surface of particles as the heating of the prepared nanoparticles was performed during the initial phase [93]. On the other hand, the surface reaction was found to proceed further upon decreasing the working temperature of the media with a low speed. The electrons in the conduction band of nanoparticles were responsible for the generation of reactive species of oxygen on the exterior surface of nanoparticles. The desorption of these reactive oxygen species was done at 80, 130, and 500 °C, respectively. Of the various types of reactive species of oxygen, it was found that the O^- species were very stable and had the tendency to react with the molecules of acetone. The sensitivity of the prepared sensor was linearly dependent on the concentration of acetone and the working temperature. In order to investigate the influence of morphology, Zhang et al. [94] used hexagonal prism-shaped nanorods of ZnO particles with planar and pyramidal tips for the sensing of acetone molecules. The fabrication of these morphologies was mainly achieved using the Teflon-lined stainless steel autoclave method. The temperature was maintained at 60 °C for around 12 h. After obtaining the precipitation, the solution was subjected to centrifugation and subsequently calcined at 550 °C for around 2 h. The addition of Cr was also done to enhance the sensitivity and response behavior of ZnO nanoparticles for acetone molecules. The crystalline size of the formed particles were around 36.97 and 39.17 for pure and Cr@ZnO nanostructures, respectively. It was found that after the addition of Cr ion as dopant, the non-uniform rods of ZnO particles were converted into hexagonal structures with planar and pyramidal tips. The hexagonal wurtzite structure of ZnO nanoparticles possessed a basal plane with (0001) plane that terminated around the lattice point of Zn metal. The other basal plane has possessed oxygen lattice point in the pyramidal structures. The sensing for acetone was found to be quite low at room temperature conditions. Therefore, external heating of the samples was required for the detection of molecules. The response behavior was also affected by the concentration of acetone molecules. The Cr doping reduced the working temperature for the sensor and acted effectively for the analysis of low concentrations of analyte. Recently, the application of bi-composites of NiO with ZnO nanoparticles has been used for the detection of ppb concentrations of acetone molecules in reaction media [95]. The formation of the bi-composite occurred by the decoration of the exterior surface of ZnO nanomaterials with NiO particles by using the solvothermal technique. The hollow structures of ZnO particles have the ability to provide large surface area for the adsorption of analyte molecules. Primarily, the sensing ability was tested by optimizing the working temperature ranges for ZnO and NiO@ZnO nanospheres. These temperature-based characteristics of as-prepared sensors were calculated at a broad range of temperature in order to search the connection among the effective temperature and the acetone response. It was found that the response of the prepared sensors to 100 ppm acetone solution followed a volcano-shaped profile relationship between acetone response and working temperature for ZnO and NiO@ZnO nanospheres-based sensors. In terms of response rate, NiO@ZnO bi-composites have shown a low operating temperature (i.e., 275 °C). On the other hand, bare ZnO particles possessed a higher operating temperature. The variation in temperature was mainly due to the inability of the acetone molecules to react with reactive oxygen species on the surface of nanoparticles at low operating temperature.

Koo et al. [96] found that the heterogeneous sensitization of ZnO nanostructures with PdO on the surface of SnO_2 nanotubes has greatly influenced the catalytic activity of the nanomaterials and enhanced the sensitivity for acetone sensing (Figure 13). The improvement in the activity was primarily interpreted by the formation of n−n type heterojunctions between the PdO@ZnO/SnO_2. Secondly, the catalytic efficiently was also affected by the presence of PdO and ZnONPs. The presence of differences in the work function further influenced the efficiency of the sensor for acetone molecules. These variations were easily captured by using the ultraviolet photoelectron spectroscopy (UPS) spectrum of surface-functionalized nanotubes of SnO_2 nanotubes (NTs) with PdO@ZnO under He I radiation with potential of 21.22 eV. The respective cut off for the binding energy was found to be 16.82 eV for SnO_2 NTs and 16.43 eV in the presence of PdO@ZnO–SnO_2 NTs. The presence of high work function for Pd-laden ZnO (5.34 eV) compared to SnO_2 NTs (4.40 eV) made the movement of the electrons from the conduction band in SnO_2 to ZnO. As a result, there was a significant shift in the energy toward the Fermi level of ZnO. Such variations caused the bending of conduction bands in an upward direction by the chemisorption of oxygen molecules. As a result, a potential barrier was initiated on the interface of PdO@ZnO/SnO_2. As an consequence, the baseline resistance value for the PdO@ZnO–SnO$_2$ NTs was amplified by ~6 kΩ as compared to bare nanotubes of SnO$_2$ NTs. Consequently, the sensing ability of the formed material was found to be higher for acetone.

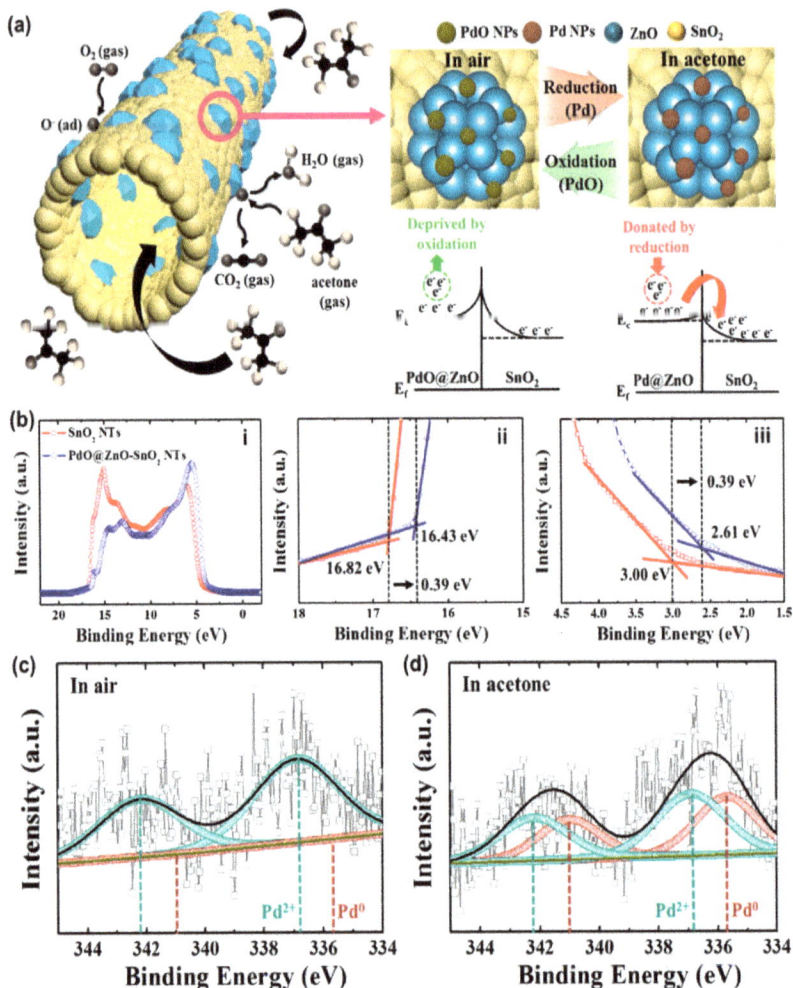

Figure 13. (**a**) Schematic illustration of acetone-sensing mechanism for PdO@ZnO–SnO$_2$ nanotubes NTs; (**b**) (i) ultraviolet photoelectron spectroscopy (UPS) spectrum of SnO$_2$ NTs and PdO@ZnO–SnO$_2$ NTs ((ii) high-binding-energy region and (iii) low-binding-energy region) and ex situ X-ray photoelectron spectroscopy (XPS) analysis using high-resolution spectra of PdO@ZnO–SnO$_2$ NTs in the vicinity of Pd 3d (**c**) in air and (**d**) in acetone after a seven-cycle sensing measurement with 5 ppm of acetone at 400 °C. Adapted figure from [96] with permission from copyright (2017), American Chemical Society (Washington, DC, USA).

In addition, the piezoelectric layer made up of Li-doped ZnO nanowires and PDMS polymer had the ability to produce an effective assimilated structure which provides a dynamic sensing material for the detection of acetone molecules at low concentrations (Figure 14).

These sensors had direct control over the additional motion and enabled the detection of instantaneous motion in the presence of external analyte. By successfully integrating these two elements in sensory

Materials **2018**, *11*, 287

devices, the respective sensing limit and sensitivity was managed to a greater extent. The as-prepared sensor displayed excellent mechanical elasticity, strength, and stretching capability, which are quite appropriate for different types of applications involving the investigation of analyte at low working temperature. It was also found that the obtained sensitivity of the sensor was mainly dependent on the surface defects present in the nanostructures. In addition, the surface area of the nanostructures also affected the interaction with the chemical molecules. Post-management during the synthesis of nanoparticles is one of best ways to adjust the defect structures on the surface of the prepared nanomaterials. The calcinations temperature also modulated the amount of defects on the surface of the prepared ZnO particles. In addition, the activation of nanomaterials under low-pressure conditions enhanced the catalytic activity of ZnO nanomaterials by providing a higher mass of oxygen vacancy (VO) defects on the exterior surface of nanoparticles. The agglomeration of nanostructures during synthesis has the tendency to decrease the efficiency of the sensor. Therefore, care was taken to control the size and agglomeration rate and luminescence during the synthesis of ZnO nanomaterials in order to obtain good chemical sensing results.

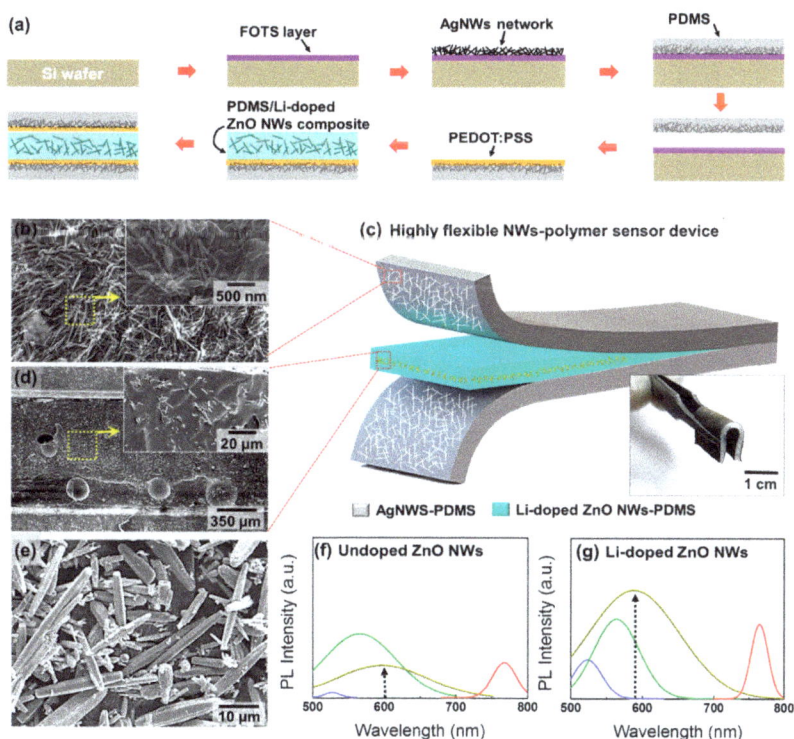

Figure 14. (**a**) Schematic illustration of the sensor fabrication process: composite layer, containing piezoelectric Li-doped ZnO NWs and PDMS, is sandwiched between two poly (3,4-ethylenedioxythiophene) polystyrene sulfonate (PEDOT:PSS)-coated Ag NW electrodes embedded in the PDMS; (**b**) SEM of the AgNW network, which is seamlessly connected on the PDMS surface. The inset shows a magnified view of the Ag NWs on the PDMS; (**c**) Schematic representation of the sensor device consisting of two main parts: resistive and piezoelectric sensing elements. The inset shows the flexibility of the device; (**d**) Cross-section of the device SEM; (**e**) SEM of as-synthesized Li-doped ZnO NWs; (**f**) photoluminescence (PL) spectra of undoped ZnO NWs, and (**g**) Li-doped ZnO NWs. The yellow emission explicitly indicates the Li-doping in ZnO. Adapted figure from [97] with permission from copyright (2017), American Chemical Society (Washington, DC, USA).

3.7. Other Chemical Sensors Based on ZnO Nanomaterials

Currently, the threat caused by the excessive use of heavy elements is a matter of concern among researchers. These heavy metals cause several types of health-associated illness, and in extreme cases cause death in human beings. There are several types of elements which are required for the proper growth of living beings. Yet, their accretions in living systems beyond tolerable limits can cause severe health predicaments due to their lethal nature. The noxious nature of a heavy metal is mainly associated with its oxidation state, concentration value, composition of the solvent media, and many other factors. Thus, there is an urgent need for a simple process to detect heavy metals. In this regard, ZnO-based nanostructures have had significant importance for providing an effective means of detecting heavy metals with great accuracy and precision. The morphology of ZnO-based nanostructures has a pronounced influence on the sensing ability of these metal ions. It has been ascertained that the nature of morphologies such as nanoflowers, cones, tubes, nails, urchins, prisms, and wires have the ability to influence the prime features of the prepared chemical sensors. These shapes have a direct effect on the limit of detection, sensitivity, reaction time, linear range, working potential, and optical activity of the produced sensor. These morphologies further affect the surface area, width, catalytic activities, electron transport characteristics, and many other properties of the developed sensor.

For instance, Bhanjana et al. [98] used well-characterized ZnO nanoparticles for the analysis of cadmium ions by using the electrochemical method. The advanced redox activities of ZnO nanoparticles provide an effective means for the analysis of these cadmium ions. The measurements were performed by using three electrode systems in which Ag/AgCl acted as reference, platinum wire worked as counter electrode, and the ZnO nanomaterials on the surface of an Au electrode acted as working electrodes, respectively. From the data, it was found that the electrochemical response for cadmium ion was reversible in nature, as CV plots displayed both oxidation and reduction peaks. The presence of ordered defects on the surface of ZnO nanomaterials makes them efficient electrocatalysts for the estimation of cadmium. Moreover, ZnO nanostructures also possess specially exposed reactive sites with advanced electronic configuration and improved electro-catalytic action. As a consequence, charge circulation at the electrode/electrolyte interface is amplified due to the development of an electrical double layer with smallest contact resistance on the interface. In addition, ZnO nanoparticles in combination with Ag nanostructures were also employed for the determination of Pb^{2+} and Cd^{2+} ions in a buffer media with the pH of the reaction media equivalent to 5 [99]. It was found that modified glassy carbon electrodes with ZnO@Ag nanoparticles displayed considerable anodic and cathodic peaks in the presence of the studied metal ions at a scan rate of 50 mV/s. Conversely, no peak was observed for the oxidation of zinc into zinc ions. The small water-soluble quantum dots (QDs) of ZnO nanodots have also had a significant role in the elimination of harmful and toxic chemicals [100]. The generation of the particles was achieved from pomegranate peels. The synthetic process was found to be quite economical and more feasible for the larger-scale synthesis of ZnO nanoparticles. The requirement of low-temperature conditions for the preparation of ZnO nanodots further made the process more economical. The prepared particles displayed good control over the optical and luminescence properties of ZnO, and provided better material for the preparation of chemical sensors for the metal ions. The developed sensor was quite selective for the estimation of Cr^{3+} ions, as it was found that out of different types of metal ions, the corresponding emission from ZnO QDs was found to be highly quenched in presence of Cr^{3+} ions. The effective interaction between the Cr^{3+} and the ZnO QDs had the ability to sense the small amount of ~2 nM of Cr^{3+} ions from the aqueous media. Additionally, a recent approach for the detection of Cu^{2+} in aqueous media was carried out by using glycol-modified ZnO nanoparticles [101]. The detection of metal ions was done by the quenching of fluorescence intensity at 351 nm in the presence of Cu^{2+} ions. These Cu^{2+} ions have the ability to encourage the aggregation of particles, and resulted in the quenching of fluorescence intensity of ZnO nanomaterials. The working range of the formed sensor was 10–200 nM, with a limit of detection of 3.33 nM. The competence of the prepared sensor was fairly high in real water samples from different sources. Ng and co-workers [102] produced a turn-off luminescent sensor for Cu^{2+} ions by using ZnO nanomaterials. The developed sensor showed good control over the selectivity and sensitivity for the metal ions under study, with the limit of detection (LOD) of ~7.68 × 10^{-7} M. The biocomposites of ZnO with ZnS nanostructures have the potential for the estimation of Cu^{2+} ions in aqueous media (Figure 15). The field site detection of metal ions was done by preparing the test papers by using ZnO@ZnS nano materials [103]. In order to investigate the response of the developed paper sensor for the selective sensing of Cu^{2+} ions from aqueous media, the solutions were prepared by using different concentrations (15, 75, 150, 300, 450, 750, and 1500 μM). The pH of the reaction media was kept acidic (pH ~4) in nature, with buffer concentration of 10 mM, respectively. For the estimation, around 20 μL of every solution was drop-casted on the paper piece of the developed sensor. The paper was kept for drying for at least 20 min, and respective photographs were taken by using a digital camera (Figure 15).

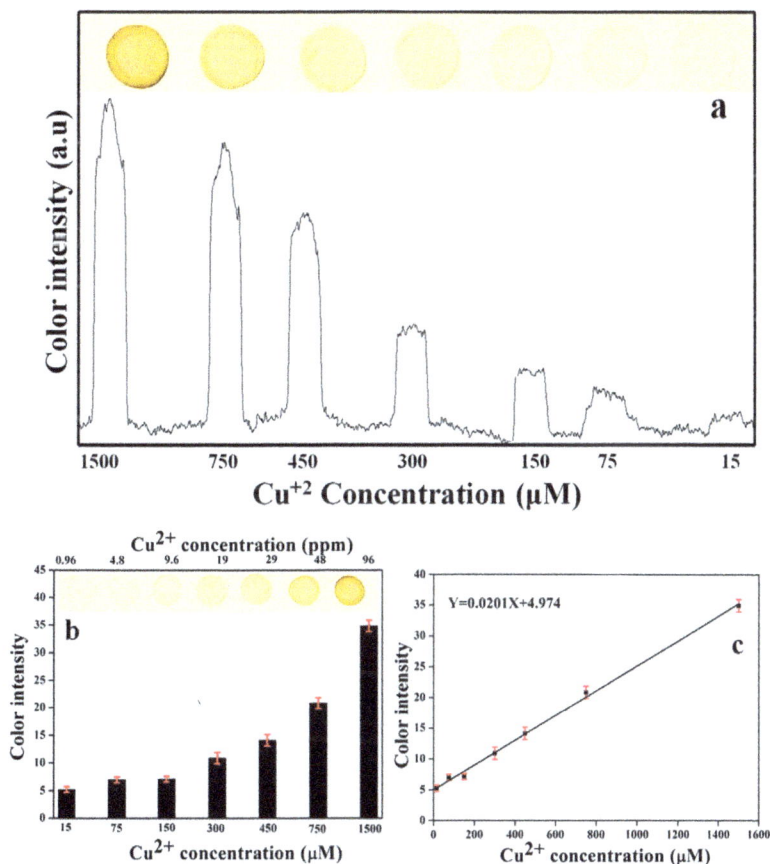

Figure 15. (a) Color intensity versus the Cu^{2+} ions concentration obtained by ImageJ with a digital photograph as an inset; (b) The color intensity versus the Cu^{2+} ions concentration; and (c) The calibration curve of the color intensity at different concentrations of the Cu^{2+} ions. Adapted figure from [103] with permission from copyright (2014), American Chemical Society (Washington, DC, USA).

In separate studies, the surface modification of ZnO nanomaterials with imine moieties had the tendency to detect Co^{2+} ions in aqueous media [104]. Shen et al. [105] employed a hierarchical mixture of ZnO with CdS nanospheres for the selective sensing of Cu^{2+} ions by using the photo-electrochemical method. ZnO nanostructures have the ability to supply an adequate quantity of light scattering centers on the surface of particles for the effective sensing of metal ions. The heterojunctions between the ZnO and CdS materials can present substantial enhancements to light absorption and charge separation. As a consequence, these biocomposites are effective for the enhancement of photocurrent intensity. It was also found that the excessive accumulation of Cu^{2+} ions on the surface CdS solution induces the effective binding of Cu^{2+} with S^{2-} and leads to the reduction of Cu^{2+} to Cu^{+} under illumination and hence detects these ions. The as-formed Cu_xS on the exterior surface of CdS nanoparticles provides a lower energy level in the particles that acts as an effectual pathway for the recombination of electrons and holes in the band gaps (Figure 16).

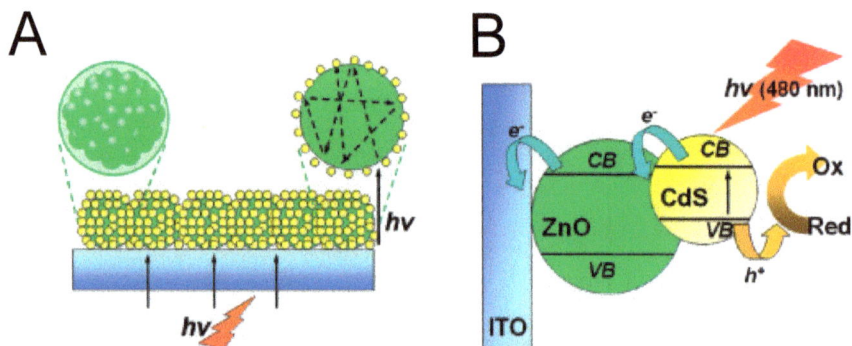

Figure 16. (**A**) Schematic of light scattering occurring on a ZnO/CdS-modified electrode; (**B**) Electron–hole pairs generation, separation, and transfer between ZnO and CdS at a ZnO/CdS-modified electrode for the sensing of Cu^{2+} ions where CB and VB are the conduction and valence band. ITO is Indium tin oxide. Adapted figure from [105] with permission from copyright (2011), American Chemical Society, (Washington, DC, USA).

The utilization of surface-enhanced Raman spectroscopy (SERS) for the sensing of Hg^{2+} ions from aqueous media was also made possible by using ZnO nanostructures [106]. Hg^{2+} is a highly noxious metal ion which is considered to be a focus in metal ion sensing. The combination of Ag ions with ZnO nanostructures make it more Raman-active and have the ability to effectively bind with Hg ions. These ions are not only detected with these nanostructures, but metal ions are also fully regenerated when external heat treatment is given to the samples. Therefore, it can be said that the ZnO nanostructures have offered the self-cleaning ability of Schottky junctions, and perform as an effective photocatalyst for harmful metal ions. Jacobsson and Edvinsson [107] have studied the characteristic properties of ZnO nanostructures by employing simple and reliable adsorption and fluorescence spectroscopic measurements. The influence of size and band gap of ZnO nanostructures were mainly associated with the available mobile trap states in as-prepared ZnO nanostructures. The luminescence properties used for the estimation of metal ions were controlled by changing the reaction parameters of the system during the particles' synthesis [108]. The presence of capping agents affected the surface defects on the surface of nanoparticles and further influenced the bioimaging role of as-formed nanoparticles.

4. Conclusions and Future Directions

The effective chemical sensors developed by using ZnO-based nanostructures has great potential in environmental remediation. Their effective surface area with biocompatible nature and manageable pore size with controlled morphologies has further boosted their range in the activities related to chemical sensing. The controlled synthesis with optimized band gap has enhanced their role in

UV-based metal ion sensors. The optimized photoluminescence properties of ZnO have provided an effective range of chemical sensors for metal ions. The development of next-generation devices with white light-emitting properties makes them effective materials in optoelectronics. The present review meticulously introduced the current advancements of the ZnO nanostructures-supported sensors with a main emphasis on chemical sensors for different analytes. The different operational factors such as effect of size, morphology, and the respective working mechanisms of nano-ZnO-based sensors along with their selectivity and sensitivity were also considered in detail.

Acknowledgments: Savita Chaudhary is grateful to DST Inspire Faculty Award [IFA-CH-17], DST Purse grant II and CAS grant for financial assistance. Ahmad Umar would like to acknowledge the support of the Ministry of Education, Kingdom of Saudi Arabia for granting the Promising Centre for Sensors and Electronic Devices (PCSED) to Najran University, Kingdom of Saudi Arabia.

Author Contributions: Savita Chaudhary and Ahmad Umar have collected the literature and wrote the review. K. K. Bhasin and Sotiris Baskoutas read, correct and revise the review

Conflicts of Interest: The authors declare no conflict of interest.

References

1. Ebrahimiasl, S.; Seifi, R.; Nahli, R.E.; Azmi Zakaria, A. Ppy/nanographene modified pencil graphite electrode nanosensor for detection and determination of herbicides in agricultural water. *Sci. Adv. Mater.* **2017**, *9*, 2045–2053.

2. Xiang, C.; Wang, Y.; Liu, H. A scientometrics review on nonpoint source pollution research. *Ecol. Eng.* **2017**, *99*, 400–408. [CrossRef]

3. Martinez, D.E.; Crondona, S.; Miglioranza, K.S.B.; Postigo, C. Groundwater pollution sources, mechanisms, and prevention. *Encycl. Anthr.* **2018**, *5*, 87–96.

4. Uma, B.B.; Uday, S.P.; Oinam, G.; Mondal, A.; Bandyopadhyay, T.K.; Tiwari, O.N. Characterization, genetic regulation and production of cyanobacterial exopolysaccharides and its applicability for heavy metal removal. *Carbohydr. Polym.* **2018**, *179*, 228–243.

5. Song, J.F.; Lin, Z.Z.; Ge, S.; Li, J.; Qiu, X.M.; Zhou, R.S.; Li, S.Z.; Guo, Z. Dispersible novel naphthalene-2,6-dicarboxylic acid monomethyl ester-based coordination polymers through *in situ* hydrolysis reaction: Highly sensitive detection of small molecules and metal ions. *Sci. Adv. Mater.* **2017**, *9*, 2054–2065.

6. Raza, W.; Ahmad, K. A highly selective Fe@ZnO modified disposable screen printed electrode based non-enzymatic glucose sensor (SPE/Fe@ZnO). *Mater. Lett.* **2018**, *212*, 231–234. [CrossRef]

7. Zhang, J.; Zhao, B.; Pan, Z.; Gu, M.; Punnoose, A. Synthesis of ZnO nanoparticles with controlled shapes, sizes, aggregations, and surface complex compounds for tuning or switching the photoluminescence. *Cryst. Growth Des.* **2015**, *15*, 3144–3149. [CrossRef]

8. Barkade, S.S.; Pinjari, D.V.; Singh, A.K.; Gogate, P.R.; Naik, J.B.; Sonawane, S.H.; Kumar, M.A.; Pandit, A.B. Ultrasound assisted miniemulsion polymerization for preparation of polypyrrole–zinc oxide (PPy/ZnO) functional latex for liquefied petroleum gas sensing. *Ind. Eng. Chem. Res.* **2013**, *52*, 7704–7712. [CrossRef]

9. Chen, C.Y.; Liu, Y.R.; Lin, S.S.; Hsu, L.J.; Tsai, S.L. Role of annealing temperature on the formation of aligned zinc oxide nanorod arrays for efficient photocatalysts and photodetectors. *Sci. Adv. Mater.* **2016**, *8*, 2197–2203. [CrossRef]

10. Ulyankina, A.; Leontyev, I.; Avramenko, M.; Zhigunov, D.; Nina Smirnova, N. Large-scale synthesis of ZnO nanostructures by pulse electrochemical method and their photocatalytic properties. *Mater. Sci. Semicond. Proc.* **2018**, *76*, 7–13. [CrossRef]

11. Niarchos, G.; Dubourg, G.; Afroudakis, G.; Tsouti, V.; Makarona, E.; Matović, J.; Crnojević-Bengin, V.; Tsamis, C. Paper-based humidity sensor coated with ZnO nanoparticles. The influence of ZnO. *Procedia Eng.* **2016**, *168*, 325–328. [CrossRef]

12. Wang, N.; Cao, X.; Wu, Q.; Zhang, R.; Wang, L.; Yin, P.; Guo, L. Hexagonal ZnO bipyramids: Synthesis, morphological evolution, and optical properties. *J. Phys. Chem. C* **2009**, *113*, 21471–21476. [CrossRef]

13. Cho, S.; Jang, J.W.; Kim, J.; Lee, J.S.; Choi, W.; Lee, K.H. Three-dimensional type II ZnO/ZnSe heterostructures and their visible light photocatalytic activities. *Langmuir* **2011**, *27*, 10243–10250. [CrossRef] [PubMed]

14. Park, H.; Alhammadi, S.; Bouras, K.; Schmerber, G.; Ferblantier, G.; Dinia, A.; Slaoui, A.; Jeon, C.W.; Park, C.; Kim, W.K. Nd-Doped SnO_2 and ZnO for application in $Cu(InGa)Se_2$ solar cells. *Sci. Adv. Mater.* **2017**, *9*, 2114–2120.

15. Khoa, N.T.; Kim, S.W.; Yoo, D.H.; Cho, S.; Kim, E.J.; Hahn, S.H. Fabrication of Au/Graphene-wrapped ZnO-nanoparticle-assembled hollow spheres with effective photoinduced charge transfer for photocatalysis. *ACS Appl. Mater. Interfaces* **2015**, *7*, 3524–3531. [CrossRef] [PubMed]

16. Guo, D.Y.; Shan, C.X.; Qu, S.N.; Shen, D.Z. Highly sensitive ultraviolet photo-detectors fabricated from ZnO quantum dots/carbon nanodots hybrid films. *Sci. Rep.* **2014**, *4*, 7469–7475. [CrossRef] [PubMed]

17. Zhang, Z.Y.; Xu, L.J.; Li, H.X.; Kong, J.L. Wavelength-tunable luminescent gold nanoparticles generated by cooperation ligand exchange and their potential application in cellular imaging. *RSC Adv.* **2013**, *3*, 59–63. [CrossRef]

18. Kim, G.E.; Noh, S.; Kang, S.H.; Kim, Y.D.; Kim, T.K. Heat treatment effect on the behavior of oxide particles in mechanically alloyed oxide dispersion strengthened powders. *Sci. Adv. Mater.* **2017**, *9*, 2126–2130.

19. Ngai, K.S.; Tan, W.T.; Zainal, Z.; Zawawi, R.M.; Juan, J.C. Electrochemical sensor based on single-walled carbon Nanotube/ZnO photocatalyst nanocomposite modified electrode for the determination of paracetamol. *Sci. Adv. Mater.* **2016**, *8*, 788–796. [CrossRef]

20. Schmidt-Mende, L.; MacManus-Driscoll, J.L. ZnO—Nanostructures, defects, and devices. *Mater. Today* **2007**, *10*, 40–48. [CrossRef]

21. Hieu, N.M.; Kim, H.; Kim, C.; Hong, S.-K.; Kim, D. A hydrogen sulfide gas sensor based on Pd-decorated ZnO nanorods. *J. Nanosci. Nanotechnol.* **2016**, *16*, 10351–10355. [CrossRef]

22. Sun, W.; Che, S.; Ying, Y.; Zhu, C.; Liu, D.; Li, W.; Jiang, L. Effects of iron(II) concentration on magnetic property and conductive mechanism of manganese zinc ferrites with different iron(III) oxide contents. *Sci. Adv. Mater.* **2016**, *8*, 2071–2076. [CrossRef]

23. Lai, L.W.; Lee, C.T. Investigation of optical and electrical properties of ZnO thin films. *Mater. Chem. Phys.* **2008**, *110*, 393–396. [CrossRef]

24. Brayner, R.; Dahoumane, S.A.; Yéprémian, C.; Djediat, C.; Meyer, M.; Couté, A.; Fiévet, F. ZnO nanoparticles: Synthesis, characterization, and ecotoxicological studies. *Langmuir* **2010**, *26*, 6522–6528. [CrossRef] [PubMed]

25. Chen, X.; Gao, Q.; Xiang, D.; Lin, H.; Han, X.; Li, X.; Du, S.; Qu, F. Controllable fabrication of porous ZnO nanofibers by electrospun with enhanced acetone sensing property. *Sci. Adv. Mater.* **2015**, *7*, 526–531. [CrossRef]

26. Chaudhary, S.; Kaur, Y.; Umar, A.; Chaudhary, G.R. 1-butyl-3-methylimidazolium tetrafluoroborate functionalized ZnO nanoparticles for removal of toxic organic dyes. *J. Mol. Liq.* **2016**, *220*, 1013–1021. [CrossRef]

27. Liu, Y.; Xu, Z.; Xu, H.; Chen, Y. Analysis of rigid-flexible coupling dynamic characteristics for power tiller. *Sci. Adv. Mater.* **2017**, *9*, 2178–2186.

28. Chaudhary, S.; Sood, A.; Mehta, S.K. Surfactant anchoring and aggregate structure at silica nanoparticles: A persuasive facade for the adsorption of azo dye. *J. Nanosci. Nanotechnol.* **2014**, *14*, 6824–6834. [CrossRef] [PubMed]

29. Ka, Y.; Jang, H.R.; Choi, W.S. Quantum dot LEDs based on solution-processed zinc oxide nano particles as electron transport layer. *Sci. Adv. Mater.* **2016**, *8*, 382–387. [CrossRef]

30. Baek, S.H.; Nam, K.H.; Park, I.K. Morphological evolution of ZnAl-layered double hydroxide nanostructures grown on Al_2O_3/Si substrate. *Sci. Adv. Mater.* **2016**, *8*, 2142–2146. [CrossRef]

31. Butler, O.T.; Cook, J.M.; Harrington, C.F.; Hill, S.J.; Rieuwerts, J.; Miles, D.L.J. Atomic spectrometry updates environmental analysis. *Anal. At. Spect.* **2006**, *21*, 217–243. [CrossRef]

32. Ibrahim, A.A.; Tiwari, P.; Al-Assiri, M.S.; Al-Salami, A.E.; Umar, A.; Kumar, R.; Kim, S.H.; Ansari, Z.A.; Baskoutas, S. Highly-sensitive picric acid chemical sensor based on ZnO nanopeanuts. *Materials* **2017**, *10*, 795. [CrossRef] [PubMed]

33. Ameen, S.; Akhtar, M.S.; Shin, H.S. Highly sensitive hydrazine chemical sensor fabricated by modified electrode of vertically aligned zinc oxide nanorods. *Talanta* **2012**, *100*, 377–383. [CrossRef] [PubMed]

34. Ibrahim, A.A.; Dar, G.N.; Zaidi, S.A.; Umar, A.; Abaker, M.; Bouzid, H.; Baskoutas, S. Growth and properties of Ag-doped ZnO nanoflowers for highly sensitive phenyl hydrazine chemical sensor application. *Talanta* **2012**, *93*, 257–263. [CrossRef] [PubMed]

35. Umar, A.; Akhtar, M.S.; Al-Hajry, A.; Al-Assiri, M.S.; Dar, G.N.; Islam, M.S. Enhanced photocatalytic degradation of harmful dye and phenyl hydrazine chemical sensing using ZnO nano-urchins. *Chem. Eng. J.* **2015**, *262*, 588–596. [CrossRef]

36. Ali, A.M.; Harraz, F.A.; Ismail, A.A.; Al-Sayari, S.A.; Algarni, H.; Al-Sehemi, A.G. Synthesis of amorphous ZnO–SiO$_2$ nanocomposite with enhanced chemical sensing properties. *Thin Solid Films* **2016**, *605*, 277–282. [CrossRef]

37. Kumar, R.; Rana, D.; Umar, A.; Sharma, P.; Chauhan, S.; Chauhan, M.S. Ag-doped ZnO nanoellipsoids: Potential scaffold for photocatalytic and sensing applications. *Talanta* **2015**, *137*, 204–213. [CrossRef] [PubMed]

38. Cho, S.; Jang, J.W.; Lee, S.H.; Lee, J.S.; Lee, K.H. A method for modifying the crystalline nature and texture of ZnO nanostructure surfaces. *Cryst. Growth Des.* **2011**, *11*, 5615–5620. [CrossRef]

39. Chishti, B.; Ansari, Z.A.; Fouad, H.; Alothman, O.Y.; Ansari, S.G. Significance of doping induced tailored zinc oxide nanoparticles: Implication on structural, morphological and optical characteristics. *Sci. Adv. Mater.* **2017**, *0*, 2202–2210.

40. Zhao, Z.; Sun, Y.; Li, P.; Sang, S.; Zhang, W.; Hu, J.; Lian, K. A sensitive hydrazine electrochemical sensor based on zinc oxide nano-wires. *J. Electrochem. Soc.* **2014**, *161*, B157–B162. [CrossRef]

41. Ni, Y.; Zhu, J.; Zhang, L.; Hong, J. Hierarchical ZnO micro/nanoarchitectures: Hydrothermal preparation, characterization and application in the detection of hydrazine. *CrystEngComm* **2010**, *12*, 2213–2218. [CrossRef]

42. Fang, B.; Zhang, C.; Zhang, W.; Wang, G. A novel hydrazine electrochemical sensor based on a carbon nanotube-wired ZnO nanoflower-modified electrode. *Electrochim. Acta* **2009**, *55*, 178–182. [CrossRef]

43. Kumar, S.; Bhanjana, G.; Dilbaghi, M.; Umar, A. Zinc oxide nanocones as potential scaffold for the fabrication of ultra-high sensitive hydrazine chemical sensor. *Ceram. Int.* **2015**, *41*, 3101–3108. [CrossRef]

44. Liu, J.; Li, Y.; Jiang, J.; Huang, X. C@ZnO nanorod array-based hydrazine electrochemical sensor with improved sensitivity and stability. *Dalton Trans.* **2010**, *39*, 8693–8697. [CrossRef] [PubMed]

45. Ahmad, R.; Tripathy, N.; Ahn, M.S.; Hahn, Y.B. Highly stable hydrazine chemical sensor based on vertically-aligned ZnO nanorods grown on electrode. *J. Colloid Int. Sci.* **2017**, *494*, 153–158. [CrossRef] [PubMed]

46. Umar, A.; Rahman, M.M.; Kim, S.H.; Hahn, Y.B. Zinc oxide nanonail based chemical sensor for hydrazine detection. *Chem. Commun.* **2008**, *2*, 166–168. [CrossRef] [PubMed]

47. Umar, A.; Rahman, M.M.; Hahn, Y.B. Ultra-sensitive hydrazine chemical sensor based on high-aspect-ratio ZnO nanowires. *Talanta* **2009**, *77*, 1376–1380. [CrossRef] [PubMed]

48. Umar, A.; Rahman, M.M.; Hahn, Y.B. Miniaturized pH sensors based on zinc oxide nanotubes/nanorods. *J. Nanosci. Nanotechnol.* **2009**, *9*, 4686–4691. [CrossRef] [PubMed]

49. Kim, D.H.; Park, M.J.; Kwon, H.I. Separate extraction of densities of interface and bulk trap states in high-mobility ZnON thin-film transistors. *J. Nanoelectron. Optoelectron.* **2017**, *12*, 1263–1266. [CrossRef]

50. Kim, S.H.; Badran, R.I.; Umar, A. Fabrication of ZnO Nanorods Based *p–n* Heterojunction Diodes and Their Electrical Behavior with Temperature. *J. Nanoelectron. Optoelectron.* **2017**, *12*, 731–735. [CrossRef]

51. Chaudhary, S.; Kaur, Y.; Umar, A.; Chaudhary, G.R. Ionic liquid and surfactant functionalized ZnO nanoadsorbent for recyclable proficient adsorption of toxic dyes from waste water. *J. Mol. Liq.* **2016**, *224*, 1294–1304. [CrossRef]

52. Han, K.N.; Li, C.A.; Bui, M.P.N.; Pham, X.H.; Seon, G.H. Control of ZnO morphologies on carbon nanotube electrodes and electrocatalytic characteristics toward hydrazine. *Chem. Commun.* **2011**, *47*, 938–940. [CrossRef] [PubMed]

53. Madhu, R.; Dinesh, B.; Chen, S.M.; Saraswathi, R.; Mani, V. An electrochemical synthesis strategy for composite based ZnO microspheres–Au nanoparticles on reduced graphene oxide for the sensitive detection of hydrazine in water samples. *RSC Adv.* **2015**, *5*, 54379–54386. [CrossRef]

54. Wu, M.; Ding, W.; Meng, J.; Ni, H.; Li, Y.; Ma, Q. Electrocatalytic behavior of hemoglobin oxidation of hydrazine based on ZnO nano-rods with carbon nanofiber modified electrode. *Anal. Sci.* **2015**, *31*, 1027–1033. [CrossRef] [PubMed]

55. Hua, J.; Zhao, Z.; Sun, Y.; Wang, Y.; Li, P.; Zhang, W.; Lian, K. Controllable synthesis of branched hierarchical ZnO nanorod arrays for highly sensitive hydrazine detection. *Appl. Surf. Sci.* **2016**, *364*, 434–441. [CrossRef]

56. Zhang, Y.; Han, T.; Wang, Z.; Zhao, C.; Li, J.; Fei, T.; Liu, S.; Lu, G.; Zhang, T. In situ formation of N-doped carbon film-immobilized Au nanoparticles-coated ZnO jungle on indium tin oxide electrode for excellent high-performance detection of hydrazine. *Sens. Actuators B* **2017**, *243*, 1231–1239. [CrossRef]

57. Park, N.K.; Park, C.J.; Kang, M.; Ryu, S.O.; Lee, T.J.; Baek, J.I.; Ryu, H.J. Reduction and oxidation behavior of ilmenite for chemical looping combustion. *Sci. Adv. Mater.* **2017**, *9*, 1998–2003.

58. Xu, T.T.; Jie, S. Synthesis of ZnO-loaded $Co_{0.85}Se$ nanocomposites and their enhanced performance for decomposition of hydrazine hydrate and catalytic hydrogenation of p-nitrophenol. *Appl. Catal. A Gen.* **2016**, *515*, 83–90. [CrossRef]

59. Dai, X.; Wildgoose, G.G.; Salter, C.; Crossley, A.; Compton, R.G. Electroanalysis using macro-, micro-, and nanochemical architectures on electrode surfaces. Bulk surface modification of glassy carbon microspheres with gold nanoparticles and their electrical wiring using carbon nanotubes. *Anal. Chem.* **2006**, *78*, 6102. [CrossRef] [PubMed]

60. Cui, Y.; Choi, E.; Shim, H.J.; Gao, Y.; Kim, K.S.; Pyo, S.G. Characterization of electroless-deposited ternary M1M2-R (M1 = Co, Ni, M2 = W, Mo, R = P, B) nano thin film for optical-sensor interconnects. *Sci. Adv. Mater.* **2017**, *9*, 2026–2031.

61. Shah, A.; Akhtar, M.; Aftab, S.; Shah, A.H.; Kraatz, H.B. Gold copper alloy nanoparticles (Au-Cu NPs) modified electrode as an enhanced electrochemical sensing platform for the detection of persistent toxic organic pollutants. *Electrochim. Acta* **2017**, *241*, 281–290. [CrossRef]

62. Thirumalraj, B.; Rajkumar, C.; Chen, S.M.; Lin, K.Y. Determination of 4-nitrophenol in water by use of a screen-printed carbon electrode modified with chitosan-crafted ZnO nanoneedles. *J. Colloid Int. Sci.* **2017**, *499*, 83–92. [CrossRef] [PubMed]

63. Divband, B.; Khatamian, M.; Kazemi Eslamian, G.R.; Darbandi, M. Synthesis of Ag/ZnO nanostructures by different methods and investigation of their photocatalytic efficiency for 4-nitrophenol degradation. *Appl. Surf. Sci.* **2013**, *284*, 80–86. [CrossRef]

64. Zhang, W.; Wang, G.; He, Z.; Hou, C.; Zhang, Q.; Wang, H.; Li, Y. Ultralong ZnO/Pt hierarchical structures for continuous-flow catalytic reactions. *Mater. Des.* **2016**, *109*, 492–502. [CrossRef]

65. Gupta, A.; Kim, B.C.; Edwards, E.; Brantley, C.; Ruffin, P. Covalent functionalization of zinc oxide nanowires for high sensitivity p-nitrophenol detection in biological systems. *Mater. Sci. Eng. B* **2012**, *177*, 1583–1588. [CrossRef]

66. Singh, K.; Ibrahim, A.A.; Umar, A.; Kumar, A.; Chaudhary, G.R.; Singh, S.; Mehta, S.K. Synthesis of CeO_2–ZnO nanoellipsoids as potential scaffold for the efficient detection of 4-nitrophenol. *Sens. Actuators B* **2014**, *202*, 1044–1050. [CrossRef]

67. Tan, S.T.; Umar, A.A.; Salleh, M.M. (001)-faceted hexagonal ZnO nanoplate thin film synthesis and the heterogeneous catalytic reduction of 4-nitrophenol characterization. *J. Alloys Compd.* **2015**, *650*, 299–304. [CrossRef]

68. Ahmad, R.; Tripathy, N.; Ahn, M.S.; Hahn, Y.B. Development of highly-stable binder-free chemical sensor electrodes for p-nitroaniline detection. *J. Colloid Int. Sci.* **2017**, *494*, 300–306. [CrossRef] [PubMed]

69. Ahmad, N.; Umar, A.; Kumar, R.; Alam, M. Microwave-assisted synthesis of ZnO doped CeO_2 nanoparticles as potential scaffold for highly sensitive nitroaniline chemical sensor. *Ceram. Int.* **2016**, *42*, 11562–11567. [CrossRef]

70. Li, M.; Chen, Y.; Yin, L.; Wang, H.; Cao, Y.; Wei, Z.; Wang, Y.; Yuan, Q.; Pu, X.; Zong, L.; et al. Formulation and stability evaluation of structure-altered paclitaxel nanosuspensions stabilized by a biocompatible amino acid copolymer. *Sci. Adv. Mater.* **2017**, *9*, 1713–1723.

71. Ibrahim, A.A.; Umar, A.; Kumar, R.; Kim, S.H.; Bumajdad, A.; Baskoutas, S. Sm_2O_3-doped ZnO beech fern hierarchical structures for nitroaniline chemical sensor. *Ceram. Int.* **2016**, *42*, 16505–16511. [CrossRef]

72. Hu, A.; Qu, S.; Wang, J.; Liang, G.; Wu, H.; Zhan, D. Synthesis and Photoluminscence Properties of Morphology- and Microstructure-Controlled S-Doped ZnO Nanostructures. *Sci. Adv. Mater.* **2017**, *9*, 316–320. [CrossRef]

73. Paraguay, F.; Miki-Yoshida, D.M.; Morales, J.; Solis, J.; Estrada, W. Influence of Al, In, Cu, Fe and Sn dopants on the response of thin film ZnO gas sensor to ethanol vapour. *Thin Solid Films* **2000**, *373*, 137–140. [CrossRef]

74. Cheng, X.L.; Zhao, H.; Huo, L.H.; Gao, S.; Zhao, J.G. ZnO nanoparticulate thin film: Preparation, characterization and gas-sensing property. *Sens. Actuators B* **2004**, *102*, 248–252. [CrossRef]

75. Nomoto, J.; Makino, H.; Yamamoto, T. Low-optical-loss transparent conductive Ga-doped ZnO films for plasmonics in the near-infrared spectral range. *Sci. Adv. Mater.* **2017**, *9*, 1815–1821.

76. Huang, J.; Shi, C.; Fu, G.; Sun, P.; Wang, X.; Gu, C. Facile synthesis of porous ZnO microbelts and analysis of their gas-sensing property. *Mater. Chem. Phys.* **2014**, *144*, 343–348. [CrossRef]

77. Faisal, M.; Khan, S.B.; Rahman, M.M.; Jamal, A.; Abdullah, M.M. Fabrication of ZnO nanoparticles based sensitive methanol sensor and efficient photocatalyst. *Appl. Surf. Sci.* **2012**, *258*, 7515–7522. [CrossRef]

78. Fahad, M.; Ye, H.; Jeon, S.; Hong, J.H.; Hong, S.K. Enhancement of reflective optical properties using photoluminescence-polymer-dispersed liquid crystal with added chiral dopant. *Sci. Adv. Mater.* **2016**, *8*, 1745–1751. [CrossRef]

79. Li, C.; Zhang, S.; Hub, M.; Xie, C. Nanostructural ZnO based coplanar gas sensor arrays from the injection of metal chloride solutions: Device processing, gas-sensing properties and selectivity in liquors applications. *Sens. Actuators B* **2011**, *153*, 415–420. [CrossRef]

80. Singh, O.; Singh, M.P.; Kohli, N.; Singh, R.C. Effect of pH on the morphology and gas sensing properties of ZnO nanostructures. *Sens. Actuators B* **2012**, *166–167*, 438–443. [CrossRef]

81. Bekermann, D.; Gasparotto, A.; Barreca, D.; Maccato, C.; Comini, E.; Sada, C.; Sberveglieri, G.; Devi, A.; Fischer, R.A. Co$_3$O$_4$/ZnO nanocomposites: From plasma synthesis to gas sensing applications. *ACS Appl. Mater. Int.* **2012**, *4*, 928–934. [CrossRef] [PubMed]

82. Gua, C.; Shanshana, L.; Huanga, J.; Shia, C.; Liu, J. Preferential growth of long ZnO nanowires and its application in gas sensor. *Sens. Actuators B* **2013**, *177*, 453–459. [CrossRef]

83. Zeng, Y.; Zhang, T.; Wang, L.; Wang, R. Synthesis and ethanol sensing properties of self-assembled monocrystalline ZnO nanorod bundles by poly(ethylene glycol)-assisted hydrothermal process. *J. Phys. Chem. C* **2009**, *113*, 3442–3448. [CrossRef]

84. Fan, F.; Feng, Y.; Tang, P.; Chen, A.; Luo, R.; Li, D. Synthesis and gas sensing performance of dandelion-like ZnO with hierarchical porous structure. *Ind. Eng. Chem. Res.* **2014**, *53*, 12737–12743. [CrossRef]

85. Wang, Z.; Tian, Z.; Han, D.; Gu, F. Highly sensitive and selective ethanol sensor fabricated with In-doped 3DOM ZnO. *ACS Appl. Mater. Int.* **2016**, *8*, 5466–5474. [CrossRef] [PubMed]

86. Thepnurat, M.; Chairuangsri, T.; Hongsith, N.; Ruankham, P.; Choopun, S. Realization of interlinked ZnO tetrapod networks for uv sensor and room-temperature gas sensor. *ACS Appl. Mater. Int.* **2015**, *7*, 24177–24184. [CrossRef] [PubMed]

87. Ameen, S.; Akhtar, M.S.; Shin, H.S. Highly dense ZnO nanowhiskers for the low level detection of p-hydroquinone. *Mater. Lett.* **2015**, *155*, 82–86. [CrossRef]

88. Freire, P.G.; Montes, R.H.O.; Romeiro, F.C.; Lemos, S.C.S.; Lima, R.C.; Richter, E.M.; Munoz, R.A.A. Morphology of ZnO nanoparticles bound to carbon nanotubes affects electrocatalytic oxidation of phenolic compounds. *Sens. Actuators B* **2016**, *223*, 557–565. [CrossRef]

89. Ahmed, J.; Rahman, M.M.; Siddiquey, I.A.; Asiri, A.M.; Hasna, M.A. Efficient hydroquinone sensor based on zinc, strontium and nickel based ternary metal oxide (TMO) composites by differential pulse voltammetry. *Sens. Actuators B* **2018**, *256*, 383–392. [CrossRef]

90. Al-Ansari, M.S.; Taqa, I.G.A.; Senouci, A.B.; Eldin, N.N.; Helal, M.; Asiado, C. Proposed formulas for estimating splitting tensile, shear and flexural strengths, and long term deflection assessment of self-compacting concrete elements. *Sci. Adv. Mater.* **2017**, *9*, 1751–1761.

91. Kumar, R.; Chauhan, M.S.; Dar, G.N.; Ansari, S.G.; Wilson, J.; Umar, A.; Chauhan, S.; Rana, D.S.; Sharma, P. ZnO nanoparticles: Efficient material for the detection of hazardous chemical. *Sens. Lett.* **2014**, *12*, 1393–1398. [CrossRef]

92. Xiao, Y.; Lu, L.; Zhang, A.; Zhang, Y.; Sun, L.; Huo, L.; Li, F. Highly enhanced acetone sensing performances of porous and single crystalline ZnO nanosheets: High percentage of exposed (100) facets working together with surface modification with Pd nanoparticles. *ACS Appl. Mater. Int.* **2012**, *4*, 3797–3804. [CrossRef] [PubMed]

93. Sahay, P.P. Zinc oxide thin film gas sensor for detection of acetone. *J. Mater. Sci.* **2005**, *40*, 4383–4385. [CrossRef]

94. Zhang, G.H.; Deng, X.Y.; Wang, P.Y.; Wang, X.L.; Chen, Y.; Ma, H.L.; Gengzang, D.J. Morphology controlled syntheses of Cr doped ZnO single-crystal nanorods for acetone gas sensor. *Mater. Lett.* **2016**, *165*, 83–86. [CrossRef]

95. Liu, C.; Zhao, L.; Wang, B.; Sun, P.; Wang, Q.; Gao, Y.; Liang, X.; Zhang, T.; Lu, G. Acetone gas sensor based on NiO/ZnO hollow spheres: Fast response and recovery, and low (ppb) detection limit. *J. Colloid Int. Sci.* **2017**, *495*, 207–215. [CrossRef] [PubMed]

96. Koo, W.T.; Jang, J.S.; Choi, S.J.; Cho, H.J.; Kim, I.D. Metal-organic framework templated catalysts: Dual sensitization of PdO–ZnO composite on hollow SnO$_2$ nanotubes for selective acetone sensors. *ACS Appl. Mater. Int.* **2017**, *9*, 18069–18077. [CrossRef] [PubMed]

97. Shin, S.H.; Park, D.H.; Jung, J.Y.; Lee, M.H.; Nah, J. Ferroelectric zinc oxide nanowire embedded flexible sensor for motion and temperature sensing. *ACS Appl. Mater. Int.* **2017**, *9*, 9233–9238. [CrossRef] [PubMed]

98. Bhanjana, G.; Dilbaghi, N.; Singh, N.K.; Kim, K.H.; Kumar, S. Zinc oxide nanopillars as an electrocatalyst for direct redox sensing of cadmium. *J. Ind. Eng. Chem.* **2017**, *53*, 192–200. [CrossRef]

99. Nagaraju, G.; Udayabhanu, H.; Prashanthb, S.S.A.; Shastri, M.; Yathish, K.V.; Anupama, C.; Rangappa, D. Electrochemical heavy metal detection, photocatalytic, photoluminescence, biodiesel production and antibacterial activities of Ag–ZnO nanomaterial. *Mater. Res. Bull.* **2017**, *94*, 54–63. [CrossRef]

100. Nam, M.; Kim, A.; Kang, K.; Choi, E.; Kwon, S.H.; Lee, S.J.; Pyo, S.G. Characterization of atomic layer deposited Al$_2$O$_3$/HfO$_2$ and Ta$_2$O$_5$/Al$_2$O$_3$ combination stacks. *Sci. Adv. Mater.* **2016**, *8*, 1958–1962. [CrossRef]

101. Geng, S.; Lin, S.M.; Li, N.B.; Luo, H.Q. Polyethylene glycol capped ZnO quantum dots as a fluorescent probe for determining copper(II) ion. *Sens. Actuators B* **2017**, *253*, 137–143. [CrossRef]

102. Ng, S.M.; Wong, D.S.N.; Phung, J.H.C.; Chua, H.S. Integrated miniature fluorescent probe to leverage the sensing potential of ZnO quantum dots for the detection of copper (II) ions. *Talanta* **2013**, *116*, 514–519. [CrossRef] [PubMed]

103. Sadollahkhani, A.; Hatamie, A.; Nur, O.; Willander, M.; Zargar, B.; Kazeminezhad, I. Colorimetric disposable paper coated with ZnO@ZnS core-shell nanoparticles for detection of copper ions in aqueous solutions. *ACS Appl. Mater. Int.* **2014**, *6*, 17694–17701. [CrossRef] [PubMed]

104. Jeong, Y.M.; Kim, J.H.; Baek, S.H. One-pot synthesis of ZnAl double hydroxide powders and their calcined oxide composites for lithium-Ion battery applications. *Sci. Adv. Mater.* **2017**, *9*, 1801–1805.

105. Shen, Q.; Zhao, X.; Zhou, S.; Hou, W.; Zhu, J.J. ZnO/CdS hierarchical nanospheres for photoelectrochemical sensing of Cu^{2+}. *J. Phys. Chem. C* **2011**, *115*, 17958–17964. [CrossRef]

106. Kandjani, A.E.; Sabri, Y.M.; Mohammadtaheri, M.; Bansal, V.; Bhargava, S.K. Detect, remove and re-use: A new paradigm in sensing and removal of Hg (II) from wastewater via SERS-active ZnO/Ag nano-asrrays. *Environ. Sci. Technol.* **2015**, *49*, 1578–1584. [CrossRef] [PubMed]

107. Qiao, L.; He, M.; Zheng, S.; Li, S. Fabrication of Blue Light-Emitting Diode with Vertical Structure on the ZnO Substrate. *Sci. Adv. Mater.* **2017**, *9*, 76–79. [CrossRef]

108. Kim, J.; Lee, G.H.; Yun, G.H.; Park, B.W.; Jeong, J.Y.; Choi, S.; Kim, S.J.; Chang, S.W.; Koh, J.H. Spin-Coated In-Doped ZnO Nanorods for Transparent Conducting Oxide Applications. *Sci. Adv. Mater.* **2017**, *9*, 1193–1196. [CrossRef]

materials

MDPI

Article

Factors Affecting the Power Conversion Efficiency in ZnO DSSCs: Nanowire vs. Nanoparticles

Myrsini Giannouli [1],*, Katerina Govatsi [2], George Syrrokostas [2], Spyros N. Yannopoulos [2] and George Leftheriotis [1]

[1] Renewable Energy and Environment Laboratory, Physics Department, University of Patras, Rion GR 26500, Greece; glefther@physics.upatras.gr
[2] Foundation for Research and Technology Hellas, Institute of Chemical Engineering Sciences (FORTH/ICE-HT), P.O. Box 1414, Rio-Patras GR-26504, Greece; govatsi@iceht.forth.gr (K.G.); gsirrokostas@yahoo.gr (G.S.); sny@iceht.forth.gr (S.N.Y.)
* Correspondence: myrgiannouli@upatras.gr; Tel.: +30 2610 997449

Received: 6 February 2018; Accepted: 7 March 2018; Published: 9 March 2018

Abstract: A comparative assessment of nanowire versus nanoparticle-based ZnO dye-sensitized solar cells (DSSCs) is conducted to investigate the main parameters that affect device performance. Towards this aim, the influence of film morphology, dye adsorption, electron recombination and sensitizer pH on the power conversion efficiency (PCE) of the DSSCs is examined. Nanoparticle-based DSSCs with PCEs of up to 6.2% are developed and their main characteristics are examined. The efficiency of corresponding devices based on nanowire arrays (NW) is considerably lower (0.63%) by comparison, mainly due to low light harvesting ability of ZnO nanowire films. The dye loading of nanowire films is found to be approximately an order of magnitude lower than that of nanoparticle-based ones, regardless of their internal surface area. Inefficient anchoring of dye molecules on the semiconductor surface due to repelling electrostatic forces is identified as the main reason for this low dye loading. We propose a method of modifying the sensitizer solution by altering its pH, thereby enhancing dye adsorption. We report an increase in the PCE of nanowire DSSCs from 0.63% to 1.84% as a direct result of using such a modified dye solution.

Keywords: dye-sensitized solar cells; ZnO semiconductor; dye loading; nanowire; pH modification

1. Introduction

Dye-sensitized solar cells (DSSCs) are regarded as a promising option to conventional solid-state semiconductor solar cells due to their low cost, ease of fabrication and environmental friendliness. DSSCs sensitized with Ruthenium-based complexes have achieved maximum power conversion efficiency (PCE) of 11.9% [1,2]. Alternatively, DSSCs with light-harvesting donor–π–acceptor (D–π–A) dyes have yielded efficiencies over 12% [3,4], while various molecularly-engineered porphyrin dyes have been synthesized [5] and resulted in a record 13% efficiency [6].

In DSSCs, the conversion of visible light to electricity is achieved through the spectral sensitization of wide bandgap semiconductors such as TiO_2, ZnO, SnO_2, NiO, etc. [7–9]. ZnO has attracted a great deal of interest in DSSCs applications, as a semiconductor with a wide band gap of 3.37 eV and a high exciton binding energy of 60 meV. In addition, ZnO has high room temperature carrier mobility (115–155 cm^2 V^{-1} s^{-1}), which enhances the performance of solar cells by reducing electron recombination [10].

Typically, the photoelectrode of DSSCs consists of a porous network of semiconductor nanoparticles. The porous nature of the photoelectrode provides a large internal surface area for chromophore anchoring to maximize dye loading (dl) and in turn, light absorption and electron generation [11]. However, the photogenerated electrons then diffuse through the nanoparticle network and will have to travel via a random path also crossing over the grain boundaries before they are collected. This electron transfer process is slow and electrons may interact with traps as they travel through the nanoparticle network and recombine with oxidizing species, thus reducing device efficiency [12,13].

DSSCs based on semiconductor nanowires are considered to have improved electron transport properties by providing a direct pathway for photogenerated electrons from the point of injection to the collecting electrode [14–16]. However, the overall PCE of such devices remains relatively low compared to nanoparticle-based ones. The reasons for the low PCE of nanowire-based DSSCs are not fully understood, but the main reason is considered to be reduced light harvesting due to low dye loading [17,18], which was also confirmed by our own findings.

Understanding the mechanisms that affect PCE in nanowire-based DSSCs is key to improving their efficiency. Towards this end, the main parameters that affect device performance, such as semiconductor morphology, light-harvesting efficiency, electron injection and electron collection, were investigated. A comparative assessment of the properties of nanoparticle and nanowire-based DSSCs was conducted to identify the underlying causes of the low nanowire-DSSC performance and improve their efficiency.

2. Methodology

2.1. ZnO Nanoparticle Films

Nanostructured ZnO films were prepared as described in [19]. Commercial ZnO nanopowder (Sigma-Aldrich, St. Louis, MO, USA) with nanoparticle diameter less than 50 nm was used to create a colloidal paste. The powder was mixed with a small amount of distilled water containing acetyl acetone (10% v/v) [20,21] to prevent the coagulation of nanoparticles and improve the porosity of the film [22]. A small amount of Triton X-100 was added to the mixture to reduce surface tension and enable even spreading of the paste [23,24].

The semiconductor oxide paste was spread on conductive glass substrates (K-glass (SnO_2:F) with sheet resistance 16.7 Ω/sq, 80% transmittance in the visible, 0.38 cm glass thickness) via a doctor blade technique and the electrodes were annealed for 30 min at 400 °C in air to enhance the electrical contact between the nanoparticles as well as between the nanoparticles and the conductive substrate [25]. The thickness of the resulting films was measured by a stylus XP-1 Ambios Technology profilometer. As the thickness of the films can affect the performance of the DSSCs, the photovoltaic characteristics of all nanoparticle-based DSSCs presented in the paper were obtained for samples of approximately the same thickness (ranging from 7 to 10 µm) [26]. We have found in our previous work [26] that ZnO films of approximately this thickness yield the optimum results in terms of device efficiency and stability. The surface area of the ZnO films was approximately 0.5 cm^2. All films had an oblong shape, with smaller width and greater length. This shape has been shown to be the most favorable for minimizing device internal series resistance and enhancing cell efficiency [27].

2.2. ZnO Nanowire Arrays

Nanostructured films based on ZnO NW arrays were developed on glass conductive substrates using wet-chemistry methods. A thin (~10 nm) ZnO seed layer was deposited by spin coating of a solution of 0.05 M Zinc acetate dihydrate (ZnAc, $Zn(CH_3COO)_2 \cdot 2H_2O$) in ethanol at a speed of 1000 rpm followed by annealing at 300 °C for 2 h. Two different approaches were employed for the growth of the NW arrays. In the first one, NW-based films were grown in an aqueous solution at 95 °C containing equal concentrations of ZnAc and hexamethylenetetramine (HMTA, $C_6H_{12}N_4$). The concentrations varied to control the NW morphology. NW arrays with narrow to moderate size distributions, and height in the range 0.5–2 µm were prepared for growth times of ~2.5 h. Renewal of the nutrient solution is necessary to achieve NW lengths up to 6–7 µm. Growth of NW arrays with larger diameter and length was also performed using the procedure described in detail elsewhere [28].

After the growth, the NW arrays were rinsed with 3D water and were heated at 300 °C for 2 h to remove undesired contamination from the surface.

The morphologies of the deposited materials were characterized by Field Emission Scanning Electron Microscopy (FE-SEM) (Zeiss SUPRA 35VP-FEG, Jena, Germany) operating at 5–20 keV. The crystal structures of the composites were investigated using X-ray diffraction (XRD) Bruker D8 diffractometer, operating at 40 kV and 40 mA, employing Cu Kα radiation (λ = 1.54056 Å).

2.3. Film Sensitization-Assessment of Dye Loading and of the Point of Zero Charge

ZnO films were sensitized in 0.3 mM solutions of the commercial dye N719 (Sigma-Aldrich, St. Louis, MO, USA) in methanol. Coating of the semiconductor surface with the dye was conducted by soaking the film in the dye solution for a time ranging from 30 min to 24 h to optimize the dying process. The optimum time the film should remain in the dye solution was investigated for all types of ZnO films considered to achieve maximum dye adsorption and device efficiency. This was found to depend on film thickness, with thicker films having to remain in the dye solution longer to achieve maximum dye loading.

Dye loading of the sensitized ZnO films was calculated by desorbing the dye in a 10 mM KOH aqueous solution and then measuring the absorbance of the solution using a UV-Vis PerkinElmer Lambda 650 spectrophotometer (Perkin Elmer, Waltham, MA, USA) [29]. The pH of the dye solution was measured using a Mettler Toledo FE20/FG2 pH Meter (Mettler Toledo, Columbus, OH, USA). The pH of the point of zero charge (pH$_{PZC}$) of ZnO films was obtained using the pH-drift method [30,31]. The point of zero charge was determined by varying the pH I of a 0.01 M NaCl 50 mL solution to pH values in the range of 2 to 12 by adding either HCl or NaOH. A semiconductor film was dipped in the solution and was then left for 48 h to reach an equilibrium pH. For each pH value, the initial pH and the final pH (after 48 h) were measured and plotted against each other. The pH at which the curve crosses the line of pH$_{final}$ = pH$_{initial}$ is the pH of the point of zero charge of the film.

2.4. DSSC Fabrication

Counter electrodes were prepared through a three-electrode electrodeposition of an aqueous hexachloroplatinic acid solution (0.002 M) on conductive glass substrates. The electrodeposition was performed using a computer-controlled Function Generator (AMEL, 586) (586, Milano, Italy), a potentiostat–galvanostat (AMEL, 2053) (2053, Milano, Italy) and a noise reducer (AMEL NR 2000) (NR 2000, Milano, Italy).

The sensitized semiconductor electrode and the platinized counter electrode were sealed together using the thermoplastic Surlyn® (DuPont, Wilmington, DE, USA) silicon. A liquid electrolyte was then inserted in the space between the two electrodes. The EL-HPE High Performance Electrolyte (Dyesol) (Queanbeyan, Australia) was used for all samples. The electrolyte was inserted into the cell with a syringe through a small aperture and the cell was then sealed with silicone. Several different devices were prepared and tested for each type of DSSCs presented in this work to ensure the reproducibility and validity of the results.

2.5. Device Characterization and Testing

Current–Voltage (I–V) curves of the cells were obtained using a Newport 96000 solar simulator (96000, Irvine, CA, USA) fitted with an AM1.5G filter (Newport AM 1.5G, Irvine, CA, USA) in conjunction with a Keithley 236 source meter. The incident irradiance was measured by a photo diode calibrated against a Melles Griot 13PE001 Broad Band Power Meter (13PE001, Rochester, NY, USA).

The main characteristics of the solar cells can be obtained from these I–V curves, namely the open circuit voltage (V_{oc}), the short circuit current (I_{sc}), the fill factor (*FF*) and the efficiency (η) of the cell.

The fill factor of each cell can be calculated according to the following equation:

$$FF = \frac{I_m V_m}{I_{SC} V_{OC}}$$ (1)

where I_m and V_m are the values of the current and the voltage for the maximum power point respectively. Finally, the energy conversion efficiency of each solar cell can be calculated using the following equation:

$$\eta = FF \frac{I_{SC} V_{OC}}{S G_T}$$ (2)

where S is the surface area of the cell and G_T is the incident light intensity.

Dark current measurements were also conducted for all DSSCs considered applying a bias voltage ramp starting from 0 V and exceeding the open circuit voltage of the cell [32]. Dark current mainly arises when triiodide ions from the electrolyte draw electrons from the semiconductor, reducing the triiodide to iodide. This occurs at the semiconductor/electrolyte interface, when there is no sensitizer adsorbed on the semiconductor surface. A secondary source of dark current is the reduction of the oxidative species of the electrolyte by the glass conductive surface. This can occur if there are pathways for the electrolyte to penetrate through the semiconductor film and reach the glass conductive surface [33]. Regardless of its origin, dark current causes electron recombination and results in the loss of photocurrent [34]. The production of dark current in a cell is also directly linked to its open circuit voltage, with high dark current reducing the open circuit voltage of the cell [35].

Open-circuit voltage decay (OCVD) measurements were conducted by stopping the illumination of the cells under open-circuit conditions and using a potentiostat–galvanostat (AMEL, 2053) to monitor the resulting decline of V_{oc} [36]. Electron lifetime (τ_n) was then determined by the reciprocal of the derivative of the decay curves normalized by the thermal voltage, using the following equation [33]:

$$\tau_n = -\frac{k_B T}{e} \left(\frac{dV_{oc}}{dt} \right)^{-1}$$ (3)

where k_B is the Boltzmann constant, T is the absolute temperature, e is the positive elementary charge, and dV_{oc}/dt is the derivative of the transient open-circuit voltage.

Incident photon to current efficiency (IPCE) spectra were obtained using a Newport setup with a 150 W Xe-lamp and a Newport (Oriel Cornerstone) monochromator. IPCE corresponds to the number of electrons measured as photocurrent in the external circuit divided by the monochromatic photon flux that strikes the cell.

The IPCE factor is given by the following equation:

$$IPCE\,(\%) = \frac{1240\,[eV\,nm] \times J_{ph}\,[mA\,cm^{-2}]}{\lambda\,[nm] \times \Phi\,[mW\,cm^{-2}]} \times 100$$ (4)

where J_{ph} is the short-circuit photocurrent density for monochromatic irradiation and λ and Φ are the wavelength and the intensity, respectively, of the monochromatic light [37,38].

3. Results and Discussion

3.1. Film Morphology

Figure 1 shows representative SEM images illustrating the morphology of a nanoparticle-based film (Figure 1a) and a nanowire-based film (Figure 1b). The nanoparticle film exhibits high porosity and very low particle agglomeration. The ZnO NWs exhibit high degree of orientation. The XRD patterns of these films are shown in Figure 1c. The length of the ZnO NWs is approximately 7–10 microns, i.e., several orders of magnitude larger than the thickness of the seed layer, which is only 10–20 nm. Therefore, the scattering intensity of the seed layer was considered negligible when obtaining the XRD patterns. The XRD data reveal that ZnO nanowires are highly crystalline and confirm the orientation of these structures normal to the substrate as only the 002 Bragg peak is visible in the corresponding XRD pattern.

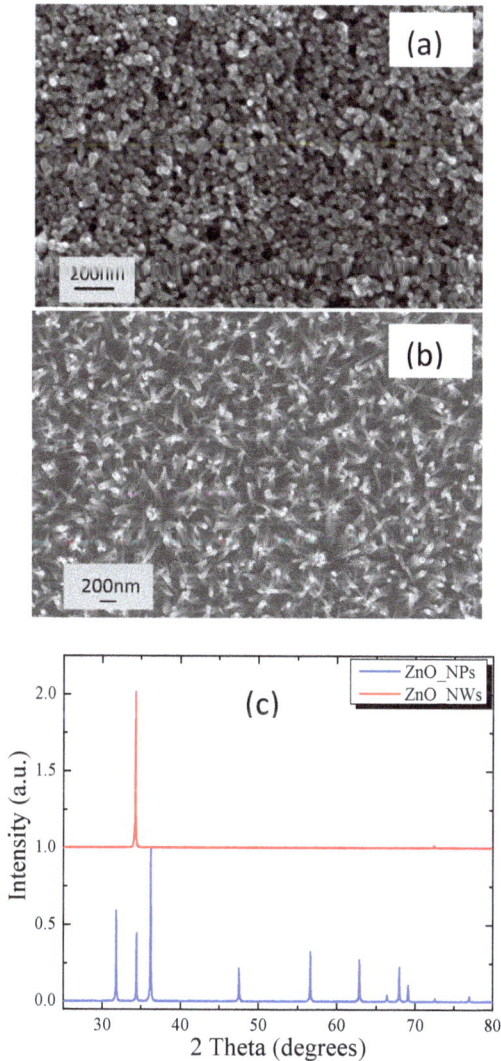

Figure 1. SEM images of: (**a**) ZnO nanoparticle film; and (**b**) a ZnO nanowire film; and (**c**) XRD patterns of a ZnO nanoparticle and a nanowire film.

Figure 2 shows step profilometer images illustrating the difference in roughness between nanowire and nanoparticle films. The thickness of nanoparticle and nanowire films is also shown in this figure. It can be observed that the film thickness of the nanoparticle films (Figure 2a) is approximately 10 μm. NW films of varying thickness were prepared and tested, by modifying the NW growth conditions. The NW array shown in Figure 2b has been prepared after renewing the nutrient solution two extra times after the initial growth.

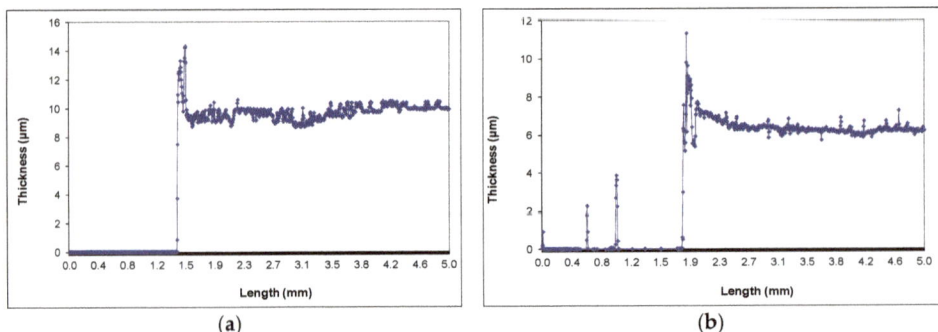

Figure 2. Morphology and thickness of nanowire based films: (**a**) ZnO nanoparticle film; and (**b**) ZnO nanowire film.

Figure 3 shows pictures of nanoparticle and nanowire-based devices sensitized with N719. Nanoparticle films sensitized with N719 have a dark red color as expected. Nanowire films on the other hand, have almost no color at all. This effect was common across all samples prepared and tested and it is an indication of very low dye adsorbance on NW films, as confirmed from dye loading measurements.

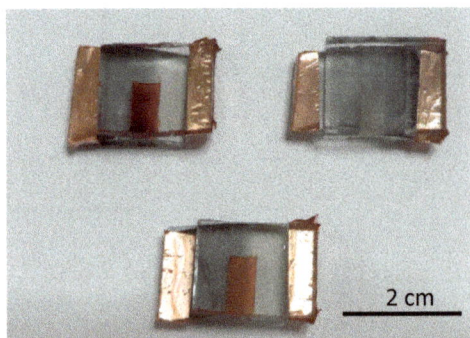

Figure 3. Images of nanoparticle (top left), nanowire (top right) and nanowire with improved dye loading (bottom) devices sensitized with N719.

3.2. Photovoltaic Characteristics

The IPCE of a cell (Equation (4)) can also be expressed in terms of the light-harvesting efficiency of the dye $LHE(\lambda)$, the quantum yield of electron injection η_{inj} and the efficiency of collecting the injected electrons η_{cc} at the back contact, according to the following equation [39]:

$$IPCE(\lambda) = LHE(\lambda) \times \eta_{inj} \times \eta_{cc} \qquad (5)$$

The IPCE values of typical nanoparticle and NW DSSCs are shown in Figure 4 as a function of the illumination wavelength. It is clear from Figure 4 that the IPCE values of nanoparticle DSSCs are considerably higher than those of nanowire cells. These results indicate that one or more of the main parameters that affect IPCE, i.e., light-harvesting efficiency, electron injection and electron collection, has far lower values for NW devices than for nanoparticle ones. In this paper, we will examine the effect of these parameters on the performance of nanoparticle and NW DSSCs and will attempt to identify the factors that are responsible for the low values in the IPCE of NW devices.

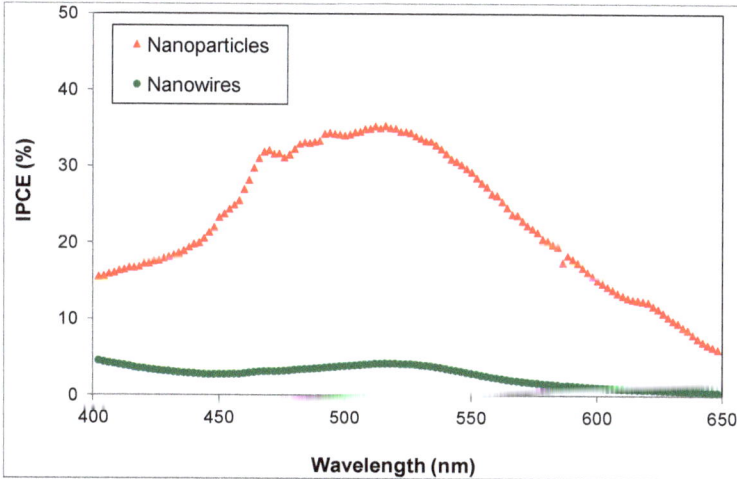

Figure 4. IPCE values of nanoparticle-based and nanowire-based DSSCs as a function of wavelength.

The J-V characteristics of ZnO nanoparticle-based and nanowire-based cells sensitized with the N719 dye are shown in Figure 5 and are summarized in Table 1. Current Density-Voltage (J-V) measurements show that the energy conversion efficiency (η) of nanoparticle-based cells sensitized with N719 was an order of magnitude higher than the efficiency of corresponding nanowire-based cells. While ZnO nanoparticle-based DSSCs yielded efficiencies over 6%, nanowire-based devices only reached PCEs of approximately 0.6%. These results are consistent with the large difference between the IPCEs of nanoparticle and NW DSSCs shown in Figure 4. In general, the short-circuit current produced by all nanowire-based devices was considerably lower than that of nanoparticle ones and their open circuit voltage was also lower. Many nanowire-based DSSCs (around 30) have been developed and tested, with varying conditions such as film thickness, time spent in the dye solution, etc. However, even with optimized device conditions, the maximum PCEs achieved were a little over 0.6%.

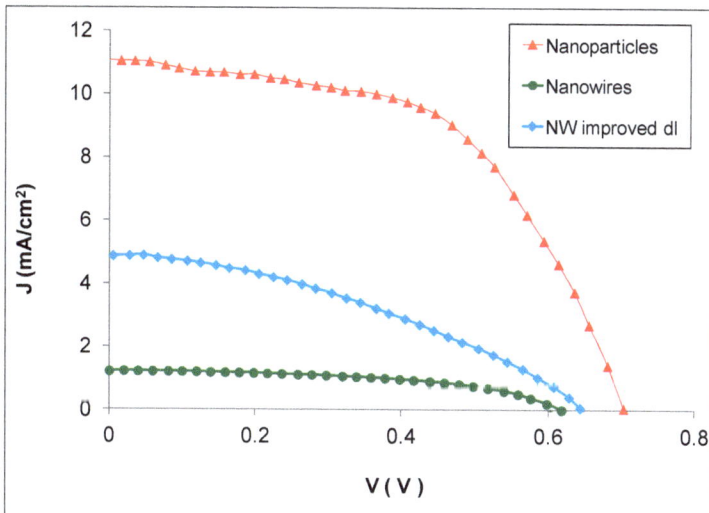

Figure 5. J-V characteristics of ZnO nanoparticle, nanowire and nanowire with improved dye loading DSSCs.

Table 1. Summary of the PV properties of the ZnO DSSCs.

Sample Type Header	V_{oc} (V)	J_{sc} (mA/cm^2)	FF	η (%)
Nanoparticles	0.70 ± 0.02	11.2 ± 1.0	0.54 ± 0.06	6.19 ± 0.60
Nanowires	0.62 ± 0.03	1.2 ± 0.4	0.53 ± 0.04	0.63 ± 0.09
Nanowires with improved dl	0.64 ± 0.02	4.8 ± 0.4	0.41 ± 0.03	1.80 ± 0.20

The efficiencies reported in this study for nanoparticle-based devices are quite high compared to corresponding values found in the literature. Keis et al. [40] report efficiencies of up to 5% for ZnO nanoparticle films sensitized with N719, while Chang et al. [41] report efficiencies of up to 5.6% for optimized devices with similar properties. He et al. [42] have obtained higher efficiencies (8%) by using air plasma to develop ZnO nanostructures with reduced charge carrier recombination. For NW-based DSSCs, Law et al. [13] report efficiencies of up to 1.5%, while Xu et al. [17] have obtained efficiencies of up to 2.1% by developing optimized ZnO NWs with length of up to 30 μm. While these efficiencies are higher than those reported in the present study, they are still lower than corresponding results for nanoparticle-based devices. The reasons for the low performance of nanowire-based DSSCs compared to nanoparticle ones are not fully understood. It has been reported in the literature that NW cells tend to have lower PCE than nanoparticle-based cells because their overall surface area does not permit high dye loading [17]. Other studies also indicate that nanowire-based DSSCs have lower *FF* than nanoparticle-based ones, due to charge recombination at the interface between the nanowires and the electrolyte [43]. In this study, the mechanisms responsible for the low PCEs of ZnO nanowire DSSCs were investigated to understand the underlying issues that prevent these devices from achieving efficiencies close to those of nanoparticle-based DSSCs.

Figure 6 shows typical dark current–voltage characteristics of nanoparticle-based cells and nanowire-based cells. For NW devices, the onset of dark current is at approximately 0.3 V and its value increases rapidly with increasing voltage. For nanoparticle-based films, on the other hand, the onset of dark current does not occur until approximately 0.55 V. The lower open circuit voltage of the nanowire DSSCs (Figure 5) is probably due to their higher dark current values.

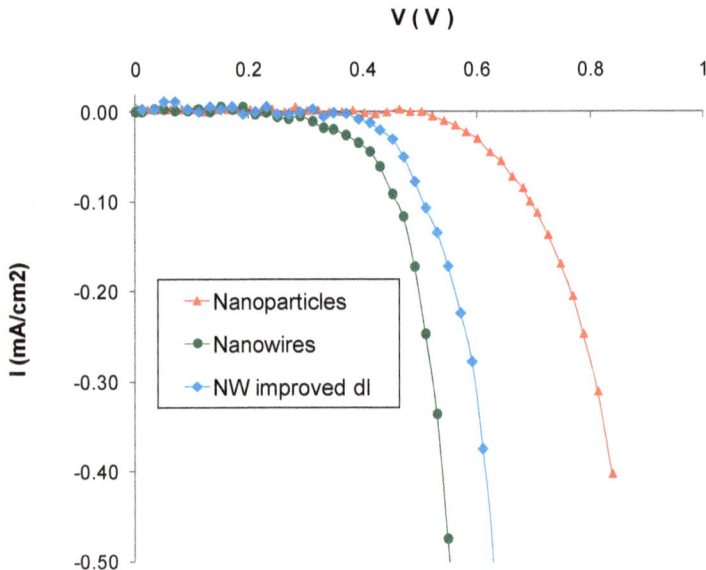

Figure 6. Dark current–voltage characteristics of ZnO nanoparticle, NW and NW with improved dye loading DSSCs.

As mentioned in Section 2, one of the causes of high dark current is the presence of semiconductor sites without dye adsorbed, that come in direct contact with the electrolyte. From Figure 3 it is apparent that dye adsorption in nanowire-based cells is not as successful as in nanoparticle ones and that is probably one of the reasons for the higher dark current of nanowire-based DSSCs. In addition, due to the porous nature of nanowire films, it is possible that there are more pathways for the electrolyte to come into contact with the glass substrate and cause charge recombination, as will be discussed below [44].

OCVD measurements can be used to determine electron recombination rates and ultimately electron lifetimes in DSSCs [45,46]. OCVD experiments measure the recombination rate of electrons injected into the conduction band of the semiconductor. Thus, the photovoltage decay measured reflects the decrease of electron concentration at the conductive glass surface, caused by charge recombination. Figure 7a shows the OCVD decay curves of typical nanoparticle and nanowire-based devices. It is apparent that the rate of open-circuit voltage decay is higher for NW devices than for nanoparticle ones. The electron lifetime with respect to the open-circuit voltage (calculated according to Equation (3)) is presented in Figure 7b and it was found to be higher for nanoparticle-based DSSCs than for nanowire cells. The results shown in Figure 7 indicate that there is higher electron recombination for nanowire DSSCs than for nanoparticle ones. It could be attributed to poor dye loading and possibly to contact of the conductive substrate with the electrolyte, as in the case of dark current.

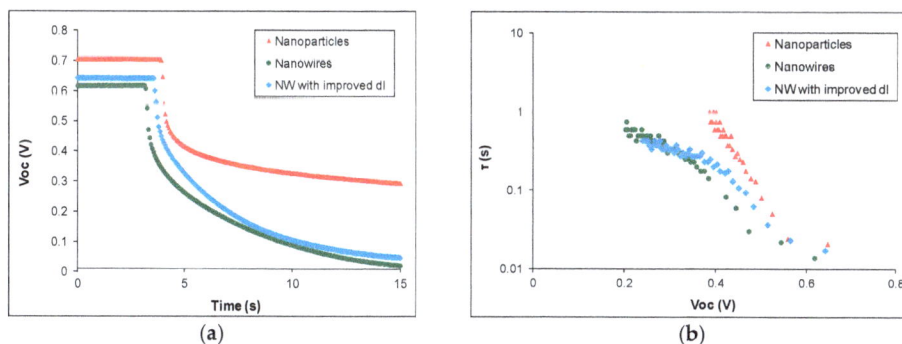

Figure 7. OCVD measurements for ZnO nanoparticle, NW and NW with improved dye loading DSSCs: (a) open-circuit voltage decay curves and (b) electron lifetime as a function of open-circuit voltage.

The dark current and OCVD results (shown in Figures 6 and 7 respectively) indicate that there are higher carrier losses for NW devices due to electron recombination. These results suggest that reduced electron collection is at least partly responsible for the low performance of NW DSSCs. It has been reported elsewhere [13] that electron injection from the dye's LUMO to the semiconductor is faster for NW DSSCs sensitized with N719 than for nanoparticle ones. It is therefore clear that the electron injection rate is not the cause of the low efficiency and IPCE of nanowire DSSCs. Below, we study the final reason for low device IPCE, namely the light-harvesting efficiency of DSSCs.

3.3. Dye Loading

Dye loading measurements were conducted, as described in Section 2, to assess the amount of dye adsorbed on each type of film. The dye loading was calculated per unit volume (film thickness × film area) to account for the difference in the surface area of the various films. However, the dye loading of the NW films was found to vary considerably depending on film morphology. NW films of different morphologies (Figure 8) were prepared and tested to optimize the dye loading process. The NW films were developed as described in [28] and each type of film was prepared according to the conditions described in Table 2.

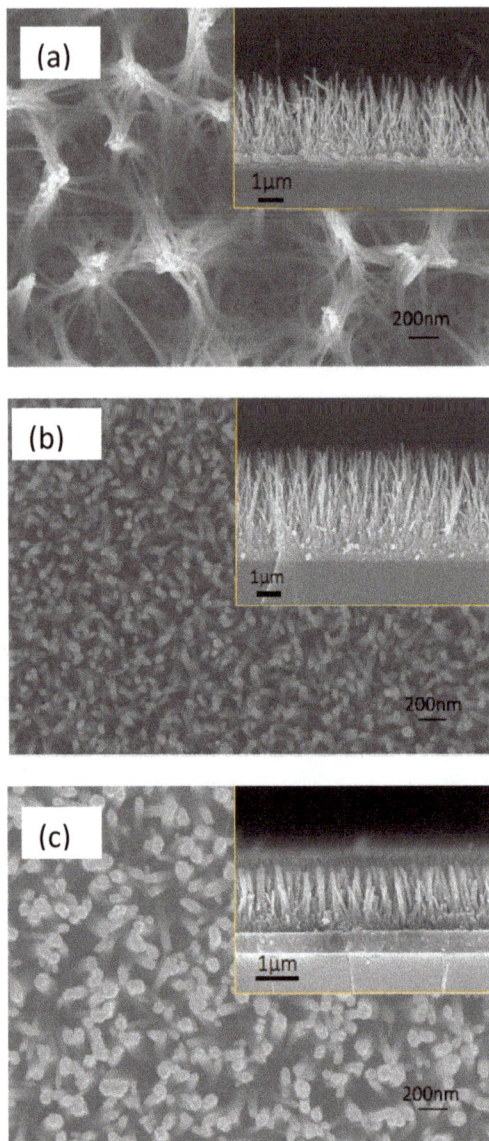

Figure 8. FE-SEM images of ZnO NW arrays used to optimize the dye loading conditions. NWs in (a–c) have average diameter 20–25, 30–40 and 80–100 nm, respectively.

The dye loading of the various NW films soaked in the standard N719 dye solution (samples a1, b1 and c1) is shown in Figure 9a. The dye loading of nanoparticle films was found to be approximately an order of magnitude higher than that of the nanowire ones with the highest dye loading measured (samples a1 and b1). This difference in the dye loading of nanoparticle and nanowire films accounts for the difference that was observed in the short-circuit current of the two types of devices and is also largely responsible for the difference in their efficiencies. The difference in the dye loading of the two types of films was also expected from visual observations of the films (Figure 3).

Table 2. Growth conditions of the three types of samples shown in Figure 8.

Sample Code	Seed Layer	Growth Conditions	
a		0.04 M Zn(NO$_3$)$_2$ 0.02 M HMTA 0.16 M PEI 0.04 M NH$_4$OH	Without renewal
	0.05 M Zinc acetate in ethanol		
b		0.04 M Zn(NO$_3$)$_2$ 0.02 M HMTA 0.16 M PEI 0.04 M NH$_4$OH	With renewal
c	0.005 M Zinc acetate in ethanol	0.05 M Zn(NO$_3$)$_2$ 0.025 M HMTA 0.08 gr PEI 0.7 M NH$_4$OH	Without renewal

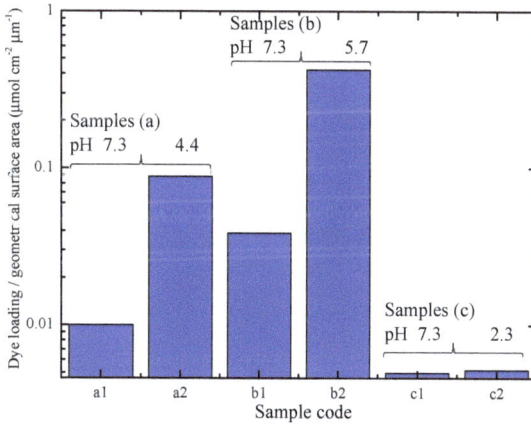

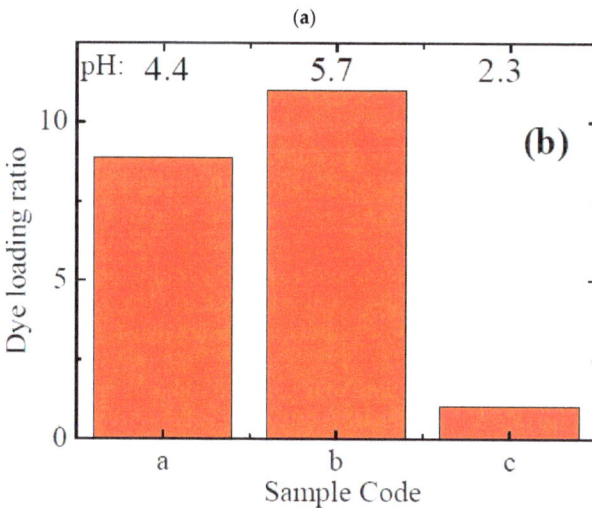

Figure 9. (a) Dye loading of NW arrays with the morphology shown in Figure 8. (b) Dye loading ratio of twin samples between the selected pH and the pH 7.3.

The relatively low dye loading of nanowire films compared to that of nanoparticle ones has also been reported elsewhere [17]. However, the reasons for the low dye absorbance of these films remain unclear. The main difference between the two types of films lies in the morphology of the ZnO nanostructures, which can also affect their surface charge. The adsorption mechanism of N719 on a ZnO surface has been shown to be predominantly electrostatically driven [47]. It has also been reported [47,48] that N719 forms more stable bonds with semiconductor films in a protonated environment. If the semiconductor surface is negatively charged on the other hand, it may repel the carboxylic groups in N719 and the dye may not be adsorbed on the semiconductor surface successfully.

To test this hypothesis, the pH of the point of zero charge (pH_{PZC}) of ZnO nanowire films was determined as described in Section 2.3 and compared to the pH_{PZC} of ZnO nanoparticle films. The values of the final against the initial pH are shown in Figure 10 and the pH at which the curve crosses the line of $pH_{final} = pH_{initial}$ is the pH of the point of zero charge of the film. In this case, the pH_{PZC} of ZnO nanowires (Figure 8a) was found in the range of 6.7–7.2 (the range represents the results of pH_{PZC} measurements in different samples). The pH_{PZC} of ZnO nanoparticles was found to be approximately 8.9 (Figure 8b), which agrees with corresponding results reported in the literature [49].

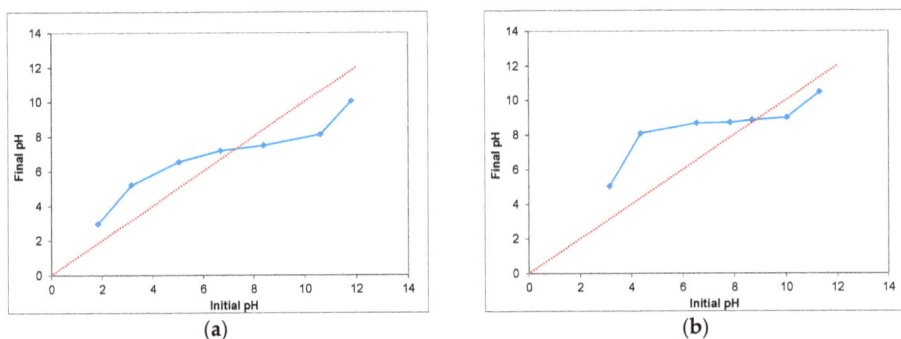

Figure 10. Determination of the pH_{PZC} of ZnO films: (**a**) nanowires; and (**b**) nanoparticles.

The high value of the pH_{PZC} of ZnO nanoparticles indicates the ZnO nanoparticle surface is positively charged, which may facilitate the adsorption process. The same is not true for ZnO nanowire films however, which may hinder the anchoring process of the dye on the semiconductor surface. The pH of the standard dye solution used was measured to be in the range of 7.3–7.7 (the range represents the pH measurements in various dye solutions), which is slightly higher than the pH_{PZC} of ZnO nanowire films. When a nanowire film is submerged in the standard dye solution, the net electrical charge on the film surface will be negative. On the other hand, the pH_{PZC} of ZnO nanoparticle films is higher than the pH of the standard dye solution, which means that the nanoparticle film/dye solution environment will be protonated. The same result can be achieved for the nanowire film/dye solution environment by lowering the pH of the dye solution below the pH_{PZC} of nanowire films. Acidic dye solutions will have a higher concentration of positive ions, which may facilitate the adsorption process for ZnO nanowire films.

To this aim, three pairs of ZnO NW films were prepared with similar morphology for each pair as shown in Figure 8. Three different pH values, i.e., 2.3, 4.4 and 5.7, were selected. One sample of each pair was soaked for 2 h in N719 solutions with pH 7.3, while the other sample of each pair was soaked for the same period in solutions of N719 with the above pH values. The dye loading of the films was estimated and the results are shown in Figure 9. Since a comparison between films with vastly different morphologies, as those shown in Figure 8, may lead to misleading conclusions we undertake only a comparison between the twin samples, i.e., one soaked in pH 7.3 and the corresponding one soaked in one of the other pH values. Figure a shows the dye loading of the samples with morphologies of type (a), (b), and (c) corresponding to the images of Figure 8. The data in Figure 9a are normalized with respect to the NW length and sample surface area. For moderate acidic pH (4.4 and 5.7), the dye loading was considerably improved compared to that at pH 7.3. The films soaked in the dye solution with PH equal to 4.4 and 5.7 were also more intensely colored, as shown in Figure 3. For the more acidic solution, i.e., at pH 2.3, there was no significant change in the dye loading with respect to that at pH 7.3.

Figure 9b shows the ratio of the dye loading achieved at the acidic pH in relation to the normal pH for all three types of NW morphologies. The gain in dye loading for the samples soaked in the solutions with pH values 2.3, 4.4 and 5.7 are ~1, ~9, and ~11, respectively. These data show best performance for pH 5.7, while even the more acidic solution (pH 4.4) still provides improved dye loading. The dye loading of nanoparticle films, however, remained higher than that of the best performing NW films (at pH 5.7) by approximately 20%.

In the case of very acidic dye solutions (pH 2.3) it was observed that there is rapid (within a few hours) degradation of the dye solution in the presence of ZnO films. Dye degradation in ZnO films occurs due to protons originating from the dye interacting with the ZnO surface and causing the dissolution of Zn surface atoms [50]. The Zn atoms then react with the protons from the dye, forming Zn^{+2}/dye complexes. This leads to the formation of inactive dye molecules which limit charge carrier injection and reduce the efficiency of the cells. The dissolution of Zn atoms depends on several factors, such as the dye concentration, the sensitization time and the pH of the dye solution [49]. Thus, to achieve optimum dye loading and efficiency, the pH of the dye should be lower than the pH_{PZC} of the film, but not so low that it causes dye degradation. It should also be noted that the samples of type (c) had lower dye loading at pH 7.3 than the samples of the other two types. This illustrates the effect of film morphology on the dye loading of the films. It can be observed from Figure 8 that the samples of type (c) had NWs with a diameter up to five times as large as that of the two other sample types. The higher diameter of the NWs leads to the formation of films with lower surface area for dye adsorption and, consequently, lower dye loading.

3.4. Nanowire DSSCs with Improved Dye Loading

The J-V characteristics of NW DSSCs with improved dye loading (pH of dye solution equal to 5.7) are shown in Figure 5, compared with those of nanoparticle-based cells and standard NW ones. The photovoltaic properties of these cells are summarized in Table 1. It is apparent from Figure 5 that the overall performance of NW cells with improved dye loading is considerably higher than that of the standard pH dye loading of NW cells. The V_{oc} of devices with improved dl is only slightly higher than that of the standard devices and it is still lower than the V_{oc} of the nanoparticle DSSCs. The J_{sc} of the devices with improved dl however, is almost four times higher than that of standard ones and that can be attributed directly to their increased dye loading. Their efficiency is also almost three times higher than the efficiency of standard nanowire DSSCs. It should be noted here that many devices of each type (approximately 30 NW, 30 nanoparticle DSSCs and over 10 NW with improved dye loading conditions) were prepared and tested to ascertain the validity of our results, and that the results shown here are representative of each type of DSSC considered. The error values shown in Table 1 reflect the variation in the J-V measurements across samples of the same type. These error values indicate that there is a high degree of reproducibility across all samples prepared and tested.

Although improving the dye loading of the devices resulted in a considerable increase in the efficiency of NW DSSCs, their performance is still not as high as that of nanoparticle devices. The dye loading of nanowire films is still lower than that of nanoparticle ones and that results in lower device current density. In addition, there are indications of high electron recombination in the case of nanowire DSSCs, as will be shown below.

Figures 6 and 7 show dark current and OCVD results, respectively, for each type of cell considered. Figure 6 indicates that the dark current of NW devices with improved dl is slightly lower compared to that of standard ones, but it is still quite high compared to that of nanoparticle DSSCs. The reduction in dark current is probably due to improved dye adsorption, which leads to lower electron recombination at the interface between the semiconductor and the electrolyte. However, the OCVD results of Figure 7 show that NW devices with improved dl have similar electron lifetimes with standard ones, which in both cases are lower than those of ZnO nanoparticles. This indicates that there is still high electron recombination, possibly due to bare sites on the glass substrate. This is understandable due to the more porous nature of the nanowire films. Their seed layer is only a few nm thick and it can be more easily permeated by the electrolyte in the gaps between the nanowires. By contrast, nanoparticle films consist of a layer of a few μm thick, which is harder for the electrolyte to penetrate.

4. Conclusions

A systematic study of the parameters that affect the performance of nanowire-based DSSCs was conducted to investigate the underlying issues that prevent these devices from achieving PCEs close to those of nanoparticle-based DSSCs. Decreased light harvesting efficiency due to low internal surface area has been broadly considered to be the main reason for the low PCE of nanowire-based DSSCs. Here, an attempt was made to further explain the reasons for the low performance of these devices. To this aim, the dye loading of nanoparticle and nanowire films was measured, regardless of their surface area, and it was found that the dl of nanoparticle films is approximately an order of magnitude higher than that of nanowire ones, thus largely accounting for the higher short-circuit current and PCE of nanoparticle DSSCs.

pH drift measurements were conducted to determine the point of zero charge of nanowire films and it was found that it is somewhat lower than the pH of the dye solution. This may cause the nanowire surface to electrostatically repel dye molecules and prevent them from successfully anchoring to the semiconductor surface. Using acidic dye solutions, which have a higher concentration of positive ions, to sensitize nanowire films led to significantly increased dye adsorption. The dye loading of nanowire films soaked in a solution of N719 with pH equal to 5.7 was found to be higher than that of corresponding films sensitized with the standard dye solution by a factor of approximately 11, and was only 20% lower than the dye loading of nanoparticle films.

The overall performance of nanowire cells with improved dye loading was considerably higher than that of the standard nanowire cells. The short-circuit current of the devices with improved dl was almost four times higher than that of standard ones, which is consistent with the increase in their dye loading. Their efficiency was also almost three times higher than the efficiency of standard nanowire DSSCs.

Electron collection in nanowire DSSCs was expected to be higher than that of corresponding nanoparticle devices, since the morphology of the nanowire films provides direct pathways for electrons to travel from the semiconductor to the collection electrode. From dark current and OCVD measurements, however, it was apparent that nanowire-based DSSCs have higher recombination losses than nanoparticle ones. Electron recombination in DSSCs occurs mainly at the interface between the semiconductor and the electrolyte. Due to the low dye loading of nanowire films, there would be many semiconductor sites without dye adsorbed. These free semiconductor sites could come in direct contact with the electrolyte, leading to a greater chance of electron recombination. Increasing the dye loading of nanowire films was found to lead to a decrease in the dark current, but not to the low levels of nanoparticle-based devices. It is possible that there is high electron recombination through contact of the electrolyte with the glass substrate, due to the more porous nature of the nanowire films.

Acknowledgments: This research has been co-financed by the European Union (European Social Fund—ESF) and Greek national funds through the Operational Program "Education and Lifelong Learning" of the National Strategic Reference Framework (NSRF)—Research Funding Program: Thales. Investing in knowledge society through the European Social Fund through the Na(Z)nowire project.

Author Contributions: M.G. was involved in the design of the experiments and was responsible for the development, characterization and testing of the DSSCs. She also conducted the dye loading and pH measurements, analyzed the experimental results and was the primary author of the paper. K.G. was involved in the synthesis and characterization of the morphology of ZnO NW arrays. G.S. was involved in the synthesis of ZnO NW arrays and to the evaluation of experimental results. S.N.Y. was involved in the design of the experiments and partly to evaluation of results and paper writing. G.L. was involved in the design of the dye loading and point of zero charge experiments. He also contributed to the evaluation of experimental results.

Conflicts of Interest: The authors declare no conflict of interest.

References

1. Green, M.A.; Hishikawa, Y.; Dunlop, E.D.; Levi, D.H.; Ebinger, J.H.; Ho-Baillie, A. Solar cell efficiency tables (version 51). *Prog. Photovolt.* **2018**, *26*, 3–12. [CrossRef]

2. Komiya, R.; Fukui, A.; Murofushi, N.; Koide, N.; Yamanaka, R.; Katayama, H. Improvement of the conversion efficiency of a monolithic type dye-sensitized solar cell module. In Proceedings of the Technical Digest, 21st International Photovoltaic Science and Engineering Conference, Fukuoka, Japan, 28 November–2 December 2011; p. 2C-5O-08.

3. Kanaparthi, R.K.; Kandhadi, J.; Giribabu, L. Metal-free organic dyes for dye-sensitized solar cells: Recent advances. *Tetrahedron* **2012**, *68*, 8383–8393. [CrossRef]

4. Yella, A.; Lee, H.W.; Tsao, H.N.; Yi, C.; Chandiran, A.K.; Nazeeruddin, M.K.; Diau, E.W.; Yeh, C.Y.; Zakeeruddin, S.M.; Grätzel, M. Porphyrin-sensitized solar cells with cobalt (II/III)-based redox electrolyte exceed 12 percent efficiency. *Science* **2011**, *4*, 629–634. [CrossRef] [PubMed]

5. Campbell, W.M.; Jolley, K.W.; Wagner, P.; Wagner, K.; Walsh, P.J.; Gordon, K.C.; Schmidt-Mende, L.; Nazeeruddin, M.K.; Wang, Q.; Grätzel, M.; et al. Highly Efficient Porphyrin Sensitizers for Dye-Sensitized Solar Cells. *J. Phys. Chem. C* **2007**, *111*, 11760–11762. [CrossRef]

6. Mathew, S.; Yella, A.; Gao, P.; Humphry-Baker, R.; Curchod, B.F.; Ashari-Astani, N.; Tavernelli, I.; Rothlisberger, U.; Nazeeruddin, M.K.; Grätzel, M. Dye-sensitized solar cells with 13% efficiency achieved through the molecular engineering of porphyrin sensitizers. *Nat. Chem.* **2014**, *6*, 242–247. [CrossRef] [PubMed]

7. Grätzel, M.; O'Regan, B. A Low-Cost, High-Efficiency Solar Cell Based on Dye-Sensitized Colloidal TiO_2 Films. *Nature* **1991**, *352*, 737–740.

8. Wang, Y.; Tian, J.; Fei, C.; Lv, L.; Liu, X.; Zhao, Z.; Cao, G. Microwave-Assisted Synthesis of SnO_2 Nanosheets Photoanodes for Dye-Sensitized Solar Cells. *J. Phys. Chem. C* **2014**, *118*, 25931–25938. [CrossRef]

9. Farré, Y.; Zhang, L.; Pellegrin, Y.; Planchat, A.; Blart, E.; Boujtita, M.; Hammarström, L.; Jacquemin, D.; Odobel, F. Second Generation of Diketopyrrolopyrrole Dyes for NiO-Based Dye-Sensitized Solar Cells. *J. Phys. Chem. C* **2016**, *120*, 7923–7940. [CrossRef]

10. Morkoc, H.; Ozgur, U. *Zinc Oxide Fundamentals, Materials and Device Technology*; Wiley-VCH: Weinheim, Germany, 2008.

11. Schlur, L.; Carton, A.; Lévêque, P.; Guillon, D.; Pourroy, G. Optimization of a New ZnO Nanorods Hydrothermal Synthesis Method for Solid State Dye Sensitized Solar Cells Applications. *J. Phys. Chem. C* **2013**, *117*, 2993–3001. [CrossRef]

12. Kao, M.C.; Chen, H.Z.; Young, S.L.; Lin, C.C.; Kung, C.Y. Structure and photovoltaic properties of ZnO nanowire for dye-sensitized solar cells. *Nanoscale Res. Lett.* **2012**, *7*, 260. [CrossRef] [PubMed]

13. Law, M.; Greene, L.; Johnson, J.; Saykally, R.; Yang, P. Nanowire dye-sensitized solar cells. *Nat. Mater.* **2005**, *4*, 455. [CrossRef] [PubMed]

14. Cheng, H.M.; Chiu, W.H.; Lee, C.H.; Tsai, S.Y.; Hsieh, W.F. Formation of Branched ZnO Nanowires from Solvothermal Method and Dye-Sensitized Solar Cells Applications. *J. Phys. Chem. C* **2008**, *112*, 16359–16364. [CrossRef]

15. Bai, Y.; Yu, H.; Li, Z.; Amal, R.; Lu, G.Q.; Wang, L. In situ growth of a ZnO nanowire network within a TiO_2 nanoparticle film for enhanced dye-sensitized solar cell performance. *Adv. Mater.* **2012**, *24*, 5850–5856. [CrossRef] [PubMed]

16. Yang, Z.; Xu, T.; Ito, Y.; Welp, U.; Kwok, W.K. Enhanced Electron Transport in Dye-Sensitized Solar Cells Using Short ZnO Nanotips on A Rough Metal Anode. *J. Phys. Chem. C* **2009**, *113*, 20521–20526. [CrossRef]

17. Xu, C.K.; Shin, P.; Cao, L.L.; Gao, D. Preferential Growth of Long ZnO Nanowire Array and Its Application in Dye-Sensitized Solar Cells. *J. Phys. Chem. C* **2010**, *114*, 125–129. [CrossRef]

18. Guerin, V.M.; Rathousky, J.; Pauporte, T. Electrochemical design of ZnO hierarchical structures for dye-sensitized solar cells. *Sol. Energy Mater. Sol. Cells* **2012**, *102*, 8–14. [CrossRef]

19. Giannouli, M. Nanostructured ZnO, TiO$_2$, and Composite ZnO/TiO$_2$ Films for Application in Dye-Sensitized Solar Cells. *Int. J. Photoenergy* **2013**, *2013*, 612095. [CrossRef]

20. Nazeeruddin, M.K.; Kay, A.; Rodicio, I.; Humpbry-Baker, R.; Miiller, E.; Liska, P.; Vlachopoulos, N.; Grätzel, M. Conversion of light to electricity by cis-X2bis(2,2'-bipyridyl-4,4'-dicarboxylate)ruthenium(II) charge-transfer sensitizers (X = Cl-, Br-, I-, CN-, and SCN-) on nanocrystalline titanium dioxide electrodes. *J. Am. Chem. Soc.* **1993**, *115*, 6382–6390. [CrossRef]

21. Smestad, G. Education and solar conversion: Demonstrating electron transfer. *Sol. Energy Mater. Sol. Cells* **1998**, *55*, 157–178. [CrossRef]

22. Pichot, F.; Pitts, R.; Gregg, B. Low-Temperature Sintering of TiO$_2$ Colloids: Application to Flexible Dye-Sensitized Solar Cells. *Langmuir* **2000**, *16*, 5626–5630. [CrossRef]

23. Liu, Y.; Wang, H.; Shen, H.; Chen, W. The 3-dimensional dye-sensitized solar cell and module based on all titanium substrates. *Appl. Energy* **2010**, *87*, 436–441. [CrossRef]

24. Van der Zanden, B.; Goossens, A. The Nature of Electron Migration in Dye-Sensitized Nanostructured TiO$_2$. *J. Phys. Chem. B* **2000**, *104*, 7171–7178. [CrossRef]

25. Shklover, V.; Nazeeruddin, M.K.; Zakeeruddin, S.M.; Barbe, C.; Kay, A.; Haibach, T.; Steurer, W.; Hermann, R.; Nissen, H.U.; Grätzel, M. Structure of Nanocrystalline TiO$_2$ Powders and Precursor to Their Highly Efficient Photosensitizer. *Chem. Mater.* **1997**, *9*, 430–439. [CrossRef]

26. Giannouli, M.; Spiliopoulou, F. Effects of the morphology of nanostructured ZnO films on the efficiency of dye-sensitized solar cells. *Renew. Energy* **2012**, *41*, 115–122. [CrossRef]

27. Lee, W.; Okada, H.; Wakahara, A.; Yoshida, A. Structural and photoelectrochemical characteristics of nanocrystalline ZnO electrode with Eosin-Y. *Ceram. Int.* **2006**, *32*, 495–498. [CrossRef]

28. Syrrokostas, G.; Govatsi, K.; Yannopoulos, S.N. High-Quality, Reproducible ZnO Nanowire Arrays Obtained by a Multiparameter Optimization of Chemical Bath Deposition Growth. *Cryst. Growth Des.* **2016**, *16*, 2140–2150. [CrossRef]

29. Han, J.; Fan, F.; Xu, C.; Lin, S.; Wei, M.; Duan, X.; Wang, Z.L. ZnO nanotube-based dye-sensitized solar cell and its application in self-powered devices. *Nanotechnology* **2010**, *21*, 405203. [CrossRef] [PubMed]

30. Benhebal, H.; Chaib, M.; Salmon, T.; Geens, J.; Leonard, A.; Lambert, S.D.; Crine, M.; Heinrichs, B. Photocatalytic degradation of phenol and benzoic acid using zinc oxide powders prepared by the sol–gel process. *Alex. Eng. J.* **2013**, *52*, 517–523. [CrossRef]

31. Lopez-Ramon, M.V.; Stoeckli, F.; Moreno-Castilla, C.; Carrasco-Marin, F. On the characterization of acidic and basic surface sites on carbons by various techniques. *Carbon* **1999**, *37*, 1215–1221. [CrossRef]

32. Giannouli, M.; Fakis, M. Interfacial Electron Transfer Dynamics and Photovoltaic Performance of TiO$_2$ and ZnO Solar Cells Sensitized with Coumarin 343. *J. Photochem. Photobiol. A* **2011**, *226*, 42–50. [CrossRef]

33. Yu, H.; Zhang, S.; Zhao, H.; Will, G.; Liu, P. An Efficient and Low-Cost TiO$_2$ Compact Layer for Performance Improvement of Dye-Sensitized Solar Cells. *Electrochim. Acta* **2009**, *54*, 1319–1324. [CrossRef]

34. Xu, W.; Dai, S.; Hu, L.; Zhang, C.; Xiao, S.; Luo, X.; Jing, W.; Wang, K. Influence of Different Surface Modifications on the Photovoltaic Performance and Dark Current of Dye-Sensitized Solar Cells. *Plasma Sci. Technol.* **2007**, *9*, 554–559.

35. Jasim, K.H. *Dye Sensitized Solar Cells—Working Principles, Challenges and Opportunities*; InTech: London, UK, 2011; ISBN 978-953-307-735-2. [CrossRef]

36. Zaban, A.; Greenshtein, M.; Bisquert, J. Determination of the electron lifetime in nanocrystalline dye solar cells by photovoltage decay measurements. *ChemPhysChem* **2003**, *4*, 859–864. [CrossRef] [PubMed]

37. Chiba, Y.; Islam, A.; Watanabe, Y.; Komiya, R.; Koide, N.; Han, L. Dye-Sensitized Solar Cells with Conversion Efficiency of 11.1%. *Jpn. J. Appl. Phys.* **2006**, *45*, 638–640. [CrossRef]

38. Hara, K.; Tachiban, Y.; Ohga, Y.; Shinpo, A.; Suga, S.; Sayama, K.; Sugihara, H.; Arakawa, H. Dye-sensitized nanocrystalline TiO$_2$ solar cells based on novel coumarin dyes. *Sol. Energy Mater. Sol. Cells* **2003**, *77*, 89–103. [CrossRef]

39. Rani, S.; Shishodia, P.K.; Mehra, R.M. Development of a dye with broadband absorbance in visible spectrum for an efficient dye-sensitized solar cell. *J. Renew. Sustain. Energy* **2010**, *2*, 043103. [CrossRef]

40. Keis, K.; Magnusson, E.; Lindström, H.; Lindquist, S.; Hagfeldt, A. A 5% efficient photoelectrochemical solar cell based on nanostructured ZnO electrodes. *Sol. Energy Mater. Sol. Cells* **2002**, *73*, 51–58. [CrossRef]

41. Chang, W.; Lee, C.; Yu, W.; Lin, C. Optimization of dye adsorption time and film thickness for efficient ZnO dye-sensitized solar cells with high at-rest stability. *Nanoscale Res. Lett.* **2012**, *7*, 688. [CrossRef] [PubMed]

42. He, Y.; Hu, J.; Xie, Y. High-efficiency dye-sensitized solar cells of up to 8.03% by air plasma treatment of ZnO nanostructures. *Chem. Commun.* **2015**, *51*, 16229–16232. [CrossRef] [PubMed]

43. Jiang, C.; Sun, X.; Lo, G.; Kwong, D. Improved dye-sensitized solar cells with a ZnO-nanoflower photoanode. *Appl. Phys. Lett.* **2007**, *90*, 263501. [CrossRef]

44. Ito, S.; Liska, P.; Comte, P.; Charvet, R.; Pechy, P.; Bach, U.; Schmidt-Mende, L.; Zakeeruddin, S.M.; Kay, A.; Nazeeruddin, M.K.; et al. Control of dark current in photoelectrochemical (TiO$_2$/I–I3-)) and dye-sensitized solar cells. *Chem. Commun.* **2005**, *34*, 4351–4353. [CrossRef] [PubMed]

45. Bisquert, A.; Zaban, M.; Greenshtein, M.; Mora-Seró, I. Determination of Rate Constants for Charge Transfer and the Distribution of Semiconductor and Electrolyte Electronic Energy Levels in Dye-Sensitized Solar Cells by Open-Circuit Photovoltage Decay Method. *J. Am. Chem. Soc.* **2004**, *126*, 13550–13559. [CrossRef] [PubMed]

46. Fabregat-Santiago, F.; Garcia-Canadas, J.; Palomares, E.; Clifford, J.N.; Haque, S.A.; Durrant, J.R.; Garcia-Belmonte, G.; Bisquert, J. The origin of slow electron recombination processes in dye-sensitized solar cells with alumina barrier coatings. *J. Appl. Phys.* **2004**, *96*, 6903–6907. [CrossRef]

47. Azpiroz, J.M.; De Angelis, F. DFT/TDDFT Study of the Adsorption of N3 and N719 Dyes on ZnO(1010) Surfaces. *J. Phys. Chem. A* **2014**, *118*, 5885–5893. [CrossRef] [PubMed]

48. Schiffmann, F.; Van deVondel, J.; Hutter, J.; Wirz, R.; Urakawa, A.; Baiker, A. Protonation-Dependent Binding of Ruthenium Bipyridyl Complexes to the Anatase(101) Surface. *J. Phys. Chem. C* **2010**, *114*, 8398–8404. [CrossRef]

49. Bahnemann, D.W. Ultrasmall Metal Oxide Particles: Preparation, Photophysical Characterization, and Photocatalytic Properties. *Isr. J. Chem.* **1993**, *33*, 115–136. [CrossRef]

50. Keis, K.; Bauer, C.; Boschloo, G.; Hagfeldt, A.; Westermark, K.; Rensmob, H.; Siegbahn, H. Nanostructured ZnO electrodes for dye-sensitized solar cell applications. *J. Photochem. Photobiol. A* **2002**, *118*, 57–64. [CrossRef]

MDPI

St. Alban-Anlage 66

4052 Basel

Switzerland

Tel. +41 61 683 77 34

Fax +41 61 302 89 18

www.mdpi.com

Materials Editorial Office

E-mail: materials@mdpi.com

www.mdpi.com/journal/materials